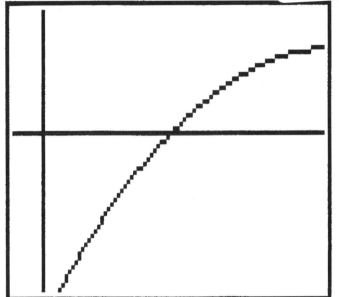

The second derivative of a function. Does the graph of the function have an inflection point? Explain why.

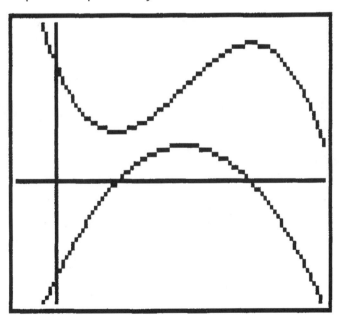

A function and one of its antiderivatives. Which is the function? Explain why.

Calculus with Applications

Also available from McGraw-Hill

Schaum's Outline Series in Mathematics & Statistics

Most outlines include basic theory, definitions and hundreds of example problems solved in step-by-step detail, and supplementary problems with answers.

Related titles on the current list include:

Advanced Calculus
Advanced Mathematics for Engineers & Scientists
Analytic Geometry
Basic Mathematics for Electricity & Electronics
Basic Mathematics with Applications to Science & Technology
Beginning Calculus
Boolean Algebra & Switching Circuits
Calculus
Calculus for Business, Economics, & the Social Sciences
College Algebra
College Mathematics
Complex Variables
Descriptive Geometry
Differential Equations

Differential Geometry
Discrete Mathematics
Elementary Algebra
Essential Computer Mathematics
Finite Differences & Difference Equations
Finite Mathematics
Fourier Analysis
General Topology
Geometry
Group Theory
Laplace Transforms
Linear Algebra
Mathematical Handbook of Formulas & Tables
Mathematical Methods for Business & Economics

Mathematics for Nurses
Matrix Operations
Modern Abstract Algebra
Numerical Analysis
Partial Differential Equations
Probability
Probability & Statistics
Real Variables
Review of Elementary Mathematics
Set Theory & Related Topics
Statistics
Technical Mathematics
Tensor Calculus
Trigonometry
Vector Analysis

Schaum's Solved Problems Books

Each title in this series is a complete and expert source of solved problems with solutions worked out in step-by-step detail.

Titles on the current list include:

3000 Solved Problems in Calculus
2500 Solved Problems in College Algebra and Trigonometry
2500 Solved Problems in Differential Equations

2000 Solved Problems in Discrete Mathematics
3000 Solved Problems in Linear Algebra
2000 Solved Problems in Numerical Analysis
3000 Solved Problems in Precalculus

Available at most college bookstores, or for a complete list of titles and prices, write to: Schaum Division
McGraw-Hill, Inc.
1221 Avenue of the Americas
New York, NY 10020

Calculus with Applications

Rosario Urso
Hillsborough Community College

McGraw-Hill, Inc.
New York   St. Louis   San Francisco   Auckland   Bogotá   Caracas
Lisbon   London   Madrid   Mexico City   Milan   Montreal   New Delhi
San Juan   Singapore   Sydney   Tokyo   Toronto

Calculus with Applications

Copyright © 1995 by McGraw-Hill, Inc. All rights reserved.
Printed in the United States of America. Except as permitted under the United States Copyright Act of 1976, no part of this publication may be reproduced or distributed in any form or by any means, or stored in a data base or retrieval system, without the prior written permission of the publisher.

 This book is printed on recycled, acid-free paper containing 10% postconsumer waste.

1 2 3 4 5 6 7 8 9 0 DOC DOC 9 0 9 8 7 6 5 4

ISBN 0-07-066651-2

This book was set in Times Roman by American Composition & Graphics, Inc.
The editors were Michael Johnson and David A. Damstra;
the design was done by A Good Thing, Inc.;
the production supervisor was Louise Karam.
The photo editor was Anne Manning.
R. R. Donnelley & Sons Company was printer and binder.

The chapter-opening art was rendered by Josette Urso. The artist lives and works in New York City.

Library of Congress Cataloging-in-Publication Data

Urso, Rosario.
    Calculus with applications / Rosario Urso.
       p.   cm.
    Includes bibliographical references and index.
    ISBN 0-07-066651-2
    1. Calculus.   I. Title.
QA303.U74   1995
515—dc20   94-1803

To Connie, my beloved wife.

Contents

Preface ix

CHAPTER 1

Functions and Limits   1

1.1 Functions   2
1.2 Finding Formulas for Functions   13
1.3 Linear Functions   22
1.4 Quadratic Functions   31
1.5 Limits at Infinity   41
1.6 Limits as $x \to c$ and Continuity   49

Important Terms   62
Review Problems   62

CHAPTER 2

Differentiation   66

2.1 Slopes of Tangents   67
2.2 The Derivative   74
2.3 Rules for Computing Derivatives   83
2.4 Approximating Change   91
2.5 More Rules for Computing Derivatives   100
2.6 Optimization on Closed Intervals   112
2.7 More on Optimization   122
2.8 More on Approximating Change   131

Important Terms   143
Review Problems   144

CHAPTER 3

More on Differentiation   149

3.1 Instantaneous Rates   150
3.2 The Second Derivative and Curve Sketching   160
3.3 More on the Second Derivative   169
3.4 Two Useful Rational Functions   175
3.5 More on Rational Functions   187
3.6 Percentage Rate of Change   199
3.7 Newton's Method for Approximating Zeros   206

Important Terms   216
Review Problems   216

CHAPTER 4

Exponential and Logarithmic Functions   221

4.1 Compound Interest   222
4.2 Exponential Functions   233
4.3 The Natural Logarithmic Function   238
4.4 Differentiation of $\ln u(x)$ 250
4.5 Differentiation of $e^{u(x)}$ 255
4.6 Two Models for Bounded Growth   264

Important Terms   275
Review Problems   275

CHAPTER 5: Integration 278

- 5.1 Antidifferentiation 279
- 5.2 The Definite Integral 291
- 5.3 Area and Definite Integrals 302
- 5.4 Integration by Substitution 313
- 5.5 Integration by Parts 326
- 5.6 Riemann Sums 333
- Important Terms 345
- Review Problems 346

CHAPTER 6: More on Integration 349

- 6.1 Simpson's Rule 350
- 6.2 Consumers' and Producers' Surpluses 360
- 6.3 Probability 369
- 6.4 Expected Value 379
- 6.5 Differential Equations for Exponential Change 386
- 6.6 Differential Equations for Bounded Change 393
- Important Terms 400
- Review Problems 401

CHAPTER 7: Differentiation of Functions of More Than One Variable 403

- 7.1 Functions of More Than One Variable 404
- 7.2 Partial Derivatives 412
- 7.3 Optimization 424
- 7.4 The Second Derivative Test 430
- 7.5 Curve Fitting 435
- 7.6 Constrained Optimization 445
- 7.7 The Method of Lagrange Multipliers 453
- Important Terms 457
- Review Problems 458

APPENDIX A: Algebra Review 460

- A.1 Real Numbers 461
- A.2 Exponents and Radicals 469
- A.3 Polynomial Expressions 477
- A.4 Rational Expressions 490

APPENDIX B: Integration by Tables 498

Answers to Odd-Numbered Section Problems and All Chapter Review Problems 504

Index 533

Preface

This book has been designed for a one-term calculus course for students in business, economics, management, and the social and biological sciences. Paced leisurely, it can be used also in a two-term version of such a course. The topics were selected and organized to allow for a maximum flexibility in the book's use.

The book's goal is to make the powerful and interesting subject of calculus as accessible as possible for the intended audience. Among the strategies used to achieve this goal, motivation is foremost: Every section is carefully motivated and includes numerous problems that demonstrate how calculus can be used in a variety of fields.

A prerequisite of intermediate algebra is assumed. In courses where college algebra is the prerequisite, many of the topics in Chapter 1 may be omitted. For classes that need additional work in algebra, an algebra review is provided in Appendix A. The review includes rules, examples, and exercises. It can be used as part of the course or as a reference.

The advent of graphing calculators has made graphical visualization more accessible than ever. Such visualization enhances understanding. Therefore, many sections include graphing calculator material that can easily be woven into the course.

Pedagogical Features

Motivation

In most cases, before defining a concept, the text presents a situation where the concept naturally arises. This approach makes students more receptive to definitions. Concepts are usually motivated through problems of particular interest to the intended audience, rather than problems whose attraction is primarily mathematical. For example, instead of introducing the definite integral through a problem in which the primary goal is to compute an area, the text introduces the definite integral through a problem in which the primary goal is to find a decrease in the size of a timberland.

Problems

Learning occurs mainly through the process of solving problems. So great care has been taken in composing the problem sets. The problem sets have the following features:

(*a.*) They are graded.
(*b.*) They are paired (odd-even) according to similarity.
(*c.*) They are coordinated with the worked-out examples in the text.
(*d.*) Practically every problem set includes a generous supply of applied problems.
(*e.*) A balance exists between the routine drill problems and the more difficult problems.
(*f.*) Review problems grouped by sections appear at the end of each chapter.

Features (*a*) through (*c*) allow instructors to base their lectures on the even (or odd) numbered problems and then assign the odd (or even) numbered problems. Thus, the design of the problem sets enhances the teachability of the book.

Verbalization

Several mathematical organizations recommend that mathematics curricula include more opportunities for students to provide verbalized (written) responses to questions. Therefore, in most problem sets, problems are included that prompt students to respond with an explanation. Problems 35 and 36 in Section 2.7 are examples of such problems, which are signified by the accompanying pencil icon.

Content Features

Few Proofs but Many Plausibility Arguments

The students to whom this book is addressed aspire to become *users* of mathematics, not *makers* of mathematics. Therefore, the text includes few mathematical proofs. Usually, a deducible statement is introduced through a concrete situation that either illustrates the statement or involves steps that parallel those in a general proof of the statement. For instance, the generalized power rule is introduced by computing the derivative of a constant multiple of a power function without using the rule. Then it is shown that the result suggests a shortcut for computing derivatives of such functions. The shortcut, of course, is the generalized power rule.

Graphing Calculator Material

Many sections include graphing calculator problems. A few sections also include a subsection that serves as a basis for some of these problems.

These problems are clearly identified with the icon . Omission of this material will not disturb the continuity of the book.

The approach to the graphing calculator material is generic and, thus, devoid of key pressing discussions. The term "graphing calculator" is also used to mean a computer loaded with graphing software.

Linear regression is introduced early (in Section 1.3). Then problems based on the visual linear regression capability of graphing calculators are included. Problem 36 in Section 4.3 is an example of such a problem.

Of course, graphing calculator problems that are not based on linear regression are also included. For example, Problems 41 and 42 in Section 2.2 ask students to use the graph of a function to approximate the points where the function is not differentiable.

Computer Material

The book includes nine Basic computer programs. The sections where these programs appear also include computer problems. The programs are clearly identified with the icon . As with the graphing calculator material, omission of this material will not disturb the continuity of the book.

EVAL1, in Section 1.1, evaluates a function at consecutive integers in its domain.

EVAL2, in Section 1.1, evaluates a function at any point in its domain.

OPT1, in Section 1.2, finds the minimum and maximum values of a function whose domain is a set of consecutive integers.

OPT2, in Section 2.6, uses the random number generator to approximate the minimum and maximum values of a continuous function on a closed interval.

NEWTON, in Section 3.7, uses Newton's Method to approximate zeros.

PROBE, in Section 3.7, finds the general location of the zeros of a continuous function on a closed interval.

INT, in Section 5.3, uses the random number generator to approximate a definite integral.

SIMP, in Section 6.1, uses Simpson's rule to approximate a definite integral.

LESQ, in Section 7.5, finds least-squares linear functions.

Two Types of Rational Functions Receive Special Attention

Many applications involve quantities that approach a fixed level either at a decreasing rate or at a rate that increases at first and decreases eventually. With certain restrictions on the coefficients, the rational functions

$$f(x) = \frac{ax + b}{cx + d}$$

and

$$f(x) = \frac{ax^2 + b}{cx^2 + d}$$

have properties that make them excellent models for such quantities. Nevertheless, these two functions seldom, if ever, receive the recognition they deserve. In this book, the importance of these functions is elevated by allocating an entire section to them.

Unambiguous Approach to Exponential Functions

Exponential functions are defined as the functions that can be expressed in the form

$$f(x) = Ae^{rx}$$

where $A > 0$ and $r \neq 0$. Such functions form the set of all positive functions that have a constant percentage rate of change.

General Comments

Chapter 1 deals primarily with algebraic functions and limits. Attention is given to finding formulas for functions in applied situations. Limits are defined intuitively without using ϵ and δ. Limits at infinity are presented before limits as x approaches a fixed number. This approach is used because to a beginner the idea of a limit is more natural in the context of the independent variable increasing without bound than in the context of the independent variable approaching a fixed number. Other topics included in Chapter 1 are average rate of change, composite functions, continuity, slopes, and several special functions used in economics.

Chapters 2 and 3 deal with differentiation of algebraic functions of one independent variable. More specifically, in Chapter 2, the derivative is interpreted as a slope and used to solve optimization problems, to approximate change, and to determine where a function increases and where it decreases. At first, derivatives are found by computing limits. But all the rules for computing derivatives of algebraic functions are eventually presented in the chapter. The idea of a function's elasticity appears in a subsection of Section 2.8.

In Chapter 3, the derivative is interpreted as a rate of change. Then the second derivative is used to determine where a function changes at an increasing rate and where it changes at a decreasing rate. In this chapter, the second derivative is also used to determine concavity. Although the students are exposed to the relationship between derivatives and graphs in Chapter 2, it is in Chapter 3 that they are first asked to use derivatives to sketch graphs. Special attention is given to two types of rational functions that have properties compatible with quantities that approach a fixed level. The topic of related rates is presented in a subsection of Section 3.1. Section 3.6 introduces percentage rate of change. The chapter concludes with a section on Newton's method.

Chapter 4 begins with compound interest. This approach motivates the exponential functions, which are defined in terms of base e. Then the natural logarithmic function is introduced and used to help solve problems that involve exponential functions. The chain rule is used to derive the rule for computing derivatives of functions of the form ln $u(x)$. The new rule is then used to derive the rule for computing derivatives of functions of the form $e^{u(x)}$. Chapter 4 ends with a section on two special functions constructed with exponential functions. These two functions are used to model quantities that experience bounded growth.

Chapters 5 and 6 deal with integration of functions of one independent variable. The first section in Chapter 5 introduces antidifferentiation and presents formulas for the antiderivatives of x^n, e^x, and $1/x$. It also presents formulas for the antiderivatives of sums and multiples of these functions. The next section uses the fundamental theorem of calculus as the definition for definite integrals and interprets definite integrals in terms of change. Then Section 5.3 interprets definite integrals in terms of area. Section 5.4 introduces integration by substitution, while Section 5.5 introduces integration by parts. (Integration by tables is presented in Appendix B.) The last section of Chapter 5 shows that a definite integral can be viewed as the limit of a sequence of Riemann sums. Definite integrals are then used to approximate Riemann sums.

Chapter 6 begins with Simpson's rule. Then definite integrals are used to find consumers' and producers' surpluses, probabilities, and expected values. Antidifferentiation is used to solve differential equations that arise where change is bounded or exponential. Improper integrals are dealt with in a subsection of Section 6.3.

In Chapter 7, partial derivatives are computed by applying the differentiation rules from earlier chapters. Such derivatives are then used to approximate change and solve optimization problems. Constrained optimization problems are solved with and without Lagrange multipliers. The method of least squares is used to fit linear and quadratic functions to data points. Chapter 7 can be studied immediately after Chapter 4.

Order of Presentation

The following sections are *not* prerequisites for other sections and, thus, may be omitted without any loss of continuity: 2.8, 3.7, 4.6, 5.5, 5.6, 6.2, 6.4, 6.5, 6.6, 7.5, and 7.7. In addition, Section 6.1 may be omitted if Section 6.2 is omitted, Section 6.3 may be omitted if Section 6.4 is omitted, and Section 7.6 may be omitted if Section 7.7 is omitted.

Student Supplements

1. A **Solutions Manual** is available at a nominal cost through the bookstore. The manual contains detailed solutions to the end-of-section odd exercises as well as solutions to all of the end-of-chapter review exercises.
2. **Environmental and Life Science Applications manual** by Anthony Barcellos (American River College)

3. **Interactive Tutorial Software** package is available for use with IBM, IBM compatibles, and Macintosh computers.
4. A **Videodisk** contains real-life simulations of several applications of calculus concepts.

Instructor Supplements

1. A unique **Computer-Generated Testing System** is available to instructors. This system allows the instructor to create tests using algorithmically generated test questions and those from a standard testbank. This testing system enables the instructor to choose questions either manually or randomly by section, question type, difficulty level, and other criteria. This system is available for IBM, IBM compatibles, and Macintosh computers.
2. A **Printed and Bound Testbank** is also available. This bank is a hard-copy listing of the questions found in the standard testbank.
3. An **Instructor's Manual** is also available. This manual contains all answers and solutions to the exercises in the text. Sample tests, transparencies, and teaching hints and suggestions are also included.
4. **Environmental and Life Science Applications manual** by Anthony Barcellos (American River College)
5. A **Videodisk** contains real-life simulations of several applications of calculus concepts.

For further information about these supplements, please contact your local McGraw-Hill College Division sales representative.

Error Check

Because of careful checking and proofing by a number of mathematics instructors, the author and publisher believe this book to be substantially error-free. For any errors remaining, the author would be grateful if they were sent to Mathematics Editor, College Division, 27th floor, McGraw-Hill, 1221 Avenue of the Americas, New York, NY 10020.

Acknowledgments

McGraw-Hill and the author would like to thank the following reviewers for their many helpful comments and suggestions: G. A. Articolo, Rutgers University, Camden; Huey Barrett, Southern Union State Junior College; Orville Bierman, University of Wisconsin, Eau Claire; E. Ray Bobo, Georgetown University; James R. Brasel, Phillips County Community College; Chris Burditt, Napa Valley College; Jim Camp, Odessa College; James Carpenter, Iona College; Larry Curnutt, Bellevue Community College; Raul Curto, University of Iowa; Ken Dunn, Dalhousie University; Gary Etgen, University of Houston; Odene Forsythe, Westark Community College; Greg Goeckel, Mohawk Valley Community College; Eleanor Goldstein, William Paterson College; Linda Holden, Indiana University; Jimmie Lawson, Louisiana State University; Daniel Lee, Southwest Texas State University; Norman Lee, Ball State University;

Steve Marsdan, Glendale Community College; Sherry Meier, Northeast Missouri State University; Stuart Mills, Louisiana State University, Shreveport; Gale Nash, Western State College; Wayne Powell, Oklahoma State University; Thomas Reifenrath, Clark College; Thomas Roe, South Dakota State University; Robert Russell, West Valley College; Daniel Scanlon, Orange Coast College; Kenneth Shiskowski, Eastern Michigan University; Jane Sieberth, Franklin University; and Lynn Wolfmeyer, Western Illinois University.

Thanks is due also to James Shields, University of Massachusetts–Amherst, for his work in preparing the *Student's Solutions Manual*.

The author thanks the able editorial and production staffs of McGraw-Hill, Inc., for generously applying their expertise to the production of this book. Special thanks go to Michael Johnson, David Damstra, Karen Minette, Anne Manning, Lester Strong, and Joseph Siegel.

The author also thanks his many friends and colleagues for their encouragement.

Finally, the author thanks his wife, Connie, for processing the manuscript almost infinitely many times, and his daughters, Josette, Lynette, and Yvette, for their unflinching support.

Rosario Urso

Calculus with Applications

CHAPTER 1
Functions and Limits

1.1 Functions

1.2 Finding Formulas for Functions

1.3 Linear Functions

1.4 Quadratic Functions

1.5 Limits at Infinity

1.6 Limits as $x \to c$ and Continuity

Important Terms

Review Problems

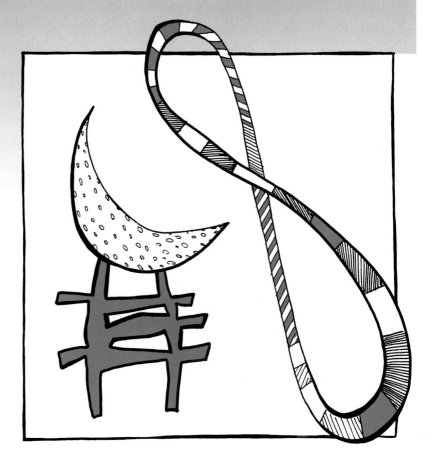

This book is mainly about *derivatives*. Derivatives are *functions*. They are defined in terms of *limits* and used to analyze other functions. It is fitting, therefore, that we begin with a chapter on functions and limits.

1.1 Functions

Many situations involve a variable quantity whose values *depend* on the values of another variable quantity. For instance, the monthly demand for a product depends on the price. In such a situation, there often exists a relatively simple mathematical formula that expresses the dependency, at least approximately. In this section, we deal with formulas that can be used in this manner. We begin with an example that involves a variable quantity whose values depend on time.

Social scientists predict that t years from now (up to $t = 10$) there will be

$$\frac{8t + 10}{t + 2}$$

million single-parent families in a Third World country. With this formula, we can predict the number of such families for any time during the next 10 years. For example, by substituting 2 for t in the formula we get

$$\frac{8(2) + 10}{2 + 2} = 6.5$$

This result tells us that 2 years from now there will be 6.5 million single-parent families in the country.

Let s represent the number of single-parent families (in millions) t years from now. Then

$$s = \frac{8t + 10}{t + 2}$$

for $0 \leq t \leq 10$. The above formula is a rule that associates with each admissible value of t exactly one corresponding value of s. For example, 6.5 is the only value the formula assigns to s when $t = 2$. Table 1.1 shows the values of s that correspond to several values of t. The table suggests that the number of single-parent families will increase during the next 10 years.

Table 1.1

t	0	1.2	2	3	4.4	6	8
s	5	6.125	6.5	6.8	7.0625	7.25	7.4

Definition 1.1

Suppose x and y are variables. A **function** is a rule that associates with each value of x *exactly one* corresponding value of y.

In the context of Definition 1.1, we say "y is a function of x according to the rule." We refer to x as the **independent variable** and to y as the **dependent variable**. The set of admissible values for the independent variable is the **domain** of the function, while the set of corresponding values of the dependent variable is the **range** of the function. In this book, we deal primarily with functions that can be expressed with mathematical formulas.

In the single-parent families situation, s is a function of t defined by the rule

$$s = \frac{8t + 10}{t + 2}$$

For this function, the independent variable is t, and the dependent variable is s. According to the social scientists, the formula for s is valid up to $t = 10$. Therefore, the domain of the function is the set of real numbers t such that $0 \leq t \leq 10$, which is the closed interval [0, 10]. (Real numbers and intervals are discussed in Section A.1 of Appendix A.) Although we will not show the details, the range consists of the closed interval [5, 7.5]. Thus, the number of single-parent families will not exceed 7.5 million during the next 10 years.

Symbols for Functions

We often use letters as names for functions. If, for instance, we use the letter f as the name of a function whose independent variable is x, the symbol $f(x)$ (read "f of x") represents the dependent variable. So there is an important difference between the symbols f and $f(x)$: f is the name of the function, while $f(x)$ represents the dependent variable. Nevertheless, in this book, we take the liberty of using the symbol $f(x)$ to represent *both* the function and the dependent variable. Which of the two it represents will be clear from the context.

According to our convention, we could use the symbol $g(n)$ (read "g of n") to represent a function whose independent variable is n. Our convention has the advantage of exhibiting the letter that represents the independent variable.

We like to represent functions with symbols that reflect the meaning of the variables. For example, since "single" begins with "s" and "time" begins with "t," we will use $s(t)$ to represent the function that predicts the number of single-parent families t years from now. Then

$$s(t) = \frac{8t + 10}{t + 2}$$

for $0 \leq t \leq 10$. According to this notation, $s(2)$, for example, represents the value of $s(t)$ when $t = 2$. Thus, $s(2) = 6.5$.

Assumption about Domains

If we remove the function

$$s(t) = \frac{8t + 10}{t + 2}$$

from its applied role and consider it from a strictly mathematical point of view, its domain is the set of all real numbers except -2. We exclude -2 from the domain because the denominator is 0 if -2 is substituted for t, and $s(-2)$, therefore, is not a real number.

From now on, we make the following assumption about domains:

> The domain of a function $f(x)$ is the set of all real numbers x such that $f(x)$ is also a real number, unless additional restrictions on x are specified or implied by an application.

Example 1.1
Find the domain of the function $f(x) = \sqrt{5 - x}$.

Solution
The square root of a negative number is not a real number. Therefore, if $x > 5$, $f(x)$ is not a real number. Otherwise, $f(x)$ is a real number. Thus, the domain of the function is the set of real numbers x such that $x \leq 5$, which is the half-open interval $(-\infty, 5]$.

As we saw, the domain of the function in Example 1.1 is the half-open interval $(-\infty, 5]$. We may require additional restrictions on the domain if we use the function to express the relationship between two variables in an application. For instance, if x represents a product's price, we would exclude negative numbers from the domain, because prices are not represented by negative numbers.

Graphs of Functions

Functions generate **ordered pairs** of real numbers. To illustrate, we return to the function

$$s(t) = \frac{8t + 10}{t + 2}$$

used in the single-parent families example. This function generates the collection of ordered pairs of real numbers of the form $(t, s(t))$, where t is in the closed interval $[0, 10]$. Since $s(2) = 6.5$, one ordered pair generated by the function $s(t)$ is $(2, 6.5)$. [*Warning:* Here $(2, 6.5)$ does not represent

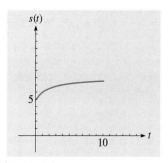

Figure 1.1
$s(t) = (8t + 10)/(t + 2)$
for t in [0, 10].

an interval.] Of course, since there are infinitely many numbers in the interval [0, 10], the function generates infinitely many ordered pairs.

The ordered pairs generated by a function can be represented by points in a **rectangular coordinate plane**. We establish such a plane by selecting a horizontal number line (**horizontal axis**) and a vertical number line (**vertical axis**) that meet at their respective origins. Then, the ordered pair (2, 6.5), for instance, can be represented by the point located 6.5 units above the point with coordinate 2 on the horizontal axis (t axis). The infinitely many points that correspond to the infinitely many ordered pairs generated by the function $s(t)$ form the curve in Figure 1.1.

We could produce the curve in Figure 1.1 by plotting several ordered pairs generated by the function $s(t)$ and then sketching a curve that passes through the points. Later, however, we will learn a method that will enable us to produce this curve more efficiently.

The **graph** of a function is the set of points in a rectangular coordinate plane that corresponds to the set of ordered pairs generated by the function. Thus, the graph of the function $s(t)$ is the curve in Figure 1.1.

Graphs reveal important information about the functions they represent. For example, the curve in Figure 1.1 reveals that the number of single-parent families will increase during the next 10 years. Actually, the curve reveals that this number will increase at a slower and slower rate. Later we will realize how the curve reveals this information.

Rectangular coordinate planes are sometimes called **cartesian planes** in recognition of the Frenchman René Descartes (1596–1650), who pioneered the use of coordinate systems.

The first number in an ordered pair of numbers is the **abscissa** of the corresponding point in a cartesian plane. The second number is the **ordinate** of the point. Thus, 2 is the abscissa and 6.5 is the ordinate of the point that represents the ordered pair (2, 6.5). If the horizontal axis is labeled the x axis and the vertical axis is labeled the y axis, the abscissa could be called the ***x* coordinate** and the ordinate could be called the ***y* coordinate**.

Constant Functions

In the single-parent families situation, the value of $s(t)$ varies whenever the value of t varies. This condition is not necessary for a rule to be a function. If the value of the dependent variable of a function does not change whenever the value of the independent variable varies, we call the function a **constant function**. An example of such a function is expressed by the formula

$$g(x) = 0.24$$

Here the value of $g(x)$ is 0.24 regardless of the value of x. The range of this constant function consists of the single number 0.24.

Change

Often we want to know how a change in the independent variable of a function affects the dependent variable. Suppose a and b are numbers in the domain of $f(x)$, and x changes from a to b. The corresponding change in $f(x)$ is

$$f(b) - f(a)$$

Such a change in $f(x)$ is an *increase* if $f(b) - f(a)$ is positive and a *decrease* if $f(b) - f(a)$ is negative.

Total Cost Functions

The next example illustrates the idea of change with a **total cost function**. Such a function expresses the total production cost as a function of the production level.

Example 1.2

During the past several months, Aqua Corporation has been producing 800 swimsuits per week. The management, however, wants to increase the weekly production level to 900 swimsuits. Let $C(x)$ represent the total weekly cost, in thousands of dollars, of producing x hundred swimsuits per week. Data suggest that $C(x)$ is close to being a function of x according to the formula

$$C(x) = 0.05x^2 + 9$$

Use this formula to determine how much the total weekly production cost increases if management increases the weekly production level from 800 swimsuits to 900 swimsuits. Figure 1.2 shows the graph of $C(x)$ for values of x in the interval $[0, +\infty)$. The figure also shows the graphical interpretation of the cost increase in question.

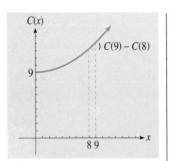

Figure 1.2 $C(x) = 0.05x^2 + 9$ for x in $[0, +\infty)$.

Solution

The total weekly cost of producing 800 (8 hundred) swimsuits per week is

$$\begin{aligned} C(8) &= 0.05(8^2) + 9 \\ &= 12.2 \end{aligned}$$

thousand dollars, while the total weekly cost of producing 900 (9 hundred) swimsuits per week is

$$\begin{aligned} C(9) &= 0.05(9^2) + 9 \\ &= 13.05 \end{aligned}$$

thousand dollars. Therefore, since

$$C(9) - C(8) = 13.05 - 12.2$$
$$= 0.85$$

the total weekly production cost increases $850 (850 = 0.85 thousand) if management increases the weekly production level from 800 swimsuits to 900 swimsuits.

In Example 1.2, $C(0) = 9$. So, the firm has a weekly cost of $9000 when it produces no swimsuits. This cost is the firm's weekly **fixed cost**.

As expected, the graph of $C(x)$ shows that the total weekly production cost increases if the weekly production level is increased. The graph also shows that additional swimsuits are progressively more expensive to produce, even at low production levels. (How does the graph show this fact?) Typically, production costs do not behave in this manner. Usually, at low production levels, additional units are progressively less expensive to produce. Thus, the graph of a typical total cost function is similar to the curve in Figure 1.3. The curve starts above the origin because of the fixed cost.

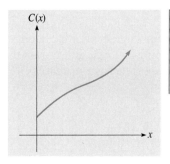

Figure 1.3 A typical total cost curve.

Average Rate of Change

Sometimes we want to determine the average amount the dependent variable of a function changes per unit increase in the independent variable.

Definition 1.2

Suppose $[a, b]$ is a closed interval contained by the domain of a function $f(x)$. The **average rate of change** of $f(x)$ with respect to x as x increases from a to b is
$$\frac{f(b) - f(a)}{b - a}$$

In the above definition, $b - a$ is the increase in x and $f(b) - f(a)$ is the corresponding change in $f(x)$. Therefore, the average rate of change tells us the average amount $f(x)$ changes *per one unit increase* in x as x increases from a to b.

In Example 1.3, we find the average rate at which a population changes during an interval of time.

Example 1.3

A conservationist projects that
$$G(x) = 0.08x^2 - 3.2x + 42$$
thousand giraffes will inhabit a certain region of Africa x years after the beginning of 1995. Find the average rate at which the giraffe population will change during the first decade of the twenty-first century.

Solution

The first decade of the twenty-first century begins 5 years after the beginning of 1995 and ends 15 years after the beginning of 1995. So we want to find the average rate of change of $G(x)$ with respect to x as x increases from 5 to 15.

Since
$$G(5) = 0.08(5^2) - 3.2(5) + 42$$
$$= 28$$

and
$$G(15) = 0.08(15^2) - 3.2(15) + 42$$
$$= 12$$

it follows that
$$G(15) - G(5) = 12 - 28$$
$$= -16$$

This result tells us $G(x)$ *decreases* 16 units as x increases from 5 to 15. Therefore, the average rate of change of $G(x)$ with respect to x as x increases from 5 to 15 is

$$\frac{G(15) - G(5)}{15 - 5} = \frac{-16}{10}$$
$$= -1.6$$

So, during the first decade of the twenty-first century, the giraffe population will *decrease* at the average rate of 1.6 thousand per year.

1.1 Problems

1. If $f(x) = 3x^2 - 4$, compute $f(5)$.

2. If $f(x) = x^3 + 2x$, compute $f(4)$.

3. If
$$y = \frac{12}{t + 3}$$
what is the value of y when $t = 1$?

4. If
$$y = \frac{5}{t + 1}$$
what is the value of y when $t = 0$?

5. Given $h(x) = \sqrt{x^2 + 5}$, find $h(2)$.

6. Given $h(x) = \sqrt{3x + 7}$, find $h(6)$.

7. Let $g(x) = 2x^3 + x$. How much does $g(x)$ change if x increases from 2 to 5?

8. Let $g(x) = x^2 - 8$. How much does $g(x)$ change if x increases from 3 to 4?

9. If $y = 36 - t^2$, what will be the change in y when t increases from 0 to 3?

10. If $y = t^2 - 16t + 64$, what will be the change in y when t increases from 1 to 4?

11. Suppose
$$f(x) = \frac{5x + 1}{x + 1}$$
How much will $f(x)$ change if x is increased from 2 to 3?

12. Suppose
$$f(x) = \frac{6x + 8}{x + 2}$$
How much will $f(x)$ change if x is increased from 2 to 6?

13. If $f(x) = 3x^2$, what is the average rate of change of $f(x)$ with respect to x as x increases from 2 to 5?

14. If $f(x) = x^3$, what is the average rate of change of $f(x)$ with respect to x as x increases from 0 to 4?

15. Given
$$y = \frac{1}{\sqrt{x}}$$
find the average rate of change of y with respect to x if x increases from 1 to 9.

16. Given
$$y = \frac{36}{\sqrt{x}}$$
find the average rate of change of y with respect to x if x increases from 4 to 9.

17. If $g(x) = x^2 + 3x$, find the average rate of change of $g(x)$ with respect to x when x increases from 0 to 6.

18. If $g(x) = x^2 - 5$, find the average rate of change of $g(x)$ with respect to x when x increases from 1 to 8.

In Problems 19 to 32, find the domain of the given function.

19. $f(x) = \dfrac{7}{x - 9}$

20. $f(x) = \dfrac{3x}{4 - x}$

21. $y = x^3 - 2x$

22. $y = 5x^2 + x - 9$

23. $h(t) = \dfrac{5 - t}{9 - t^2}$

24. $h(t) = \dfrac{t - 8}{t^2 - 25}$

25. $z = \sqrt{7 - t}$

26. $z = \sqrt{t - 2}$

27. $f(x) = \dfrac{1}{\sqrt{x - 4}}$

28. $f(x) = \dfrac{4x}{\sqrt{5 - x}}$

29. $g(x) = \sqrt{16 - x^2}$

30. $g(x) = \sqrt{25 - x^2}$

31. $h(x) = \dfrac{3x}{\sqrt{4 - x^2}}$

32. $h(x) = \dfrac{5}{\sqrt{36 - x^2}}$

33. **Microbiology** The population of a colony of microbes, which presently is 2000, will increase at a rate that depends on the size of the population. A microbiologist believes that when there are x thousand microbes in the colony, the population will increase at the rate of
$$r(x) = 0.24x - 0.02x^2$$
thousand microbes per day. At what rate will the population increase when there are 5000 (5 thousand) microbes in the colony?
Comment: The microbiologist believes the population will approach 12,000 but will never quite reach 12,000. Therefore, since the population is presently 2000, the domain of the function $r(x)$ is the half-open interval [2, 12).

34. **Physiology** For a particular vein, let $v(x)$ represent the rate (in centimeters per second) at which blood x centimeters from the center of the vein flows. Taking into account what is known as Poisenille's law, a physiologist assumes $v(x)$ is a function of x according to the formula
$$v(x) = 40,500(0.000081 - x^2)$$
At what rate does blood 0.005 of a centimeter from the center of the vein flow?
Comment: The inner radius of the vein is 0.009 of a centimeter. Therefore, the domain of the function $v(x)$ is the half-open interval [0, 0.009).

35. **Two-Income Families** In a European country, the number of families that depend on two incomes is expected to increase. The government projects that t years from now the number of such families will be
$$S(t) = \frac{3t + 3}{2t + 4}$$
million. According to this projection, how much will the number of such families increase during the next 4 years?

36. **Car Phones** A telephone company predicts that t years from now there will be
$$g(t) = \frac{36t + 8}{t + 1}$$
thousand cars equipped with telephones in a southwestern metropolitan area. According to this prediction, how much will the number of such cars increase during the next 3 years?

37. **Depreciation** An accountant believes that t years from now a machine will be worth
$$V(t) = \frac{96}{t + 2}$$
thousand dollars. How much will the machine depreciate during the 4-year period that begins 2 years from now?

38. **Average Production Cost** A firm that manufactures umbrellas finds that increasing the monthly production level decreases the *average cost* of producing each umbrella. To be more specific, the firm

finds that when it produces x thousand umbrellas per month, the average cost of producing each umbrella is

$$A(x) = \frac{5x + 12}{x}$$

dollars. How much will the average cost of producing each umbrella decrease if the firm increases the monthly production level from 4000 to 6000 umbrellas?

39. **Average Velocity** A baseball is dropped from the top of a tall building. Let $d(t)$ represent the number of feet it falls during the first t seconds. Then $d(t)$ is close to being a function of t according to the formula

$$d(t) = 16t^2$$

Use this formula to find the average rate of change of $d(t)$ with respect to t as t increases from 2 to 5. In other words, find the *average velocity* at which the ball falls during the 3-second interval of time that begins 2 seconds after it is dropped.

Comment: It takes 6 seconds for the ball to strike the ground. Therefore, the domain of $d(t)$ is the closed interval $[0, 6]$.

40. **Embryology** The weight of a human fetus (in kilograms) is proportional to the cube of its age (in weeks). Thus, if $w(x)$ represents the weight of an x-week-old fetus, there exists a constant k such that $w(x) = kx^3$. Suppose for a particular fetus $k = 0.00005$, so that

$$w(x) = 0.00005x^3$$

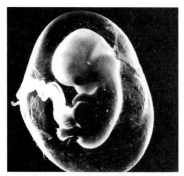

A human fetus after eight weeks.
(Dr. M. A. Ansary/Science Photo Library/Photo Researchers)

Find the average rate at which the weight increases during the interval of time that begins when the fetus becomes 20 weeks old and ends when it becomes 30 weeks old.

41. **Steel Industry Jobs** In Orange Waters County, the number of jobs provided by the steel industry is expected to decrease. County officials project that t months from now the number of such jobs will be

$$J(t) = \frac{24t + 90}{t + 2}$$

thousand. At what average rate will the number of jobs decrease during the next 6 months?

42. **Hematology** Let $N(t)$ represent the concentration of white blood cells in the blood of a typical t-year-old person. It is reasonable to assume that

$$N(t) = \frac{15t + 22}{2t + 2}$$

if the concentration is measured in thousands of white blood cells per cubic millimeter of blood. Find the average rate at which this concentration decreases during a typical person's first 10 years of life.

43. Suppose the average rate of change of $f(x)$ with respect to x is *zero* as x increases from a to b. Can we conclude that the value of $f(x)$ remains constant as x increases from a to b? Why?

44. Suppose the value of $f(x)$ remains constant as x increases from a to b. Can we conclude that the average rate of change of $f(x)$ with respect to x is *zero* as x increases from a to b? Why?

Graphing Calculator Problems

45. **Population** Set the viewing screen on your graphing calculator to $[0, 75] \times [0, 50]$ with a scale of 5 on both axes. Then draw the graph of the function in Example 1.3. The graph will show that the giraffe population will decrease for a while and then increase. Use the trace function to estimate when the population will be least. Approximately how large will the population be when it is least?

46. **Microbiology** Set the viewing screen on your graphing calculator to $[2, 12] \times [0, 1]$ with a scale of 1 on the x axis and a scale of 0.1 on the y axis. Then draw the graph of the function in Problem 33. The graph will show that the growth rate of the microbe population will increase at first and then decrease. Use the trace function to estimate what the

population will be when it is growing most rapidly. Approximately how rapidly will the population be growing when it is growing most rapidly?

47. **Profit** A manufacturing firm finds that its monthly profit is

$$P(x) = -2x^3 + 15x^2 - 25$$

thousand dollars when it sells x tons of its product per month. Set the viewing screen on your graphing calculator to $[0, 10] \times [-50, 130]$ with a scale of 1 on the x axis and a scale of 20 on the y axis. Then draw the graph of the above profit function. The graph will show that the firm loses money if its sales level is too low or too high. (A loss is indicated by a negative profit.) Use the trace function to estimate the sales levels at which the firm breaks even. (The firm **breaks even** when its profit is zero.)

48. **Market Equilibrium** Suppose consumers buy x tons of a product monthly when it sells for

$$f(x) = \frac{14}{x + 2}$$

thousand dollars per ton. Also, suppose the producers supply the market with x tons of the product monthly when it sells for

$$g(x) = 0.8x$$

thousand dollars per ton. Set the viewing screen on your graphing calculator to $[0, 5] \times [-2, 8]$ with a scale of 1 on both axes. Then draw the graphs of the above two functions so that they appear simultaneously. The graphs will show that there is a price at which the amount sold monthly is equal to the amount supplied monthly. (At such a price, the market is in **equilibrium**.) Use the trace function to estimate this price.

Computer Problems

The Basic program EVAL1 listed below computes the values of a function $f(x)$ at consecutive integer values of x in the domain. Execution of the program prompts you to enter only the least and greatest integers in the set of consecutive integers of your choosing. After you enter these two integers, the program prints the value of $f(x)$ (rounded to two decimal places) at *each* integer in the set of consecutive integers.

As listed, EVAL1 applies to the function

$$f(x) = x^2 - 6x + 10$$

However, you can make the program applicable to other functions by simply editing line 50.

The Program EVAL1 and a Sample Run

```
LIST
10   REM                          EVAL1
20   REM
30   REM A PROGRAM THAT EVALUATES A FUNCTION AT CONSECUTIVE INTEGERS IN ITS DOMAIN
40   REM
50   DEF FNF(X)=X^2-6*X+10
60   PRINT "ENTER THE LEAST AND GREATEST INTEGERS."
70   INPUT A,B
80   PRINT
90   FOR X=A TO B STEP 1
100  PRINT
110  PRINT "F(";X;") =";
120  PRINT USING "#####.##";FNF(X)
130  NEXT X
140  END
Ok
RUN
ENTER THE LEAST AND GREATEST INTEGERS.
? -6,9
```

F(-6) = 82.00
F(-5) = 65.00
F(-4) = 50.00
F(-3) = 37.00
F(-2) = 26.00
F(-1) = 17.00
F(0) = 10.00
F(1) = 5.00
F(2) = 2.00
F(3) = 1.00
F(4) = 2.00
F(5) = 5.00
F(6) = 10.00
F(7) = 17.00
F(8) = 26.00
F(9) = 37.00
Ok

The next Basic program, EVAL2, computes the values of a function $f(x)$ at *any* value of x in the domain. Execution of the program prompts you to enter an x value. After you enter an x value, the program prints the corresponding $f(x)$ value rounded to two decimal places. Then it tells you to enter Y, if you want the value of $f(x)$ at another x value, or N, if you want to exit the program. If you enter Y, the program prompts you to enter another x value, and so on.

As listed, EVAL2 applies to the function

$$f(x) = 3x^2 - 250$$

But you can make the program applicable to other functions by editing line 50.

The Program EVAL2 and a Sample Run

```
LIST
10   REM                          EVAL2
20   REM
30   REM A PROGRAM THAT EVALUATES A FUNCTION AT ANY POINT IN ITS DOMAIN
40   REM
50   DEF FNF(X)=3*X^2-250
60   PRINT "ENTER AN X VALUE."
70   INPUT X
80   PRINT
90   PRINT "F(";X;") =";
100  PRINT USING "#####.##";FNF(X)
110  PRINT
120  PRINT
130  PRINT "IF YOU WANT TO EVALUATE F(X) AT ANOTHER X VALUE, ENTER Y."
140  PRINT "OTHERWISE, ENTER N."
150  INPUT A$
160  PRINT
170  IF A$="Y" THEN 60
180  END
Ok
RUN
ENTER AN X VALUE.
? 8

F(8) = -58.00

IF YOU WANT TO EVALUATE F(X) AT ANOTHER X VALUE, ENTER Y.
OTHERWISE, ENTER N.
? Y
```

ENTER AN X VALUE.
? 4.63

F(4.63) = -185.69

IF YOU WANT TO EVALUATE F(X) AT ANOTHER X VALUE, ENTER Y.
OTHERWISE, ENTER N.
? Y

ENTER AN X VALUE.
? -17

F(-17) = 617.00

IF YOU WANT TO EVALUATE F(X) AT ANOTHER X VALUE, ENTER Y.
OTHERWISE, ENTER N.
? N
Ok

49. The domain of the function

$$f(x) = 0.04x^3 - 1.12x^2 + 6.4x + 5$$

is the closed interval [0, 20]. Use the program EVAL1 to compute the ordinates of the points on the graph of $f(x)$ whose abscissas are integers. Then exhibit the graph of $f(x)$ by sketching a curve that passes through these points.

50. The domain of the function

$$f(x) = 0.025x^3 - 0.675x^2 + 4.05x$$

is the closed interval [1, 18]. Use the program EVAL1 to compute the ordinates of the points on the graph of $f(x)$ whose abscissas are integers. Then exhibit the graph of $f(x)$ by sketching a curve that passes through these points.

51. Given

$$f(x) = \frac{x^2}{x + 5}$$

use the program EVAL2 to compute the following values:

a. $f(-3)$ b. $f(6.2)$

c. $f(3.25)$ d. $f(-9)$

52. Given

$$f(x) = \sqrt{x + 14}$$

use the program EVAL2 to compute the following values:

a. $f(9)$ b. $f(-4.2)$

c. $f(-0.8)$ d. $f(20)$

1.2 Finding Formulas for Functions

In this section we derive formulas for functional relationships. The first example involves a total cost function and a demand function. We have already introduced the idea of a total cost function in Section 1.1.

Demand Functions

The demand for a product is the product's sales volume. A **demand function** either expresses the demand for a product as a function of the price, or it expresses the price as a function of the demand, as is the case in the example that follows. Either way, typically, the value of the dependent variable *decreases* if the value of the independent variable increases.

Example 1.4

The management of Shirts by Rubin finds that the total weekly cost of producing x shirts per week is

$$C(x) = 6x + 10{,}000$$

dollars. The firm can sell x shirts weekly if it sells each shirt for

$$D(x) = -0.01x + 36$$

dollars. According to the management, these formulas are valid as long as x is restricted to the nonnegative integers less than or equal to 3600. Moreover, if the firm produces x shirts weekly, the firm sets the price of each shirt at $-0.01x + 36$ dollars. Thus, if the firm produces x shirts weekly, it sells x shirts weekly.

a. Find a formula that expresses the weekly revenue obtained from selling the shirts as a function of the weekly production level. (**Revenue** is income without taking cost into account.)
b. Find a formula that expresses the weekly profit earned from selling the shirts as a function of the weekly production level. (**Profit** takes cost into account.)
c. Find a formula that expresses the average cost of producing each shirt as a function of the weekly production level.

Solution for (*a*)

When the firm produces x shirts weekly, it sells x shirts weekly at $D(x)$ dollars per shirt. Thus, the weekly revenue obtained when the firm produces x shirts weekly is

$$\begin{aligned} R(x) &= x \cdot D(x) \\ &= x(-0.01x + 36) \\ &= -0.01x^2 + 36x \end{aligned}$$

dollars. Since the formula for $D(x)$ is valid only for values of x restricted to the nonnegative integers not greater than 3600, the domain of the function $R(x)$ is this same set of integers.

Solution for (*b*)

The profit is the difference between the revenue and the cost. Thus, the weekly profit earned when the firm produces (and sells) x shirts weekly is

$$\begin{aligned} P(x) &= R(x) - C(x) \\ &= (-0.01x^2 + 36x) - (6x + 10{,}000) \\ &= -0.01x^2 + 30x - 10{,}000 \end{aligned}$$

dollars. The domain of the function $P(x)$ is also the set of nonnegative integers less than or equal to 3600.

Solution for (c)

The average cost of producing each shirt is the total production cost divided by the number of shirts the firm produces. Therefore, the average cost of producing each shirt when the firm produces x shirts weekly is

$$A(x) = \frac{C(x)}{x}$$

$$= \frac{6x + 10,000}{x}$$

$$= 6 + \frac{10,000}{x}$$

dollars. The domain of the function $A(x)$ is the set of positive integers less than or equal to 3600. We exclude 0 from the domain of the function $A(x)$ because division by 0 is not allowed. Besides, it does not make sense to talk about the average cost of producing each shirt when no shirts are produced.

Revenue, Profit, and Average Cost Functions

Suppose $C(x)$ is the total cost of producing x units of a product during a specified time period and $D(x)$ is the price of the product at which x units are sold during the time period. Suppose, also, that whenever x units are produced, the price is set at $D(x)$. In Example 1.4, we illustrated that these two functions can be used to derive the following functions:

a. The function

$$R(x) = x \cdot D(x)$$

gives the revenue obtained when x units are produced and is called a **revenue function**.

b. The function

$$P(x) = R(x) - C(x)$$

gives the profit earned when x units are produced and is called a **profit function**.

c. The function

$$A(x) = \frac{C(x)}{x}$$

gives the average cost of producing each unit when x units are produced and is called an **average cost function.**

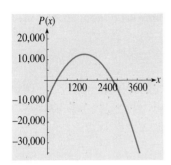

Figure 1.4
$P(x) = -0.01x^2 + 30x - 10{,}000$
for x in $[0, 3600]$.

Since the domain of the profit function $P(x)$ we derived in Example 1.4 is a set of integers, its graph is a collection of isolated points. However, if we assume the domain of this function is the closed interval $[0, 3600]$, rather than just the integers in this interval, the graph is the curve in Figure 1.4. The curve is below the x axis for small and large values of x. Thus, if the weekly production level is too low or too high, the weekly profit is negative, which means the company loses money.

Figure 1.4 shows that the graph of the profit function $P(x)$ has a *highest point*. The x coordinate (abscissa) of the highest point is the weekly production level at which the weekly profit is *maximum*. Later we will learn how to apply calculus to the formula for $P(x)$ to determine this production level.

In Example 1.5, we derive a function whose graph has a *lowest point*.

Example 1.5

A baker plans to bake 1470 heart-shaped cakes for St. Valentine's Day using cake pans that cost $3 each. Her oven has room for 100 cakes and consumes $2.50 worth of electricity to bake each batch. Express the combined cost of the pans and the electricity as a function of the number of pans bought and used. (Each pan is reusable.)

> **Solution**
>
> Let x represent the number of pans bought and used. Then
>
> $$\frac{1470}{x}$$
>
> represents the number of batches baked. Since the oven consumes $2.50 worth of electricity to bake each batch,
>
> $$2.50\left(\frac{1470}{x}\right)$$
>
> represents the cost (in dollars) of the electricity. Therefore, since each cake pan costs $3, the combined cost of the pans and the electricity is
>
> $$C(x) = 3x + 2.50\left(\frac{1470}{x}\right)$$
> $$= 3x + \frac{3675}{x}$$
>
> dollars if x pans are bought and used. Since the oven has room for 100 cakes, the domain of the function $C(x)$ is the integers from 1 to 100.

In Example 1.5, the number of batches baked must be, in reality, an integer. But $1470/x$ fails to be an integer for some values of x. Therefore, the function we derived, at best, approximates the relationship between the cost and the number of pans.

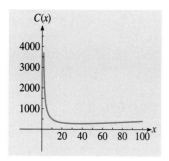

Figure 1.5
$C(x) = 3x + 3675/x$
for x in [1, 100].

If we assume the domain of the function in Example 1.5 is the closed interval [1, 100], rather than just the integers in this interval, the graph is the curve in Figure 1.5. The x coordinate of the *lowest point* on the curve is the number of pans that should be bought and used to *minimize* the combined cost of the pans and the electricity. Later, using calculus, we can show that this number of pans is 35.

New functions can be formed by adding, subtracting, multiplying, or dividing other functions. For instance, the function

$$C(x) = 3x + \frac{3675}{x}$$

may be viewed as the function obtained by adding the functions $3x$ and $3675/x$. Similarly, the function

$$s(t) = \frac{8t + 10}{t + 2}$$

may be viewed as the function obtained by dividing the function $8t + 10$ by the function $t + 2$. There is another way we can form new functions from other functions.

Composite Functions

Consider the function

$$f(x) = \sqrt{5 - x}$$

We can think of this function as the function formed by substituting $5 - x$ for x in the formula $\sqrt{x}$. Therefore, in view of the next definition, we can think of $f(x)$ as a composite function.

Definition 1.3

Suppose $g(x)$ and $u(x)$ are functions. The **composite** of $g(x)$ and $u(x)$ is the function formed by substituting $u(x)$ for x in the formula for $g(x)$ and is represented by $g[u(x)]$.

Thus, if we let

$$g(x) = \sqrt{x}$$

and

$$u(x) = 5 - x$$

then

$$g[u(x)] = \sqrt{u(x)} = \sqrt{5 - x}$$

So we can think of the function $f(x) = \sqrt{5 - x}$ as the composite of $g(x)$ and $u(x)$.

The domain of a composite function $g[u(x)]$ is the set of real numbers x such that x is in the domain of $u(x)$ and the corresponding value of $u(x)$ is in the domain of $g(x)$.

Example 1.6

Let
$$f(x) = \frac{5}{(x^2 + 9)^3}$$
Find two functions $g(x)$ and $u(x)$ whose composite is $f(x)$.

Solution

Let
$$g(x) = \frac{5}{x^3}$$
and
$$u(x) = x^2 + 9$$
Then
$$g[u(x)] = \frac{5}{[u(x)]^3}$$
$$= \frac{5}{(x^2 + 9)^3}$$

So the composite of $g(x)$ and $u(x)$ is $f(x)$.

In Example 1.6, if we instead let
$$g(x) = \frac{5}{x}$$
and
$$u(x) = (x^2 + 9)^3$$
we also would be able to conclude that
$$g[u(x)] = \frac{5}{(x^2 + 9)^3}$$

The order in which the composite of two functions is formed usually matters. For instance, in Example 1.6, we saw that if
$$g(x) = \frac{5}{x^3}$$
and
$$u(x) = x^2 + 9$$
then
$$g[u(x)] = \frac{5}{(x^2 + 9)^3}$$
But
$$u[g(x)] = [g(x)]^2 + 9$$
$$= \left(\frac{5}{x^3}\right)^2 + 9$$
$$= \frac{25}{x^6} + 9$$

1.2 Problems

1. **Average Production Cost** Suppose the total monthly cost of producing x tons of a product per month is
$$C(x) = 0.02x^2 + 18$$
hundred dollars. Find a formula that expresses the average cost of producing each ton as a function of the monthly production level.

2. **Average Production Cost** Suppose the total weekly cost of producing x barrels of a commodity per week is
$$C(x) = 0.01x^2 + 5$$
hundred dollars. Find a formula that expresses the average cost of producing each barrel as a function of the weekly production level.

3. **Revenue** The owners of a coffee mill find they can sell x pounds of their highest grade Colombian coffee daily if they sell it for
$$D(x) = \frac{2000}{x + 500}$$
dollars per pound. Whenever the mill processes x pounds of the coffee daily, the owners set the price at $D(x)$ dollars per pound. Find a formula that expresses the daily revenue obtained from selling the coffee as a function of the amount processed daily.

4. **Revenue** The owners of a bake shop know they can sell x pounds of cookies daily if they sell them for
$$D(x) = \frac{1200}{x + 400}$$
dollars per pound. Whenever the shop bakes x pounds of cookies daily, the owners set the price at $D(x)$ dollars per pound. Find a formula that expresses the daily revenue obtained from selling the cookies as a function of the amount baked daily.

5. **Profit** The total weekly cost of producing x pounds of a flea powder per week is
$$C(x) = 1.2x + 1400$$
dollars. The company that produces the powder can sell x pounds of the powder weekly at
$$D(x) = -0.002x + 6$$
dollars per pound. (These formulas, the company claims, are valid up to $x = 3000$.) Whenever the company produces x pounds of the powder weekly, the company sets the price at $D(x)$ dollars per pound. Find a formula that expresses the weekly profit earned from selling the powder as a function of the weekly production level.

6. **Profit** Sam, the owner of Sandwiches by Turedo, finds that the total daily cost of making x sandwiches per day is
$$C(x) = 0.8x + 100$$
dollars. Sam can sell x sandwiches daily at
$$D(x) = -0.02x + 5$$
dollars per sandwich. (Sam claims that these formulas are valid up to $x = 250$.) Whenever Sam makes x sandwiches per day, he sets the price of each sandwich at $D(x)$ dollars. Find a formula that expresses the daily profit Sam earns from selling the sandwiches as a function of the number of sandwiches he makes each day.

7. **Production Cost** A ceramist has decided to make 300 copies of a vase she designed. Her kiln has room for 25 molds and consumes $12 worth of electricity each time it is fired. The cost of building each mold is $9, and each mold is reusable. Find a formula that expresses the combined cost of making the molds and firing the kiln as a function of the number of molds that are built and used. What is the domain of this function?

8. **Delivery Cost** The ceramist in Problem 7 plans to use her truck to deliver the 300 vases to the warehouse of a department store. Special reusable crates for the vases have to be built at a cost of $9 per crate. Each crate holds one vase, and the truck has room for 50 crates. Therefore, the truck must make several trips, each trip costing $27. Find a formula that expresses the combined cost of crating and delivering the 300 vases as a function of the number of crates built and used. What is the domain of this function?

9. **Volume** A rectangular cardboard is 20 feet long and 12 feet wide. Equal squares are cut from each of the four corners. (See the figure for this problem.) The resulting flaps are folded up to form a topless box. Let x represent the length of each side of the

squares. Find a formula that expresses the volume of the box as a function of x. What is the domain of this function?

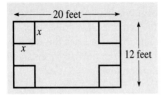

10. **Volume** Do Problem 9 for a cardboard 17 feet long and 9 feet wide.

11. **Surface Area** A container company has agreed to manufacture containers for Great Taste Foods. Each container must be a right circular cylinder with a volume of 20π cubic inches. Therefore, the amount of material required for each container may be viewed as a function of the radius of the base. Find a formula for this function and tell what the domain is.
    Hint: First express the height of each container in terms of the radius by using the fact the volume is given by the formula $V = \pi r^2 h$. (See the figure for this problem.)

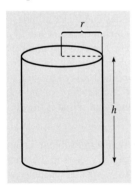

12. **Volume** A carpenter wants to build a closed rectangular box with a square base. The box must have a surface area of 72 square feet. Therefore, the volume of the box may be viewed as a function of the length of each side of the square base. Find a formula for this function and tell what the domain is.
    Hint: First express the height of the box in terms of the length of each side of the square base by making use of the fact the surface area is given by the formula $S = 2x^2 + 4xy$. (See the figure for this problem.)

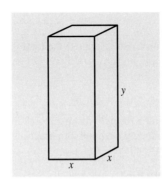

In Problems 13 to 18, find a formula for the composite of $g(x)$ and $u(x)$. In other words, find a formula for $g[u(x)]$.

13. $g(x) = x^5$, $u(x) = x - 9$

14. $g(x) = x^3$, $u(x) = 2x + 1$

15. $g(x) = \dfrac{1}{x}$, $u(x) = \sqrt{3x + 5}$

16. $g(x) = \dfrac{4}{x}$, $u(x) = (x - 1)^7$

17. $g(x) = \dfrac{2}{x^3}$, $u(x) = 4x + 6$

18. $g(x) = \dfrac{1}{\sqrt{x}}$, $u(x) = 3x + 8$

In Problems 19 to 24, find two functions $g(x)$ and $u(x)$ such that $f(x) = g[u(x)]$.

19. $f(x) = (3x + 4)^5$

20. $f(x) = (x^2 + 1)^8$

21. $f(x) = \dfrac{1}{x^2 + 7x}$

22. $f(x) = \dfrac{2}{4x - 6}$

23. $f(x) = \dfrac{9}{\sqrt{x - 11}}$

24. $f(x) = \dfrac{1}{(x^3 + x)^2}$

 Graphing Calculator Problems

25. Set the viewing screen on your graphing calculator to $[-7.5, 7.5] \times [-5, 5]$ with a scale of 1 on both axes. Then draw the graphs of the following functions:
$$g(x) = x^2$$
$$h(x) = (x - 3)^2$$
$$i(x) = (x + 5)^2$$

Now suppose $f(x)$ is *any* function and c is *any* positive real number.

a. The graphs of $g(x)$ and $h(x)$ suggest that there is a simple relationship between the graph of the composite function $f(x - c)$ and the graph of $f(x)$. What is the suggested relationship?

b. The graphs of $g(x)$ and $i(x)$ suggest that there is a simple relationship between the graph of the composite function $f(x + c)$ and the graph of $f(x)$. What is the suggested relationship?

26. Set the viewing screen on your graphing calculator to $[-10, 10] \times [-10, 10]$ with a scale of 1 on both axes. Then draw the graphs of the following functions:

$$g(x) = 0.5x^2 + 3$$
$$h(x) = -(0.5x^2 + 3)$$

Now suppose $f(x)$ is *any* function. The graphs of $g(x)$ and $h(x)$ suggest that there is a simple relationship between the graph of $f(x)$ and the graph of the composite function $-f(x)$. What is the suggested relationship?

Computer Problems

The Basic program OPT1 listed below finds the minimum and maximum values of a function $f(x)$ whose domain is a set of consecutive integers. Execution of the program prompts you to enter only the least and greatest integers in the domain. After you enter these two integers, the program prints the minimum and maximum values of $f(x)$. It also prints the values of x at which these values of $f(x)$ occur.

As listed, OPT1 applies to the function

$$f(x) = -x^2 + 16x + 80$$

However, you can make the program applicable to other functions by merely editing line 60.

The Program OPT1 and a Sample Run

```
LIST
10   REM                              OPT1
20   REM
30   REM A PROGRAM THAT FINDS THE MINIMUM AND MAXIMUM VALUES OF A FUNCTION WHOSE
40   REM DOMAIN IS A SET OF CONSECUTIVE INTEGERS
50   REM
60   DEF FNF(X)= X^2+16*X+80
70   PRINT "ENTER THE LEAST AND GREATEST INTEGERS IN THE DOMAIN."
80   INPUT A,B
90   PRINT
100  LET MN=FNF(A)
110  LET L=A
120  LET MX=FNF(B)
130  LET G=B
140  FOR I=A TO B STEP 1
150  IF FNF(I)>=MN THEN 180
160  LET MN=FNF(I)
170  LET L=I
180  IF FNF(I)<=MX THEN 210
190  LET MX=FNF(I)
200  LET G=I
210  NEXT I
220  PRINT "THE MINIMUM VALUE OF F(X) IS";MN; "AND IT OCCURS AT X=";L;"."
230  PRINT "THE MAXIMUM VALUE IS";MX;"AND IT OCCURS AT X=";G;"."
240  END
Ok
RUN
```

ENTER THE LEAST AND GREATEST INTEGERS IN THE DOMAIN.
? 3,20

THE MINIMUM VALUE OF F(X) IS 0 AND IT OCCURS AT X=20.
THE MAXIMUM VALUE IS 144 AND IT OCCURS AT X=8.
Ok

27. **Maximum Revenue** In Example 1.4a, we found that when Shirts by Rubin produces x shirts per week, the weekly revenue is

$$R(x) = -0.01x^2 + 36x$$

dollars, where x is a nonnegative integer less than or equal to 3600. Use the program OPT1 to find the weekly production level that maximizes the weekly revenue. How much revenue does the company obtain at that production level?

28. **Maximum Profit** In Example 1.4b, we found that when Shirts by Rubin produces x shirts per week, the weekly profit is

$$P(x) = -0.01x^2 + 30x - 10{,}000$$

dollars, where x is a nonnegative integer less than or equal to 3600. Use the program OPT1 to find the weekly production level that maximizes the weekly profit. How much profit does the company earn at that production level?

29. **Optimal Firing Batch** Use the program OPT1 to find the number of molds the ceramist in Problem 7 should build and use to minimize the combined cost of building the molds and firing the kiln.

30. **Optimal Baking Batch** Use the program OPT1 to find the number of cake pans the baker in Example 1.5 should buy and use to minimize the combined cost of the pans and the electricity.

1.3 Linear Functions

Calculus is primarily used in situations where a quantity varies at a *variable rate*. In this section, however, we focus on functions that are appropriate models for quantities that vary at a *constant rate*. We begin with a problem about the American population.

The American population is getting older. At the beginning of 1990, the median age of the population was 33. But the United States Bureau of the Census predicts that the median age will increase to 36 by the beginning of the year 2000. (The **median age** of a population is the age for which the number of people younger is equal to the number of people older.)

Let $A(x)$ represent the median age of the American population x years after the beginning of 1990, where x is in the interval $[0, 20]$. We assume that $A(x)$ is the function whose graph is the line segment in Figure 1.6.

The line segment in Figure 1.6 includes the points $(0, 33)$ and $(10, 36)$. So our assumption about the function $A(x)$ is compatible with the median age being 33 at the beginning of 1990 and 36 at the beginning of the year 2000.

In what follows, we derive a formula for $A(x)$. Then we can compute the median age for any time during the 20-year period from the beginning of 1990 to the beginning of 2010.

Suppose c is *any* number greater than zero in the interval $[0, 20]$. The average rate of change of $A(x)$ with respect to x as x increases from 0 to c is

$$\frac{A(c) - A(0)}{c - 0} = \frac{A(c) - 33}{c}$$

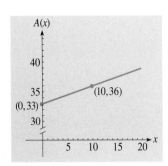

Figure 1.6 The graph of $A(x)$ for x in $[0, 20]$.

In particular, the average rate of change of $A(x)$ with respect to x as x increases from 0 to 10 is

$$\frac{A(10) - A(0)}{10 - 0} = \frac{36 - 33}{10}$$
$$= 0.3$$

(So, during the nineties, the median age will increase at the average rate of 0.3 of a year per year.) Since the graph of $A(x)$ is a line segment, $A(x)$ always changes the same amount whenever x increases 1 unit, regardless of the level from which x increases. Thus, the average rate of change of $A(x)$ with respect to x is 0.3, regardless of the interval on which x increases. Therefore,

$$\frac{A(c) - 33}{c} = 0.3$$
$$A(c) - 33 = 0.3c \quad \text{multiplying by } c$$
$$A(c) = 0.3c + 33 \quad \text{adding 33}$$

Since c is any number greater than 0 in the interval [0, 20], the above result tells us that

$$A(x) = 0.3x + 33$$

for any value of x in the half-open interval (0, 20]. According to this formula, $A(0) = 33$, which agrees with the median age being 33 at the beginning of 1990. So, actually, the above formula gives the value of $A(x)$ for any value of x in the closed interval [0, 20], not just those greater than 0.

Now, we can find the median age of the American population for any time during the 20-year period from the beginning of 1990 to the beginning of 2010. For example, by substituting 8 for x in the formula $A(x) = 0.3x + 33$, we get

$$A(8) = 0.3(8) + 33$$
$$= 35.4$$

Therefore, the median age will be 35.4 at the beginning of 1998. (The beginning of 1998 is 8 years after the beginning of 1990.)

The function

$$A(x) = 0.3x + 33$$

is an example of a linear function.

Definition 1.4

A **linear function** is a function that can be expressed in the form

$$f(x) = mx + b$$

where m and b are constants.

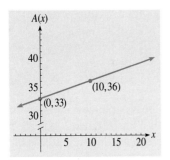

Figure 1.7 The graph of $A(x) = 0.3x + 33$ with no restrictions on the domain.

If the domain is not restricted, the graph of a linear function is a nonvertical line. So, if we allow the domain of the function $A(x) = 0.3x + 33$ to be the set of *all* real numbers, rather than just the interval [0, 20], the graph would be the *entire* line that passes through the points (0, 33) and (10, 36), rather than just the segment shown in Figure 1.6. Figure 1.7 shows the graph of $A(x) = 0.3x + 33$ with no restrictions on the domain.

The Slope of a Line

Since the graph of a linear function is either a nonvertical line or a portion of a nonvertical line, the value of the dependent variable always changes the same amount whenever the value of the independent variable increases 1 unit. In other words, the value of the dependent variable changes at a *constant rate* as the value of the independent variable increases. This result makes Definition 1.5 possible.

Definition 1.5

Suppose $f(x)$ is a linear function whose graph is an entire line. The **slope** of the line is the constant rate at which $f(x)$ changes as x increases.

If the slope is positive, $f(x)$ increases whenever x increases; if the slope is negative, $f(x)$ decreases whenever x increases.

In our aging population example, $A(x)$ increases at the constant rate of 0.3 of a year per year. Therefore, according to Definition 1.5, if the domain is the set of real numbers, the slope of the line that represents the function $A(x)$ is 0.3. Moreover, since $A(0) = 33$, the line crosses the vertical axis at the point (0, 33). Recalling that

$$A(x) = 0.3x + 33$$

we see that we have illustrated Theorem 1.1.

Theorem 1.1

The graph of a function that can be expressed in the form

$$f(x) = mx + b$$

is the line whose slope is m and that crosses the vertical axis at the point (0, b).

In Theorem 1.1, the domain of $f(x)$ is the set of all real numbers. Using the theorem, we can immediately conclude, for example, that the graph of the linear function

$$f(x) = -4x + 5$$

is the line whose slope is -4 and that crosses the vertical axis at the point $(0, 5)$. Since the slope is -4, $f(x)$ *decreases* at the constant rate of 4 units per 1 unit increase in x.

Figure 1.8 shows four lines that have different slopes. The slope of line L_1 is 3, the slope of line L_2 is $\frac{1}{2}$, the slope of line L_3 is 0, and the slope of line L_4 is -2.

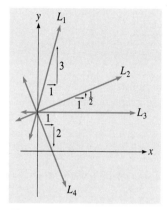

Figure 1.8 Four lines with different slopes.

Horizontal Lines

Line L_3 in Figure 1.8 illustrates the following result:

> The lines with zero slopes are the horizontal lines.

Constant functions can be expressed in the form

$$f(x) = 0x + b$$

Therefore, the graph of a constant function is a horizontal line.

Vertical Lines

On a vertical line every point has the same abscissa. Therefore, a vertical line can be characterized by writing

$$x = c$$

where c is the common abscissa of the points on the line. For example,

$$x = 6$$

characterizes the vertical line that crosses the horizontal axis at the point $(6, 0)$. Figure 1.9 shows this vertical line.

> The idea of slope is left undefined for vertical lines.

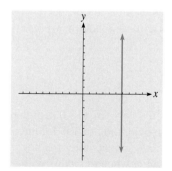

Figure 1.9 The vertical line $x = 6$.

A Formula for Finding Slopes

The validity of Theorem 1.2 is based primarily on Definition 1.5.

Theorem 1.2

If (x_1, y_1) and (x_2, y_2) are any two points on a nonvertical line, the slope of the line is

$$m = \frac{y_2 - y_1}{x_2 - x_1}$$

We use Theorem 1.2 in Example 1.7.

Example 1.7

Find the slope of the line that passes through the points (3, 1) and (7, 9). Then find a formula for the linear function whose graph is this line.

Solution

Since the line passes through the points (3, 1) and (7, 9), its slope is

$$\frac{9-1}{7-3} = 2$$

Let $f(x)$ be the linear function whose graph is the line in question. Then there exist constants m and b such that

$$f(x) = mx + b$$

In fact, since the slope of the line is 2, $m = 2$ and so

$$f(x) = 2x + b$$

Since the graph of $f(x)$ passes through the points (3, 1) and (7, 9), $f(3) = 1$ and $f(7) = 9$. To determine the value of b, we can either use $f(3) = 1$ or $f(7) = 9$. If we use $f(7) = 9$,

$$9 = 2 \cdot 7 + b$$

which implies

$$b = -5$$

Hence,

$$f(x) = 2x - 5$$

Figure 1.10 shows the graph of this function.

Since the slope of the line in Figure 1.10 is 2, $f(x)$ *increases* at the constant rate of 2 units per 1 unit increase in x.

We have seen that if $f(x)$ is a linear function, then $f(x)$ changes at a constant rate as x increases. The converse of this result is also valid. That is, if $f(x)$ changes at a constant rate, then $f(x)$ is a linear function. We use this result in Example 1.8.

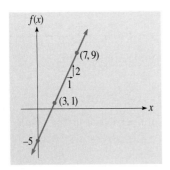

Figure 1.10 $f(x) = 2x - 5$.

Example 1.8

A machine that depreciates at a constant rate had a value of $18,000 at the beginning of 1990 and a value of $14,400 at the beginning of 1993. Find a formula that gives its value, in thousands of dollars, x years after the beginning of 1990. Then use the formula to determine when the value will be $6000.

Solution

Let $V(x)$ represent the value (in thousands of dollars) x years after the beginning of 1990. Since the value decreases at a constant rate, $V(x)$ is a linear function. So there exist constants m and b such that

$$V(x) = mx + b$$

Since the value was $18,000 ($18 thousand) at the beginning of 1990, $V(0) = 18$. Thus, $b = 18$, and so

$$V(x) = mx + 18$$

The beginning of 1993 was 3 years after the beginning of 1990. Therefore, since the value was $14,400 ($14.4 thousand) at the beginning of 1993, $V(3) = 14.4$. Thus,

$$14.4 = m \cdot 3 + 18$$

which implies

$$m = -1.2$$

Therefore,

$$V(x) = -1.2x + 18$$

is a formula for the value of the machine x years after the beginning of 1990.

Now we find the value of x for which $V(x) = 6$:

$$6 = -1.2x + 18$$
$$1.2x = 12$$
$$x = 10$$

So the machine's value will be $6,000 at the beginning of 2000 (10 years after the beginning of 1990). Figure 1.11 shows the graph of $V(x)$.

In Example 1.8, since $V(x)$ represents the value of a machine, $V(x)$ cannot be negative. In consideration of this requirement, we restrict the domain to the closed interval [0, 15]. (See Figure 1.11.) The fact $m = -1.2$ tells us the machine's value *decreases* at the constant rate of $1200 ($1.2 thousand) per year.

Another way of saying the dependent variable of a function changes at a constant rate is to say that it *changes linearly*. So we could say the machine in Example 1.8 depreciates linearly. We use this terminology in the problems at the end of this section.

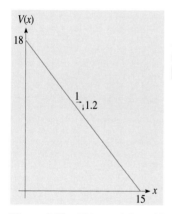

Figure 1.11 $V(x) = -1.2x + 18$ for x in [0, 15].

Regression Lines

Sometimes we use data about two nonlinearly related variables to derive a linear function that *approximates* the relationship between the two variables. In the following situation, such a derivation is feasible.

The price of a liter of Momiles, a gasoline additive that improves mileage, has changed several times. Table 1.2 shows the monthly demand for each price.

Table 1.2

Price per liter, dollars	3	5	8	9	12	14
Monthly demand, millions of liters	6	3.8	3.6	2.2	1	1.2

Figure 1.12 depicts the information in Table 1.2 as **data points** in a cartesian plane. The prices are plotted on the horizontal axis (the *x* axis) and the corresponding demands are plotted on the vertical axis (the *y* axis).

The demand is not a linear function of the price, because there is no line that passes through all six of the data points in Figure 1.12. Nevertheless, Figure 1.13 suggests that it would be reasonable to *approximate* the relationship with a linear function.

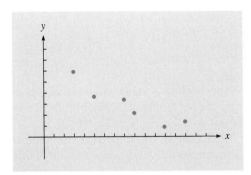

Figure 1.12 The information in Table 1.2 depicted as data points.

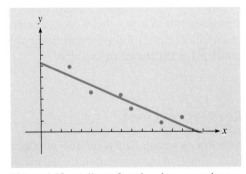

Figure 1.13 A linear function that approximates the relationship between Momiles' demand and its price.

Let $D(x)$ represent the linear function whose graph is the line (segment) in Figure 1.13. We want to make the function $D(x)$ a good approximation for the actual relationship between the demand and the price. One way of accomplishing this result is to position the line in Figure 1.13 so that the sum of the squares of the vertical distances between the data points and the line is as small as possible. Such a line is then defined as the **regression line** for the six data points. The corresponding function $D(x)$ is the **least-squares linear function** that approximates the relationship between the demand and the price. A graphing calculator problem at the end of this section (Problem 37) will ask you to find this function.

In a situation where the data points are collinear, the regression line is, in fact, the line that passes through all of the data points.

In Chapter 7, we will use calculus to find regression lines for data points.

1.3 Problems

The graph of each function in Problems 1 to 8 is a line. Find the slope of the line and the point where the line crosses the vertical axis. Then sketch the line in the cartesian plane, showing the coordinates of at least two points on the line.

1. $f(x) = 3x + 1$
2. $f(x) = -1.5x + 6$
3. $f(x) = -2.5x + 4$
4. $f(x) = x - 3$
5. $f(x) = 2$
6. $f(x) = 4$
7. $f(x) = -x$
8. $f(x) = x$

In Problems 9 to 18, find the slope of the line that passes through the given pair of points. Then find a formula for the linear function whose graph is the line.

9. (0.5, 6) and (3, 1)
10. (1, 5) and (4, 14)
11. (2, 3) and (5, 15)
12. (2, 7) and (6, 1)
13. (0, 1) and (4, 3)
14. (0, 7) and (2, 3)
15. (4, 2) and (9, 2)
16. (3, 3) and (8, 3)
17. (0, 0) and (8, 2)
18. (0, 0) and (4, 3)

19. Let $f(x)$ be a linear function for which $f(0) = 15$ and $f(3) = 3$. Find a formula for $f(x)$. Then find $f(1)$.

20. Let $f(x)$ be a linear function for which $f(0) = 0$ and $f(8) = 1$. Find a formula for $f(x)$. Then find $f(6)$.

21. If $f(x)$ is a linear function for which $f(1) = 7$ and $f(4) = 22$, find a formula for $f(x)$. Then find $f(9)$.

22. If $f(x)$ is a linear function for which $f(3) = 7$ and $f(5) = 5$, find a formula for $f(x)$. Then find $f(8)$.

23. **Production** Let $S(x)$ represent the quantity of tomatoes, in millions of cartons, the farmers of a region in Arizona produce yearly when their tomatoes sell for x cents per pound. Suppose, for x in the interval [15, 75], that $S(x)$ is a linear function for which $S(30) = 10$ and $S(36) = 14$. Find a formula for $S(x)$. Then use the formula to determine how many cartons of tomatoes the farmers produce yearly when their tomatoes sell for 60 cents per pound.

24. **Demand** Let $D(x)$ represent the quantity of leaf lettuce, in thousands of crates, the people of a region in Kentucky buy monthly when leaf lettuce sells for x cents per pound. Suppose, for x in the interval [60, 120], that $D(x)$ is a linear function for which $D(80) = 250$ and $D(90) = 225$. Find a formula for $D(x)$. Then use the formula to determine how many crates of the lettuce the people buy monthly when the lettuce sells for $1 (100 cents) per pound.

25. **Energy Production Cost** In a developing country, the present cost of producing each megawatt hour of electrical energy is $70. Experts predict that in 5 years this cost will be $66. Suppose the cost decreases at a constant rate during the next 20 years. Find a formula for the cost t years from now. Then use the formula to determine when it will cost $64.40 to produce each megawatt hour of electrical energy.

26. **Velocity** An object is projected straight upward with an initial velocity of 256 feet per second. After 2 seconds, its velocity is 192 feet per second. Assume the velocity changes at a constant rate. Find a formula that gives the velocity t seconds after the

object is projected. Then use the formula to determine when the velocity is 96 feet per second.

27. **Consumption Rate** According to a U.S. Department of Agriculture report, on the average, each person was consuming fresh vegetables at the rate of 146.2 pounds per year at the beginning of 1981 and at the rate of 150.9 pounds per year at the beginning of 1982. Assume the consumption rate increases linearly. Find a formula for the consumption rate t years after the beginning of 1981. Then use the formula to determine the consumption rate at the beginning of 1998.

28. **Jogging** According to a Gallup Poll, at the beginning of 1977 joggers, on the average, were jogging 1.6 miles per outing. At the beginning of 1984, they were averaging 2.5 miles per outing. Assume the average number of miles jogged per outing increases linearly. Find a formula for this average t years after the beginning of 1977. Then use the formula to determine the average at the beginning of 1998.

29. **Marketing** Canadian Summer, a new mouthwash, will soon be introduced in a community whose population is 300,000. Let $R(x)$ represent the rate, in thousands per day, at which the community will be introduced to the mouthwash when x thousand people are already introduced to it. The introduction will be accompanied by frequent advertisements that will be continued until 270,000 people are introduced to the product. Taking this strategy into account, a marketing expert predicts that while the advertising campaign is in force, $R(x)$ will be a linear function for which $R(25) = 5.5$ and $R(100) = 4$. Find a formula for $R(x)$. Then use the formula to determine how rapidly the community will be introduced to the product when 250,000 (250 thousand) people are already introduced to it.

30. **Oxygen Uptake** The maximal oxygen uptake of an individual is the maximum rate at which the individual can take in oxygen from the atmosphere. At about age 25, this rate begins to decline. Let $M(x)$ represent the maximal oxygen uptake, in liters per minute, of a typical physically active x-year-old male, where x is in the interval [25, 80]. It is reasonable to assume that $M(x)$ is a linear function for which $M(30) = 4.5$ and $M(60) = 3$. Use this assumption to derive a formula for $M(x)$. Then use the formula to compute the maximal oxygen uptake of a typical active 40-year-old male.

31. **Treadmill Time** The maximum treadmill time of a typical physically fit person is 11 minutes at age 40 and 9.5 minutes at age 50. Assume the maximum treadmill time decreases linearly from age 40 to age 80. Find a formula that expresses the maximum treadmill time as a function of age. Then use the formula to find the maximum treadmill time of a typical physically fit 62-year-old person.

32. **Heart Rate** A woman's maximal heart rate (in beats per minute) was 216 at age 5 and 196 at age 25. It is reasonable to assume that the woman's maximal heart rate is a linear function of her age. Use this assumption to derive a formula that expresses her maximal heart rate as a function of her age. Then use the formula to find the woman's maximal heart rate at age 37.

33. **Mixture** Two grades of dried fruit are mixed to make a batch worth $45. Let x represent the number of pounds of the lower-grade fruit used, and let y represent the number of pounds of the higher-grade fruit used. Then y is a linear function of x. Suppose the lower-grade fruit is worth $3 per pound and the higher-grade fruit is worth $5 per pound. Find a formula that expresses y as a function of x. What is the domain of this function?

34. **Temperature** The temperature C of a substance in degrees Celsius is a linear function of its temperature F in degrees Fahrenheit. Find a formula for this functional relationship. Then use the formula to determine the temperature of a flask of water according to the Celsius scale if its temperature according to the Fahrenheit scale is 70 degrees.

    Hint: According to the Fahrenheit scale, water freezes at 32 degrees and boils at 212 degrees. According to the Celsius scale, it freezes at 0 degrees and boils at 100 degrees.

35. Why did we leave the idea of slope undefined for vertical lines?

36. Suppose [a, b] is a closed interval contained by the domain of a function $f(x)$. Explain why the *average rate of change* of $f(x)$ with respect to x as x increases from a to b is equal to the *slope* of the line that passes through the points $(a, f(a))$ and $(b, f(b))$.

Graphing Calculator Problems

37. **Demand** As we have seen, Table 1.2 contains information about the demand for a gasoline additive

called Momiles. Figure 1.12 depicts the information as data points in a cartesian plane.

a. Set the viewing screen on your graphing calculator to [0, 16] × [0, 7] with a scale of 1 on both axes, and draw the regression line for the data points in Figure 1.12. Then use the trace function to approximate the monthly demand when the price is $4.25 per liter.

b. The least-squares linear function that corresponds to the regression line you drew in part (*a*) can be expressed in the form

$$D(x) = mx + b$$

Press the appropriate keys on your graphing calculator to find the values of *m* and *b*. Then use the resulting formula for $D(x)$ to approximate the monthly demand when the price is $4.25 per liter. Your answer should be close to the answer you obtained in part (*a*) using the trace function.

38. **Population** The following table shows the population of a country on five different dates:

Date, Jan. 1	1986	1987	1990	1991	1993
Population, millions	42.2	44	44.3	45	47

a. Set the viewing screen on your graphing calculator to [0, 20] × [40, 55] with a scale of 1 on both axes. Then think of the population as a function of the number of years after the beginning of 1985, and plot the data points that correspond to the information in the above table.

b. Draw the regression line for the data points you plotted in part (*a*). Then use the trace function to estimate what the population will be at the beginning of 1999. (1999 is 14 years after the beginning of 1985.)

c. The least-squares linear function that corresponds to the regression line you drew in part (*b*) can be expressed in the form

$$P(x) = mx + b$$

Press the appropriate keys on your graphing calculator to find the values of *m* and *b*. Then use the resulting formula for $P(x)$ to estimate what the population will be at the beginning of 1999. Your answer should be close to the answer you obtained in part (*b*) using the trace function.

1.4 Quadratic Functions

The **maximum value** of a function is the ordinate of the *highest* point on the graph of the function. Similarly, the **minimum value** is the ordinate of the *lowest* point on the graph. So, if the graph does not have a highest point, the function does not have a maximum value, and if the graph does not have a lowest point, it does not have a minimum value.

In Chapter 2, we will use calculus to find maximum and minimum values of functions. In this section, however, we deal with functions whose maximum and minimum values can be found without calculus. We begin with a business problem.

Several months ago, a company began using a fleet of five trucks to sell frozen yogurt in the streets of a city. Since then, each truck has been producing a monthly profit of $600. Management believes that if it increases the fleet, each additional truck will cause the monthly profit produced by *every* truck to decrease $40. The city permits the company to use at most 18 trucks. In what follows, we will determine the level to which the fleet should be increased so that it produces the highest possible total monthly profit.

Let $P(x)$ represent the total monthly profit, in *hundreds* of dollars, if the fleet is increased to x trucks. [We want to find the value of x for which $P(x)$ is maximum.] Since the fleet already has five trucks,

$$x - 5$$

represents the number of trucks added to the fleet. Since each additional truck causes the monthly profit per truck to decrease $40 ($0.4 hundred) and since the monthly profit per truck is $600 ($6 hundred) if no trucks are added, each one of the x trucks in the expanded fleet will produce a monthly profit of

$$6 - 0.4(x - 5)$$

hundred dollars. Therefore, the entire expanded fleet of x trucks will produce a total monthly profit of

$$P(x) = x[6 - 0.4(x - 5)]$$
$$= -0.4x^2 + 8x$$

hundred dollars.

Management does not plan to decrease the size of the fleet. Therefore, since the city permits the company to use at most 18 trucks, the domain of the function $P(x)$ is the set of integers from 5 to 18. The problem, then, is to find which of these integers is the x coordinate (abscissa) of the highest point on the graph of $P(x)$. We could find this integer by substituting each of the integers from 5 to 18 in the formula $P(x) = -0.4x^2 + 8x$. But we will use an alternative approach.

The function

$$P(x) = -0.4x^2 + 8x$$

is an example of a quadratic function.

Definition 1.6

A **quadratic function** is a function that can be expressed in the form

$$f(x) = ax^2 + bx + c$$

where a, b, and c are constants and a is not zero.

For the function $P(x) = -0.4x^2 + 8x$, $a = -0.4$, $b = 8$, and $c = 0$. As we will see, quadratic functions arise in many situations.

Parabolas

If the domain is the set of all real numbers, the graph of a quadratic function is a curve called a **parabola**. Figure 1.14 shows the parabola that represents the quadratic function

$$P(x) = -0.4x^2 + 8x$$

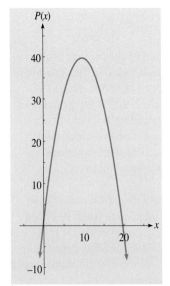

Figure 1.14 The parabola that represents $P(x) = -0.4x^2 + 8x$.

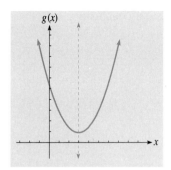

Figure 1.15 The parabola that represents $g(x) = 0.5x^2 - 3x + 5.5$.

when its domain is the set of all real numbers.

A parabola that represents a quadratic function has either a highest point or a lowest point, called the **vertex**. Such a parabola is symmetrical with respect to the vertical line that passes through the vertex. So we call this vertical line the **axis of symmetry**.

The vertex of the parabola in Figure 1.14 is the highest point on the parabola. Figure 1.15 shows the parabola that represents the quadratic function

$$g(x) = 0.5x^2 - 3x + 5.5$$

The vertex of this parabola is the lowest point on the parabola. The dashed line is the axis of symmetry. If we fold the page along the axis of symmetry, the portions of the parabola on opposite sides of the axis of symmetry would coincide.

We can determine whether the vertex is the highest or lowest point on a parabola that represents a function of the form

$$f(x) = ax^2 + bx + c$$

by the sign of a. The function $P(x) = -0.4x^2 + 8x$ illustrates that if a is *negative*, the vertex is the *highest* point. The function $g(x) = 0.5x^2 - 3x + 5.5$ illustrates that if a is *positive*, the vertex is the *lowest* point.

Finding the Vertex of a Parabola

Suppose

$$f(x) = ax^2 + bx + c$$

where $a \neq 0$. Let v denote the x coordinate of the vertex of the corresponding parabola. We will find a formula for v. For now, we assume $v \neq 0$. Note that

$$f(0) = c$$

Since the axis of symmetry is the vertical line that crosses the x axis at $x = v$,

$$f(2v) = c$$

also. (See Figure 1.16.) Thus,

$$a(2v)^2 + b(2v) + c = c$$
$$4av^2 + 2bv + c = c$$
$$4av^2 + 2bv = 0 \quad \text{subtracting } c$$
$$2av + b = 0 \quad \text{dividing by } 2v$$
$$v = \frac{-b}{2a} \quad \text{solving for } v$$

It can be shown that if $v = 0$, then $b = 0$. So the formula

$$v = \frac{-b}{2a}$$

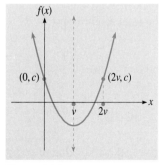

Figure 1.16
$f(x) = ax^2 + bx + c$, where $a > 0$.

is valid even if $v = 0$. Theorem 1.3 summarizes our conclusion.

Theorem 1.3

The vertex of a parabola that represents a function of the form
$$f(x) = ax^2 + bx + c$$
is the point whose x coordinate is
$$\frac{-b}{2a}$$

As we said, the parabola in Figure 1.15 represents the function
$$g(x) = 0.5x^2 - 3x + 5.5$$
For this function, $a = 0.5$ and $b = -3$. So, according to the above theorem, the x coordinate of the vertex of the parabola is
$$\frac{-b}{2a} = \frac{-(-3)}{2(0.5)}$$
$$= 3$$

Maximum and Minimum Values of Quadratic Functions

Theorem 1.3 implies that if
$$f(x) = ax^2 + bx + c$$
and the domain of $f(x)$ includes $-b/2a$, the maximum value of $f(x)$ (if a is negative) or the minimum value of $f(x)$ (if a is positive) occurs at
$$x = \frac{-b}{2a}$$

Now we can complete the solution of our truck fleet problem. We found that if the fleet is expanded to x trucks, the total monthly profit is
$$P(x) = -0.4x^2 + 8x$$
hundred dollars. For this function, $a = -0.4$ and $b = 8$. So
$$\frac{-b}{2a} = \frac{-8}{2(-0.4)}$$
$$= 10$$

Since 10 is in the domain of $P(x)$ and a is negative, 10 is the value of x that maximizes $P(x)$. So the fleet should be increased to 10 trucks for the total monthly profit to be as high as possible.

In Example 1.9, we find the minimum value of a quadratic function.

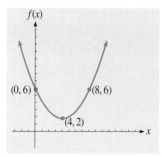

Figure 1.17
$f(x) = 0.25x^2 - 2x + 6$.

Example 1.9

Suppose the domain of

$$f(x) = 0.25x^2 - 2x + 6$$

is the interval [0, 10]. Find the minimum value of $f(x)$.

Solution

For the given function, $a = 0.25$ and $b = -2$. Therefore,

$$\frac{-b}{2a} = \frac{-(-2)}{2(0.25)}$$
$$= 4$$

Since 4 is in the interval [0, 10] and a is positive, the minimum value of $f(x)$ occurs at $x = 4$. Since

$$f(4) = 0.25(4^2) - 2 \cdot 4 + 6$$
$$= 2$$

the minimum value of $f(x)$ is 2.

If the domain is the set of *all* real numbers, the graph of the function in Example 1.9 is the parabola in Figure 1.17. The vertex of the parabola is the point (4, 2).

In sketching a parabola, we start by plotting at least three points that lie on the parabola. One point should be the vertex, and two points should be on opposite sides of the axis of symmetry, preferably equally distant from the axis of symmetry. Then we sketch the parabola that passes through the plotted points. We used this strategy to sketch the parabola in Figure 1.17. There we started by plotting the points (0, 6), (4, 2), and (8, 6).

The Zeros of Quadratic Functions

The **zeros** of a function $f(x)$ are the values of x for which $f(x) = 0$.

A parabola (or a portion of a parabola) can cross the horizontal axis in *at most two places*. Thus, if $f(x)$ is a quadratic function, there are at most two values of x (among the reals) for which $f(x) = 0$. Sometimes we find such values of x by factoring. As an example, consider the quadratic function

$$f(x) = x^2 - 9x + 14$$

By factoring, we obtain

$$x^2 - 9x + 14 = (x - 2)(x - 7)$$

This result reveals that $f(x) = 0$ when $x = 2$ or $x = 7$. Thus, if the domain contains 2 and 7, the graph of $f(x)$ crosses the x axis at $x = 2$ and at $x = 7$.

Theorem A.5 in Section A.3 of Appendix A implies an alternative method for finding the zeros of a quadratic function. The method is as follows:

Suppose $f(x)$ is a quadratic function. Then $f(x)$ can be expressed in the form
$$f(x) = ax^2 + bx + c$$
where $a \neq 0$. If $b^2 - 4ac \geq 0$, Theorem A.5 implies that the zeros of $f(x)$ are
$$x = \frac{-b + \sqrt{b^2 - 4ac}}{2a}$$
and
$$x = \frac{-b - \sqrt{b^2 - 4ac}}{2a}$$

Note that if $b^2 - 4ac = 0$, both of the above formulas simplify to $-b/2a$. So, if $b^2 - 4ac = 0$, $f(x)$ has only one real zero, namely,
$$x = \frac{-b}{2a}$$

However, if $b^2 - 4ac < 0$, then $f(x)$ has no real zeros, because the square root of a negative number is not a real number. The formula
$$\frac{-b \pm \sqrt{b^2 - 4ac}}{2a}$$
is known as the **quadratic formula**, and the value of $b^2 - 4ac$ is called the **discriminant**.

Example 1.10

For each of the following quadratic functions, use the quadratic formula to find the values of x for which $f(x) = 0$.

a. $f(x) = -2x^2 + 9x - 4$
b. $f(x) = 9x^2 - 12x + 4$
c. $f(x) = x^2 - 6x + 10$

Solution for (a)

For the function $f(x) = -2x^2 + 9x - 4$, we have $a = -2$, $b = 9$, and $c = -4$. Thus, the discriminant is
$$b^2 - 4ac = 9^2 - 4(-2)(-4)$$
$$= 49$$

Since the discriminant is positive, there exist two real values of x for which $f(x) = 0$. These values are
$$x = \frac{-9 + \sqrt{49}}{2(-2)}$$
$$= 0.5$$
and
$$x = \frac{-9 - \sqrt{49}}{2(-2)}$$
$$= 4$$

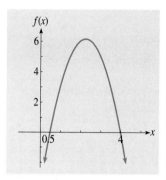

Figure 1.18
$f(x) = -2x^2 + 9x - 4$.

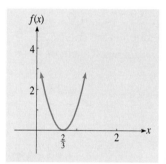

Figure 1.19
$f(x) = 9x^2 - 12x + 4$.

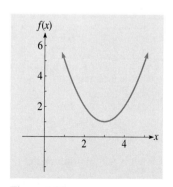

Figure 1.20
$f(x) = x^2 - 6x + 10$.

(See Figure 1.18.)

Solution for (b)

For the function $f(x) = 9x^2 - 12x + 4$, we have $a = 9$, $b = -12$, and $c = 4$. Thus, the discriminant is

$$b^2 - 4ac = (-12)^2 - 4(9)(4)$$
$$= 0$$

Therefore, there is only one real value of x for which $f(x) = 0$. This value is

$$x = \frac{-(-12)}{2(9)}$$
$$= \frac{2}{3}$$

(See Figure 1.19.)

Solution for (c)

For the function $f(x) = x^2 - 6x + 10$, we have $a = 1$, $b = -6$, and $c = 10$. Thus, the discriminant is

$$b^2 - 4ac = (-6)^2 - 4(1)(10)$$
$$= -4$$

Since the discriminant is negative, there is no value of x (among the real numbers) for which $f(x) = 0$. (See Figure 1.20.)

Example 1.10 illustrates that a quadratic function has two real zeros if its discriminant is positive, one real zero if its discriminant is zero, and no real zeros if its discriminant is negative.

Polynomial Functions

Quadratic functions are polynomial functions.

Definition 1.7

A **polynomial function** of degree n is a function that can be expressed in the form

$$f(x) = a_n x^n + a_{n-1} x^{n-1} + a_{n-2} x^{n-2} + \cdots + a_1 x + a_0$$

where the exponents $n, n-1, n-2, \ldots, 1$ are positive integers and the coefficients $a_n, a_{n-1}, a_{n-2}, \ldots, a_1, a_0$ are real numbers with $a_n \neq 0$.

Nonzero constant functions are considered polynomial functions of degree 0. The constant function

$$i(x) = 13$$

therefore, is a polynomial function of degree 0. The constant function

$$j(x) = 0$$

is called the **zero polynomial function**, but *no degree* is assigned to it. The following are examples of polynomial functions:

$f(x) = 5x^4 - x^2 + 0.5x$	degree 4
$f(x) = x^3 + 4x - 7$	degree 3
$f(x) = -9$	degree 0
$f(x) = 0$	no degree
$f(x) = x^2 - 6x$	degree 2
$f(x) = 3x^7$	degree 7
$f(x) = 5x + 8$	degree 1

The **leading coefficient** of a polynomial function is the coefficient of the highest power of x that appears. Thus, the leading coefficient of

$$f(x) = 2x^4 + 7x^3 - 9$$

is 2, and the leading coefficient of

$$g(x) = 8x - x^3$$

is -1.

The polynomial functions of degree 2 are the quadratic functions, and the polynomial functions of degree less than 2 (including the zero polynomial function) are the linear functions.

1.4 Problems

For each quadratic function in Problems 1 to 10, find the coordinates of the vertex of the corresponding parabola. Then sketch the parabola.

1. $f(x) = 0.2x^2 - 2x + 6$
2. $f(x) = -0.25x^2 + 3x + 1$
3. $f(x) = -0.125x^2 + x + 2$
4. $f(x) = x^2 - 4x + 7$
5. $f(x) = -x^2 + 6x$
6. $f(x) = -x^2 + 8x$
7. $f(x) = -\frac{1}{3}x^2 + 12$
8. $f(x) = 0.5x^2 + 3$
9. $f(x) = x^2$
10. $f(x) = 2x^2$

11. Suppose the domain of

$$f(x) = x^2 - 6x + 13$$

is the interval $[2, 7]$. Find the value of x at which $f(x)$ is minimum.

12. Suppose the domain of

$$f(x) = -3x^2 + 24x + 2$$

is the interval $(0, 6)$. Find the value of x at which $f(x)$ is maximum.

13. If the domain of

$$f(x) = -2x^2 + 36x - 10$$

is the interval $(0, +\infty)$, what is the maximum value of $f(x)$?

14. If the domain of

$$f(x) = x^2 - 14x + 57$$

is the interval $[0, +\infty)$, what is the minimum value of $f(x)$?

For each quadratic function in Problems 15 to 20, use the quadratic formula to find the real values of x at which $f(x) = 0$.

15. $f(x) = 12x^2 - 5x - 2$
16. $f(x) = 25x^2 - 10x + 1$
17. $f(x) = 64x^2 - 48x + 9$
18. $f(x) = -3x^2 + 35x + 12$
19. $f(x) = -x^2 + 8x - 17$
20. $f(x) = x^2 - 12x + 40$

21. **Inflation Rate** The inflation rate in an Asian country is presently 6.4 percent. Let $I(x)$ represent the rate x years from now. Economists believe that $I(x)$ will decrease for a while and then begin an upward trend. More specifically, they believe that

$$I(x) = 0.06x^2 - 0.6x + 6.4$$

for x in the interval [0, 8]. When will the inflation rate be least during the next 8 years? Sketch the graph of $I(x)$ for x in [0, 8].

22. **Average Production Cost** Let $A(x)$ represent the average cost, in dollars, of producing each barrel of a product when x thousand barrels is produced weekly. The management of the firm that produces the product assumes

$$A(x) = 0.07x^2 - 0.42x + 5$$

for x in the interval [1, 10]. What is the weekly production level at which the average cost of producing each barrel is least? Sketch the graph of $A(x)$ for x in [1, 10].

23. **Profit** Each week, a firm sells whatever amount of a product it produces. Thus, the firm's weekly profit from selling the product may be viewed as a function of the firm's weekly production level. Let $P(x)$ represent the weekly profit, in hundreds of dollars, the firm earns when it produces x tons of the product per week. If x is either too small or too large, $P(x)$ is negative, which means the firm loses money. To be more specific,

$$P(x) = -x^2 + 13x - 22$$

a. Find the weekly production levels at which the firm breaks even. In other words, find the weekly production levels at which the firm's profit is zero.

b. Find the weekly production level at which the profit is greatest. What is the weekly profit at this production level?

24. **Vertical Motion** A batter popped a baseball straight up in such a way that t seconds later its height was

$$H(t) = -16t^2 + 96t + 4$$

feet.

a. When did the ball reach its highest point? How high was the ball at that time?

b. When did the ball strike the ground? (Round the answer to the nearest hundredth of a second.)

25. **Profit** The price of a special fertilizer is $2000 per ton. At this price, whatever amount is produced is immediately sold. So the monthly profit earned from selling the fertilizer may be viewed as a function of the monthly production level. The total monthly cost of producing x tons of the fertilizer per month is

$$C(x) = 0.02x^2 + 3$$

thousand dollars.

a. Derive a formula that expresses the monthly profit as a function of the monthly production level.

b. Find the monthly production level that maximizes the profit. How much is the profit at this production level?

c. Find the monthly production levels at which the company breaks even. That is, find the monthly production levels at which the profit is zero. (Round your answers to the nearest tenth of a ton.)

26. **Profit** Do Problem 25 again. This time, however, assume that whatever amount of the fertilizer is produced is immediately sold at $1600 ($1.6 thousand) per ton and that the total monthly cost of producing x tons per month is

$$C(x) = 0.04x^2 + 5$$

thousand dollars.

27. **Market Equilibrium** Market researchers in a southeastern state find that when the price of shrimp averages x dollars per pound, the monthly amount consumers buy (demand) is

$$D(x) = \frac{44.625}{x + 2}$$

million pounds and the monthly amount shrimpers supply is

$$S(x) = 0.5x$$

million pounds. The researchers believe these formulas are valid only for values of x in the interval $[4, 12]$. At what price does the monthly demand equal the monthly supply?

28. **Market Equilibrium** Do Problem 27 again. This time, however, assume that

$$D(x) = -0.5x + 6$$

and

$$S(x) = 0.04x^2$$

29. **Optimal Workload** A company assigns each of its sales representatives at least 12 accounts but not more than 20. One of the representatives finds that when she has 12 accounts, each account brings in $200 per month in commission. If she has more accounts, the amount of attention she gives to each account decreases. Thus, each account brings in less commission. Suppose each additional account causes *every* account to bring in $10 less per month in commission. Find a formula that expresses the representative's total monthly commission as a function of the number of accounts she has. What is the domain of this function? How many accounts should the representative have for her total monthly commission to be maximum?

30. **Optimal Enrollment** Josette has agreed to teach a watercolor class at a community art center if the enrollment is at least 15 but not more than 30. If the enrollment is 15, she will charge every student a fee of $70. Otherwise, for each student in excess of 15, she will lower the fee of *every* student $2. For example, if the enrollment is 18, she will charge every student a fee of $64. Find a formula that expresses the total amount Josette will receive from the fees as a function of the enrollment. What is the domain of this function? What enrollment maximizes the total amount she will receive from the fees?

31. **Optimal Dimension** Exactly 207 feet of fencing will be used to enclose a rectangular plot of land adjacent to an existing wall. (See the figure for this problem.) Let x represent the length of each of the two equal sides perpendicular to the wall. Find a formula that expresses the area of the plot as a function of x. Then find the value of x that maximizes the area.

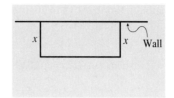

32. **Optimal Time to Sell** A farmer has 5200 crates of a tropical fruit that requires canning within 30 days. Presently, the farmer can sell the fruit to the cannery for $8 a crate. The price the cannery pays for each crate, however, will increase during the next 30 days at the rate of $0.25 per day. Pests ruin the fruit at the rate of 65 crates per day. Find a formula for the total amount of money the farmer will receive for the fruit x days from now. Then determine when the farmer should sell the fruit to the cannery.

 Graphing Calculator Problems

33. Set the viewing screen on your graphing calculator to $[-5, 5] \times [-1, 6]$ with a scale of 1 on both axes. Then draw the graphs of the following functions:

$$f(x) = x^2$$
$$g(x) = 2x^2$$
$$h(x) = 4x^2$$

The graphs of these functions suggest that there is a simple relationship between the graph of $f(x) = x^2$ and the graph of any function of the form

$$i(x) = ax^2$$

where $a > 1$. What is the suggested relationship?

34. Set the viewing screen on your graphing calculator to $[-5, 5] \times [-1, 6]$ with a scale of 1 on both axes. Then draw the graphs of the following functions:

$$f(x) = x^2$$
$$g(x) = 0.5x^2$$
$$h(x) = 0.25x^2$$

The graphs of these functions suggest that there is a simple relationship between the graph of $f(x) = x^2$ and the graph of any function of the form

$$i(x) = ax^2$$

where $0 < a < 1$. What is the suggested relationship?

1.5 Limits at Infinity

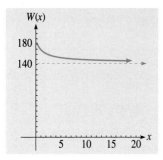

Figure 1.21
$W(x) = (280x + 540)/(2x + 3)$ for $x \geq 0$.

Many applications involve functions for which the values of the dependent variable approach a fixed number as the values of the independent variable increase. In this section, we deal with such functions. We begin with a situation from the life sciences.

A physician recently placed one of her patients on a special diet. She predicts that after x months of dieting, the patient will weigh

$$W(x) = \frac{280x + 540}{2x + 3}$$

pounds. According to this formula, the patient's weight will approach a certain level while dieting.

Table 1.3 displays the values of $W(x)$ that correspond to several values of x, while Figure 1.21 shows the graph of the function $W(x)$ for $x \geq 0$. The table and the figure both suggest that as x increases, the values of

$$\frac{280x + 540}{2x + 3}$$

get closer and closer to 140. In fact, as x increases without bound, the values of $(280x + 540)/(2x + 3)$ eventually get and stay arbitrarily close to 140. So the patient's weight will approach 140 pounds while dieting.

Table 1.3

x	0	3	6	18	42	72	96	120	160
$W(x)$	180	153.3	148	143.1	141.4	140.8	140.6	140.5	140.4

To obtain additional evidence that the patient's weight will approach 140 pounds while dieting, we divide x into the numerator and the denominator of the formula for $W(x)$. Accordingly, we get

$$\frac{280x + 540}{2x + 3} = \frac{280 + 540/x}{2 + 3/x}$$

provided $x \neq 0$. Note that as x increases, $540/x$ and $3/x$ both get closer and closer to 0. Thus, as x increases, the values of

$$\frac{280x + 540}{2x + 3}$$

get closer and closer to

$$\frac{280}{2}$$

which is equal to 140.

The diet problem illustrates the concept defined below. In the definition, we use the phrase "x approaches positive infinity" (abbreviated $x \to +\infty$) to mean that x increases without bound.

Definition 1.8

Suppose L is a number related as follows to a function $f(x)$:

> As x increases without bound, the values of $f(x)$ eventually get and stay arbitrarily close to L.

Then L is the **limit of $f(x)$ as x approaches positive infinity.** This statement is expressed by writing

$$\lim_{x \to +\infty} f(x) = L$$

Since 140 is related to the function $(280x + 540)/(2x + 3)$ as described in Definition 1.8, we can say that 140 is the limit of $(280x + 540)/(2x + 3)$ as x approaches positive infinity. We can express our statement by writing

$$\lim_{x \to +\infty} \left(\frac{280x + 540}{2x + 3} \right) = 140$$

Our statement does not require the values of $(280x + 540)/(2x + 3)$ to eventually reach and stay equal to 140. In fact, as x approaches positive infinity, the values of this function are never equal to 140.

To say L is the limit of $f(x)$ as x approaches *negative* infinity is similar to saying L is the limit of $f(x)$ as x approaches *positive* infinity. Specifically, it means that as x *decreases* without bound, the values of $f(x)$ eventually get and stay arbitrarily close to L. This statement is expressed by writing

$$\lim_{x \to -\infty} f(x) = L$$

Nonexistence of a Limit

It may happen that a function $f(x)$ is such that there is *no* number L for which

$$\lim_{x \to +\infty} f(x) = L$$

In this case, we say the limit of $f(x)$ as x approaches positive infinity *does not exist*. For example,

$$\lim_{x \to +\infty} \frac{x^2}{x + 1}$$

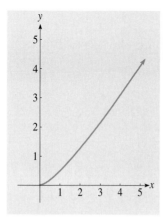

Figure 1.22 The graph of $x^2/(x+1)$ for $x \geq 0$.

does not exist. To show why this limit does not exist, we divide the numerator and the denominator by x^2, which is the highest power of x in the formula. Accordingly, we get

$$\frac{x^2}{x+1} = \frac{1}{1/x + 1/x^2}$$

provided $x \neq 0$. Note that as x increases, $1/x$ and $1/x^2$ both get smaller and smaller. Thus,

$$\frac{1}{1/x + 1/x^2}$$

gets larger and larger. In fact, it increases without bound. So

$$\lim_{x \to +\infty} \frac{x^2}{x+1}$$

indeed does not exist. Figure 1.22 supports our conclusion.

In general, if $f(x)$ increases without bound as x approaches positive infinity,

$$\lim_{x \to +\infty} f(x)$$

does not exist. This situation is expressed by writing

$$\lim_{x \to +\infty} f(x) = +\infty$$

In particular, then, our conclusion regarding the function $x^2/(x+1)$ can be expressed by writing

$$\lim_{x \to +\infty} \frac{x^2}{x+1} = +\infty$$

Similarly, if $f(x)$ *decreases* without bound as x approaches positive infinity,

$$\lim_{x \to +\infty} f(x)$$

does not exist. This situation is expressed by writing

$$\lim_{x \to +\infty} f(x) = -\infty$$

A function $f(x)$ can fail to have a limit as x increases without bound, even though the values of $f(x)$ neither increase without bound nor decrease without bound. Figure 1.23 shows the graph of such a function. Note that the values of $f(x)$ oscillate between 1 and 3 as x increases.

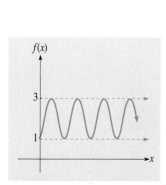

Figure 1.23 A function without a limit as $x \to +\infty$.

Properties of Limits

The following properties help us evaluate limits as $x \to +\infty$:

Properties of Limits at Infinity

Suppose $\lim_{x \to +\infty} f(x)$ and $\lim_{x \to +\infty} g(x)$ both exist, and k is a constant. Then

P1 $\quad \lim_{x \to +\infty} k = k$

P2 $\quad \lim_{x \to +\infty} [f(x) \pm g(x)] = \lim_{x \to +\infty} f(x) \pm \lim_{x \to +\infty} g(x)$

P3 $\quad \lim_{x \to +\infty} [f(x) \cdot g(x)] = \left(\lim_{x \to +\infty} f(x) \right) \left(\lim_{x \to +\infty} g(x) \right)$

P4 $\quad \lim_{x \to +\infty} \dfrac{f(x)}{g(x)} = \dfrac{\lim_{x \to +\infty} f(x)}{\lim_{x \to +\infty} g(x)}$, provided $\lim_{x \to +\infty} g(x) \neq 0$

P5 $\quad \lim_{x \to +\infty} \dfrac{k}{h(x)} = 0$, if $\lim_{x \to +\infty} h(x) = +\infty$ (or $-\infty$)

P6 $\quad$ If $\lim_{x \to +\infty} f(x) \neq 0$ and $\lim_{x \to +\infty} g(x) = 0$, then $\lim_{x \to +\infty} \dfrac{f(x)}{g(x)}$ does not exist

The above properties remain valid if $x \to +\infty$ is replaced by $x \to -\infty$. Properties P1 to P4 can be paraphrased as follows.

P1—The limit of a constant is the constant itself.
P2—The limit of a sum (or difference) is the sum (or difference) of the limits.
P3—The limit of a product is the product of the limits.
P4—The limit of a quotient is the quotient of the limits, provided the limit of the divisor (denominator) is not zero.

Note that if n is a positive integer, $\lim_{x \to +\infty} x^n = +\infty$. Thus, according to property P5, we have the following result:

$$\lim_{x \to +\infty} \dfrac{k}{x^n} = 0$$

Note also that

$$\lim_{x \to +\infty} k \cdot f(x) = \left(\lim_{x \to +\infty} k \right) \left(\lim_{x \to +\infty} f(x) \right) \qquad \text{P3}$$

$$= k \left(\lim_{x \to +\infty} f(x) \right) \qquad \text{P1}$$

So properties P1 and P3 imply the following result:

$$\lim_{x \to +\infty} k \cdot f(x) = k \left(\lim_{x \to +\infty} f(x) \right)$$

That is, the limit of a constant times a function is the constant times the limit of the function (provided the limit of the function exists).

We use some of the limit properties in Example 1.11.

Example 1.11

Find

$$\lim_{x \to +\infty} \frac{4x}{x^2 + 7}$$

Solution

We begin by dividing the numerator and the denominator by the highest power of x in the formula, which is x^2. We get

$$\frac{4x}{x^2 + 7} = \frac{4/x}{1 + 7/x^2}$$

provided $x \neq 0$. Therefore,

$$\lim_{x \to +\infty} \frac{4x}{x^2 + 7} = \lim_{x \to +\infty} \frac{4/x}{1 + 7/x^2}$$

$$= \frac{\lim_{x \to +\infty} (4/x)}{\lim_{x \to +\infty} (1 + 7/x^2)} \quad \text{P4}$$

$$= \frac{\lim_{x \to +\infty} (4/x)}{\lim_{x \to +\infty} 1 + \lim_{x \to +\infty} (7/x^2)} \quad \text{P2}$$

$$= \frac{0}{1 + 0} \quad \text{P1 and P5}$$

$$= 0$$

Thus

$$\lim_{x \to +\infty} \frac{4x}{x^2 + 7} = 0$$

Rational Functions

The functions

$$W(x) = \frac{280x + 540}{2x + 3}$$

$$f(x) = \frac{x^2}{x+1}$$

and
$$g(x) = \frac{4x}{x^2+7}$$

are examples of rational functions.

Definition 1.9

A **rational function** is a function that can be expressed in the form

$$f(x) = \frac{P(x)}{Q(x)}$$

where $P(x)$ and $Q(x)$ are polynomial functions.

In the rational function

$$W(x) = \frac{280x + 540}{2x + 3}$$

the leading coefficient of the numerator is 280 and the leading coefficient of the denominator is 2. Note that $280/2 = 140$, which, as we saw, is the limit of $W(x)$ as x approaches positive infinity. Thus, since the degree of the numerator is equal to the degree of the denominator (they are both of degree 1), the function $W(x)$ illustrates Theorem 1.4.

Theorem 1.4

Suppose $f(x)$ is a rational function for which the degree of the numerator is *equal* to the degree of the denominator. Then

$$\lim_{x \to +\infty} f(x) = \frac{a}{b} \quad \text{and} \quad \lim_{x \to -\infty} f(x) = \frac{a}{b}$$

where a and b are, respectively, the *leading coefficients of* the numerator and denominator.

Example 1.12 makes use of Theorem 1.4.

Example 1.12

Find

$$\lim_{x \to +\infty} \frac{13x^2 - 104x + 658}{2x^2 - 16x + 182}$$

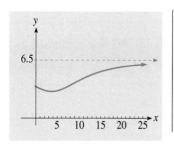

Figure 1.24 The function in Example 1.12 for $x \geq 0$.

Solution

This function is a rational function for which the degree of the numerator is equal to the degree of the denominator. So we can use Theorem 1.4. The leading coefficient of the numerator is 13 and the leading coefficient of the denominator is 2. Consequently,

$$\lim_{x \to +\infty} \frac{13x^2 - 104x + 658}{2x^2 - 16x + 182} = \frac{13}{2} = 6.5$$

Figure 1.24 shows the graph of the function in Example 1.12 for $x \geq 0$. A proof of Theorem 1.5 involves steps similar to those in Example 1.11.

Theorem 1.5

If $f(x)$ is a rational function for which the degree of the numerator is *less* than the degree of the denominator, then

$$\lim_{x \to +\infty} f(x) = 0 \quad \text{and} \quad \lim_{x \to -\infty} f(x) = 0$$

We use Theorem 1.5 in Example 1.13.

Example 1.13

Find

$$\lim_{x \to -\infty} \frac{5x + 8}{x^2}$$

Solution

Since the degree of the numerator is less than the degree of the denominator, we use Theorem 1.5 and obtain

$$\lim_{x \to -\infty} \frac{5x + 8}{x^2} = 0$$

Earlier we showed that the limit as x approaches positive infinity does not exist for the function $x^2/(x + 1)$. Thus, we illustrated the following result:

If $f(x)$ is a rational function for which the degree of the numerator is *greater* than the degree of the denominator, then neither

$$\lim_{x \to +\infty} f(x) \quad \text{nor} \quad \lim_{x \to -\infty} f(x)$$

exist.

1.5 Problems

In Problems 1 to 10, if the limit exists, find it. Otherwise, write "does not exist."

1. $\lim\limits_{x \to +\infty} \dfrac{4x^2 + x - 5}{3x^2 + 7}$

2. $\lim\limits_{x \to -\infty} \dfrac{2x^3 + 1}{7x^3 + 2x + 8}$

3. $\lim\limits_{x \to -\infty} \dfrac{5x + 9}{x^2 + 6}$

4. $\lim\limits_{x \to +\infty} \dfrac{x^2 + 5x + 1}{x^3 + 2}$

5. $\lim\limits_{x \to +\infty} \dfrac{4x^3}{x^3 + 8}$

6. $\lim\limits_{x \to +\infty} \dfrac{x}{5x + 2}$

7. $\lim\limits_{x \to -\infty} \dfrac{3x^2}{4x + 1}$

8. $\lim\limits_{x \to +\infty} \dfrac{x^3}{2x + 13}$

9. $\lim\limits_{x \to +\infty} \dfrac{7}{x + 9}$

10. $\lim\limits_{x \to -\infty} \dfrac{1}{x^2 + 2}$

11. **Life Expectancy** In a developing country, the life expectancy for males is currently 68 years. An actuary predicts that x years from now the life expectancy will be

$$L(x) = \dfrac{222x + 136}{3x + 2}$$

years. What level will it approach as time passes?

12. **Unemployment Rate** The unemployment rate in a southeastern community is currently 12 percent. Officials predict that x months from now it will be

$$U(x) = \dfrac{16x + 36}{2x + 3}$$

percent. Find the level it will approach as time passes.

13. **Prime Rate** In a European country, the prime lending rate is presently 12 percent. An economist predicts that x months from now it will be

$$I(x) = \dfrac{18x + 12}{2x + 1}$$

percent. Approximate the long-term expectation for the prime lending rate.

14. **Demand** The monthly demand for a new product is presently 150 tons. The management of the firm that produces the product projects that x years from now the monthly demand will be

$$D(x) = \dfrac{275x^2 + 600}{x^2 + 4}$$

tons. Approximate the level the monthly demand will reach after a long period of time.

15. **Supply** Suppose the producers of a commodity supply the market with

$$S(x) = \dfrac{120x^2}{4x^2 + 53}$$

thousand barrels of the commodity per week when its selling price is x cents per liter. What level will the weekly supply approach as the price increases?

16. **Public Health** A communicable disease was recently introduced into a community. Although the disease is not serious, any individual who contracts it becomes a carrier for life. Health officials believe that x months from now the percentage of carriers will be

$$D(x) = \dfrac{120x^2 + 43}{2x^2 + 30}$$

Approximate what the percentage of carriers will be eventually.

Graphing Calculator Problems

17. Set the viewing screen on your graphing calculator to $[0, 24] \times [0, 16]$ with a scale of 1 on both axes. Then draw the graphs of

$$g(x) = \dfrac{18x + 5}{2x + 5}$$

and

$$y = 9$$

so that they appear simultaneously. Explain how the graphs are compatible with Theorem 1.4.

18. Set the viewing screen on your graphing calculator to $[0, 15] \times [0, 10]$ with a scale of 1 on both axes. Then draw the graphs of

$$f(x) = \dfrac{6x + 16}{3x + 2}$$

and

$$y = 2$$

so that they appear simultaneously. Explain how the graphs are compatible with Theorem 1.4.

1.6 Limits as $x \to c$ and Continuity

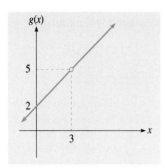

Figure 1.25
$g(x) = (x^2 - x - 6)/(x - 3)$.

Table 1.4

x	2	2.5	2.9	2.99
$g(x)$	4	4.5	4.9	4.99

Table 1.5

x	4	3.5	3.1	3.01
$g(x)$	6	5.5	5.1	5.01

In Section 1.5, we examined the behavior of functions as the independent variable *increases without bound*. In this section, we examine the behavior of functions as the independent variable *approaches a fixed number*.

We begin with the function

$$g(x) = \frac{x^2 - x - 6}{x - 3}$$

The domain of this function is the set of all real numbers except 3. We exclude 3 from the domain because the denominator is 0 when $x = 3$. Nevertheless, we want to know how $g(x)$ behaves as x approaches 3.

First, we must clarify two things. Whenever we say "x approaches a real number c," we mean x gets arbitrarily close to c without actually reaching c. Also, if we say "a quantity behaves in a certain way as x approaches c," we mean the quantity behaves that way *regardless of the side from which x approaches c*, unless we state otherwise.

We now return to the function $g(x) = (x^2 - x - 6)/(x - 3)$. Table 1.4 displays the values of $g(x)$ for several values of x *less* than 3, while Table 1.5 displays the values of $g(x)$ for several values of x *greater* than 3. Figure 1.25 shows the graph of the function $g(x)$, which is a line with a puncture at the point (3, 5). The two tables and the figure suggest that as x approaches 3, the values of

$$\frac{x^2 - x - 6}{x - 3}$$

get closer and closer to 5. In fact, as x approaches 3, the values of $(x^2 - x - 6)/(x - 3)$ eventually get and stay arbitrarily close to 5.

With the function $g(x)$ we have illustrated the concept noted in Definition 1.10.

Definition 1.10

Suppose $f(x)$ is a function and c is a number for which the following is true:

As x approaches c, the values of $f(x)$ eventually get and stay arbitrarily close to a number L.

Then L is the **limit of $f(x)$ as x approaches** c. This statement is expressed by writing

$$\lim_{x \to c} f(x) = L$$

In Definition 1.10, the number c may or may not be in the domain of the function $f(x)$. Also, if a number L such as described in the definition does not exist, we say the limit of $f(x)$ as x approaches c *does not exist*.

According to the above definition, we can say that 5 is the limit of $(x^2 - x - 6)/(x - 3)$ as x approaches 3 and write

$$\lim_{x \to 3} \frac{x^2 - x - 6}{x - 3} = 5$$

Our statement means that as x approaches 3 *from either side of 3 and without actually reaching 3*, the values of

$$\frac{x^2 - x - 6}{x - 3}$$

eventually get and stay arbitrarily close to 5. So our statement does not require the values of $(x^2 - x - 6)/(x - 3)$ to eventually reach and stay equal to 5. In fact, Figure 1.25 shows that the values of this function are never equal to 5, but are close to 5 for values of x close to 3.

Properties of Limits

The following properties help us evaluate limits as $x \to c$. Some of these properties are similar to some of the limit properties listed in Section 1.5.

Properties of Limits as $x \to c$

Suppose $\lim_{x \to c} f(x)$ and $\lim_{x \to c} g(x)$ both exist. Also, suppose k is any constant and n is a positive integer. Then

L1 $\lim_{x \to c} k = k$

L2 $\lim_{x \to c} [f(x) \pm g(x)] = \lim_{x \to c} f(x) \pm \lim_{x \to c} g(x)$

L3 $\lim_{x \to c} [f(x) \cdot g(x)] = \left(\lim_{x \to c} f(x)\right)\left(\lim_{x \to c} g(x)\right)$

L4 $\lim_{x \to c} \frac{f(x)}{g(x)} = \frac{\lim_{x \to c} f(x)}{\lim_{x \to c} g(x)}$, provided $\lim_{x \to c} g(x) \neq 0$

L5 $\lim_{x \to c} kx^n = kc^n$

L6 $\lim_{x \to c} \sqrt[n]{f(x)} = \sqrt[n]{\lim_{x \to c} f(x)}$, provided $\lim_{x \to c} f(x)$ is not negative if n is even

Limits of Polynomial Functions

We use some of the above limit properties in Example 1.14.

Example 1.14
Find
$$\lim_{x \to 5} (x^2 - 7x + 4)$$

Solution

$$\lim_{x \to 5} (x^2 - 7x + 4) = \lim_{x \to 5} x^2 - \lim_{x \to 5} 7x + \lim_{x \to 5} 4 \quad \text{L2}$$
$$= \lim_{x \to 5} x^2 - \lim_{x \to 5} 7x + 4 \quad \text{L1}$$
$$= 5^2 - 7 \cdot 5 + 4 \quad \text{L5}$$
$$= -6$$

If we substitute 5 for x in $x^2 - 7x + 4$, we also obtain -6. According to Theorem 1.6, this result is not a coincidence! The theorem can be proved with steps similar to those in Example 1.14.

Theorem 1.6

If $P(x)$ is a polynomial function, then
$$\lim_{x \to c} P(x) = P(c)$$

Theorem 1.6 tells us that the limit of a polynomial function can be found by simply evaluating the function at the number that x approaches.

Limits of Rational Functions

We use Theorem 1.6 in Example 1.15.

Example 1.15
Find
$$\lim_{x \to 4} \frac{x^2 + 1}{5x + 2}$$

Solution

Since $x^2 + 1$ and $5x + 2$ are polynomial functions, using Theorem 1.6,

$$\lim_{x \to 4} (x^2 + 1) = 4^2 + 1 = 17$$

and

$$\lim_{x \to 4} (5x + 2) = 5 \cdot 4 + 2 = 22$$

So, according to property L4,

$$\lim_{x \to 4} \frac{x^2 + 1}{5x + 2} = \frac{17}{22}$$

If we substitute 4 for x in $(x^2 + 1)/(5x + 2)$, we also obtain 17/22. This result illustrates Theorem 1.7.

Theorem 1.7

If $P(x)$ and $Q(x)$ are polynomial functions, then

$$\lim_{x \to c} \frac{P(x)}{Q(x)} = \frac{P(c)}{Q(c)}$$

provided $Q(c) \neq 0$. That is, if $R(x)$ is a rational function, then

$$\lim_{x \to c} R(x) = R(c)$$

provided the denominator does not become zero when c is substituted for x.

We use Theorem 1.7 in Example 1.16.

Example 1.16

Find

$$\lim_{x \to 6} \frac{x^2 - 7x + 6}{x^2 - 36}$$

Solution

Note that $x^2 - 36$ is 0 when $x = 6$. So we cannot apply Theorem 1.7 to the given formula. But, for $x \neq 6$,

$$\frac{x^2 - 7x + 6}{x^2 - 36} = \frac{(x - 6)(x - 1)}{(x - 6)(x + 6)}$$

$$= \frac{x - 1}{x + 6}$$

Since $x + 6 \neq 0$ when $x = 6$, we apply Theorem 1.7 to $(x - 1)/(x + 6)$ and get

$$\lim_{x \to 6} \frac{x-1}{x+6} = \frac{6-1}{6+6}$$

$$= \frac{5}{12}$$

Therefore, $\lim_{x \to 6} \frac{x^2 - 7x + 6}{x^2 - 36} = \frac{5}{12}$ also.

Next, we discuss one-sided limits.

One-Sided Limits

The statement "L is the limit of $f(x)$ as x approaches c *from the left*" is expressed by writing

$$\lim_{x \to c^-} f(x) = L$$

This statement means that as x approaches c through values *less* than c, the corresponding values of $f(x)$ eventually get and stay arbitrarily close to L. Similarly, the statement "L is the limit of $f(x)$ as x approaches c *from the right*" is expressed by writing

$$\lim_{x \to c^+} f(x) = L$$

The meaning of this statement is similar to the meaning of $\lim_{x \to c^-} f(x) = L$. As expected, all of the properties of limits (including Theorems 1.6 and 1.7) are valid when $x \to c$ is replaced by $x \to c^-$ or by $x \to c^+$.

What we have said about limits (including one-sided limits) implies Theorem 1.8.

Theorem 1.8

If $\lim_{x \to c^-} f(x) = \lim_{x \to c^+} f(x) = L$, then $\lim_{x \to c} f(x) = L$.

This theorem tells us that if the limits from the left and right both exist and are equal, then the limit exists. Moreover, the value of the limit is the common value of the limits from the left and right.

We use Theorem 1.8 in Example 1.17.

Example 1.17

Find

$$\lim_{x \to 2} h(x)$$

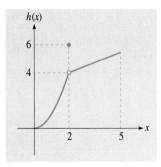

Figure 1.26 The graph of $h(x)$ for x in $[0, 5]$.

if $$h(x) = \begin{cases} x^2 & \text{when } 0 \leq x < 2 \\ 6 & \text{when } x = 2 \\ 0.5x + 3 & \text{when } 2 < x \leq 5 \end{cases}$$

[Figure 1.26 shows the graph of $h(x)$.]

Solution

We begin by finding the limit from the left. Since $h(x) = x^2$ when $x < 2$,

$$\lim_{x \to 2^-} h(x) = \lim_{x \to 2^-} x^2$$
$$= 2^2 \qquad \text{Theorem 1.6}$$
$$= 4$$

(Figure 1.26 supports this conclusion.) Now we find the limit from the right. Since $h(x) = 0.5x + 3$ when $x > 2$,

$$\lim_{x \to 2^+} h(x) = \lim_{x \to 2^+} (0.5x + 3)$$
$$= 0.5(2) + 3 \qquad \text{Theorem 1.6}$$
$$= 4$$

(Figure 1.26 also supports this conclusion.) Thus, according to Theorem 1.8,

$$\lim_{x \to 2} h(x) = 4$$

In Example 1.17, $h(2) = 6$. So the limit of $h(x)$ as x approaches 2 is not the same number as the value of $h(2)$.

In Figure 1.26, the point $(2, 6)$ is part of the graph, but the point $(2, 4)$ is not part of the graph.

Nonexistence of Limits

The inverse of Theorem 1.8, which is stated in Theorem 1.9, is also valid.

Theorem 1.9

$\lim_{x \to c} f(x)$ does not exist if at least one of the following occurs:

1. $\lim_{x \to c^-} f(x) \neq \lim_{x \to c^+} f(x)$.

2. $\lim_{x \to c^-} f(x)$ does not exist.

3. $\lim_{x \to c^+} f(x)$ does not exist.

Theorem 1.9 is used in Examples 1.18 and 1.19.

Example 1.18

Video Village, a company that rents movies on video cassettes, allows its customers to keep a cassette up to 4 days for a flat fee of $6. Customers who keep a cassette more than 4 days pay a $2 penalty plus $0.75 for each day beyond the first 4 days. Let $R(x)$ represent the cost of keeping a cassette x days.

a. Find a formula for $R(x)$.
b. Show that $\lim_{x \to 4} R(x)$ does not exist.

Solution for (a)

Since customers can keep a cassette up to 4 days for a flat fee of $6,

$$R(x) = 6$$

when $0 < x \leq 4$. Since customers who keep a cassette more than 4 days pay a $2 penalty plus $0.75 for each day beyond the first 4 days,

$$R(x) = 6 + 2 + (x - 4)0.75$$
$$= 0.75x + 5$$

when $x > 4$. Summarizing,

$$R(x) = \begin{cases} 6 & \text{when } 0 < x \leq 4 \\ 0.75x + 5 & \text{when } x > 4 \end{cases}$$

Solution for (b)

Figure 1.27 shows the graph of $R(x)$. Since $R(x) = 6$ when $x \leq 4$,

$$\lim_{x \to 4^-} R(x) = \lim_{x \to 4^-} 6$$
$$= 6 \quad \text{limit property L1}$$

(Figure 1.27 supports this conclusion.) Since $R(x) = 0.75x + 5$ when $x > 4$,

$$\lim_{x \to 4^+} R(x) = \lim_{x \to 4^+} (0.75x + 5)$$
$$= 0.75(4) + 5 \quad \text{Theorem 1.6}$$
$$= 8$$

(Figure 1.27 also supports this conclusion.) Therefore,

$$\lim_{x \to 4^-} R(x) \neq \lim_{x \to 4^+} R(x)$$

So, according to Theorem 1.9, $\lim_{x \to 4} R(x)$ does not exist.

In Example 1.18, note that $R(4) = 6$, not 8.

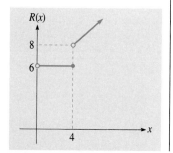

Figure 1.27 The graph of $R(x)$ for x in $(0, +\infty)$.

Example 1.19

Let
$$j(x) = \frac{10}{5 - x}$$

Show that $\lim_{x \to 5} j(x)$ does not exist.

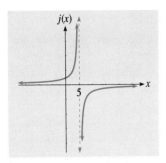

Figure 1.28 $j(x) = 10/(5 - x)$.

Solution

Figure 1.28 shows the graph of $j(x)$. From the graph we see that as x approaches 5 from the left, $j(x)$ gets larger and larger. In fact, as x approaches 5 from the left, $j(x)$ increases without bound. Thus, $\lim_{x \to 5^-} j(x)$ does not exist. So, according to Theorem 1.9, we can conclude that $\lim_{x \to 5} j(x)$ also does not exist.

In Example 1.19, we could have shown that $\lim_{x \to 5} j(x)$ does not exist by observing that $\lim_{x \to 5^+} j(x)$ does not exist.

With the function
$$j(x) = \frac{10}{5 - x}$$

we illustrated the following result:

Theorem 1.10

If
$$R(x) = \frac{P(x)}{Q(x)}$$

is a rational function and c is a number for which $P(c) \neq 0$ but $Q(c) = 0$, then
$$\lim_{x \to c} R(x)$$

does not exist.

We use this result in Example 1.20.

Example 1.20

Show that
$$\lim_{x \to 9} \frac{5x^2 - 45x}{x^2 - 18x + 81}$$

does not exist.

Solution

The numerator and the denominator are both zero when $x = 9$. Thus, according to Theorem A.8 in Section A.3 of Appendix A, $x - 9$ is a factor of the numerator and the denominator. In fact, for $x \neq 9$,

$$\frac{5x^2 - 45x}{x^2 - 18x + 81} = \frac{5x(x - 9)}{(x - 9)(x - 9)}$$
$$= \frac{5x}{x - 9}$$

Since $5x/(x - 9)$ is a rational function whose numerator is not zero when $x = 9$ and whose denominator is zero when $x = 9$, according to Theorem 1.10,

$$\lim_{x \to 9} \frac{5x}{x - 9}$$

does not exist. Thus,

$$\lim_{x \to 9} \frac{5x^2 - 45x}{x^2 - 18x + 81}$$

does not exist also.

Continuity

Figure 1.29 shows the graph of a function that is *continuous* at each number in its domain, which is an interval.

Definition 1.11

A function $f(x)$ is **continuous** at $x = c$ if all three of the following conditions are met:

1. c is in the domain of $f(x)$.
2. $\lim_{x \to c} f(x)$ exists.
3. $\lim_{x \to c} f(x) = f(c)$.

If S is a set of real numbers and $f(x)$ is continuous at each number in S, we say that "$f(x)$ is continuous on S."

The function

$$g(x) = \frac{x^2 - x - 6}{x - 3}$$

is discontinuous (not continuous) at $x = 3$ because condition (1) and, thus, condition (3) are not met. (See Figure 1.25.) The function

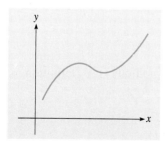

Figure 1.29 A continuous function.

$$h(x) = \begin{cases} x^2 & \text{when } 0 \leq x < 2 \\ 6 & \text{when } x = 2 \\ 0.5x + 3 & \text{when } 2 < x \leq 5 \end{cases}$$

is discontinuous at $x = 2$ because condition (3) is not met. (See Figure 1.26.) The function

$$R(x) = \begin{cases} 6 & \text{when } 0 < x \leq 4 \\ 0.75x + 5 & \text{when } x > 4 \end{cases}$$

is discontinuous at $x = 4$ because condition (2) and, thus, condition (3) are not met. (See Figure 1.27.) The function

$$j(x) = \frac{10}{5 - x}$$

is discontinuous at $x = 5$ because conditions (1) and (2) [and, thus, condition (3)] are not met. (See Figure 1.28.)

The function $g(x)$ is continuous everywhere except at $x = 3$. Since the domain of $g(x)$ is all the real numbers except 3, we can say $g(x)$ is continuous on its domain. Similarly, we can say $j(x)$ is continuous on its domain, which is the set of all real numbers except 5. The graphs of $g(x)$ and $j(x)$ (Figures 1.25 and 1.28) show that it is possible for a function to be continuous on its domain and yet have a *disconnected* graph. However, if the domain is an interval, we have the following result:

> If a function is continuous on its domain and the domain is an *interval*, then the graph of the function can be traced without lifting the pen from the paper at any point along the way.

The function in Figure 1.29 illustrates the above result.

Example 1.21

Let

$$f(x) = \begin{cases} 9 - x^2 & \text{when } 0 \leq x < 2 \\ 0.5x^2 + 3 & \text{when } 2 \leq x \leq 4 \end{cases}$$

Determine whether $f(x)$ is continuous at $x = 2$. [Figure 1.30 shows the graph of $f(x)$.]

Solution

The number 2 is in the domain of $f(x)$. Thus, condition (1) of Definition 1.11 is met. Since $f(x) = 9 - x^2$ when $x < 2$,

$$\lim_{x \to 2^-} f(x) = \lim_{x \to 2^-} (9 - x^2)$$
$$= 5 \qquad \text{Theorem 1.6}$$

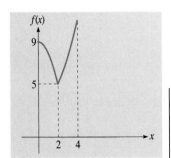

Figure 1.30 The graph of $f(x)$ for x in $[0, 4]$.

Since $f(x) = 0.5x^2 + 3$ when $x \geq 2$,

$$\lim_{x \to 2^+} f(x) = \lim_{x \to 2^+} (0.5x^2 + 3)$$
$$= 5 \quad \text{Theorem 1.6}$$

Thus, according to Theorem 1.8, $\lim_{x \to 2} f(x)$ exists, and so condition (2) is met. In fact, according to Theorem 1.8,

$$\lim_{x \to 2} f(x) = 5$$

Therefore, since

$$f(2) = 0.5(2^2) + 3$$
$$= 5$$

also, condition (3) is met. Since all three of the conditions for continuity are met, $f(x)$ is continuous at $x = 2$.

Although the function in Example 1.21 is continuous at $x = 2$, Figure 1.30 shows that its graph is *not smooth* at the point whose x coordinate is 2. Recall that if $P(x)$ is a polynomial function, then

$$\lim_{x \to c} P(x) = P(c)$$

Also if $R(x)$ is a rational function, then

$$\lim_{x \to c} R(x) = R(c)$$

provided the denominator does not become 0 when c is substituted for x. These results imply Theorem 1.11.

Theorem 1.11

Polynomial functions and rational functions are continuous everywhere on their domains.

1.6 Problems

In Problems 1 to 22, if the limit exists, find it. Otherwise, write "does not exist."

1. $\lim_{x \to 7} \dfrac{x^2 - 5x - 14}{x - 7}$

2. $\lim_{x \to 1} \dfrac{x^2 + 3x - 4}{x - 1}$

3. $\lim_{x \to 4} (3x^2 + x - 2)$

4. $\lim_{x \to 2} (x^3 - 4x + 1)$

5. $\lim_{x \to 3} x^2 \sqrt{5x + 10}$

6. $\lim_{x \to 5} 2x \sqrt{3x^2 + 6}$

7. $\lim_{x \to 1} \dfrac{5x}{4x - 4}$

8. $\lim_{x \to 3} \dfrac{8x}{x^2 - 9}$

9. $\lim_{x \to 2} \dfrac{7x}{x^2 - 1}$

10. $\lim_{x \to 4} \dfrac{x^2}{x^2 - 9}$

11. $\lim_{x \to 9} \dfrac{x^2 - 81}{x^2 - 4x + 5}$

12. $\lim_{x \to 6} \dfrac{x - 6}{x^2 - 2x + 3}$

13. $\lim_{x \to 5} \dfrac{x-5}{x^2-10x+25}$

14. $\lim_{x \to 7} \dfrac{3x-21}{x^2-14x+49}$

15. $\lim_{x \to 0} \dfrac{(6+x)^2-36}{x}$

16. $\lim_{x \to 0} \dfrac{(3+x)^2-9}{x}$

17. $\lim_{x \to 8} \dfrac{x^2-5x-24}{x^2-13x+40}$

18. $\lim_{x \to 6} \dfrac{x^2+x-42}{x^2-7x+6}$

19. $\lim_{x \to 5} f(x)$ if $f(x) = \begin{cases} 0.4x^2 & \text{when } 0 \le x \le 5 \\ 2x+1 & \text{when } 5 < x \le 7 \end{cases}$

20. $\lim_{x \to 9} f(x)$ if $f(x) = \begin{cases} \sqrt{x} & \text{when } 0 \le x \le 9 \\ \tfrac{1}{3}x - 2 & \text{when } 9 < x \le 12 \end{cases}$

21. $\lim_{x \to 4} f(x)$ if $f(x) = \begin{cases} \sqrt{x} & \text{when } 0 \le x < 4 \\ \tfrac{1}{8}x^2 & \text{when } 4 \le x \le 10 \end{cases}$

22. $\lim_{x \to 6} f(x)$ if $f(x) = \begin{cases} 7 - 0.5x & \text{when } 0 \le x < 6 \\ \tfrac{2}{3}x & \text{when } 6 \le x \le 8 \end{cases}$

In Problems 23 to 40, determine whether the given function is continuous at the designated point. If it is not continuous, tell which of the three conditions for continuity is/are not met.

23. $f(x) = \dfrac{x^2-16}{x-4}$ at $x = 4$

24. $f(x) = \dfrac{x^2-4x-5}{x-5}$ at $x = 5$

25. $f(x) = \dfrac{x^2+2x-15}{x-3}$ at $x = 6$

26. $f(x) = \dfrac{x^2-36}{x-6}$ at $x = 7$

27. $f(x) = x^3 - 5x + 1$ at $x = 13$

28. $f(x) = 4x^2 + x$ at $x = 8$

29. $f(x) = \dfrac{x^2+6x-7}{x^2-3x+2}$ at $x = 1$

30. $f(x) = \dfrac{x^2-6x+8}{x^2+x-6}$ at $x = 2$

31. $f(x) = \begin{cases} \dfrac{19}{x-3} & \text{when } x \ne 3 \\ 5 & \text{when } x = 3 \end{cases}$ at $x = 3$

32. $f(x) = \begin{cases} \dfrac{4x}{x-6} & \text{when } x \ne 6 \\ 1 & \text{when } x = 6 \end{cases}$ at $x = 6$

33. $f(x) = \begin{cases} x^2 - 4x + 5 & \text{when } 0 \le x < 3 \\ x - 1 & \text{when } x \ge 3 \end{cases}$ at $x = 3$

34. $f(x) = \begin{cases} 9 - x^2 & \text{when } 0 \le x \le 2 \\ 2.5x & \text{when } x > 2 \end{cases}$ at $x = 2$

35. $f(x) = \begin{cases} 6 - 0.25x & \text{when } 0 \le x \le 4 \\ 0.2x + 1 & \text{when } 4 < x \le 7 \end{cases}$ at $x = 4$

36. $f(x) = \begin{cases} x^2 + 3 & \text{when } 0 \le x < 9 \\ \sqrt{x} & \text{when } 9 \le x \le 16 \end{cases}$ at $x = 9$

37. $f(x) = \sqrt{5x + 16}$ at $x = 4$

38. $f(x) = \sqrt{2x^2 + 14}$ at $x = 5$

39. $f(x) = \begin{cases} x^2 - 2x + 2 & \text{when } 0 \le x < 2 \\ 1 & \text{when } x = 2 \\ 3 - 0.5x & \text{when } 2 < x \le 10 \end{cases}$ at $x = 2$

40. $f(x) = \begin{cases} 2x + 3 & \text{when } 0 \le x < 1 \\ 7 & \text{when } x = 1 \\ 8 - 3x & \text{when } 1 < x \le 12 \end{cases}$ at $x = 1$

In Problems 41 to 48, sketch the graph of a function $f(x)$ that satisfies the listed conditions.

41. The domain of $f(x)$ is $[2, 5]$.
$f(x)$ is continuous on $[2, 5]$.

42. The domain of $f(x)$ is $[0, 4]$.
$f(x)$ is continuous on $[0, 4]$.

43. The domain of $f(x)$ is $[0, 7]$.
$\lim_{x \to 4^-} f(x)$ and $\lim_{x \to 4^+} f(x)$ both exist, but their values are not equal.

44. The domain of $f(x)$ is $[1, 6]$.
$\lim_{x \to 5^-} f(x)$ and $\lim_{x \to 5^+} f(x)$ both exist, but their values are not equal.

45. The domain of $f(x)$ is $[0, 5]$.
Neither $\lim_{x \to 2^-} f(x)$ nor $\lim_{x \to 2^+} f(x)$ exist.

46. The domain of $f(x)$ is $[0, 8]$.
Neither $\lim_{x \to 6^-} f(x)$ nor $\lim_{x \to 6^+} f(x)$ exist.

47. The domain of $f(x)$ is $[3, 10]$.
$\lim_{x \to 7} f(x)$ exists, but its value is not equal to $f(7)$.

48. The domain of $f(x)$ is $[0, 9]$.
$\lim_{x \to 3} f(x)$ exists, but its value is not equal to $f(3)$.

49. **Production Cost** The fixed weekly cost in the production of a product is $9000. Up to 20 tons of the product can be produced weekly at a cost of $400 per ton. If the weekly production level exceeds 20 tons, each extra ton costs $800 to produce. Let $C(x)$ represent the total weekly cost of producing x tons of the product per week.

   a. Find a formula for $C(x)$.

   b. Sketch the graph of $C(x)$ for $0 \le x \le 40$.
   (Let each unit on the horizontal axis represent 5 tons, and let each unit on the vertical axis represent $5000.)

   c. Determine whether the function $C(x)$ is continuous at $x = 20$.

50. **Rental Fee** Rent and Mow, a company that rents lawn mowers, rents one of its mowers for $10 plus $5 per hour up to 4 hours. After the first 4 hours, the cost per additional hour is only $2. Let $R(x)$ represent the rental fee when the mower is rented for x hours.

   a. Find a formula for $R(x)$.

   b. Sketch the graph of $R(x)$ for $0 < x \le 10$.
   (Let each unit on the vertical axis represent $5.)

   c. Determine whether the function $R(x)$ is continuous at $x = 4$.

51. **Production Cost** Let $C(x)$ represent the total monthly cost of producing x tons of a product per month. The figure shows the graph of $C(x)$ for $0 \le x \le 9$ for this problem. Give a feasible explanation for the abrupt increase in the production cost when the production level is increased beyond 5 tons.

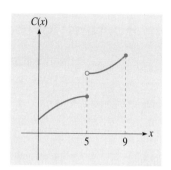

52. **Population** Let $P(x)$ represent the beaver population in a Canadian forest x years after the beginning of 1990. The figure shows the graph of $P(x)$ for $0 \le x \le 10$ for this problem. Give a feasible explanation for the abrupt decrease in the population 3 years after the beginning of 1990.

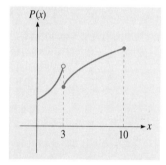

53. Suppose $f(x)$ is a function that is continuous on an interval. Also, suppose a and b are numbers in the interval. If $f(a)$ and $f(b)$ have opposite signs, can we conclude that $f(x)$ has a *zero* between a and b? Why?

   Recall: A zero of $f(x)$ is a value of x for which $f(x) = 0$.

54. Suppose $f(x)$ is a function that is continuous on its domain, and its domain is an interval. Also, suppose 3 and 7 are the only zeros of $f(x)$.

   a. If $f(x)$ is positive for a value of x that is less than 3, can we conclude that $f(x)$ is positive for *all* values of x that are less than 3? Why?

   b. If $f(x)$ is negative for a value of x that is between 3 and 7, can we conclude that $f(x)$ is negative for *all* values of x that are between 3 and 7? Why?

Chapter 1 Important Terms

Section 1.1

function (2)
independent and dependent
   variables (3)
domain and range (3)
$f(x)$ notation (3)
ordered pair of real numbers (4)
rectangular coordinate plane (5)
horizontal and vertical axes (5)
cartesian plane (5)
graph of a function (5)
abscissa and ordinate (5)
x and y coordinates (5)
constant function (5)
change in $f(x)$ resulting from a
   change in x (6)
total cost function (6)
fixed cost (7)
average rate of change (7)

Section 1.2

demand and revenue functions
   (13–15)

profit function (14–15)
average cost function (15)
composite function (17)

Section 1.3

linear function (23)
slope of a line (24)
horizontal and vertical lines (25)
data point (28)
regression line (27–29)
least-squares linear function (29)

Section 1.4

maximum and minimum values of a
   function (31)
quadratic function (32)
parabola (32)
vertex of a parabola (33)
axis of symmetry of a parabola
   (33)
zeros of a function (35)
quadratic formula (36)

discriminant (36)
polynomial function (37)
leading coefficient (38)

Section 1.5

limit of $f(x)$ as x approaches infinity
   (42)
rational function (46)

Section 1.6

limit of $f(x)$ as x approaches a
   number (49)
one-sided limit (53)
continuity and discontinuity of a
   function (57)

Chapter 1 Review Problems

Section 1.1

1. If
$$f(x) = \frac{3x + 25}{x + 4}$$
how much does $f(x)$ change if x increases from 1 to 6?

2. Suppose
$$g(x) = 0.04x^3 + 2$$
Find the average rate of change of $g(x)$ with respect to x as x increases from 3 to 7.

3. Find the domain of each function.

   a. $h(x) = \dfrac{x - 5}{x^2 - 36}$ b. $y = \sqrt{t - 4}$

   c. $f(t) = \dfrac{3}{\sqrt{16 - t^2}}$ d. $f(x) = 3x^4 - 9x + 4$

4. **Oxygen Uptake** The oxygen uptake of an individual is the rate at which the individual takes in oxygen from the atmosphere. Suppose the oxygen uptake of a woman is
$$R(t) = \frac{7t + 1}{2t + 5}$$
liters per minute t minutes after she begins pedaling on a stationary bicycle. How much does her oxygen uptake increase during the first 3 minutes of pedaling?

5. **Depreciation** An accountant predicts that a machine used in extracting juice from oranges will be worth

$$V(x) = \frac{480}{x+3}$$

thousand dollars x years from now. Find the average rate at which the machine will depreciate during the 6-year period that begins 1 year from now.

Section 1.2

6. **Business** Seemore Video sells x video cassette recorders per week if it sets the price of each recorder at

$$D(x) = \frac{2100}{x+260}$$

hundred dollars. The company sets the price of each recorder at $D(x)$ hundred dollars whenever it produces x recorders per week. Thus, the weekly sales level is the same as the weekly production level. The total weekly cost of producing x recorders per week is

$$C(x) = 3x + 20$$

hundred dollars.

a. Find a formula that expresses the weekly revenue obtained from selling the recorders as a function of the weekly production level.

b. Find a formula that expresses the weekly profit earned from selling the recorders as a function of the weekly production level.

c. Find a formula that expresses the average cost of producing each recorder as a function of the weekly production level.

7. **Delivery Cost** Henry plans to use his truck to deliver 250 showcases he built for a department store. Special reusable crates have to be built for the showcases at a cost of $30 per crate. The truck has space for at most 40 crates, and each crate holds only one showcase. Thus, the truck has to make several trips, each trip costing $65. Find a formula that expresses the combined cost of building the crates and delivering the showcases as a function of the number of crates built and used. What is the domain of this function?

8. If $g(x) = \sqrt{x}$ and $u(x) = 5x - 1$, find formulas for $g[u(x)]$ and $u[g(x)]$.

9. Suppose

$$f(x) = \frac{8}{(x^2+1)^3}$$

Find two functions $g(x)$ and $u(x)$ such that $f(x) = g[u(x)]$.

Section 1.3

The graph of each function in Problems 10 to 13 is a line. Find the slope of the line and the point where it crosses the vertical axis. Then sketch the line in the cartesian plane, showing the coordinates of at least two points on the line.

10. $f(x) = -3x + 2$ 11. $f(t) = 0.5t$

12. $g(x) = x - 4$ 13. $f(x) = 6$

In Problems 14 to 17, find the slope of the line that passes through the given pair of points. Then find a formula for the linear function whose graph is the line.

14. (2, 9) and (6, 7)   15. (0, 2) and (5, 10)

16. (3, 5) and (8, 5)   17. (0, 0) and (9, 9)

18. Suppose $f(x)$ is a linear function for which $f(0) = 3$ and $f(2) = 6$. Find a formula for $f(x)$. Then find $f(8)$.

19. Suppose $g(x)$ decreases linearly as x increases. What is the value of $g(10)$ if $g(1) = 9$ and $g(6) = 2$?

20. **Demand** The makers of Fast Lane jogging shoes find that the monthly demand for the shoes is 72,000 (72 thousand) pairs when their price is $65 and 56,000 (56 thousand) pairs when their price is $85. Suppose the monthly demand is a linear function of the price for prices that range from $50 to $100 per pair. Find a formula for this function. Then use the formula to find the monthly demand when the price is $90 per pair.

21. **Physical Activity** In a European country, the percentage of adults that exercise at least 4 hours per week increased from 28 percent at the beginning of 1985 to 40 percent at the beginning of 1991. Let $E(x)$ represent this percentage x years after the beginning of 1985, where x is in the interval [0, 20]. Assume that $E(x)$ increases at a constant rate. Find a formula for $E(x)$. Then use the formula to determine what percent of the adults will exercise at least 4 hours per week at the beginning of the year 2002.

Section 1.4

22. For each of the following quadratic functions, find the coordinates of the vertex of the corresponding parabola. Then sketch the parabola.

   a. $f(x) = 0.4x^2 - 4x + 13$

   b. $f(x) = -0.5x^2 + 8x - 22$

23. If the domain of $f(x) = x^2 - 14x + 61$ is the interval $[0, +\infty)$, what is the minimum value of $f(x)$?

24. Suppose the domain of $f(x) = -0.2x^2 + 2.4x + 18$ is the interval $(0, 10)$. Find the value of x at which $f(x)$ is maximum.

25. For each of the following quadratic functions, find the real values of x at which $f(x) = 0$.

   a. $f(x) = -8x^2 + 43x - 15$

   b. $f(x) = 4x^2 - 36x + 81$

   c. $f(x) = x^2 - 8x + 20$

26. **Entomology** The population of a colony of fleas, which is presently 6000, is expected to increase. Let $R(x)$ represent the rate, in thousands of fleas per day, at which the colony will grow when it consists of x thousand fleas. According to findings by Pierre-François Verhulst, a nineteenth century Belgian who studied population growth patterns, it is reasonable to assume $R(x)$ is a quadratic function. In fact, an entomologist assumes

$$R(x) = -0.05x^2 + 2x$$

   for x in the interval $[6, 40)$. What will the size of the population be when it increases most rapidly?

27. **Profit** The price of a certain chemical is $240 ($2.4 hundred) per barrel. At this price, whatever amount is produced is immediately sold. So the weekly profit earned from selling the chemical may be viewed as a function of the weekly production level. The total weekly cost of producing x barrels of the chemical per week is

$$C(x) = 0.08x^2 + 6$$

   hundred dollars.

   a. Derive a formula that expresses the weekly profit as a function of the weekly production level.

   b. Find the weekly production level that maximizes the profit. How much is the profit at this production level?

   c. Find the weekly production levels at which the company breaks even. That is, find the weekly production levels at which the profit is zero. (Round your answers to the nearest tenth of a barrel.)

28. **Optimal Workload** The management of a restaurant assigns each waiter between 5 and 12 tables. One of the waiters finds that when he has five tables, each table brings in $52 per week in tips. If he has more tables, the amount of attention he gives to each table decreases. Thus, each table brings in less per week in tips. Suppose each additional table causes *every* table to bring in $4 less per week in tips. Derive a formula that expresses the waiter's weekly earnings from tips as a function of the number of tables he has. What is the domain of this function? How many tables should the waiter have for his weekly earnings from tips to be maximum?

Section 1.5

29. If the limit exists, find it. Otherwise, write "does not exist."

   a. $\lim\limits_{x \to +\infty} \dfrac{12x^3 + x - 3}{x^3 + 5}$

   b. $\lim\limits_{x \to +\infty} \dfrac{3x + 8}{3x^2 + 2}$

   c. $\lim\limits_{x \to +\infty} \dfrac{x^3 - 1}{4x^2 + 3x + 7}$

30. **Plastic Waste** In a European country, growing public awareness of the environmental damage attributable to plastic waste is causing the demand for plastic products to decline. Thus, the plastic industry is gradually decreasing the production of plastic products. At the beginning of 1990, the industry was producing plastic products at the rate of 546 billion cubic centimeters per year. Suppose that x years after this date the industry produces such products at the rate of

$$P(x) = \dfrac{75x + 546}{3x + 1}$$

   billion cubic centimeters per year. Determine the level the production rate approaches as time passes.

Section 1.6

31. If the limit exists, find it. Otherwise, write "does not exist."

a. $\lim_{x \to 3} \dfrac{x^2 + 5x - 24}{x - 3}$

b. $\lim_{x \to 5} (x^2 - 3x + 1)$

c. $\lim_{x \to 2} \dfrac{3x}{x^2 + 7}$

d. $\lim_{x \to 1} 5x\sqrt{2x^2 + 7}$

e. $\lim_{x \to 4} \dfrac{x^2 - 16}{x^2 - 3x - 4}$

f. $\lim_{x \to 6} \dfrac{2x - 12}{x^2 + 5}$

g. $\lim_{x \to 5} \dfrac{3x}{x^2 - 25}$

h. $\lim_{x \to 2} \dfrac{3x - 6}{x^2 - 4x + 4}$

i. $\lim_{x \to 0} \dfrac{(4 + x)^2 - 16}{x}$

j. $\lim_{x \to 0} \dfrac{(2 + x)^3 - 8}{x}$

k. $\lim_{x \to 7} f(x)$ if $f(x) = \begin{cases} 0.2x^2 & \text{when } x \leq 7 \\ \sqrt{x - 3} & \text{when } x > 7 \end{cases}$

l. $\lim_{x \to 6} f(x)$ if $f(x) = \begin{cases} 40 - 0.7x^2 & \text{when } x < 6 \\ 0.3x^2 + 4 & \text{when } x \geq 6 \end{cases}$

32. Determine whether the given function is continuous at the designated point. If it is not continuous, tell which of the three conditions for continuity is/are not met.

a. $f(x) = x^3 - 4x + 3$ at $x = 5$

b. $f(x) = \dfrac{x^2 + x - 6}{x - 2}$ at $x = 2$

c. $f(x) = \dfrac{4x}{x - 1}$ at $x = 3$

d. $f(x) = \dfrac{x^2 - 2x - 3}{x - 6}$ at $x = 6$

e. $f(x) = \sqrt{3x + 1}$ at $x = 5$

f. $f(x) = \dfrac{\sqrt{x}}{x - 5}$ at $x = 9$

g. $f(x) = \begin{cases} 2x + 1 & \text{when } x < 4 \\ 13 & \text{when } x = 4 \\ x^2 - 7 & \text{when } x > 4 \end{cases}$ at $x = 4$

h. $f(x) = \begin{cases} 0.5x^3 & \text{when } x \leq 2 \\ 4x + 7 & \text{when } x > 2 \end{cases}$ at $x = 2$

i. $f(x) = \begin{cases} 5x - 17 & \text{when } x < 4 \\ \sqrt{x + 5} & \text{when } x \geq 4 \end{cases}$ at $x = 4$

33. For each of the following, sketch the graph of a function $f(x)$ that satisfies the listed conditions.

a. The domain of $f(x)$ is $[1, 6]$.
   $f(x)$ is continuous on $[1, 6]$.

b. The domain of $f(x)$ is $[0, 8]$.
   $\lim_{x \to 3^-} f(x)$ and $\lim_{x \to 3^+} f(x)$ both exist, but their values are not equal.

c. The domain of $f(x)$ is $[2, 9]$.
   Neither $\lim_{x \to 5^-} f(x)$ nor $\lim_{x \to 5^+} f(x)$ exist.

d. The domain of $f(x)$ is $[0, 7]$.
   $\lim_{x \to 4} f(x)$ exists, but its value is not equal to $f(4)$.

34. **Rental Fee** Rent and View, a company that rents movies on video cassettes, allows its customers to keep a cassette up to 3 days for a flat fee of $4. The fee for any additional time beyond the first 3 days is $2 per day. Let $R(x)$ represent the rental fee if a customer keeps a cassette x days.

a. Find a formula for $R(x)$.

b. Sketch the graph of $R(x)$ for $0 < x \leq 7$.

c. Determine whether the function $R(x)$ is continuous at $x = 3$.

CHAPTER 2

Differentiation

2.1 Slopes of Tangents
2.2 The Derivative
2.3 Rules for Computing Derivatives
2.4 Approximating Change
2.5 More Rules for Computing Derivatives
2.6 Optimization on Closed Intervals
2.7 More on Optimization
2.8 More on Approximating Change

Important Terms

Review Problems

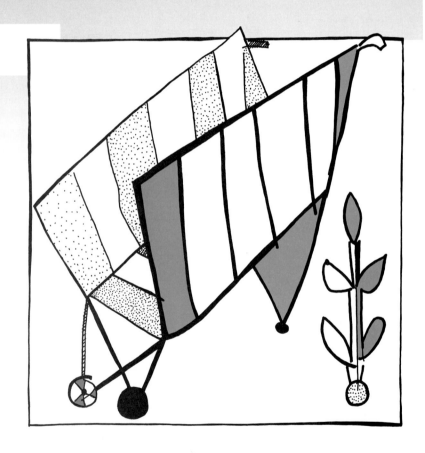

Many people have contributed to the development of calculus. But the discipline crystalized mainly through the independent efforts of Isaac Newton (1642–1727) and Gottfried Wilhelm Leibniz (1646–1716).

According to the approach we use in this book, the core of calculus is the *derivative*. In this chapter, we think of this concept as a function that gives the slopes of the lines tangent to a curve. In Chapter 3, we will see that the derivative has another important interpretation.

2.1 Slopes of Tangents

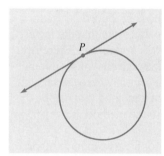

Figure 2.1 A line tangent to a circle at a point *P*.

A line is **tangent** to a circle if it touches the circle at one point only, as illustrated in Figure 2.1. This criterion for tangency, however, does not apply to all curves. In this section, we present a criterion for tangency that applies to curves that represent functions. We also show how the slope of a tangent to such a curve can be obtained using the formula for the function and the point of tangency.

The Idea of a Line Tangent to a Curve

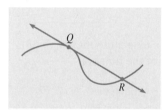

Figure 2.2 A line tangent to a curve that is not a circle.

Imagine a particle moving along the path of the curve in Figure 2.2. If at the very instant the particle reaches point *Q* it somehow loses its traction with the curve, it would leave the curve along the path of the line shown in the figure *regardless of the direction from which it approaches point Q*. (This result is due to the *law of inertia*.) We single out this line and say that it is tangent to the curve at point *Q*. We make this statement even though the line passes through the curve at point *R*. The line, however, is not tangent to the curve at point *R*, because if the particle loses its traction with the curve when it reaches point *R*, the linear path it would follow as it leaves the curve would not be that of the line shown in the figure.

Figure 2.3 shows that it is possible for a line to meet a curve at only one point and yet not be tangent to the curve. The reason is not that the line crosses the curve. Figure 2.4 shows a line tangent to a curve at a point (labeled *M*) where it crosses the curve.

Figure 2.5 shows a curve that fails to have a tangent at a point. Imagine a particle moving from left to right along the path of the curve shown in the figure. If the particle loses its traction with the curve when it reaches point *N*, it would leave the curve along the path of the line shown in the figure. However, if the particle is moving instead from right to left when it loses its traction with the curve at point *N*, it would not leave the curve along the path of the line shown in the figure. Therefore, the curve does not have a tangent at point *N*.

The curve in Figure 2.5 is *not smooth* at point *N*. In this book, we deal primarily with functions whose graphs are smooth curves. Such curves have a tangent at every point.

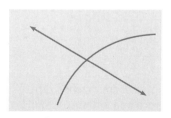

Figure 2.3 A line not tangent to a curve.

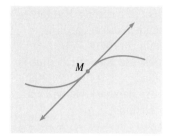

Figure 2.4 A line tangent to a curve where it crosses the curve.

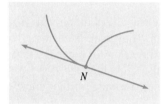

Figure 2.5 A curve that does not have a tangent at a point.

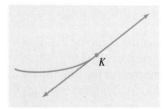

Figure 2.6 A line tangent to a curve at an endpoint.

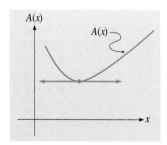

Figure 2.7 A curve with a horizontal tangent at its lowest point.

Figure 2.6 illustrates that *curves can have tangents at endpoints*. In such cases, we have to alter the moving particle description of tangency because the particle can approach the point of tangency from one direction only.

According to our description of the concept of tangency, a line has a tangent at every point. Moreover, at every point the tangent is the line itself.

Horizontal Tangents

There are many situations where the idea of a line tangent to a curve is useful. For example, let $A(x)$ represent the average cost of producing each barrel of a sweet wine when x barrels of the wine is produced per year. Suppose the graph of $A(x)$ is the curve in Figure 2.7. Note that the *lowest point* on the curve is the point where *the tangent is horizontal*. Therefore, the yearly production level that minimizes the average cost of producing each barrel is the x coordinate of the point on the curve where the tangent is horizontal. In the next section, we will present a method that can be used to find the x coordinate of this point. To use the method, however, we must have a formula for $A(x)$.

Figures 2.8 and 2.9 show that it is possible for a curve to have a horizontal tangent at a point that is *not* the lowest point on the curve.

Finding Slopes of Tangents

We now illustrate how to find the slope of a line tangent to the graph of a function. The curve in Figure 2.10 is the graph of the function

$$g(x) = 5x - x^2$$

for x in the closed interval $[0, 5]$. The line labeled T is tangent to the curve at the point $(1, g(1))$. The line labeled S is not tangent to the curve. We will use line S to help find the slope of line T.

The method we used in Chapter 1 to find the slope of a line requires the coordinates of two points on the line. As long as h is not zero, point $(1 + h, g(1 + h))$ does not coincide with point $(1, g(1))$. (When h is positive, $1 + h$ is to the right of 1, as in Figure 2.10. When h is negative, $1 + h$ is to the left of 1.) So, as long as h is not zero,

$$\text{Slope of line } S = \frac{g(1 + h) - g(1)}{(1 + h) - 1}$$
$$= \frac{g(1 + h) - g(1)}{h}$$

We now examine how the slope of line S behaves as h approaches zero.

As h approaches zero (from either side of zero and without actually reaching zero), the point $(1 + h, g(1 + h))$ moves along the path of the curve and gets closer and closer to the point $(1, g(1))$. This activity

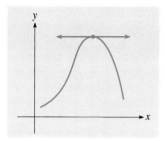

Figure 2.8 A curve with a horizontal tangent at its highest point.

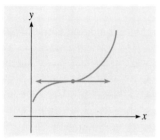

Figure 2.9 A curve with a horizontal tangent at a point that is neither a low nor a high point.

causes line S to pivot on the point $(1, g(1))$ and approach line T. Thus, as h approaches zero, the slope of line S gets closer and closer to the slope of line T. (See Figure 2.11.)

More precisely, as h approaches zero, the slope of line S eventually gets and stays arbitrarily close to the slope of line T. Therefore, since

$$\text{Slope of line } S = \frac{g(1+h) - g(1)}{h}$$

we conclude that

$$\text{Slope of line } T = \lim_{h \to 0} \frac{g(1+h) - g(1)}{h}$$

We will compute this limit *after* we simplify $[g(1+h) - g(1)]/h$ to a form in which h is not a factor of the denominator.

Since $g(x) = 5x - x^2$, we have that

$$\frac{g(1+h) - g(1)}{h} = \frac{[5(1+h) - (1+h)^2] - [5 \cdot 1 - 1^2]}{h}$$

$$= \frac{5 + 5h - 1 - 2h - h^2 - 5 + 1}{h}$$

$$= \frac{3h - h^2}{h}$$

$$= \frac{h(3 - h)}{h}$$

$$= 3 - h$$

Now we compute the limit:

$$\lim_{h \to 0} \frac{g(1+h) - g(1)}{h} = \lim_{h \to 0} (3 - h)$$

$$= 3 \qquad \text{Theorem 1.6 in Section 1.6}$$

Thus, the slope of line T is 3.

With the function $g(x) = 5x - x^2$ we illustrated Theorem 2.1.

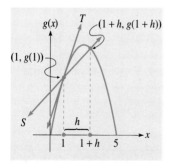

Figure 2.10 The graph of $g(x) = 5x - x^2$ and its tangent at $(1, g(1))$.

Theorem 2.1

Suppose c is the abscissa of a point on the graph of a function $f(x)$ where the tangent is nonvertical. (See Figure 2.12.) Then

$$\lim_{h \to 0} \frac{f(c+h) - f(c)}{h}$$

exists and its value is the slope of the tangent.

In Theorem 2.1, if the point of tangency is the left endpoint of the graph, we would compute the limit as h approaches 0 from the *right* only. Similarly, if the point of tangency is the right endpoint of the graph, we

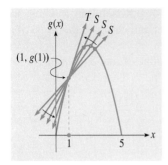

Figure 2.11 Line S approaching line T.

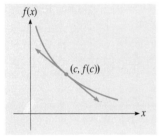

Figure 2.12 A nonvertical tangent.

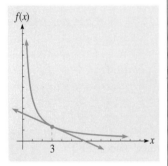

Figure 2.13 The tangent to the graph of $f(x) = 4/x$ at the point where $x = 3$.

would compute the limit as h approaches 0 from the *left* only. Such restrictions ensure that $c + h$ is in the domain of $f(x)$.

We use Theorem 2.1 in Example 2.1.

Example 2.1

Find the slope of the tangent to the graph of

$$f(x) = \frac{4}{x}$$

at the point whose x coordinate is 3. (See Figure 2.13.)

Solution

According to Theorem 2.1, the slope of the tangent in question is the limit of the quotient

$$\frac{f(3 + h) - f(3)}{h}$$

as h approaches zero. We will compute this limit after we simplify the quotient to the point where h is not a factor of the denominator:

$$\frac{f(3 + h) - f(3)}{h} = \frac{\frac{4}{3 + h} - \frac{4}{3}}{h} \quad \text{since } f(x) = 4/x$$

$$= \frac{3 \cdot 4 - (3 + h)4}{3(3 + h)h} \quad \text{multiplying by } 3(3+h)/3(3+h)$$

$$= \frac{-4h}{(9 + 3h)h}$$

$$= \frac{-4}{9 + 3h}$$

So

$$\lim_{h \to 0} \frac{f(3 + h) - f(3)}{h} = \lim_{h \to 0} \frac{-4}{9 + 3h}$$

$$= \frac{-4}{9} \quad \text{Theorem 1.7 in Section 1.6}$$

Therefore, the slope of the tangent to the graph of $f(x) = 4/x$ at the point where $x = 3$ is $-4/9$.

Approximating Tangents with a Graphing Calculator

A curve and a tangent to the curve are practically identical near the point of tangency. So we can use a line through two points on a curve to approximate a tangent to the curve if the two points are near the point of tangency. In Example 2.2, we use this approach and a graphing calculator to approximate a tangent.

Example 2.2

Let $f(x) = 0.06x^3 + 2$
Then $f(2) = 0.06(2^3) + 2$
$= 2.48$

So the point (2, 2.48) is on the graph of $f(x)$.

a. Use a graphing calculator to verify that the graph of $f(x)$ is practically a line segment near the point (2, 2.48).
b. Use the trace function to select two points on the graph of $f(x)$ near the point (2, 2.48).
c. Set the viewing screen on the graphing calculator to [0, 9] × [0, 6] with a scale of 1 on both axes. Then draw the graph of $f(x)$. Also, draw the regression line for the two points selected in part (b) so that the regression line appears simultaneously with the graph of $f(x)$. (Regression lines are discussed in Section 1.3.)
d. What is the slope of the regression line drawn in part (c)?

Solution for (a)

We begin by "zooming in" on the point (2, 2.48). As a first attempt, we set the viewing screen to [1.94, 2.06] × [2.44, 2.52] with a scale of 0.01 on both axes. Then we draw the graph of $f(x)$. Figure 2.14 shows that our viewing screen selection is a good selection, because on this viewing screen the graph of $f(x)$ is practically a line segment.

Solution for (b)

Any two points on the portion of the graph of $f(x)$ that we drew in part (a) will be near the point (2, 2.48). So, applying the trace function to this portion of the graph of $f(x)$, we can select, for example, the point with coordinates

$x = 1.942553191$
$y = 2.439814971$

and the point with coordinates

$x = 2.054893617$
$y = 2.52061812$

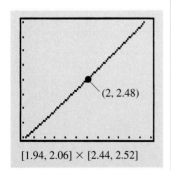

[1.94, 2.06] × [2.44, 2.52]

Figure 2.14 $f(x) = 0.06x^3 + 2$ near the point (2, 2.48).

as our two points near the point (2, 2.48).

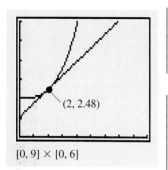

Figure 2.15 A line that approximates a tangent to $f(x) = 0.06x^3 + 2$.

Solution for (c)

Figure 2.15 shows the graph of $f(x)$ and the regression line for the two points we selected in part (b). The regression line approximates the tangent to the graph of $f(x)$ at the point (2, 2.48).

Solution for (d)

We can get the slope of the regression line we drew in part (c) by pressing the appropriate keys on the graphing calculator. The slope, truncated at three decimal places, is

$$m = 0.719$$

Therefore, since the regression line approximates the tangent to the graph of $f(x)$ at the point (2, 2.48), the slope of this tangent is approximately 0.719.

Using Theorem 2.1, we could show that the actual slope of the tangent to the graph of

$$f(x) = 0.06x^3 + 2$$

at the point (2, 2.48) is 0.72. So the approximation we obtained in Example 2.2 using a graphing calculator is quite accurate.

In part (a) of Example 2.2, we selected a viewing screen in which the point (2, 2.48) is at the center of the screen. We also selected a vertical range whose length is two-thirds the length of the horizontal range. This ratio preserves the actual shape of the graph.

In Section 2.2, we will introduce a procedure for finding where the graph of a function has horizontal tangents. The procedure will enable us to find the high and low points on the graph of a function.

2.1 Problems

In Problems 1 to 10, sketch the line that is tangent to the given curve at the point indicated by the dot.

1.

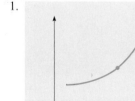

2.

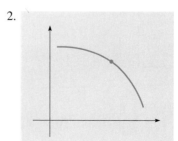

Sec. 2.1 Slopes of Tangents 73

3.
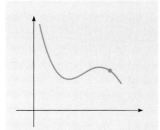

4.
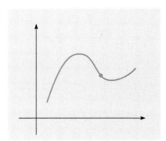

5.
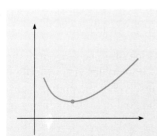

6.
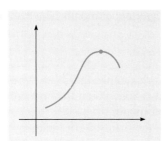

7.
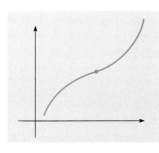

8.
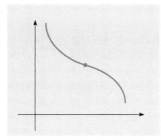

9.
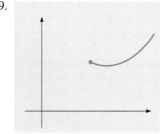

10.
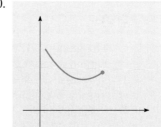

In Problems 11 to 22, find the slope of the tangent to the graph of the function at the given point.

11. $f(x) = 4x^2$ $(3, f(3))$
12. $f(x) = x^2$ $(5, f(5))$
13. $f(x) = x^3$ $(4, f(4))$
14. $f(x) = 2x^3$ $(1, f(1))$
15. $f(x) = 5x^2 + 1$ $(2, f(2))$
16. $f(x) = x^2 + 4$ $(3, f(3))$
17. $f(x) = x^2 - 8x$ $(3, f(3))$
18. $f(x) = 10x - 2x^2$ $(4, f(4))$
19. $f(x) = 196x - 12x^3$ $(2, f(2))$
20. $f(x) = 169x - 3x^3$ $(5, f(5))$
21. $f(x) = 5/x$ $(2, f(2))$
22. $f(x) = 1/x$ $(6, f(6))$

23. Find the linear function whose graph is the tangent to the graph of
$$f(x) = 0.25x^2 + 4$$
at the point where $x = 6$.

24. Find the linear function whose graph is the tangent to the graph of
$$f(x) = 8x - x^2$$
at the point where $x = 5$.

Graphing Calculator Problems

25. Set the viewing screen on your graphing calculator to $[0, 9] \times [0, 6]$ with a scale of 1 on both axes. Then draw the graph of
$$f(x) = |0.25x^2 - 3| + 2$$
You will see a point where the graph has no tangent. Use the trace function to approximate the coordinates of this point.

26. Set the viewing screen on your graphing calculator to $[0, 9] \times [0, 6]$ with a scale of 1 on both axes. Then draw the graph of
$$f(x) = \sqrt[3]{x - 5} + 4$$
You will see a point where the graph has a vertical tangent. Use the trace function to approximate the x coordinate of this point.

In Problems 27 and 28, use an approach like in Example 2.2 to approximate the slope of the tangent to the graph of the function at the point with the given x coordinate.

27. $f(x) = 5 - 0.12x^2$ $x = 4$

28. $f(x) = \dfrac{10}{x + 2}$ $x = 3$

2.2 The Derivative

In Section 2.1, we showed how to find the slope of a line tangent to the graph of a function. In this section, we show how to derive a formula that expresses such a slope as a function of the abscissa of the point of tangency. We begin with a situation in which the goal is to find a firm's optimal production level.

A chemical firm recently began producing a new insulating material. Let $P(x)$ represent the monthly profit (in thousands of dollars) the firm earns if it produces x hundred tons of the material per month. The management believes that

$$P(x) = 256x - 3x^3$$

for values of x in the interval $[2, 8]$. Note, for example, that $P(3) = 687$, which tells us that the firm's monthly profit is $687,000 if the firm produces 300 tons of the insulating material per month.

The curve in Figure 2.16 is the graph of the firm's profit function. The graph shows that as x is increased, $P(x)$ at first increases but eventually decreases. Thus, the monthly profit is low if the firm produces either too little or too much of the material. We want to find the monthly production level at which the monthly profit is *greatest*. That is, we want to find the value of x at which $P(x)$ is *maximum*. To put it still another way, we want to find the value of x for which $(x, P(x))$ is the *highest* point on the graph of the profit function.

We cannot solve the problem by finding the x coordinate of the vertex of a parabola, because $P(x)$ is not a quadratic function. Nor can we rely entirely on Figure 2.16. For example, the figure suggests that the highest

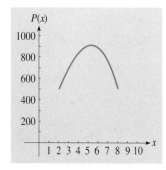

Figure 2.16 $P(x) = 256x - 3x^3$ for x in $[2, 8]$.

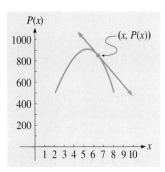

Figure 2.17 A tangent to the graph of $P(x)$.

point on the graph of $P(x)$ is possibly the point where $x = 5$. But we shall soon see that this conclusion would not be correct. Figure 2.16 does correctly suggest that the highest point on the graph of $P(x)$ is the point where the tangent is horizontal, and hence its slope is zero. Therefore, we will solve the problem using an approach based on this observation.

In Figure 2.17, the point $(x, P(x))$ is an *arbitrary fixed point* on the graph of the profit function. The line labeled T is tangent to the graph at this point. Naturally, the slope of line T depends on the value of x. We will derive a formula for the slope of line T in terms of x. Then we will find the value of x for which the slope of line T is zero. This value of x is the monthly production level that maximizes the profit, because, as we observed, the highest point on the graph is the point where the tangent is horizontal.

Let $m(x)$ represent the slope of line T. Then, according to Theorem 2.1 in Section 2.1,

$$m(x) = \lim_{h \to 0} \frac{P(x+h) - P(x)}{h}$$

We will compute this limit *after* we simplify $[P(x + h) - P(x)]/h$ to the point where h is not a factor of the denominator.

Since $P(x) = 256x - 3x^3$, it follows that

$$\frac{P(x+h) - P(x)}{h} = \frac{[256(x+h) - 3(x+h)^3] - [256x - 3x^3]}{h}$$

$$= \frac{256x + 256h - 3(x^3 + 3x^2h + 3xh^2 + h^3) - 256x + 3x^3}{h}$$

$$= \frac{256h - 9x^2h - 9xh^2 - 3h^3}{h}$$

$$= \frac{h(256 - 9x^2 - 9xh - 3h^2)}{h}$$

$$= 256 - 9x^2 - 9xh - 3h^2$$

Since $(x, P(x))$ is a fixed point on the graph of the profit function, the value of x is fixed. So we can think of $256 - 9x^2 - 9xh - 3h^2$ as a polynomial function whose independent variable is h. Therefore, we can use Theorem 1.6 in Section 1.6 to compute the limit of this function as h approaches zero:

$$\lim_{h \to 0} \frac{P(x+h) - P(x)}{h} = \lim_{h \to 0} (256 - 9x^2 - 9xh - 3h^2)$$

$$= 256 - 9x^2 \qquad \text{Theorem 1.6 in Section 1.6}$$

Thus, a formula for the slope of line T is

$$m(x) = 256 - 9x^2$$

Since $(x, P(x))$ is an arbitrary fixed point on the graph of the profit function, the above formula is valid for *any* value of x in the interval $[2, 8]$. So we can use the formula to obtain the slope of *any* tangent to the graph of the profit function. For example, by substituting 5 for x in the formula, we get

$$m(5) = 256 - 9 \cdot 5^2$$
$$= 31$$

This result tells us that the slope of the tangent at the point $(5, P(5))$ is 31.

In Figure 2.16, it appears that the slope of the tangent at the point $(5, P(5))$ should be much less than 31. In fact, it appears that the slope could possibly be zero. We get this false impression because the graph of $P(x)$ is flattened considerably by the relatively large scale we chose for the vertical axis.

We now find the x coordinate of the point on the graph of $P(x)$ at which the tangent is horizontal. So we set the formula for the slopes of the tangents equal to zero and solve for x:

$$256 - 9x^2 = 0$$
$$-9x^2 = -256$$
$$x^2 = \frac{256}{9}$$
$$x = \pm\frac{16}{3}$$

Thus, the graph of the profit function has a horizontal tangent at the point whose x coordinate is $\frac{16}{3}$. (We disregard $-\frac{16}{3}$ because the domain of the profit function is the interval [2, 8].) As we saw in Figure 2.16, the highest point on the graph is the point where the tangent is horizontal. Therefore, since $\frac{16}{3}$ is approximately 5.33, the chemical firm should produce about 533 (5.33 hundred) tons of the insulating material monthly to maximize the profit.

With the profit problem, we illustrated the fundamental concept of this book, which is stated in Definition 2.1.

Definition 2.1

The **derivative** of a function $f(x)$ is a function $f'(x)$ (read "f prime of x") whose formula can be derived by computing

$$\lim_{h \to 0} \frac{f(x+h) - f(x)}{h}$$

It is common to say that $f(x)$ is **differentiable** at those values of x for which the above limit exists. It is also common to refer to

$$\frac{f(x+h) - f(x)}{h}$$

as the **difference quotient** of $f(x)$.

In solving the profit problem, we computed

$$\lim_{h \to 0} \frac{P(x + h) - P(x)}{h}$$

and obtained the function

$$m(x) = 256 - 9x^2$$

Thus, according to Definition 2.1, the derivative of the function $P(x)$ is $m(x) = 256 - 9x^2$. This result can be expressed by writing

$$P'(x) = 256 - 9x^2$$

The Connection between Derivatives and Slopes of Tangents

The validity of Theorem 2.2 is based on Theorem 2.1 in Section 2.1 and Definition 2.1.

Theorem 2.2

Suppose c is the abscissa of a point on the graph of a function $f(x)$ where the tangent is nonvertical. Then $f(x)$ is differentiable at $x = c$ and $f'(c)$ is the slope of the tangent.

The converse of Theorem 2.2 is also valid.

Converse of Theorem 2.2

Suppose a function $f(x)$ is differentiable at $x = c$. Then the graph has a nonvertical tangent at the point $(c, f(c))$ and the slope of the tangent is $f'(c)$.

This result is used in Example 2.3.

Example 2.3

Let $C(t)$ be the temperature, in degrees Celsius, at a ski resort t hours from now. The weather bureau predicts that

$$C(t) = \tfrac{1}{3}t^2 - 3t$$

up to $t = 10$. Figure 2.18 shows that the lowest point on the graph of this function is the point where the tangent is horizontal. Use the derivative to determine when the minimum temperature will occur during the next 10 hours. What will the temperature be at that time?

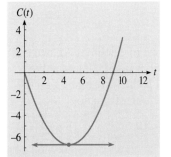

Figure 2.18 $C(t) = \tfrac{1}{3}t^2 - 3t$ for t in $[0, 10]$.

Solution

First we find the derivative of $C(t)$. According to Definition 2.1, a formula for this derivative can be found by computing the limit of the difference quotient

$$\frac{C(t+h) - C(t)}{h}$$

as h approaches zero. We will compute this limit after we simplify the difference quotient to the point where h does not appear as a factor of the denominator:

$$\frac{C(t+h) - C(t)}{h} = \frac{[\tfrac{1}{3}(t+h)^2 - 3(t+h)] - [\tfrac{1}{3}t^2 - 3t]}{h}$$

$$= \frac{\tfrac{1}{3}t^2 + \tfrac{2}{3}th + \tfrac{1}{3}h^2 - 3t - 3h - \tfrac{1}{3}t^2 + 3t}{h}$$

$$= \frac{\tfrac{2}{3}th + \tfrac{1}{3}h^2 - 3h}{h}$$

$$= \frac{h(\tfrac{2}{3}t + \tfrac{1}{3}h - 3)}{h}$$

$$= \tfrac{2}{3}t + \tfrac{1}{3}h - 3$$

Now we compute the limit. For each fixed value of t in the interval $[0, 10]$,

$$\lim_{h \to 0} \frac{C(t+h) - C(t)}{h} = \lim_{h \to 0} (\tfrac{2}{3}t + \tfrac{1}{3}h - 3)$$

$$= \tfrac{2}{3}t - 3 \quad \text{Theorem 1.6 in Section 1.6}$$

Thus, for each value of t in $[0, 10]$, the derivative of $C(t)$ is

$$C'(t) = \tfrac{2}{3}t - 3$$

Next, we set the formula for the derivative equal to 0 and solve for t:

$$\tfrac{2}{3}t - 3 = 0$$
$$\tfrac{2}{3}t = 3$$
$$t = 4.5$$

Thus, $C'(t) = 0$ when $t = 4.5$. Using the converse of Theorem 2.2, we can therefore conclude that the graph of $C(t)$ has a horizontal tangent at the point whose t coordinate is 4.5. Since this point is the lowest point on the graph, the minimum temperature will occur 4.5 hours from now. Since

$$C(4.5) = \tfrac{1}{3}(4.5^2) - 3(4.5)$$
$$= -6.75$$

the minimum temperature will be 6.75 degrees below zero.

The function $C(t)$ in the above example is a quadratic function. Therefore, we could have used an approach based on Theorem 1.3 in Section 1.4 to determine when the minimum temperature will occur.

The Leibniz Notation for Derivatives

If the dependent variable of a function is y and the independent variable is x, we can represent the derivative of the function by

$$\frac{dy}{dx}$$

(read "dee y dee x"). We will not think of this symbol as a quotient. Nevertheless, it does remind us that a derivative is the limit of a quotient. This notation was introduced by Leibniz. We use it in Example 2.4.

Example 2.4

Find a formula for dy/dx, if

$$y = 2\sqrt{x}$$

Then find the slope of the tangent to the graph of this function at the point (9, 6). (See Figure 2.19.)

Solutions

First we simplify the difference quotient of the given function:

$$\frac{2\sqrt{x+h} - 2\sqrt{x}}{h} = \frac{(2\sqrt{x+h} + 2\sqrt{x})(2\sqrt{x+h} - 2\sqrt{x})}{(2\sqrt{x+h} + 2\sqrt{x})h}$$

$$= \frac{4(x+h) - 4x}{(2\sqrt{x+h} + 2\sqrt{x})h}$$

$$= \frac{4h}{(2\sqrt{x+h} + 2\sqrt{x})h}$$

$$= \frac{4}{2\sqrt{x+h} + 2\sqrt{x}}$$

Now we compute the limit of the difference quotient as h approaches zero. For each fixed *positive* value of x,

$$\lim_{h \to 0} \frac{4}{2\sqrt{x+h} + 2\sqrt{x}} = \frac{4}{2\sqrt{x} + 2\sqrt{x}} \quad \text{limit properties}$$

$$= \frac{4}{4\sqrt{x}}$$

$$= \frac{1}{\sqrt{x}}$$

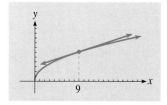

Figure 2.19 The tangent to the graph of $y = 2\sqrt{x}$ at the point where $x = 9$.

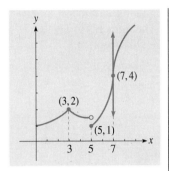

Figure 2.20 A function that is not differentiable at $x = 3, 5,$ and 7.

Thus, according to Definition 2.1, the derivative of $y = 2\sqrt{x}$ is

$$\frac{dy}{dx} = \frac{1}{\sqrt{x}}$$

for each positive value of x. In particular, when $x = 9$,

$$\frac{dy}{dx} = \frac{1}{\sqrt{9}}$$
$$= \frac{1}{3}$$

So, according to the converse of Theorem 2.2, the slope of the tangent to the graph of $y = 2\sqrt{x}$ at the point $(9, 6)$ is $1/3$.

The Connection between Differentiable and Continuous

The curve in Figure 2.20 does not have tangents at the points $(3, 2)$ and $(5, 1)$. So, according to the converse of Theorem 2.2, the corresponding function is not differentiable at $x = 3$ and at $x = 5$. The function is also not differentiable at $x = 7$, because the curve has a vertical tangent at the point where $x = 7$.

Note that the function in Figure 2.20 is discontinuous at $x = 5$. Since the function is not differentiable at $x = 5$, we have illustrated that a function cannot be differentiable at a point where it is not continuous. In other words, we have illustrated the following result:

> If a function is differentiable at $x = c$, then the function is continuous at $x = c$.

The converse of the above result is not valid. So it is possible for a function to be continuous at a point without the function being differentiable at the point. For example, the function in Figure 2.20 is continuous at $x = 3$ and at $x = 7$. But, as we observed, the function is not differentiable at these two points.

Comment about the Problem Set

Several problems at the end of this section contain a statement informing us that the highest (or lowest) point on a graph is the point where the tangent is horizontal. We can confirm such a statement by drawing the graph with a graphing calculator. Whether or not we confirm the statement, we will use the derivative of the corresponding function to find the abscissa of the point where the tangent is horizontal, as in Example 2.3. This abscissa, of course, is the value of the independent variable where the dependent variable attains its maximum (or minimum) value.

Approximating Derivatives with a Graphing Calculator

Example 2.2 in Section 2.1 illustrates how we can use a graphing calculator to approximate the slope of a tangent. Since the slope of a tangent is the value of a derivative, we can use the same approach to approximate the value of a derivative. Two problems at the end of this section require that we use this approach. (See Problems 39 and 40.)

2.2 Problems

In Problems 1 to 20, find a formula for the derivative of the function by computing the limit of a difference quotient as h approaches zero. Then use the formula to find the slope of the tangent at the given point.

1. $f(x) = 8x^2$
   $(0.25, 0.5)$

2. $f(x) = x^2$
   $(3, 9)$

3. $f(x) = x^3$
   $(1, 1)$

4. $f(x) = 4x^3$
   $(2, 32)$

5. $f(x) = 3x^2 - 4$
   $(5, 71)$

6. $f(x) = x^2 + 5$
   $(0.5, -5.25)$

7. $f(x) = x^2 - 6x$
   $(2, -8)$

8. $f(x) = 7x^2 - x$
   $\left(\frac{1}{14}, -\frac{1}{28}\right)$

9. $f(x) = 2x^3 + x^2$
   $\left(\frac{1}{6}, \frac{1}{54}\right)$

10. $f(x) = x^3 + 8x^2$
    $(-1, 7)$

11. $f(x) = x^2 - 3x + 1$
    $(1.5, -1.25)$

12. $f(x) = 3x^2 + x - 4$
    $(0, -4)$

13. $f(x) = 4x^3 + x + 3$
    $(1, 8)$

14. $f(x) = x^3 - 15x + 9$
    $(2, -13)$

15. $f(x) = x^4$
    $(-2, 16)$

16. $f(x) = 3x^4$
    $(0, 0)$

17. $f(x) = 2\sqrt{x}$
    $(9, 6)$

18. $f(x) = \sqrt{x}$
    $(4, 2)$

19. $f(x) = \dfrac{1}{x}$
    $(5, 0.2)$

20. $f(x) = \dfrac{3}{x}$
    $(2, 1.5)$

In Problems 21 to 24, use the derivative of the function to find the x coordinates of the points where the graph has horizontal tangents.

21. $y = x^2 - 6x + 10$

22. $y = x^3 - 75x$

23. $y = 108x - x^3$

24. $y = -x^2 + 8x$

25. Let the domain of
$$f(x) = 27x - x^3$$
be the set of nonnegative real numbers. Then the highest point on the graph of $f(x)$ is the point where the tangent is horizontal. Find the value of x (among the nonnegative real numbers) where $f(x)$ is maximum.

26. The highest point on the graph of
$$f(x) = 7x - x^2$$
is the point where the tangent is horizontal. Use the derivative to find the value of x where $f(x)$ is maximum.

27. The lowest point on the graph of
$$f(x) = x^2 - 8x + 25$$
is the point where the tangent is horizontal. Use the derivative to find the value of x where $f(x)$ is minimum.

28. Let the domain of
$$f(x) = x^3 - 48x + 130$$
be the set of nonnegative real numbers. Then the lowest point on the graph of $f(x)$ is the point where the tangent is horizontal. Find the value of x (among the nonnegative real numbers) where $f(x)$ is minimum.

29. **Demography** Let $A(t)$ represent the median age of the people in a South American country t years from now. Demographers project that
$$A(t) = \tfrac{3}{64}t^2 - \tfrac{3}{4}t + 31$$
for t in the interval $[0, 20]$. The lowest point on the graph of this function is the point where the tangent is horizontal. Use the derivative to determine when the median age will be minimum. What will the median age be at that time?

30. **Demand** In a Third World country, the average monthly income per household is only $400. However, it is increasing. Let $D(x)$ represent the country's monthly demand for powdered milk (in tons) when the average monthly income per household is x hundred dollars. Officials believe that $D(x)$ will at first increase as x increases. But, since powdered milk is a money-saving substitute for regular milk, the officials project that $D(x)$ will eventually decrease as x continues to increase. Suppose

$$D(x) = -\tfrac{1}{3}x^3 + 5x^2$$

for values of x in the interval $[4, 14]$. The highest point on the graph of this function is the point where the tangent is horizontal. For what average monthly income per household will the monthly demand for powdered milk be greatest? What will the monthly demand be when it is greatest?

31. **Optimal Production Level** Each day a company sells whatever amount of papaya juice it produces. Thus, we can think of the company's daily profit earned from selling the juice as a function of the daily production level. Suppose the daily profit is

$$P(x) = -\tfrac{1}{3}x^3 + 9x^2 + 63x - 1200$$

dollars when the company produces x barrels of the juice per day. The highest point on the graph of this function is the point where the tangent is horizontal. Determine the daily production level that maximizes the profit.

A papaya tree. (Richard Rowan's Collection/Photo Researchers)

32. **Least Average Cost** The management of a tobacco processing firm assumes that the average cost of producing each ounce of Choco-Chew (a chocolate flavored chewing tobacco) is

$$A(x) = \tfrac{1}{45}x^2 - 2x + 61$$

cents when the weekly production level is x tons. The lowest point on the graph of this function is the point where the tangent is horizontal. Use the derivative to determine the weekly production level at which the average cost of producing each ounce is least.

33. **Law of Inertia** From a tabletop 3 feet high, a marble rolls off with a horizontal speed of 8 feet per second. Then its height y (in feet) above the floor is a quadratic function of its horizontal distance x (also in feet) from the edge of the table according to the formula

$$y = -0.25x^2 + 3$$

The path of the marble is a portion of a parabola as shown in the figure. Under normal conditions the marble strikes the floor at a point $2\sqrt{3}$ feet from the base of the table. But suppose the force of gravity somehow ceases to exist beginning at the instant when the position of the marble is $(1, 2.75)$. Then, according to the law of inertia, the marble would leave its parabolic path and travel along the line that is tangent to the parabolic path at the point $(1, 2.75)$. The figure shows this linear path. If the force of gravity ceases as described, at a point how far from the base of the table would the marble strike the floor?

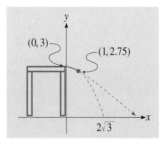

34. Show that the function

$$f(x) = |x|$$

fails to have a derivative at $x = 0$. In other words, show that there is no number equal to

$$\lim_{h \to 0} \frac{f(0+h) - f(0)}{h}$$

35. Suppose $f(x)$ is the function whose graph is the curve in the figure for this problem. Explain how the graph shows that $f'(x)$ decreases as x increases.

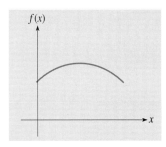

36. Suppose $f(x)$ is the function whose graph is the curve in the figure for this problem. Explain how the graph shows that there is a value of x where $f'(x)$ is maximum.

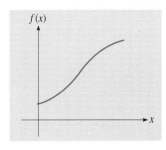

37. Explain why the derivative of a linear function must be a constant function.

38. If two different functions have the same derivative, how are the graphs of the two functions related?

Graphing Calculator Problems

In Problems 39 and 40, use your graphing calculator to approximate the derivative of the function at the given value of x.

Hint: See Example 2.2 in Section 2.1.

39. $f(x) = \dfrac{12x}{x+2}$ \quad $x = 6$

40. $f(x) = \dfrac{x+24}{x+3}$ \quad $x = 2$

41. Set the viewing screen on your graphing calculator to $[0, 9] \times [0, 6]$ with a scale of 1 on both axes. Then draw the graph of

$$f(x) = \sqrt[3]{x-6} + 3$$

The graph will show that there is a value of x where $f(x)$ is not differentiable. Use the trace function to approximate this value of x.

42. Set the viewing screen on your graphing calculator to $[0, 9] \times [0, 6]$ with a scale of 1 on both axes. Then draw the graph of

$$f(x) = |0.3x^2 - 2| + 1$$

The graph will show that there is a value of x where $f(x)$ is not differentiable. Use the trace function to approximate this value of x.

2.3 Rules for Computing Derivatives

In Section 2.2, we saw that the derivative of a function $f(x)$ is a function $f'(x)$ whose formula can be obtained by computing

$$\lim_{h \to 0} \frac{f(x+h) - f(x)}{h}$$

But the computation of such a limit is often laborious. Fortunately, there are rules that we can use to obtain derivatives with less effort. In this section, we present several such rules. We begin with a rule for constant functions.

The Derivative of a Constant Function

Let $f(x) = c$, where c is a constant. Applying the definition of the derivative (Definition 2.1), for each fixed value of x,

$$f'(x) = \lim_{h \to 0} \frac{f(x+h) - f(x)}{h}$$

$$= \lim_{h \to 0} \frac{c - c}{h}$$

$$= \lim_{h \to 0} \frac{0}{h}$$

$$= \lim_{h \to 0} 0$$

$$= 0$$

(We could have obtained this result without the above calculations. How?)
   We just established Theorem 2.3, the rule for computing the derivative of a constant function.

Theorem 2.3 (Constant Rule)

If c is a real number, the derivative of

$$f(x) = c$$

is

$$f'(x) = 0$$

In other words, the derivative of a constant function is zero. We use this rule in Example 2.5.

Example 2.5

Find the derivative of

$$f(x) = 0.73$$

Solution

Using the constant rule,

$$f'(x) = 0$$

We now turn to functions that can be expressed in the form cx^n.

The Derivative of a Function of the Form cx^n

The function

$$f(x) = 7x^3$$

is of the form cx^n, where $c = 7$ and $n = 3$. For each fixed value of x,

$$\begin{aligned} f'(x) &= \lim_{h \to 0} \frac{7(x+h)^3 - 7x^3}{h} \\ &= \lim_{h \to 0} \frac{7(x^3 + 3x^2h + 3xh^2 + h^3) - 7x^3}{h} \\ &= \lim_{h \to 0} \frac{21x^2h + 21xh^2 + 7h^3}{h} \\ &= \lim_{h \to 0} \frac{(21x^2 + 21xh + 7h^2)h}{h} \\ &= \lim_{h \to 0} (21x^2 + 21xh + 7h^2) \\ &= 21x^2 \quad \text{Theorem 1.6 in Section 1.6} \end{aligned}$$

Thus, for each value of x, the derivative of $f(x) = 7x^3$ is

$$f'(x) = 21x^2$$

Note that $21x^2 = 3 \cdot 7x^{3-1}$. This result suggests the rule given in Theorem 2.4.

Theorem 2.4 (Power Rule)

If c is a real number and n is a rational number, the derivative of

$$f(x) = cx^n$$

is

$$f'(x) = ncx^{n-1}$$

In Section 2.5, we will present a more general version of this rule.

Using what is known in algebra as the **binomial theorem**, the power rule can be proved for the case where n is any positive integer. Such a proof parallels our derivation of the derivative of $f(x) = 7x^3$.

We use the power rule in Example 2.6.

Example 2.6

Find the derivatives of the following functions.

a. $g(x) = \dfrac{12}{\sqrt{x}}$

b. $y = 5x$

c. $f(x) = x^9$

> **Solution for (a)**
>
> Since
>
> $$\frac{12}{\sqrt{x}} = 12x^{-1/2}$$

we can write
$$g(x) = 12x^{-1/2}$$
Thus, $g(x)$ is of the form cx^n, with $c = 12$ and $n = -1/2$. Using the power rule,
$$\begin{aligned}g'(x) &= -6x^{-3/2} \\ &= \frac{-6}{x^{3/2}} \\ &= \frac{-6}{(\sqrt{x})^3}\end{aligned}$$

Solution for (b)

The formula for y is of the form cx^n, with $c = 5$ and $n = 1$. So, according to the power rule,
$$\begin{aligned}\frac{dy}{dx} &= 5x^0 \\ &= 5 \cdot 1 \\ &= 5\end{aligned}$$

Solution for (c)

The formula for $f(x)$ is of the form cx^n, with $c = 1$ and $n = 9$. Therefore, applying the power rule,
$$f'(x) = 9x^8$$

Next, we consider functions formed by adding or subtracting other functions.

The Derivative of a Sum or Difference

In Section 2.2, we found, after much effort, that the derivative of
$$P(x) = 256x - 3x^3$$
is
$$P'(x) = 256 - 9x^2$$

The power rule tells us that 256 is the derivative of $256x$ and $9x^2$ is the derivative of $3x^3$. Thus, with the function $P(x)$, we illustrated the rule given in Theorem 2.5.

Theorem 2.5 (Sum and Difference Rule)

If $u(x)$ and $v(x)$ are differentiable functions, the derivative of
$$f(x) = u(x) + v(x)$$
is
$$f'(x) = u'(x) + v'(x)$$
and the derivative of
$$g(x) = u(x) - v(x)$$
is
$$g'(x) = u'(x) - v'(x)$$

In other words, the derivative of a sum is the sum of the derivatives and the derivative of a difference is the difference of the derivatives. This rule extends to sums and differences of *more than two* functions. You will be asked to prove this rule in the problem set at the end of this section.

Before we continue our discussion of the sum and difference rule, we introduce an alternative notation for derivatives. The symbol

$$D_x(f(x))$$

is sometimes used to represent the derivative of a function $f(x)$. For example, the derivative of

$$f(x) = 6x^4$$

can be represented by

$$D_x(6x^4)$$

which, according to the power rule, is $24x^3$.

Using the above notation, we can express the sum and difference rule as follows:

$$D_x(u(x) + v(x)) = D_x(u(x)) + D_x(v(x))$$

and

$$D_x(u(x) - v(x)) = D_x(u(x)) - D_x(v(x))$$

We use the sum and difference rule in the remaining examples of this section.

Example 2.7

Find the derivative of

$$y = (x^2 + 5)^3$$

Solution

First note that

$$(x^2 + 5)^3 = x^6 + 3x^4 \cdot 5 + 3x^2 \cdot 25 + 125$$
$$= x^6 + 15x^4 + 75x^2 + 125$$

Therefore, we can write

$$y = x^6 + 15x^4 + 75x^2 + 125$$

So $dy/dx = D_x(x^6) + D_x(15x^4) + D_x(75x^2) + D_x(125)$ sum rule
$= D_x(x^6) + D_x(15x^4) + D_x(75x^2) + 0$ constant rule
$= 6x^5 + 60x^3 + 150x$ power rule

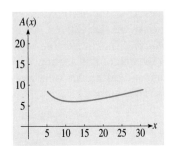

Figure 2.21
$A(x) = 0.25x + 36/x$
for $5 \leq x \leq 30$.

We need a special rule for computing derivatives of functions like the one in Example 2.7. Imagine, for example, how much work it would take to expand $(x^2 + 5)^{74}$. In Section 2.5, we will present a rule that allows us to compute the derivative of this function without having to expand its formula.

Example 2.8 involves an average cost function.

Example 2.8

The average cost of producing each ton of a special fertilizer is

$$A(x) = 0.25x + \frac{36}{x}$$

hundred dollars if x tons of the fertilizer is produced weekly. Figure 2.21 shows that the lowest point on the graph of $A(x)$ is the point where the tangent is horizontal. How many tons of the fertilizer should be produced weekly to minimize the average cost of producing each ton?

Solution

Applying the power rule,

$$D_x(0.25x) = 0.25$$

and

$$D_x\left(\frac{36}{x}\right) = D_x(36x^{-1}) = -36x^{-2} = \frac{-36}{x^2}$$

Therefore, applying the sum and difference rule,

$$A'(x) = 0.25 + \frac{-36}{x^2}$$
$$= 0.25 - \frac{36}{x^2}$$

Now we set the derivative equal to 0 and solve for x:

$$0.25 - \frac{36}{x^2} = 0$$
$$0.25 = \frac{36}{x^2}$$
$$0.25x^2 = 36$$
$$x^2 = 144$$
$$x = \pm 12$$

Thus, $A'(x) = 0$ at $x = 12$. (We disregard -12 because the value of x cannot be a negative number in our application.) So the graph of the average cost function has a horizontal tangent at the point where $x = 12$. Since this point is the lowest point on the graph, 12 tons of the fertilizer should be produced weekly to minimize the average cost of producing each ton.

2.3 Problems

In Problems 1 to 36, find a formula for $f'(x)$ by applying the rules introduced in this section.

1. $f(x) = -4x$
2. $f(x) = 0.4x$
3. $f(x) = 0.75$
4. $f(x) = -9$
5. $f(x) = 5x^3$
6. $f(x) = -3x^4$
7. $f(x) = -x^5$
8. $f(x) = x^6$
9. $f(x) = x^2$
10. $f(x) = -4x^2$
11. $f(x) = \dfrac{1}{x}$
12. $f(x) = -\dfrac{8}{x}$
13. $f(x) = -\dfrac{4}{x^2}$
14. $f(x) = \dfrac{1}{x^3}$
15. $f(x) = 2\sqrt{x}$
16. $f(x) = 6\sqrt[3]{x}$
17. $f(x) = -\dfrac{1}{\sqrt[3]{x}}$
18. $f(x) = \dfrac{4}{\sqrt{x}}$
19. $f(x) = 3x + 8$
20. $f(x) = 0.5x - 7$
21. $f(x) = x^2 - 9x + 4$
22. $f(x) = 3x^2 + x$
23. $f(x) = 2x^3 + 0.5x$
24. $f(x) = x^3 - 7x - 1$
25. $f(x) = x^3 + 7x^2 - 5x$
26. $f(x) = 4x^3 - x^2 + 6x$
27. $f(x) = x - \dfrac{2}{x}$
28. $f(x) = 5x + \dfrac{1}{x}$
29. $f(x) = 8x + \sqrt{x} + 5$
30. $f(x) = \sqrt[3]{x} + 7$
31. $f(x) = 3\sqrt[3]{x} + \dfrac{1}{x^2}$
32. $f(x) = \sqrt{x} - \dfrac{2}{x^3}$
33. $f(x) = x^3 + \dfrac{1}{\sqrt{x}}$
34. $f(x) = x^2 - \dfrac{12}{\sqrt[3]{x}}$
35. $f(x) = (x^3 - 5)^2$
36. $f(x) = (x + 2)^3$

In Problems 37 to 46, use the derivative of the function to find the x coordinates of the points where the graph has horizontal tangents.

37. $y = x^2 - 12x + 3$
38. $y = 3x^2 + 18x - 7$
39. $y = 2x^3 - 3x^2 - 120x$
40. $y = x^3 - 12x^2 + 21x$
41. $y = x^3 - 10x^2 + 12x$
42. $y = 4x^3 + 27x^2 + 24x$
43. $y = x + \dfrac{16}{x}$
44. $y = 4x + \dfrac{36}{x}$
45. $y = 24x - 4(\sqrt{x})^3$
46. $y = 60x - 8(\sqrt{x})^3$

47. Let the domain of
$$f(x) = 5x + \dfrac{45}{x}$$
be the set of positive real numbers. Then the lowest point on the graph of $f(x)$ is the point where the tangent is horizontal. Find the value of x (among the positive real numbers) where $f(x)$ is minimum.

48. Let the domain of
$$f(x) = x + \dfrac{81}{x}$$
be the set of positive real numbers. Then the lowest point on the graph of $f(x)$ is the point where the tangent is horizontal. Find the value of x (among the positive real numbers) where $f(x)$ is minimum.

49. The highest point on the graph of
$$f(x) = 6x - 2(\sqrt{x})^3$$
is the point where the tangent is horizontal. Find the value of x where $f(x)$ is maximum.

50. The highest point on the graph of
$$f(x) = 27x - 6(\sqrt{x})^3$$
is the point where the tangent is horizontal. Find the value of x where $f(x)$ is maximum.

51. **Optimal Speed** When a tractor trailer is driven at a speed of x miles per hour (up to $x = 90$), it gets
$$m(x) = -0.0028x^2 + 0.28x$$

miles from each gallon of fuel. The highest point on the graph of this function is the point where the tangent is horizontal. Use the derivative to find the speed at which the tractor trailer gets the best mileage. What is the mileage at that speed?

52. **Population Growth Rate** A farmer has decided to allow a flock of turkeys to multiply before marketing them. The growth rate of the flock will at first increase. But, due to certain constraining factors, the growth rate will eventually begin to decrease. Let $r(x)$ represent the growth rate (in thousands of turkeys per year) when the flock consists of x thousand turkeys. The farmer projects that

$$r(x) = -0.08x^2 + 1.92x$$

for x in the interval $[1, 20]$. The highest point on the graph of this function is the point where the tangent is horizontal. Use the derivative to find the size of the flock when its growth rate is greatest. What is the greatest growth rate?

53. **House Prices** Realtors in a southern city believe that the average price of a new house will decline for a while and then begin to increase. They predict that x months from now the average price will be

$$f(x) = \tfrac{4}{3}(\sqrt{x})^3 - 6x + 95$$

thousand dollars. The lowest point on the graph of this function is the point where the tangent is horizontal. When will the average price be lowest?

54. **Insect Population** After a fruit grower sprays an orchard with a biodegradable insecticide, the population of a particular kind of insect gradually declines. However, the insects are not entirely eliminated. In fact, the insects eventually begin to multiply. The grower believes that each acre has

$$p(t) = \tfrac{1}{3}(\sqrt{t})^3 - 2t + 15$$

thousand insects t days after spraying. The lowest point on the graph of this function is the point where the tangent is horizontal. When is the insect population least?

55. **Worker Productivity** Before fatigue sets in, the rate at which postal workers sort packages increases gradually. After fatigue sets in, however, their sorting rate begins to decline. Suppose after working x hours the average postal worker sorts

$$r(x) = 64x - \tfrac{16}{27}x^3 + 5$$

packages per hour, where x is in the interval $[0, 8]$. The highest point on the graph of this function is the point where the tangent is horizontal. When is the average postal worker most productive? At what rate does the average worker sort packages at that time?

56. **Consumer Debt** Presently, the consumer debt in a South American country is $19 billion. Economists believe this debt will decline for a while and then begin to increase. Suppose x months from now the debt is

$$d(x) = \tfrac{27}{4900}x^3 - \tfrac{81}{100}x + 19$$

billion dollars. The lowest point on the graph of this function is the point where the tangent is horizontal. When will the consumer debt be lowest? What will be the lowest debt?

57. In general, it is not true that the derivative of the product of two functions is the product of their derivatives. Use the function $f(x) = 7x$ to verify this fact. Note that this function can be regarded as the product of the functions $g(x) = 7$ and $h(x) = x$.

58. In general, it is not true that the derivative of the quotient of two functions is the quotient of their derivatives. Use the function $f(x) = 4/x^2$ to verify this fact. Note that this function can be regarded as the quotient of the functions $g(x) = 4$ and $h(x) = x^2$.

59. Suppose $u(x)$ and $v(x)$ are differentiable functions. Use the definition of the derivative to prove that the derivative of

$$f(x) = u(x) + v(x)$$

is

$$f'(x) = u'(x) + v'(x)$$

60. Suppose $u(x)$ and $v(x)$ are differentiable functions. Use the definition of the derivative to prove that the derivative of

$$f(x) = u(x) - v(x)$$

is

$$f'(x) = u'(x) - v'(x)$$

Graphing Calculator Problems

61. Set the viewing screen on your graphing calculator to $[0, 12] \times [-7, 16]$ with a scale of 5 on both axes. Then draw the graph of

$$f(x) = 0.5x^2 - 6x + 14$$

Also, draw the graph of $f'(x)$ so that it appears simultaneously with the graph of $f(x)$.

a. You will see that there is an interval such that $f'(x)$ is negative for each value of x in the interval. How does $f(x)$ behave as x increases on this interval?

b. You will also see that there is an interval such that $f'(x)$ is positive for each value of x in the interval. How does $f(x)$ behave as x increases on this interval?

c. Make a conjecture based on the results in parts (a) and (b).

62. Do Problem 61 again. This time let

$$f(x) = 1.5x - 0.125x^3$$

and set the viewing screen on your graphing calculator to $[0, 3] \times [-2, 3]$ with a scale of 1 on both axes.

63. Let

$$f(x) = 14\sqrt{x}$$

for $0 \leq x \leq 12$. Set the viewing screen on your graphing calculator to $[0, 12] \times [0, 8]$ with a scale of 1 on both axes and draw the graph of $f'(x)$. Also, draw the graph of

$$g(x) = \frac{f(x + 0.4) - f(x)}{0.4}$$

so that it appears simultaneously with the graph of $f'(x)$. Then explain why the graph of $g(x)$ is almost identical to the graph of $f'(x)$.

64. Do Problem 63 again. This time let

$$f(x) = 0.08x^3 - 1.44x^2 + 10.64x$$

2.4 Approximating Change

Suppose $f(x)$ is a function that is differentiable at $x = c$. As we have seen, $f'(c)$ can be viewed as the slope of the tangent to the graph of $f(x)$ at the point where $x = c$. In this section, we show that $f'(c)$ can also be viewed as an approximation for the amount $f(x)$ changes if x increases from c to $c + 1$. We begin with a problem that involves the training of workers.

Suppose after receiving x hours of training (up to $x = 30$), each worker on an assembly line is able to install

$$P(x) = -0.01x^3 + 0.5x^2$$

meters of piping per day. Presently, each worker receives 10 hours of training. Therefore, each worker is able to install

$$P(10) = -0.01 \cdot 10^3 + 0.5 \cdot 10^2$$
$$= 40$$

meters of piping per day. However, the management is considering a 1-hour increase in training time. Consequently, it wants to know what effect this increase would have on productivity.

If the worker training time is increased to 11 hours, each worker would then be able to install

$$P(11) = -0.01 \cdot 11^3 + 0.5 \cdot 11^2$$
$$= 47.19$$

meters of piping per day. So, if the training time is increased from 10 hours to 11 hours, the number of meters of piping each worker is able to install per day would increase by

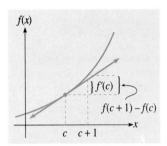

Figure 2.22 The graph of $f(x)$ and its tangent at $(c, f(c))$.

$$P(11) - P(10) = 47.19 - 40 = 7.19$$

In other words, $P(x)$ would increase by 7.19 if x is increased from 10 to 11.

Now, if we compute the derivative of $P(x) = -0.01x^3 + 0.5x^2$, we obtain

$$P'(x) = -0.03x^2 + x$$

Thus,
$$P'(10) = -0.03 \cdot 10^2 + 10 = 7$$

Since $P(11) - P(10) = 7.19$, we see that $P'(10)$ is a fairly close *approximation* for the amount $P(x)$ would increase if x is increased from 10 to 11. We have just illustrated the following fact:

> Suppose $f(x)$ is a function that is differentiable at $x = c$. In many cases, $f'(c)$ is a close approximation for the amount $f(x)$ changes if x increases from c to $c + 1$.

The actual amount $f(x)$ changes if x increases from c to $c + 1$ is

$$f(c + 1) - f(c)$$

Figure 2.22 depicts $f(c + 1) - f(c)$ and $f'(c)$ as lengths of line segments. The figure shows that the quality of the approximation of $f(c + 1) - f(c)$ by $f'(c)$ depends on how sharply the graph of $f(x)$ bends near the point $(c, f(c))$. If the graph of $f(x)$ does not bend too sharply near that point, $f'(c)$ is a good approximation for $f(c + 1) - f(c)$.

To realize why $f'(c)$ can be depicted as in Figure 2.22, let $T(x)$ be the linear function whose graph is the line in the figure. Since this line is tangent to the graph of $f(x)$ at the point where $x = c$, the slope of the line is $f'(c)$. Thus, the change in $T(x)$ is always $f'(c)$ whenever x increases one unit. In particular, the change in $T(x)$ is $f'(c)$ when x increases from c to $c + 1$.

In Example 2.9, we use a life science problem to illustrate the idea of approximating change with derivatives.

Example 2.9

A biologist is studying a rare tropical rodent. She finds that the percent of the rodents that live at least x months is

$$L(x) = -0.25x^2 + 100$$

for $0 \leq x \leq 20$. (See Figure 2.23.) Use the derivative of $L(x)$ to approximate the percent of the rodents that die as their age increases from 14 months to 15 months.

Solution

The derivative of $L(x)$ is

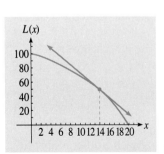

Figure 2.23 The graph of $L(x)$ and its tangent at $(14, L(14))$.

$$L'(x) = -0.5x$$

In particular, $L'(14) = -7$. Thus, $L(x)$ decreases approximately 7 units as x increases from 14 to 15. So approximately 7 percent of the rodents die as their age increases from 14 months to 15 months.

In Example 2.9, the exact percent of the rodents that die as their age increases from 14 months to 15 months is 7.25. This result is obtained by computing $L(15) - L(14)$.

Frequently, management wants to compare the change in the total production cost with the change in the revenue if the production level is increased one unit. In what follows, we discuss approximations for such changes.

Marginal Cost and Marginal Revenue

Let $C(x)$ be the total cost of producing x units of a product during a specified period of time.

> The **marginal cost** at production level x is $C'(x)$.

The marginal cost at production level x *approximates* the increase in the total production cost if the production level is increased 1 unit (from x to $x + 1$). Unless the total cost function is linear, the additional cost of producing an extra unit of the product depends on the production level x.

Now let us consider revenue. Let $R(x)$ be the revenue obtained from the sale of x units of a product during a specified period of time.

> The **marginal revenue** at selling level x is $R'(x)$.

The marginal revenue at selling level x *approximates* the change in the revenue if the selling level is increased 1 unit.

Examples 2.10 and 2.11 deal with marginal cost and marginal revenue.

Example 2.10

Suppose the total monthly cost of producing x kilograms of an experimental drug per month is

$$C(x) = \frac{1}{27}x^3 - x^2 + 11x + 5$$

thousand dollars. Find the marginal cost at production level 9. Then interpret the answer.

Solution

The formula for the marginal cost function is

$$C'(x) = \tfrac{1}{9}x^2 - 2x + 11$$

Since $C'(9) = 2$, the marginal cost is $2000 at production level 9. This result means that the total monthly production cost would increase approximately $2000 if the monthly production level were increased from 9 kilograms to 10 kilograms. In other words, it would cost approximately $2000 to produce the 10th kilogram.

Example 2.11

Each month, the experimental drug in Example 2.10 is priced so that the amount produced during the month is sold that month. Thus, the monthly revenue obtained from selling the drug is a function of its monthly production level. Suppose the monthly revenue is

$$R(x) = 7x - \tfrac{1}{3}x^2$$

thousand dollars when x kilograms is produced per month. (This formula is valid up to $x = 21$.) Use the concept of marginal revenue to determine if it would be profitable to increase the monthly production level from 9 kilograms to 10 kilograms.

Solution

The formula for the marginal revenue function is

$$R'(x) = 7 - \tfrac{2}{3}x$$

Since $R'(9) = 1$, the marginal revenue is $1000 at production level 9. This result tells us that the monthly revenue would increase approximately $1000 if the monthly production level were increased from 9 kilograms to 10 kilograms. In Example 2.10 we found that it would cost approximately $2000 to produce the 10th kilogram. Therefore, it would *not* be profitable to increase the monthly production level from 9 kilograms to 10 kilograms.

The production level of a product depends on both the size of the work force and the amount of capital used in the production operation (e.g., equipment). If one of these two variables is held fixed, the production level is viewed as a function of the other variable.

Often, management wants to compare the change in the production level if the *labor force* is increased 1 unit with the change in the production level if the *capital* is increased 1 unit. In what follows, we discuss how such changes can be approximated.

Marginal Productivity of Labor and Marginal Productivity of Capital

Let $Q(w)$ be the quantity of a product that can be produced during a specified period of time when w units of labor is used.

The **marginal productivity of labor** is $Q'(w)$ when w units of labor is used.

This marginal productivity *approximates* the increase in the production level if the labor force is increased 1 unit (from w to $w + 1$).

Now we will view production level as a function of capital. Let $Q(k)$ be the quantity of a product that can be produced during a specified period of time when k units of capital is used.

The **marginal productivity of capital** is $Q'(k)$ when k units of capital is used.

This marginal productivity *approximates* the increase in the production level if the capital outlay is increased 1 unit. We use the idea of marginal productivity in Example 2.12.

Example 2.12

Rite Paint Company produces Rust-X (a rust inhibitor) with 6400 hours of labor per week and a capital outlay of $27 million. Let $Q(w)$ be the number of barrels of the inhibitor the company can produce per week with w hundred hours of labor and a capital outlay of $27 million. Also, let $Q(k)$ be the number of barrels the company can produce per week with 6400 hours of labor and a capital outlay of k million. The management finds that

$$Q(w) = 2700(\sqrt[3]{w})^2$$

and

$$Q(k) = 14{,}400 \sqrt[3]{k}$$

Use the idea of marginal productivity to determine which has the greater impact on the weekly production level: an increase in the weekly labor force of 100 hours (from $w = 64$ to $w = 65$) or an increase in the capital outlay of $1 million (from $k = 27$ to $k = 28$)?

Solution

Since $2700(\sqrt[3]{w})^2 = 2700w^{2/3}$, we can write

$$Q(w) = 2700w^{2/3}$$

So, using the power rule, the formula for the marginal productivity of labor function is

$$Q'(w) = 1800w^{-1/3}$$
$$= \frac{1800}{\sqrt[3]{w}}$$

Hence, the marginal productivity of labor when 6400 (64 hundred) hours of labor is used is

$$Q'(64) = \frac{1800}{\sqrt[3]{64}}$$
$$= 450$$

This result tells us the weekly production level increases approximately 450 barrels if the weekly labor force is increased 100 (1 hundred) hours. Since $14{,}400\sqrt[3]{k} = 14{,}400k^{1/3}$, we can write

$$Q(k) = 14{,}400k^{1/3}$$

So, according to the power rule, the formula for the marginal productivity of capital function is

$$Q'(k) = 4800k^{-2/3}$$
$$= \frac{4800}{(\sqrt[3]{k})^2}$$

Thus, the marginal productivity of capital when $27 million of capital is used is

$$Q'(27) = \frac{4800}{(\sqrt[3]{27})^2}$$
$$= 533\tfrac{1}{3}$$

This result tells us the weekly production level increases approximately 533.3 barrels if the capital outlay is increased $1 million. Therefore, an increase in the capital outlay of $1 million has a greater impact on the weekly production level than an increase in the weekly labor force of 100 hours.

Approximating Derivatives with a Graphing Calculator

Suppose $f(x)$ is a function whose domain is an interval. Also, suppose a formula for $f(x)$ is unknown to us. If we know the values of

$$f(x) \quad \text{and} \quad f(x+1)$$

for *several* values of x, we sometimes can derive a function $g(x)$ that approximates the derivative of $f(x)$. The derivation of such a function is based on the assumption that at each value of x in the domain of $f(x)$,

$$f(x+1) - f(x)$$

is a close approximation for $f'(x)$. We use this idea in Example 2.13.

Example 2.13

Let $C(x)$ represent the total monthly cost of producing x units of a product per month. Table 2.1 shows the values of $C(x)$ for several values of x.

Table 2.1

x	2	3	5	6	9	10	12	13	14	15
$C(x)$	41.2	42.9	47.5	50.3	64.9	70	83.2	90.7	99.3	107.5

Find a function that approximates the corresponding marginal cost function.

Solution

We can use the information in Table 2.1 to compute the values of

$$C(x+1) - C(x)$$

for $x = 2, 5, 9, 12, 13$, and 14. Table 2.2 shows these values of $C(x+1) - C(x)$.

Table 2.2

x	2	5	9	12	13	14
$C(x+1) - C(x)$	1.7	2.8	5.1	7.5	8.6	8.2

Now, thinking of $C(x+1) - C(x)$ as a function of x, we plot the data points that correspond to the information in Table 2.2. We plot these points using a graphing calculator whose viewing screen is set to $[0, 18] \times [0, 12]$ with a scale of 1 on both axes. Figure 2.24 suggests that we should approximate the function $C(x+1) - C(x)$ with the least-squares linear function for the data points. Such a least-squares function can be expressed in the form

$$g(x) = mx + b$$

(Least-squares linear functions are discussed in Section 1.3.)

By pressing the appropriate keys on the graphing calculator and truncating at one decimal place, we get $m = 0.6$ and $b = 0.1$. Therefore, we can use the linear function

$$g(x) = 0.6x + 0.1$$

to approximate the function $C(x+1) - C(x)$.

Since $C(x+1) - C(x)$ approximates $C'(x)$ and $C'(x)$ is the marginal cost function, we can use the function

$$g(x) = 0.6x + 0.1$$

to also approximate the marginal cost function.

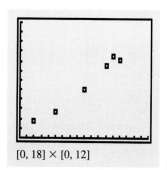

$[0, 18] \times [0, 12]$

Figure 2.24 The information in Table 2.2 depicted as data points.

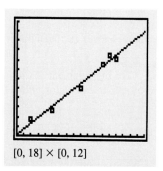

$[0, 18] \times [0, 12]$

Figure 2.25 The regression line for the data points in Figure 2.24.

The graph of the linear function $g(x)$ in Example 2.13 is the regression line for the data points in Figure 2.24. Figure 2.25 shows this regression line and the data points.

2.4 Problems

In Problems 1 to 10, use the derivative to approximate the amount $f(x)$ changes if x increases 1 unit as indicated.

1. $f(x) = \frac{1}{6}x^2 + x + \frac{3}{2}$ x increases from 3 to 4
2. $f(x) = \frac{1}{12}x^2 - x + 3$ x increases from 5 to 6
3. $f(x) = 80 - \frac{1}{100}x^3$ x increases from 10 to 11
4. $f(x) = \frac{1}{60}x^3$ x increases from 6 to 7
5. $f(x) = 2\sqrt{x}$ x increases from 16 to 17
6. $f(x) = 4(\sqrt{x})^3$ x increases from 25 to 26
7. $f(x) = \frac{1}{2x^2}$ x increases from 4 to 5
8. $f(x) = \frac{7}{x}$ x increases from 8 to 9
9. $f(x) = \frac{8}{\sqrt{x}}$ x increases from 9 to 10
10. $f(x) = \frac{2}{3(\sqrt{x})^3}$ x increases from 4 to 5

11. **Mortality** Health officials have been studying the mortality patterns of the population of an island. As a result of their findings, they believe that the probability an x-year-old islander will die of ischemic heart disease is

$$m(x) = 0.00005x^2 - 0.0045x + 0.10325$$

for x in the interval [45, 90]. Use the derivative of $m(x)$ to approximate the amount this probability increases as an islander's age increases from 70 to 71.

12. **Production Cost** The management of The Cheese House finds that the average cost of producing each pound of an expensive Romano cheese is

$$A(x) = 2 + \frac{75}{x}$$

dollars if x hundred pounds of the cheese is produced daily. Use the derivative of $A(x)$ to approximate the amount the average cost changes if the daily production level is increased from 1000 (10 hundred) pounds to 1100 (11 hundred) pounds.

13. **Demand** When the price of Yummy Bar, a low-calorie candy bar, is x cents per bar,

$$D(x) = 10 + \frac{14}{\sqrt{x}}$$

million bars are sold per month, where x is in the interval [10, 80]. Use the derivative of $D(x)$ to approximate the decrease in the monthly demand for Yummy Bar if the price is increased from 25 cents to 26 cents per bar.

14. **Unemployment Benefits** The government of a country assumes that when the unemployment rate is x percent,

$$B(x) = 8\sqrt{x}$$

billion dollars is spent per year on unemployment benefits. Use the derivative of $B(x)$ to approximate the increase in this yearly expenditure if the unemployment rate increases from 9 percent to 10 percent.

15. **Marginal Cost** Let $C(x)$ represent the total weekly cost (in hundreds of dollars) of producing x liters of a cologne per week. Suppose $C(x)$ is a function of x according to

$$C(x) = \frac{1}{24}x^3 - x^2 + 10x + 3$$

Compute the marginal cost at production level 12. Then interpret the result.

16. **Marginal Revenue** Each week, D-Bug (an insecticide) is priced so that the amount produced during the week is sold that week. Hence, the weekly revenue obtained from selling the insecticide can be viewed as a function of the weekly production level. Suppose the weekly revenue is

$$R(x) = 20x - 0.5x^2$$

hundred dollars when x tons is produced per week. (This formula is valid up to $x = 40$.) Compute the marginal revenue at production level 18. Then interpret your answer.

17. **Marginal Productivity** The management of Calcutex, a manufacturer of graphing calculators, believes that

$$Q(w) = 864 - \frac{288}{w}$$

graphing calculators can be produced weekly with w units of labor, provided $w \geq 1$. Find the marginal productivity of labor when 12 units of labor is used. Then interpret the answer.

18. **Marginal Productivity** Let $Q(k)$ represent the number of kilograms of a commodity that can be produced weekly with a capital outlay of $\$k$ million. Suppose

$$Q(k) = 2k + 42\sqrt{k}$$

Find the marginal productivity of capital when $49 million of capital is used. Then interpret your result.

19. **Least Marginal Cost** Suppose the total monthly cost of producing x units of a fabric per month is given by

$$C(x) = \tfrac{1}{60}x^3 - x^2 + 22x + 15$$

Let $MC(x)$ represent the marginal cost at production level x. As is often the case in production situations, at low production levels, $MC(x)$ decreases if x is increased. At high production levels, however, $MC(x)$ increases if x is increased. The lowest point on the graph of the marginal cost function is the point where the tangent is horizontal. At what monthly production level is the marginal cost least?

20. **Minimum Average Cost** Suppose the total weekly cost of producing x units of a product per week is given by

$$C(x) = 0.5x^2 + 8$$

Then the average cost of producing each unit is given by

$$A(x) = \frac{C(x)}{x} = \frac{0.5x^2 + 8}{x} = 0.5x + \frac{8}{x}$$

when x units are produced per week.

a. Find the weekly production level at which the average cost of producing each unit is equal to the marginal cost.

b. In this production situation, at low production levels, $A(x)$ decreases if x is increased. At high production levels, however, $A(x)$ increases if x is increased. The lowest point on the graph of the average cost function is the point where the tangent is horizontal. At what weekly production level is the average cost of producing each unit minimum? How does this production level compare with the production level found in part (a)?

21. **Optimal Production Level** Each day, a commodity is priced so that the amount produced during the day is sold that day. Thus, the daily revenue and the daily profit obtained from selling the commodity may be viewed as functions of the daily production level. If x units per day are produced (up to $x = 1596$), the daily revenue is

$$R(x) = -\tfrac{1}{6}x^2 + 266x$$

thousand dollars and the total daily production cost is

$$C(x) = \tfrac{1}{15}x^3 - 5x^2 + 126x + 20$$

thousand dollars. The production level that maximizes the profit is the production level at which the marginal revenue equals the marginal cost. Find the production level that maximizes the profit.

22. **Propensity to Consume** The derivative of a function that expresses how expenditure depends on income is referred to as a **marginal propensity to consume** function. Suppose a community spends

$$E(x) = 0.75x + 0.25\sqrt{x}$$

billion dollars per week when its weekly income is x billion dollars, provided $x \geq 1$. Although the expenditure increases if the income increases, the marginal propensity to consume decreases. The government has decided to gradually raise the income of the community through various tax strategies. However, it plans to stop when the marginal propensity to consume is reduced to 0.775 of a billion dollars. Up to what level, then, will the government raise the income of the community?

23. Suppose $f(x)$ is a function that satisfies the following two conditions: (a) $f(x)$ increases as x increases; (b) $f'(x)$ decreases as x increases. Then for any number c in the domain, $f'(c)$ is *greater* than the actual amount $f(x)$ increases as x increases from c to $c + 1$. Explain why.

   Hint: Sketch a curve that could be the graph of $f(x)$.

24. Suppose $f(x)$ is a linear function. Then for any number c in the domain, $f'(c)$ is the actual change $f(x)$ undergoes as x increases from c to $c + 1$. Explain why.

Graphing Calculator Problems

25. **Average and Marginal Costs** The average cost function that corresponds to the total cost function in Problem 19 is

$$A(x) = \frac{C(x)}{x}$$
$$= \frac{1}{60}x^2 - x + 22 + \frac{15}{x}$$

Set the viewing screen on your graphing calculator to [0, 48] × [0, 32] with a scale of 4 on both axes. Then draw the graphs of the average cost function and the marginal cost function so that the graphs appear simultaneously. Use the trace function to approximate the production level at which the average cost of producing each unit is equal to the marginal cost. How does this production level compare with the production level at which the average cost of producing each unit is least?

26. **Average and Marginal Costs** Do Problem 25 again. This time, however, use the total cost function in Example 2.10.

27. Let

$$f(x) = 0.025x^3 - 0.225x^2 + 1.5x$$

and $\quad d(x) = f(x + 1) - f(x)$

Set the viewing screen on your graphing calculator to [0, 9] × [0, 6] with a scale of 1 on both axes. Then draw the graphs of $f'(x)$ and $d(x)$ so that the graphs appear simultaneously. Explain why the two graphs are almost identical.

28. The table below shows the values of a function $f(x)$ for several values of x. Use an approach like in Example 2.13 to find a function that approximates the derivative of $f(x)$.

x	2	3	4	6	7	10	11	14	15	17	18
$f(x)$	49	58.3	66.7	82.9	90.2	110	116	130.8	134.7	142.2	145.4

2.5 More Rules for Computing Derivatives

In this section, we present rules for computing derivatives of the following kinds of functions:

1. Functions that are constant multiples of *powers* of other functions
2. Functions that are *products* of other functions
3. Functions that are *quotients* of other functions

We begin with the function

$$f(x) = 5(x^2 + 4)^3$$

Using the formula

$$(a + b)^3 = a^3 + 3a^2b + 3ab^2 + b^3$$

we can expand $5(x^2 + 4)^3$, and thus express the function $f(x)$ as

$$f(x) = 5x^6 + 60x^4 + 240x^2 + 320$$

Then we can compute its derivative using the rules presented in Section 2.3 and get

$$f'(x) = 30x^5 + 240x^3 + 480x$$

We could use the same approach to compute the derivative of

$$f(x) = 3(5x^2 - 8)^9$$

But expanding $3(5x^2 - 8)^9$ requires a great deal of work.

The Derivative of a Function of the Form $c[u(x)]^n$

The functions $f(x) = 5(x^2 + 4)^3$ and $f(x) = 3(5x^2 - 8)^9$ are constant multiples of power functions.

Definition 2.2

A **power function** is a function that can be expressed in the form

$$f(x) = [u(x)]^n$$

where $u(x)$ is a function and n is a rational number.

The function

$$f(x) = (x^3 + 4x + 1)^5$$

is an example of a power function. Here $u(x) = x^3 + 4x + 1$ and $n = 5$. Another example is

$$f(x) = x^4$$

In this case, $u(x) = x$ and $n = 4$. Since

$$\frac{1}{\sqrt{x+7}} = (x+7)^{-1/2}$$

the function

$$f(x) = \frac{1}{\sqrt{x+7}}$$

is still another example of a power function. Here $u(x) = x + 7$ and $n = -\frac{1}{2}$.

As we illustrated with the function $f(x) = 5(x^2 + 4)^3$, we can compute the derivative of a constant multiple of a power function using the rules presented in Section 2.3, provided we can expand the formula for the function. However, if n (the power) is large, this approach is not practical. Indeed, if n is not a positive integer, this approach is impossible. Evidently, we need a technique for computing the derivatives of these functions that does not involve expanding.

Let us take a closer look at the derivative of

$$f(x) = 5(x^2 + 4)^3$$

which, as we saw, is

$$f'(x) = 30x^5 + 240x^3 + 480x$$

Factoring,

$$\begin{aligned} 30x^5 + 240x^3 + 480x &= 30x(x^4 + 8x^2 + 16) \\ &= 30x[(x^2)^2 + 8x^2 + 16] \\ &= 30x(x^2 + 4)^2 \\ &= 3 \cdot 5(x^2 + 4)^2 2x \end{aligned}$$

So we can alternatively express the derivative of $f(x)$ as

$$f'(x) = 3 \cdot 5(x^2 + 4)^2 2x$$

This formula is related to $5(x^2 + 4)^3$ in an interesting way. In particular, the factor $2x$ is the derivative of $x^2 + 4$. This result suggests the rule given in Theorem 2.7.

Theorem 2.7 (Generalized Power Rule)

Suppose $u(x)$ is a differentiable function, c is a real number, and n is a rational number. Then the derivative of

$$f(x) = c[u(x)]^n$$

is

$$f'(x) = nc[u(x)]^{n-1} u'(x)$$

Theorem 2.4 in Section 2.3 is the special case of the above rule where $u(x) = x$. We apply the rule in Examples 2.14 and 2.15.

Example 2.14

Find the derivative of

$$f(x) = 3(5x^2 - 8)^9$$

Solution

Using the generalized power rule,

$$\begin{aligned} f'(x) &= 27(5x^2 - 8)^8 10x \\ &= 270x(5x^2 - 8)^8 \end{aligned}$$

The next example involves a function of the form $c[u(x)]^n$ with $c = 1$ and $n = -1$.

Example 2.15

Recently, some of the inhabitants of an island in the Indian Ocean were exposed to a virus that causes abdominal cramps lasting about a day. Health officials predict that x weeks from now

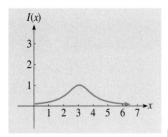

Figure 2.26
$I(x) = 1/(x^2 - 6x + 10)$ for $x \geq 0$.

$$I(x) = \frac{1}{x^2 - 6x + 10}$$

thousand islanders will be experiencing the cramps. Figure 2.26 shows that the highest point on the graph of $I(x)$ is the point where the tangent is horizontal. When will the number of islanders experiencing the cramps be greatest?

Solution

Since

$$\frac{1}{x^2 - 6x + 10} = (x^2 - 6x + 10)^{-1}$$

we can express the given function as

$$I(x) = (x^2 - 6x + 10)^{-1}$$

Then, applying the generalized power rule, the derivative of $I(x)$ is

$$I'(x) = (-1)(x^2 - 6x + 10)^{-2}(2x - 6)$$
$$= \frac{6 - 2x}{(x^2 - 6x + 10)^2}$$

Now we set the derivative of $I(x)$ equal to zero and solve for x:

$$\frac{6 - 2x}{(x^2 - 6x + 10)^2} = 0$$

$6 - 2x = 0$ multiplying by $(x^2 - 6x + 10)^2$
$6 = 2x$
$3 = x$

Thus, $I'(x) = 0$ at $x = 3$. This result tells us the graph of the function $I(x)$ has a horizontal tangent at the point whose x coordinate is 3. Since this point is the highest point on the graph, the number of islanders experiencing the cramps will be greatest 3 weeks from now.

Next, we consider functions formed by multiplying other functions.

The Derivative of a Product

In Section 2.3, we saw that the derivative of the sum of two or more functions is the sum of their derivatives. We also saw that the derivative of the difference of two functions is the difference of their derivatives. However, the derivative of the product of two or more functions is *not* the product of their derivatives. We can illustrate this fact with the function

$$h(x) = 3x^2(5x + 1)$$

If we multiply the factors $3x^2$ and $5x + 1$, we can express this function as

$$h(x) = 15x^3 + 3x^2$$

Therefore, its derivative is

$$h'(x) = 45x^2 + 6x$$

If we compute the derivatives of the factors $3x^2$ and $5x + 1$, we get $6x$ and 5, respectively. The product of these two derivatives is $30x$, which is *not* equal to $45x^2 + 6x$. So the derivative of $h(x) = 3x^2(5x + 1)$ is not the product of the derivatives of $3x^2$ and $5x + 1$.

The rule stated in Theorem 2.8 enables us to compute the derivative of a product without first having to multiply the factors as we did in the above example.

Theorem 2.8 (Product Rule)

If $u(x)$ and $v(x)$ are differentiable functions, the derivative of

$$f(x) = u(x)v(x)$$

is

$$f'(x) = u(x)v'(x) + v(x)u'(x)$$

In other words, the derivative of the product of two functions is the first function times the derivative of the second function, plus the second function times the derivative of the first function.

Using the product rule, the derivative of

$$h(x) = 3x^2(5x + 1)$$

is

$$h'(x) = (3x^2)5 + (5x + 1)6x$$

This result agrees with our earlier result since

$$(3x^2)5 + (5x + 1)6x = 45x^2 + 6x$$

In Example 2.16, we use both the product rule and the generalized power rule.

Example 2.16

Find the derivative of

$$f(x) = x^2(3x + 1)^5$$

Solution

According to the product rule,

$$f'(x) = (x^2)D_x(3x + 1)^5 + (3x + 1)^5 D_x(x^2)$$

Using the generalized power rule,

$$D_x(3x + 1)^5 = 5(3x + 1)^4 \, 3$$
$$= 15(3x + 1)^4$$

and
$$D_x(x^2) = 2x$$

Thus,
$$\begin{aligned} f'(x) &= (x^2)15(3x+1)^4 + (3x+1)^5 2x \\ &= (3x+1)^4[15x^2 + (3x+1)2x] \quad \text{factoring} \\ &= (3x+1)^4(21x^2 + 2x) \\ &= x(3x+1)^4(21x+2) \quad \text{factoring} \end{aligned}$$

Example 2.16 demonstrates that when we use the generalized power rule, we usually can factor the derivative without too much effort.

Now suppose
$$f(x) = c \cdot u(x)$$
where c is a real number and $u(x)$ is a differentiable function. According to the product rule,
$$f'(x) = c \cdot u'(x) + u(x) \cdot D_x(c)$$
But the constant rule tells us
$$D_x(c) = 0$$
Therefore
$$f'(x) = c \cdot u'(x)$$

We have just derived the rule given in Theorem 2.9.

Theorem 2.9 (Constant Multiple Rule)

If c is a real number and $u(x)$ is a differentiable function, the derivative of
$$f(x) = c \cdot u(x)$$
is
$$f'(x) = c \cdot u'(x)$$

In other words, the derivative of a constant times a function is the constant times the derivative of the function. This rule is a special case of the product rule. We will use it in Chapter 4.

We now turn to functions formed by dividing two other functions.

The Derivative of a Quotient

As was the case with the derivative of the product of two or more functions, the derivative of the quotient of two functions is *not* the quotient of their derivatives. We can illustrate this fact with the function
$$g(x) = \frac{6x^2 + 1}{2x}$$

Note that

$$\frac{6x^2 + 1}{2x} = \frac{6x^2}{2x} + \frac{1}{2x}$$
$$= 3x + 0.5x^{-1}$$

So the above function can alternatively be expressed as

$$g(x) = 3x + 0.5x^{-1}$$

Thus, using rules from Section 2.3, its derivative is

$$g'(x) = 3 - 0.5x^{-2}$$
$$= 3 - \frac{0.5}{x^2}$$

Now, note that the derivative of $6x^2 + 1$ is $12x$ and the derivative of $2x$ is 2. The quotient of these two derivatives is

$$\frac{12x}{2} = 6x$$

which is certainly *not* equal to $3 - 0.5/x^2$. So the derivative of

$$g(x) = \frac{6x^2 + 1}{2x}$$

is not the quotient of the derivatives of $6x^2 + 1$ and $2x$.

The rule given in Theorem 2.10 allows us to compute the derivative of a quotient without first having to find an alternative formula as we did in the above example.

Theorem 2.10 (Quotient Rule)

If $u(x)$ and $v(x)$ are differentiable functions, the derivative of

$$f(x) = \frac{u(x)}{v(x)}$$

is

$$f'(x) = \frac{v(x)u'(x) - u(x)v'(x)}{[v(x)]^2}$$

In other words, the derivative of the quotient of two functions is the bottom function times the derivative of the top function, minus the top function times the derivative of the bottom function, all over the square of the bottom function.

Using the quotient rule, the derivative of

$$g(x) = \frac{6x^2 + 1}{2x}$$

is

$$g'(x) = \frac{(2x)D_x(6x^2 + 1) - (6x^2 + 1)D_x(2x)}{(2x)^2}$$

$$= \frac{(2x)12x - (6x^2 + 1)2}{4x^2}$$

$$= \frac{12x^2 - 2}{4x^2}$$

$$= \frac{12x^2}{4x^2} - \frac{2}{4x^2}$$

$$= 3 - \frac{0.5}{x^2}$$

which is what we obtained earlier.

In Example 2.17, we use both the quotient rule and the generalized power rule.

Example 2.17

Let

$$f(x) = \left(\frac{5x - 2}{2x + 4}\right)^6$$

Use the derivative of $f(x)$ to approximate the amount $f(x)$ changes if x increases from 2 to 3.

Solution

The given function has the form

$$f(x) = c[u(x)]^n$$

where $c = 1$, $n = 6$, and $u(x) = (5x - 2)/(2x + 4)$. So we begin by applying the generalized power rule (Theorem 2.7):

$$f'(x) = 6\left(\frac{5x - 2}{2x + 4}\right)^5 D_x\left(\frac{5x - 2}{2x + 4}\right) \quad \text{generalized power rule}$$

$$= 6\left(\frac{5x - 2}{2x + 4}\right)^5 \frac{(2x + 4)5 - (5x - 2)2}{(2x + 4)^2} \quad \text{quotient rule}$$

$$= 6\left(\frac{5x - 2}{2x + 4}\right)^5 \frac{24}{(2x + 4)^2}$$

$$= \frac{144}{(2x + 4)^2}\left(\frac{5x - 2}{2x + 4}\right)^5$$

In particular,

$$f'(2) = \frac{144}{8^2}\left(\frac{8}{8}\right)^5$$

$$= 2.25$$

So $f(x)$ increases approximately 2.25 units if x increases from 2 to 3.

We could use the quotient rule (along with the generalized power rule) to find the derivative of

$$f(x) = \frac{5}{(6x+1)^3}$$

It is easier, however, to first rewrite this function as

$$f(x) = 5(6x+1)^{-3}$$

and then use the generalized power rule to obtain

$$f'(x) = -15(6x+1)^{-4} 6$$
$$= \frac{-90}{(6x+1)^4}$$

Similarly, although we could use the quotient rule to find the derivative of

$$y = \frac{8}{x^3}$$

it is better to first rewrite this function as

$$y = 8x^{-3}$$

and then use the power rule to obtain

$$\frac{dy}{dx} = -24x^{-4}$$
$$= \frac{-24}{x^4}$$

Theorem 2.6 in Section 2.3 and the quotient rule imply Theorem 2.11.

Theorem 2.11

Rational functions are differentiable at every point where the denominator is not zero.

The Chain Rule

The generalized power rule is a *special case* of the following rule:

Chain Rule

If $g(x)$ and $u(x)$ are differentiable functions, the derivative of the composite function

$$f(x) = g[u(x)]$$

is

$$f'(x) = g'[u(x)]u'(x)$$

In the above rule, suppose

$$g(x) = cx^n$$

where c is a real number and n is a rational number. Then

$$f(x) = c[u(x)]^n$$

and, since $g'(x) = ncx^{n-1}$,

$$g'[u(x)] = nc[u(x)]^{n-1}$$

Therefore, $$f'(x) = nc[u(x)]^{n-1}u'(x)$$

Thus, the generalized power rule is the special case of the chain rule where $g(x) = cx^n$.

We will use the chain rule in future chapters.

2.5 Problems

In Problems 1 to 44, find the derivative of the given function.

1. $y = (4x^5 + 9)^7$
2. $y = (8x - 6)^5$
3. $y = (x^4 - 5x)^3$
4. $y = (3x^2 + x)^2$
5. $y = 5(2x - 7)^6$
6. $y = 0.4(4x^3 + 7)^5$
7. $y = (x^3 - 4x^2 + 5x)^2 - x^2$
8. $y = 0.25(7x^5 + 3x - 8)^4 + 7x^2$
9. $y = \dfrac{6}{x^3 + 7}$
10. $y = \dfrac{-1}{5x^2 - x}$
11. $y = \dfrac{-3}{(2x^4 - 9x)^5}$
12. $y = \dfrac{2}{(x^3 - 2x + 9)^3}$
13. $y = \sqrt[3]{2x^4 + 5x - 2}$
14. $y = \sqrt{6x - 9}$
15. $y = \dfrac{1}{\sqrt{9x + 8}}$
16. $y = \dfrac{-7}{\sqrt[4]{x^3 + x^2}}$
17. $y = (x^4 + 3x - 9)(5x^2 + 7)$
18. $y = (3x - 1)(2x^3 - x + 8)$
19. $y = 4x(5x^2 + x)^7$
20. $y = (x^3 - 5x)^4(7x + 1)$
21. $y = x^2(6x + 3)^5 + \dfrac{7}{x}$
22. $y = 4x^3(9x - 2)^6 - \dfrac{1}{x^2}$
23. $y = (x^2 + 9x)^5(5x^2 - 3x)$
24. $y = (8x + 3)(3x^4 + 8)^3$
25. $y = x\sqrt{x^4 - 7}$
26. $y = 8x\sqrt{5x^2 + 9x}$
27. $y = (5x - 2)^3(x^2 + 1)^3$
28. $y = (3x^4 + 5)^2(x - 4)^6$
29. $y = (x^3 + 2x + 4)^5(7x + 3)^2$
30. $y = (x^2 + x)^7\sqrt{3x^2 - 4x}$
31. $y = \dfrac{x^5}{3x + 9}$
32. $y = \dfrac{7x}{x^3 - 7}$
33. $y = \dfrac{7x - 2}{5x + 1} + \sqrt{x}$
34. $y = \dfrac{x^3 + 5}{4x^2 - 1} + \dfrac{8}{\sqrt{x}}$
35. $y = \dfrac{x^2 + 2x}{x^4 + 3x - 5}$
36. $y = \dfrac{4x^3 - x + 2}{x^2 + 5x}$
37. $y = \dfrac{x^2 + 3x}{7x^2 - 2}$
38. $y = \dfrac{x^3 + x}{x^3 + x^2}$
39. $y = \dfrac{(2x + 1)^3}{x + 7}$
40. $y = \dfrac{x^2 - 5x}{(x - 3)^4}$
41. $y = \left(\dfrac{x^3 - 7}{x^3 + 5}\right)^4$
42. $y = \sqrt{\dfrac{x^2 - 4}{x^2 + 5}}$
43. $y = \dfrac{7x + 5}{\sqrt{x + 3}}$
44. $y = \dfrac{\sqrt{2x + 1}}{(x + 7)^3}$

45. The lowest point on the graph of

$$f(x) = \sqrt{x^2 - 8x + 17}$$

is the point where the tangent is horizontal. Find the value of x where $f(x)$ is minimum.

46. The highest point on the graph of

$$f(x) = \dfrac{2}{3x^2 - 9x + 8}$$

is the point where the tangent is horizontal. Find the value of x where $f(x)$ is maximum.

47. Let the domain of
$$f(x) = \frac{2x - 9}{(x + 7)^4}$$
be the set of nonnegative real numbers. Then the highest point on the graph of $f(x)$ is the point where the tangent is horizontal. Find the value of x (among the nonnegative reals) where $f(x)$ is maximum.

48. Suppose the domain of
$$f(x) = (2x - 9)(4x + 6)^3$$
is the set of nonnegative real numbers. Then the lowest point on the graph of $f(x)$ is the point where the tangent is horizontal. Find the value of x (among the nonnegative reals) where $f(x)$ is minimum.

In Problems 49 to 52, use the derivative to approximate the amount $f(x)$ changes if x increases 1 unit as indicated.

49. $f(x) = \frac{1}{6}(x^2 - 62)^3$ x increases from 8 to 9

50. $f(x) = \frac{7}{x^2 + 5}$ x increases from 2 to 3

51. $f(x) = \frac{3x}{x^2 + 1}$ x increases from 4 to 5

52. $f(x) = 3x\sqrt{4x + 1}$ x increases from 6 to 7

53. **Marginal Revenue** Each week, an alloy is priced so that the amount produced during the week is sold that week. Hence, the weekly revenue obtained from selling the alloy may be viewed as a function of the weekly production level. Suppose the weekly revenue is
$$R(x) = \frac{1}{90}x(81 - 3x)^3$$
dollars when x tons is produced per week, where x is in the interval [0, 27]. Find the marginal revenue at production level 7. Then interpret your answer.

54. **Marginal Cost** The management of a bakery assumes the total weekly cost of producing x tons of wheat bread per week is
$$C(x) = \sqrt{7x + 39}$$
thousand dollars. Find the marginal cost at production level 15. Then interpret the result.

55. **Related Production Levels** A company produces two kinds of house siding: aluminum and vinyl. Let $f(x)$ represent the number of tons of aluminum siding it can produce daily when it produces x tons of vinyl siding daily. Suppose
$$f(x) = \sqrt{9 - 0.25x}$$
for $0 \le x \le 36$. Use the derivative of $f(x)$ to approximate the amount the daily production level of aluminum siding would have to be decreased if the daily production level of vinyl siding is increased from 32 tons to 33 tons.

56. **Average Production Cost** Let $A(x)$ represent the average cost (in hundreds of dollars) of producing each barrel of apple juice when x barrels is produced per day. Suppose
$$A(x) = \frac{\sqrt{2x + 7}}{x}$$
for $x \ge 1$. Use the derivative of $A(x)$ to approximate the decrease in the average cost of producing each barrel if the daily production level is increased from 9 barrels to 10 barrels.

57. **Daily Reading Volume** In a southern European country, the average number of words each adult reads daily was 12,000 at the beginning of 1993. Suppose that x years later this average is
$$R(x) = \frac{8x + 12}{x + 1}$$
thousand. Use the derivative of $R(x)$ to approximate the amount the daily average will change during the 1-year period from the beginning of 1997 to the beginning of 1998.

58. **Illiteracy** About 27 million Americans were functionally illiterate at the beginning of 1990. Suppose that x years later the number of illiterate Americans is
$$I(x) = \frac{16x + 54}{x + 2}$$
million. Use the derivative of $I(x)$ to approximate the amount by which the number of illiterate Americans will change during the 1-year period from the beginning of 1998 to the beginning of 1999.

59. **Pollution Removal** Let $P(x)$ represent the percentage of a pollutant that can be removed from a lake if the capital outlay is x million dollars. Suppose
$$P(x) = \frac{400x}{4x + 9}$$

a. Use the derivative of $P(x)$ to approximate the additional percentage of the pollutant that can be removed from the lake if the capital outlay is increased from 4 million dollars to 5 million dollars.

b. Although $P(x)$ increases if x is increased, $P'(x)$ decreases. The county commissioners have decided to gradually increase the capital outlay until $P'(x)$ is reduced to 1. How much, then, will they eventually invest in the project?

Comment: Since $P'(x)$ approximates the additional percentage of the pollutant that can be removed by increasing the capital investment from x million dollars to $x + 1$ million dollars, the commissioners, evidently, are willing to increase the investment as long as the returns are sufficiently large.

60. **Physical Training**   Let $W(x)$ represent the weight (in hundreds of pounds) Frank can lift if he spends x hours per day training. Suppose

$$W(x) = \frac{16x + 2}{3x + 1}$$

a. Use the derivative of $W(x)$ to approximate the additional weight Frank can lift if he increases his daily training time from 1 hour to 2 hours.

b. Although $W(x)$ increases if x is increased, $W'(x)$ decreases. Frank has decided to gradually increase his daily training time until $W'(x)$ is reduced to 0.1. How many hours per day, then, will he eventually spend training?

Comment: Since $W'(x)$ approximates the additional weight Frank can lift if he increases his daily training time from x hours to $x + 1$ hours, Frank, evidently, is willing to increase his daily training time as long as the returns are sufficiently large.

61. **Optimal Production Level**   A bakery immediately sells whatever amount of Pan de Higos (a fig cake) it produces. Thus, the monthly profit earned from selling the cake may be viewed as a function of its monthly production level. Suppose the monthly profit is

$$P(x) = \frac{800x}{x + 2} - 4x - 600$$

thousand dollars when the bakery produces x tons of the cake monthly, where $x \geq 0$. The highest point on the graph of this function is the point where the tangent is horizontal. Find the monthly production level at which the profit is maximum. What is the profit at this production level?

62. **Maximum Yield**   Let $Y(x)$ represent the number of tons of a hybrid wheat a farmer can produce if he uses x tons of a special fertilizer. The yield is low if he uses either too little or too much of the fertilizer. Suppose

$$Y(x) = \frac{35}{x^2 - 46x + 530}$$

The highest point on the graph of this function is the point where the tangent is horizontal. Find the amount of fertilizer that the farmer should use to maximize the yield. If he uses this amount of fertilizer, how many tons of the wheat will he produce?

63. **Mortgage Rates**   The interest rate for conventional mortgages will decline for a while and then begin to increase. Economists predict that x months from now the interest rate will be

$$I(x) = 14 - \frac{10}{x^2 - 10x + 30}$$

percent. The lowest point on the graph of this function is the point where the tangent is horizontal. When will the interest rate be lowest? What will the interest rate be at that time?

64. **Optimal Labor Force**   Suppose the daily production level of a product is

$$P(W) = \frac{7W^2}{W^2 + 64}$$

barrels if W units of labor is used per day. Then the average number of barrels produced by each unit of labor is

$$A(W) = \frac{P(W)}{W} = \frac{7W}{W^2 + 64}$$

a. How many units of labor should be used per day to make the average productivity of each unit of labor equal to the marginal productivity of labor?

b. The highest point on the graph of the average productivity function is the point where the tangent is horizontal. Find the number of units of labor that should be used per day to maximize the average productivity of each unit of labor. How does the size of this labor force compare with the size of the labor force you found in part (*a*)?

65. Explain how to find the derivative of the quotient of two functions without using the quotient rule or the definition of a derivative.

66. We said that the power rule (Theorem 2.4 in Section 2.3) is the special case of the generalized power rule (Theorem 2.7) where $u(x) = x$. Explain why we can make this claim.

Graphing Calculator Problems

67. Set the viewing screen on your graphing calculator to $[0, 6] \times [-0.5, 2]$ with a scale of 1 on both axes. Then draw the graph of

$$f(x) = \frac{6x}{x^2 + 3}$$

Also, draw the graph of $f'(x)$ so that it appears simultaneously with the graph of $f(x)$.

a. You will see that there is an interval such that $f'(x)$ is positive for each value of x in the interval. How does $f(x)$ behave as x increases on this interval?

b. You will see that there is an interval such that $f'(x)$ is negative for each value of x in the interval. How does $f(x)$ behave as x increases on this interval?

68. Do Problem 67 again. This time, however, let

$$f(x) = 5 - \frac{24x}{x^2 + 9}$$

and set the viewing screen on your graphing calculator to $[0, 6] \times [-3, 5]$ with a scale of 1 on both axes.

2.6 Optimization on Closed Intervals

Recall that the **maximum value** of a function is the ordinate of the *highest* point on the graph of the function. Similarly, the **minimum value** is the ordinate of the *lowest* point on the graph. In this section, we discuss how to find the maximum and minimum values of differentiable functions whose domains are closed intervals.

Let $g(x)$ represent the function whose graph is the curve in Figure 2.27. Note that the curve has a nonvertical tangent at every point. So, according to Theorem 2.2 in Section 2.2, $g(x)$ has a derivative at each x in the closed interval $[3, 11]$. In other words, $g(x)$ is differentiable everywhere on the interval $[3, 11]$.

Note also that the graph of $g(x)$ has a highest point and a lowest point. The highest point is one of the endpoints. The lowest point is a point where the tangent is horizontal. Therefore, the derivative of $g(x)$ is zero at the x coordinate of the lowest point.

The function $g(x)$ illustrates Theorem 2.12.

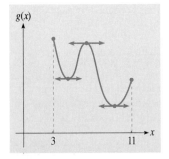

Figure 2.27 A differentiable function whose domain is a closed interval.

Theorem 2.12

Suppose $f(x)$ is a function that is differentiable everywhere on its domain. If the domain is a *closed interval*, $f(x)$ has a maximum value and a minimum value. The maximum value occurs at an endpoint or at a point (in the domain) where $f'(x) = 0$. Similarly, the minimum value occurs at an endpoint or at a point where $f'(x) = 0$.

If the domain of a function $f(x)$ is not a closed interval, then $f(x)$ may fail to have a maximum or minimum value, even if $f(x)$ is differentiable everywhere on its domain. For example, suppose

$$f(x) = 3\sqrt{x}$$

Figure 2.28 shows that if we make the domain of this function the open interval $(1, +\infty)$, then the graph has neither a highest nor a lowest point. So $f(x)$ has neither a maximum nor a minimum value. And yet $f(x)$ is differentiable everywhere on the interval $(1, +\infty)$. The graph would have a lowest point if we make the domain the half-open interval $[1, +\infty)$, rather than the open interval $(1, +\infty)$. In this case, the lowest point would be $(1, 3)$. But the graph would still not have a highest point.

To find the maximum or minimum value of a differentiable function $f(x)$ whose domain is a closed interval, we compute $f(x)$ at the endpoints and at the points in the domain where the derivative is zero. Then we compare these values of $f(x)$ to determine which is largest and which is smallest. We use this approach in Examples 2.18 and 2.19.

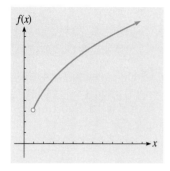

Figure 2.28 $f(x) = 3\sqrt{x}$ for x in $(1, +\infty)$.

Example 2.18

In a South American country, 1500 steel workers are presently unemployed. If x months from now the number of unemployed steel workers is

$$U(x) = \frac{30}{x^2 - 8x + 20}$$

thousand, will it ever reach 9000 during the next 10 months?

Solution

The discriminant of $x^2 - 8x + 20$ is negative, namely,

$$(-8)^2 - 4(1)(20) = -16$$

Therefore, the rational function $U(x)$ has a denominator that is never 0. So, according to Theorem 2.11 in Section 2.5, $U(x)$ is differentiable everywhere. In particular, $U(x)$ is differentiable on the closed interval $[0, 10]$. Thus, according to Theorem 2.12, $U(x)$ attains a maximum value as x varies on the closed interval $[0, 10]$. We will find this maximum value. Then we will compare it to 9000.

Since

$$\frac{30}{x^2 - 8x + 20} = 30(x^2 - 8x + 20)^{-1}$$

we can write

$$U(x) = 30(x^2 - 8x + 20)^{-1}$$

Then, using the generalized power rule, we obtain

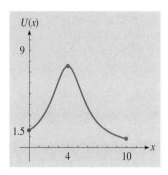

Figure 2.29
$U(x) = 30/(x^2 - 8x + 20)$ for x in $[0, 10]$.

$$U'(x) = -30(x^2 - 8x + 20)^{-2}(2x - 8)$$
$$= \frac{240 - 60x}{(x^2 - 8x + 20)^2}$$

The only value of x in the interval $[0, 10]$ at which $U'(x) = 0$ is 4. So, according to Theorem 2.12, the maximum value $U(x)$ attains as x varies on the interval $[0, 10]$ must occur at $x = 0$, $x = 4$, or $x = 10$. Substituting these three numbers in the formula for $U(x)$, we get

$$U(0) = 1.5$$
$$U(4) = 7.5$$

and
$$U(10) = 0.75$$

So 7.5 is the maximum value $U(x)$ attains as x varies on the interval $[0, 10]$. Therefore, during the next 10 months, the number of unemployed steel workers will never exceed 7500 and, hence, will never reach 9000. Figure 2.29 shows the graph of $U(x)$.

Sometimes it is convenient to assume the domain of a function is an interval, even though, in reality, the domain is a set of integers. In Example 2.19, we make such an assumption.

Example 2.19

The owners of a potato processing plant permit nearby farmers to rent their machines during weekends. The rental fee is $50 per machine plus $30 per hour for the use of the plant, regardless of the number of machines rented. Suppose 20 machines are available. How many machines should a farmer rent to process 25,000 bushels of potatoes if each machine processes 75 bushels per hour?

Comment: The cost of the electricity the machines consume depends on the number of bushels processed—not on the number of machines used. Therefore, although the farmer is billed for this cost, it is not necessary to account for it in solving the problem.

Solution

First we will find a formula that expresses the combined cost of renting the machines and using the plant as a function of the number of machines rented. Since each machine processes 75 bushels per hour, x machines process $75x$ bushels per hour. Thus, the number of hours it takes x machines to process 25,000 bushels is

$$\frac{25{,}000}{75x}$$

Since the rental fee is $50 per machine plus $30 per hour for the use of the plant, the combined cost of renting x machines and using the plant is

$$C(x) = 50x + 30\left(\frac{25{,}000}{75x}\right)$$
$$= 50x + \frac{10{,}000}{x}$$

dollars. Since only 20 machines are available, the domain of this function is the set of positive integers from 1 to 20. However, so that we can use calculus, we temporarily assume the domain is the closed interval [1, 20].

The derivative of $C(x)$ is

$$C'(x) = 50 - \frac{10{,}000}{x^2}$$

We set this formula equal to 0 and solve for x:

$$50 - \frac{10{,}000}{x^2} = 0$$
$$50 = \frac{10{,}000}{x^2}$$
$$50x^2 = 10{,}000$$
$$x^2 = 200$$
$$x = \pm\sqrt{200}$$

Thus, $C'(x) = 0$ at $x = \sqrt{200}$ and at $x = -\sqrt{200}$. But, of these two numbers, only $\sqrt{200}$ is in the interval [1, 20].

Note that $14 < \sqrt{200} < 15$. Therefore, in view of Theorem 2.12 and the fact the number of machines must be a positive integer, the value of x at which $C(x)$ is minimum must be 1, 14, 15, or 20. Since

$$C(1) = 10{,}050 \qquad C(15) = 1416\tfrac{2}{3}$$
$$C(14) = 1414\tfrac{2}{7} \qquad C(20) = 1500$$

we see that $C(x)$ is minimum at $x = 14$. So the farmer should rent 14 machines.

In Example 2.19, even though $\sqrt{200}$ is closer to 14 than to 15, we considered 15 as a possibility for the value of x that minimizes $C(x)$. We considered 15 because the point $(15, C(15))$ could possibly be lower than the point $(14, C(14))$ on the graph of $C(x)$.

In reality, the domain of the function $C(x)$ in Example 2.19 is the positive integers from 1 to 20. Therefore, we could have found the minimum value of $C(x)$ by substituting each one of these 20 positive integers in the formula for $C(x)$. Using calculus, however, we only had to substitute 4 of the 20 positive integers in the formula for $C(x)$.

A differentiable function whose domain is a closed interval may have a derivative that is nowhere zero on the interval. In this case, the maximum and minimum values of the function occur at endpoints. Figure 2.30 shows such a function.

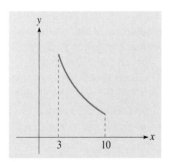

Figure 2.30 A differentiable function whose derivative is nowhere zero on the closed interval [3, 10].

Optimization with a Graphing Calculator

As we have seen, many problems involve finding values of x where a derivative $f'(x)$ is 0. So far, we have managed to find such values of x using basic algebraic methods. In some cases, however, an algebraic approach is not practical. Example 2.20 illustrates such a case.

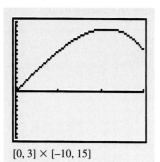

[0, 3] × [−10, 15]

Figure 2.31 The graph of $f(x)$ for x in [0, 3].

Example 2.20

Suppose

$$f(x) = -0.0625x^4 - 0.5x^3 + 9.25x$$

for values of x in the interval [0, 3]. Find the value of x where $f(x)$ is maximum.

Solution

We set the viewing screen on a graphing calculator to [0, 3] × [−10, 15] with a scale of 1 on both axes. Then we draw the graph of $f(x)$. Figure 2.31 shows that the highest point on the graph of $f(x)$ is the point where the tangent is horizontal. Therefore, $f(x)$ is maximum at the value of x (in the interval [0, 3]) where $f'(x) = 0$.

The derivative of $f(x)$ is

$$f'(x) = -0.25x^3 - 1.5x^2 + 9.25$$

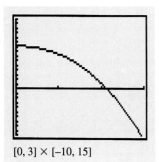

[0, 3] × [−10, 15]

Figure 2.32 $f'(x) = -0.25x^3 - 1.5x^2 + 9.25$ for x in [0, 3].

Using an algebraic approach, we would have difficulty finding the value of x where $f'(x) = 0$. So we will use a graphical approach to find an *approximation* for this value of x. Our approach involves "zooming in" on the point where the graph of $f'(x)$ crosses the x axis.

Using the setting on the viewing screen we used for $f(x)$, we begin by drawing the graph of $f'(x)$. Figure 2.32 shows that the graph of $f'(x)$ crosses the x axis at a point between 2 and 3.

Now we set the viewing screen to [2, 3] × [−1, 1] with a scale of 0.1 on both axes and again draw the graph of $f'(x)$. Figure 2.33 shows that the graph of $f'(x)$ crosses the x axis at a point between 2.1 and 2.2. So 2.1 approximates the x intercept of the graph of $f'(x)$ with an error less than 0.1.

Now we set the viewing screen to [2.1, 2.2] × [−0.1, 0.1] with a scale of 0.01 on both axes and again draw the graph of $f'(x)$. Figure 2.34 shows that the graph of $f'(x)$ crosses the x axis at a point between 2.13 and 2.14. Therefore, 2.13 approximates the x intercept of the graph of $f'(x)$ with an error less than 0.01.

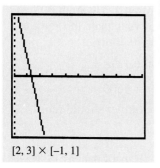

[2, 3] × [−1, 1]

Figure 2.33 The x intercept of $f'(x)$ between 2.1 and 2.2.

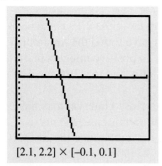

[2.1, 2.2] × [−0.1, 0.1]

Figure 2.34 The x intercept of $f'(x)$ between 2.13 and 2.14.

Now we set the viewing screen to [2.13, 2.14] × [−0.01, 0.01] with a scale of 0.001 on both axes and again draw the graph of $f'(x)$. Figure 2.35 shows that the graph of $f'(x)$ crosses the x axis at a point between 2.132 and 2.133. Hence, 2.132 approximates the x intercept of the graph of $f'(x)$ with an error less than 0.001.

Now we set the viewing screen to [2.132, 2.133] × [−0.001, 0.001] with a scale of 0.0001 on both axes and again draw the graph of $f'(x)$. Figure 2.36 shows that the graph of $f'(x)$ crosses the x axis at a point between 2.1329 and 2.1330. Therefore, 2.1329 approximates the x intercept of the graph of $f'(x)$ with an error less than 0.0001.

By continuing the above process, we could obtain more accurate approximations for the x intercept of the graph of $f'(x)$ than 2.1329. However, we will settle for 2.1329.

The x intercept of the graph of $f'(x)$ is the value of x where $f'(x) = 0$. Therefore, since $f(x)$ is maximum at this value of x, we conclude that $f(x)$ is maximum at a value of x that differs from 2.1329 by less than 0.0001.

In Example 2.20, we could approximate the value of x where $f(x)$ is maximum by applying the trace function to the graph of $f(x)$. This approach, however, would not provide us with information regarding the accuracy of the approximation.

The functions in the problem set for this section are differentiable on their domains. Moreover, the domains are closed intervals. In Section 2.7, we solve optimization problems involving functions whose domains are not closed intervals.

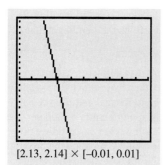

[2.13, 2.14] × [−0.01, 0.01]

Figure 2.35 The x intercept of $f'(x)$ between 2.132 and 2.133.

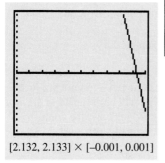

[2.132, 2.133] × [−0.001, 0.001]

Figure 2.36 The x intercept of $f'(x)$ between 2.1329 and 2.1330.

2.6 Problems

In Problems 1 to 14, the given function is differentiable on its domain, which is the accompanying closed interval. Find the highest and lowest points on the graph of the function.

1. $f(x) = x^3 - 9x^2 + 15x + 32$ [0, 6]
2. $f(x) = -x^3 + 15x^2 - 63x + 135$ [2, 10]
3. $f(x) = -x^3 + 6x^2 + 15x$ [1, 6]
4. $f(x) = x^3 - 12x^2 - 60x + 800$ [0, 8]
5. $f(x) = 12(\sqrt{x})^3 - 72x + 400$ [9, 25]
6. $f(x) = 48x - 4(\sqrt{x})^3$ [1, 100]
7. $f(x) = 120 - \dfrac{108}{x} - 3x$ [1, 4]
8. $f(x) = 5x + \dfrac{320}{x}$ [4, 10]
9. $f(x) = \dfrac{15}{x^2 - 12x + 37}$ [0, 10]
10. $f(x) = 8 - \dfrac{18}{x^2 - 16x + 70}$ [0, 10]
11. $f(x) = \sqrt{x^2 - 12x + 40}$ [2, 8]
12. $f(x) = 7 - \sqrt{x^2 - 6x + 10}$ [1, 9]
13. $f(x) = \dfrac{9x}{x^2 + 4}$ [1, 6]
14. $f(x) = \dfrac{16x}{x^2 + 25}$ [1, 6]

15. **Unemployment Rate** In a Third World country, the unemployment rate is presently 9 percent. Economists predict that x months from now the rate will be

$$U(x) = 9 - \frac{42x}{x^2 + 49}$$

percent. When during the next 12 months will the rate be lowest? How low will the rate be at that time?

16. **Worker Productivity** In a factory, the work day is 6 hours long for each worker. Management finds that the average worker produces

$$r(x) = \frac{15}{x^2 - 4x + 5}$$

units per hour after being on the job x hours. When is the average worker most productive? At what rate does the average worker produce when most productive?

17. **Medicine** Let $S(x)$ represent a patient's blood sugar level (in milligrams per deciliter) x hours after the patient receives a dose of an experimental insulin. The patient's doctor assumes

$$S(x) = 6x + \frac{600}{x + 3}$$

for x in the closed interval $[0, 24]$. When during the 24-hour period after the patient receives the insulin is the patient's blood sugar level lowest? How low is the patient's blood sugar level at that time?

18. **Demography** Demographers predict that x years from now the population of a country will include

$$f(x) = 4(\sqrt{x})^3 - 12x + 60$$

million teenagers. When during the next 9 years will the number of teenagers be least? How many teenagers will be in the country at that time?

19. **Interest Rate** Suppose x months from now the prime interest rate is

$$I(x) = 1.8x - 0.4(\sqrt{x})^3 + 12$$

percent, where x is in the closed interval $[0, 16]$. When will the rate peak during the next 16 months? What will be the rate at that time?

20. **Home Affordability** In an Asian country, 90 million families can afford to purchase homes worth at least $50,000. If x months from now the number of such families is

$$f(x) = \sqrt{x^2 - 26x + 8100}$$

million, when during the next 30 months will this number be least? What will be the number at that time?

21. **Optimal Work Force** A manufacturer wants to produce 810,000 units of a new product. The workers who will be assigned to the task require special training, which costs the manufacturer $1600 per worker. Once trained, each worker can produce 50 units of the product per hour. To produce the product, a special machine has to be rented. The rental fee for the machine is $200 per hour while in operation. If at most 60 workers can be trained, how many workers should be assigned to the task so the combined cost of training the workers and renting the machine is least?

    Suggestion: See Example 2.19.

22. **Optimal Work Force** A company wants to produce 640,000 units of a product it has not produced before. The workers who will be assigned to the task require special training, which costs the company $400 per worker. Once trained, each worker can produce 25 units of the product per hour. To produce the product, a special machine has to be rented. The rental fee for the machine is $100 per hour while in operation. If at most 95 workers can be trained, how many workers should be assigned to the task so the combined cost of training the workers and renting the machine is least?

    Suggestion: See Example 2.19.

23. **Optimal Production Levels** The management of Continental Video finds that the total monthly cost of producing x video cassette recorders per month is

    $$C(x) = 117x + 50{,}000$$

    dollars. The company can sell x recorders per month whenever it sets the price of each recorder at

    $$D(x) = -0.01x^2 + 1200$$

    dollars. The management claims these formulas are valid for values of x in the closed interval $[100, 300]$. If the firm produces x recorders per month, it sets the price of each recorder at $D(x)$ dollars.

    a. Find the monthly production level that maximizes the revenue obtained from selling the recorders.

       Suggestion: See Example 1.4 in Section 1.2.

b. Find the monthly production level that maximizes the profit earned from selling the recorders. At this production level, what is the selling price and what is the monthly profit?

c. If the cost of producing each recorder increases $11.37, what is the monthly production level that maximizes the profit earned from selling the recorders? At this production level, what is the selling price and what is the monthly profit?

24. **Optimal Production Levels** The management of a firm assumes that the total weekly cost of producing x kilograms of a commodity per week is

$$C(x) = \tfrac{1}{60}x^2 + 75$$

dollars. The firm can sell x kilograms of the commodity per week whenever it sells each kilogram for

$$D(x) = -\tfrac{1}{60}x + 40$$

dollars. The management claims these formulas are valid for values of x in the closed interval [300, 700]. If the firm produces x kilograms of the commodity per week, it sells each kilogram for $D(x)$ dollars.

a. Use calculus to find the weekly production level that maximizes the revenue obtained from selling the commodity.
   Suggestion: See Example 1.4 in Section 1.2.

b. Use calculus to find the weekly production level that maximizes the profit earned from selling the commodity. At this production level, what is the selling price and what is the weekly profit?

c. Suppose the government requires the firm to pay a tax of $13 per kilogram produced. Use calculus to determine the weekly production level that maximizes the profit earned from selling the commodity. At this production level, what is the selling price and what is the weekly profit?

25. **Optimal Number of Flower Stands** Several months ago a nationwide floral company opened a sidewalk flower stand in a city. The flower stand has been earning a daily profit of $300. The owner of the company believes that if additional stands are opened in the city, each additional stand would cause the daily profit per stand to decrease $20 (0.2 of a hundred dollars). The city allows a maximum of 12 stands. Use calculus to determine how many stands would maximize the total daily profit.

   Suggestion: See the opening example in Section 1.4.

26. **Optimal Work Load** The management of a restaurant assigns each waitress between 7 and 15 tables. One waitress finds that when she is assigned seven tables, each table brings in $30 per week in tips. If she is assigned more tables, the amount of attention she gives to each table decreases. Thus, each table brings in less per week in tips. Suppose each additional table causes *every* table to bring in $2 less per week in tips. Use calculus to determine how many tables she should be assigned to maximize her weekly earnings from tips.
   Suggestion: See the opening example in Section 1.4.

27. **Optimal Delivery Batch** The owner of a cabinet shop plans to use his truck to deliver 841 showcases designed and built exclusively for a nearby department store. Special crates have to be built for the showcases at a cost of $25 per crate. The truck has space for at most 55 crates, and each crate holds only one showcase. Thus, the truck must make several trips. It costs $49 to make each trip. If each crate is reusable, how many crates should be built (and used) to minimize the combined cost of building the crates and delivering the 841 showcases?
   Suggestion: See Example 1.5 in Section 1.2.

28. **Optimal Firing Batch** Helen wants to make 1200 copies of a vase she designed. Her kiln has space for at most 24 molds and consumes $6 worth of electricity each time it is fired. It costs $32 to build each mold. If each mold is reusable, how many molds should Helen build (and use) to minimize the combined cost of the molds and the electricity?
   Suggestion: See Example 1.5 in Section 1.2.

29. **Optimal Route** An injured person has to be transported by a boat from point A, located 40 miles from shore, to a point D on the shore. See the figure for this problem. From point D, the patient will be transported by an ambulance to a hospital located at point B on the shore. The boat travels at 30 miles per hour, and the ambulance travels at 50 miles per hour. Suppose triangle AFD is a right triangle. Determine how far point D should be from point F to minimize the time it takes to transport the patient to the hospital.

Chap. 2 Differentiation

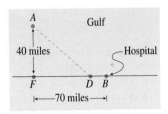

Suggestion: Let x represent the distance between point F and point D. Then find a formula that expresses the transportation time as a function of x.

30. **Optimal Route** A builder plans to use a dump truck to transport a load of dirt from point A to point B via point D. See the figure for this problem. Point B is located in a prairie 20 miles from the paved road. On the paved road, the truck gets 10 miles per gallon of fuel. But on the prairie it only gets 6 miles per gallon of fuel. Suppose triangle BFD is a right triangle. Determine how far point D should be from point F to minimize the amount of fuel consumed in transporting the dirt.

Suggestion: Let x represent the distance between point D and point F. Then find a formula that expresses the fuel consumption as a function of x.

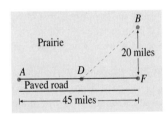

31. **Optimal Tapping Point** An existing gas main will be tapped at a single point to supply gas to two factories. See the figure for this problem. Point D represents where the existing main will be tapped. Suppose triangles AFD and BGD are right triangles. Determine how far point D should be from point F to minimize the amount of pipe needed.

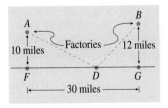

32. **Optimal Docking Point** A ferryboat soon will make one round trip per day to each of two islands from a single docking point on the mainland. See the figure for this problem. Point D represents the docking point on the mainland. Suppose triangles AFD and BGD are right triangles. Determine how far point D should be from point F to minimize the total distance the ferryboat will travel each day.

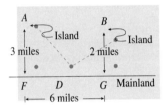

 Graphing Calculator Problems

33. Set the viewing screen on your graphing calculator to $[0, 5] \times [-2, 4]$ with a scale of 1 on both axes. Then draw the graph of

$$f(x) = 0.03x^4 - 0.19x^3 + 0.38x^2 - 1.25x + 4$$

You will see that if the domain is the closed interval $[0, 5]$, the lowest point on the graph is the point where the tangent is horizontal. Use the zooming-in approach we used in Example 2.20 to approximate the value of x where $f(x)$ is minimum. Continue zooming in until the difference between the approximation and the actual value is less than 10^{-8}.

34. Set the viewing screen on your graphing calculator to $[0, 2] \times [-2, 4]$ with a scale of 1 on both axes. Then draw the graph of

$$f(x) = 0.25x^4 - x^3 + 3x$$

You will see that if the domain is the closed interval $[0, 2]$, the highest point on the graph is the point where the tangent is horizontal. Use the zooming-in approach we used in Example 2.20 to approximate the value of x where $f(x)$ is maximum. Continue zooming in until the difference between the approximation and the actual value is less than 10^{-6}.

Computer Problems

The Basic program OPT2 listed below approximates the minimum and maximum values of a continuous function $f(x)$ whose domain is a closed interval. The program compares the values of $f(x)$ at randomly selected values of x and prints the minimum and maximum of these values. It also prints the values of x at which these values of $f(x)$ occur. Execution of the program prompts you to enter the endpoints of the interval and the number of points to be randomly selected from the interval. (The more points are selected the better the accuracy of the result.) It also prompts you to *seed* the set of randomly selected points with an integer between $-32{,}768$ and $32{,}767$. (Different seeds produce different sets of random points.)

As listed, OPT2 applies to the function

$$f(x) = x^3 - 12x + 19$$

But you can easily make the program applicable to other continuous functions by editing line 60.

The Program OPT2 and a Sample Run

```
LIST
10   REM                              OPT2
20   REM
30   REM A PROGRAM THAT USES THE RANDOM NUMBER GENERATOR TO APPROXIMATE THE
40   REM MINIMUM AND MAXIMUM VALUES OF A CONTINUOUS FUNCTION ON A CLOSED INTERVAL
50   REM
60   DEF FNF(X)=X^3-12*X+19
70   PRINT "ENTER THE ENDPOINTS OF THE INTERVAL."
80   INPUT A,B
90   PRINT
100  PRINT "ENTER THE NUMBER OF POINTS TO BE RANDOMLY SELECTED FROM THE"
110  PRINT "INTERVAL. (SEVERAL HUNDRED POINTS ARE USUALLY NEEDED.)"
120  INPUT N
130  PRINT
140  RANDOMIZE
150  PRINT
160  LET MN=FNF(A)
170  LET L=A
180  LET MX=FNF(B)
190  LET G=B
200  FOR I=1 TO N
210  LET X=(B-A)*RND+A
220  IF FNF(X)>=MN THEN 250
230  LET MN=FNF(X)
240  LET L=X
250  IF FNF(X)<=MX THEN 280
260  LET MX=FNF(X)
270  LET G=X
280  NEXT I
290  PRINT "THE MINIMUM VALUE OF F(X) ON THE CLOSED INTERVAL FROM";A;"TO";B;
300  PRINT "IS APPROXIMATELY";MN;"AND IT OCCURS AT ABOUT X=";L;".THE MAXIMUM"
310  PRINT "VALUE IS APPROXIMATELY";MX;"AND IT OCCURS AT ABOUT X =";G;"."
320  END
Ok
```

```
RUN
ENTER THE ENDPOINTS OF THE INTERVAL.
? 0,3

ENTER THE NUMBER OF POINTS TO BE RANDOMLY SELECTED FROM THE
INTERVAL.(SEVERAL HUNDRED POINTS ARE USUALLY NEEDED.)
?500
Random number seed (-32768 to 32767)? 9

THE MINIMUM VALUE OF F(X) ON THE CLOSED INTERVAL FROM 0 TO 3
IS APPROXIMATELY 3.00002 AND IT OCCURS AT ABOUT X=2.00182. THE MAXIMUM
VALUE IS APPROXIMATELY 18.95336 AND IT OCCURS AT ABOUT X=3.887057E-03.
Ok
```

35. **Unemployment Rate** Use the program OPT2 to approximate the lowest unemployment rate expected during the next 12 months in the country in Problem 15. Approximately when is the lowest unemployment rate expected?

36. **Worker Productivity** Use the program OPT2 to approximate the highest production rate the average worker achieves during a work day in the factory in Problem 16. Approximately when is the highest production rate achieved?

37. **Optimal Route** Use the program OPT2 to approximate the time it takes to transport the injured person in Problem 29 to the hospital using the best possible route.

38. **Optimal Route** Use the program OPT2 to approximate the amount of fuel the truck in Problem 30 consumes if it uses the best possible route to transport the dirt from point A to point B.

2.7 More on Optimization

In Section 2.6, we found the maximum and minimum values of functions whose domains are *closed* intervals. In this section, we find the maximum and minimum values of functions whose domains are *not closed* intervals. First, however, we discuss how the sign of the derivative of a function can be used to analyze the function.

Analyzing a Function Using the Sign of Its Derivative

Let $g(x)$ represent the function whose graph is the curve in Figure 2.37. The figure shows that the tangents have negative slopes at points whose x coordinates are less than 3. Therefore, $g'(x)$ is negative when $x < 3$. The figure also shows that the tangents have positive slopes at points whose x coordinates are greater than 3. So $g'(x)$ is positive when $x > 3$. Note that $g(x)$ *decreases* as x increases towards 3, but *increases* as x increases beyond 3.

The function $g(x)$ suggests Theorem 2.13.

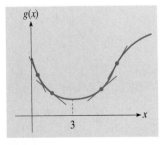

Figure 2.37 The graph of $g(x)$ and several of its tangents.

Theorem 2.13

If $f'(x)$ is negative for each x in an interval, then $f(x)$ decreases as x increases on the interval. If $f'(x)$ is positive for each x in an interval, then $f(x)$ increases as x increases on the interval.

From now on, for the sake of simplicity, whenever we say $f(x)$ decreases on an interval, we mean $f(x)$ decreases *as x increases* on the interval. Similarly, whenever we say $f(x)$ increases on an interval, we mean $f(x)$ increases *as x increases* on the interval.

We use Theorem 2.13 in Example 2.21.

Example 2.21

Let
$$f(x) = \frac{4}{(x-2)^2}$$

Determine where $f(x)$ increases and where it decreases.

Solution

Since
$$\frac{4}{(x-2)^2} = 4(x-2)^{-2}$$

we can express the given function as
$$f(x) = 4(x-2)^{-2}$$

Applying the generalized power rule, we get
$$f'(x) = -8(x-2)^{-3}$$
$$= \frac{-8}{(x-2)^3}$$

Since the denominator is zero when $x = 2$, $f(x)$ is not differentiable at $x = 2$. For $x < 2$, the numerator and denominator of $-8/(x-2)^3$ are both negative. Thus, for $x < 2$, $f'(x)$ is positive. Therefore, according to Theorem 2.13, $f(x)$ increases on the interval $(-\infty, 2)$.

For $x > 2$, the numerator of $-8/(x-2)^3$ is again negative. (In fact, the numerator is always negative.) But now the denominator is positive. Thus, for $x > 2$, $f'(x)$ is negative. Therefore, according to Theorem 2.13, $f(x)$ decreases on the interval $(2, +\infty)$. Figure 2.38 shows the graph of $f(x)$.

The **sign chart** in Figure 2.39 shows the information we gathered in Example 2.21 about the derivative of the function

$$f(x) = \frac{4}{(x-2)^2}$$

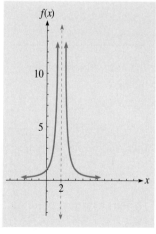

Figure 2.38 $f(x) = 4/(x-2)^2$

$f'(x)$	positive	negative
	2	

Figure 2.39 The sign chart for $f'(x) = -8/(x-2)^3$.

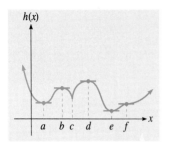

Figure 2.40 A curve that has two high points and three low points.

Now that we know how to determine where a function increases and where it decreases, we can find the high and low points on a graph.

High and Low Points on a Graph

Let $h(x)$ be the function whose graph is the curve in Figure 2.40. The points with x coordinates b and d are **high points** on the graph. The points with x coordinates a, c, and e are **low points**. In fact, the point with x coordinate e is the lowest point. However, the graph does not have a highest point.

In Figure 2.40, at the high and low points on the curve, the tangents are either horizontal or fail to exist. (The tangent fails to exist at the point where $x = c$.) Although the tangent is horizontal at the point where $x = f$, this point is neither a high point nor a low point. These observations lead us to Theorem 2.14.

Theorem 2.14

Excluding endpoints, the high and low points on the graph of a function are among the points where the derivative is either zero or fails to exist.

Keep in mind that a derivative may be zero at a point that is neither a high point nor a low point. This situation, as we observed, is illustrated in Figure 2.40 by the point where $x = f$.

Critical Points

The numbers in the domain of a function where the derivative is either zero or fails to exist are important enough to be labeled.

Definition 2.3

The **critical points** of a function $f(x)$ are the values of x in the domain where $f'(x) = 0$ or $f'(x)$ fails to exist.

If c is a critical point of $f(x)$, we refer to $(c, f(c))$ as a critical point on the graph of $f(x)$.

Theorem 2.14 tells us that the high and low points (excluding endpoints) on the graph of a function are among the critical points. But remember that it is possible for a critical point to be neither a high point nor a low point.

In this book, we usually work with functions whose critical points are points where the derivative is zero, rather than points where the derivative fails to exist.

In Example 2.22, we use Theorem 2.13 and Theorem 2.14 to locate the high and low points on the graph of a function.

Example 2.22

Find the high and low points on the graph of the function

$$f(x) = \tfrac{1}{24}x^4 - \tfrac{4}{9}x^3 + x^2 + 8$$

Solution

Since $f(x)$ is a polynomial function, it is differentiable everywhere. Therefore, the only critical points of $f(x)$ are the numbers where the derivative is zero. The derivative of $f(x)$ is

$$\begin{aligned} f'(x) &= \tfrac{1}{6}x^3 - \tfrac{4}{3}x^2 + 2x \\ &= \tfrac{1}{6}x(x^2 - 8x + 12) \\ &= \tfrac{1}{6}x(x - 2)(x - 6) \end{aligned}$$

So $f'(x) = 0$ only when $x = 0, 2,$ or 6. Thus, 0, 2, and 6 are the critical points. According to Theorem 2.14, the high and low points can occur only at the points with these numbers as x coordinates.

$\tfrac{1}{6}x$	negative	positive	positive	positive
$x-2$	negative	negative	positive	positive
$x-6$	negative	negative	negative	positive
$f'(x)$	negative	positive	negative	positive
	0		2	6

Figure 2.41 The sign chart for $f'(x) = \tfrac{1}{6}x(x-2)(x-6)$.

The sign chart in Figure 2.41 analyzes the sign of $f'(x)$ in terms of the factors of $f'(x)$. The chart shows that $f'(x)$ is negative when $x < 0$ and when $2 < x < 6$. The chart also shows that $f'(x)$ is positive when $0 < x < 2$ and when $x > 6$.

Using Theorem 2.13, we can conclude that $f(x)$ decreases on the interval $(-\infty, 0)$ and on the interval $(2, 6)$. We can also conclude that $f(x)$ increases on the interval $(0, 2)$ and on the interval $(6, +\infty)$.

Therefore, the low points on the graph of $f(x)$ are the points where $x = 0$ and where $x = 6$. Since $f(0) = 8$ and $f(6) = 2$, the low points are $(0, 8)$ and $(6, 2)$. The only high point is the point where $x = 2$. Since $f(2) = \tfrac{82}{9}$, the high point is $(2, \tfrac{82}{9})$. Figure 2.42 shows the graph of $f(x)$.

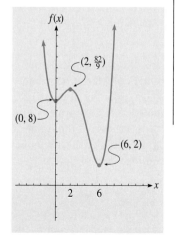

Figure 2.42
$f(x) = \tfrac{1}{24}x^4 - \tfrac{4}{9}x^3 + x^2 + 8$

A Comment on Terminology

In this book, we call the ordinate of the highest point on the graph of a function the maximum value of the function. Similarly, we call the ordinate of the lowest point on the graph the minimum value of the function. In many books, what we call the maximum value is called the **absolute**

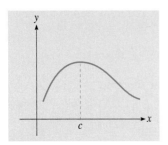

Figure 2.43 A function whose derivative is positive for $x < c$ and negative for $x > c$.

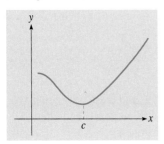

Figure 2.44 A function whose derivative is negative for $x < c$ and positive for $x > c$.

maximum value, and what we call the minimum value is called the **absolute minimum value**. In such books, the ordinate of a high point on the graph of a function is called a **relative maximum value**, and the ordinate of a low point on the graph is called a **relative minimum value**.

Using the above terminology, the function in Example 2.22 has two relative minimum values, namely, 8 and 2. (See Figure 2.42.) In fact, 2 is the absolute minimum value of the function. The function has one relative maximum value, namely, $\frac{89}{9}$, but no absolute maximum value.

Optimization on Intervals That Are Not Closed

Suppose c is a critical point of a function $f(x)$ whose domain is an interval which does not necessarily contain its endpoints. Also, suppose $f'(x)$ is positive for all $x < c$ and negative for all $x > c$. Then, according to Theorem 2.13, $f(x)$ increases as x increases towards c and decreases as x increases beyond c. So the graph of $f(x)$ has a highest point, namely, the point $(c, f(c))$. This conclusion implies that $f(x)$ has a maximum value at $x = c$, and the maximum value is $f(c)$. (See Figure 2.43.) Similarly, if $f'(x)$ is negative for all $x < c$ and positive for all $x > c$, then $f(x)$ has a minimum value at $x = c$ and the minimum value is $f(c)$. (See Figure 2.44.) Theorem 2.15 summarizes this paragraph.

Theorem 2.15

Suppose c is a critical point of a function $f(x)$ whose domain is an interval. If $f'(x)$ is positive for $x < c$ and negative for $x > c$, then $f(x)$ has a *maximum* value at $x = c$. If $f'(x)$ is negative for $x < c$ and positive for $x > c$, then $f(x)$ has a *minimum* value at $x = c$.

In Example 2.23, we use Theorem 2.15 to solve an inventory problem.

Example 2.23

During the coming year, a hospital will use 256 liters of Mercurochrome at a constant rate. The hospital will place equal-size orders of at most 128 liters that arrive at the moment the supply is exhausted. The supplier charges a delivery fee of $4 per order, regardless of the size of the order. It costs the hospital $2 to store each liter for 1 year. How many liters should the hospital order each time to minimize the combined cost of delivery and storage?

Solution

First we will find a formula that expresses the combined cost of delivery and storage for the year as a function of the number of liters ordered each time. If x liters is ordered each time, $256/x$ orders are placed during the year. Thus, since the delivery fee is $4 per order, the delivery cost for the year is

$$\left(\frac{256}{x}\right)4 = \frac{1024}{x}$$

dollars. Since the Mercurochrome is used at a constant rate, the average number of liters in storage during the year is $x/2$. (We will be able to prove this fact after we complete Chapter 5.) Thus, since it costs $2 to store each liter for one year, the total storage cost for the year is

$$\left(\frac{x}{2}\right)2 = x$$

dollars. Therefore, if x liters is ordered each time, the combined cost of delivery and storage for the year is

$$C(x) = x + \frac{1024}{x}$$

dollars. Since each order consists of at most 128 liters, the domain of this function is the half-open interval $(0, 128]$.

The derivative of $C(x)$ is

$$C'(x) = 1 - \frac{1024}{x^2}$$

Setting this formula equal to zero and solving for x, we get

$$1 - \frac{1024}{x^2} = 0$$

$$1 = \frac{1024}{x^2}$$

$$x^2 = 1024$$

$$x = \pm 32$$

Thus, among the values of x in $(0, 128]$, $C'(x) = 0$ only at $x = 32$.

For values of x in $(0, 128]$, $C'(x)$ is negative when $x < 32$ and positive when $x > 32$. [For example, $C'(10) = -9.24$, while $C'(40) = 0.36$.] So, according to Theorem 2.15, $C(x)$ has a minimum value at $x = 32$. Therefore, the hospital should order 32 liters each time to minimize the combined cost of delivery and storage. Figure 2.45 shows the graph of $C(x)$.

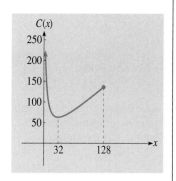

Figure 2.45
$C(x) = x + 1024/x$
for x in $(0, 128]$.

2.7 Problems

In Problems 1 to 16, find the intervals on which $f(x)$ increases and the intervals on which it decreases. Then identify the high and low points on the graph.

1. $f(x) = x^3 - 15x^2 + 63x - 9$
2. $f(x) = 2x^3 - 12x^2 - 72x$
3. $f(x) = -x^3 - 6x^2 + 15x$
4. $f(x) = -x^3 + 18x^2 - 96x + 5$
5. $f(x) = \frac{1}{3}x^3 - x^2 + x + 2$
6. $f(x) = -3x^3 + 27x^2 - 81x$

7. $f(x) = -x^4 + 16x^3 - 70x^2 + 1100$
8. $f(x) = \frac{1}{4}x^4 - \frac{10}{3}x^3 + 12x^2$
9. $f(x) = -x^3 + x^2 - 5x$
10. $f(x) = x^3 + 4.5x^2 + 9x + 7$
11. $f(x) = x^4 - 8x^3 + 18x^2 + 9$
12. $f(x) = -x^4 + 16x^3 - 72x^2 + 440$
13. $f(x) = \dfrac{x^2 + 3x - 12}{x^2}$
14. $f(x) = \dfrac{3x - 9}{x^2 - 10x + 25}$
15. $f(x) = \dfrac{2x^3}{x - 6}$
16. $f(x) = \dfrac{x - 5}{x^2 - 9}$

17. **Acid Rain** The pH of a substance is a number between 0 and 14 that measures the acidity or alkalinity of the substance. A substance whose pH is less than 7 is acidic. A substance whose pH is greater than 7 is alkaline. In a region of North America, the pH of rainwater is presently 6. Environmentalists predict that x years from now the pH of the region's rainwater will be

$$A(x) = 6 - \dfrac{12x}{x^2 + 9}$$

When will the pH be lowest? When the pH is lowest, what will it be?

The effect of acidic rainwater on a German forest. (Gernot Huber/Woodfin Camp)

18. **Consumer Debt** Presently, the consumer debt in a country is approximately $5.8 billion. Economists predict that x years from now the debt will be

$$D(x) = \sqrt{x^2 - 10x + 34}$$

billion dollars. When will the debt be lowest? When the debt is lowest, what will it be?

19. **Optimal Fertilization Level** Let $Y(x)$ represent the number of flats of strawberries each acre of land produces if x sacks of a special fertilizer is applied to each acre. The yield is low if either too little or too much of the fertilizer is applied. Suppose

$$Y(x) = \dfrac{800}{x^2 - 12x + 40}$$

How much of the fertilizer should be applied to maximize the yield?

20. **Household Income** Although the average income per household in a country is only $500 per month, it is increasing. Let $D(x)$ represent the number of million pounds of powdered eggs sold monthly when the average monthly income per household is x hundred. There are indications $D(x)$ will at first increase as x increases. However, since powdered eggs are a money-saving substitute for fresh eggs, $D(x)$ is expected to eventually decrease as x continues to increase. Suppose

$$D(x) = \dfrac{256x}{x^2 + 64}$$

when $x \geq 5$. For what average monthly income per household will the monthly demand for powdered eggs be greatest?

21. **Car Affordability** In a country, 80 million people can presently afford to purchase automobiles costing at least $10,000. Suppose that x months from now the number of such people is

$$f(x) = \tfrac{2}{3}(\sqrt{x})^3 - 4x + 80$$

million. When will this number be least?

22. **Drug Concentration** A physician believes that x hours after a patient swallows a dose of an experimental drug, the patient's bloodstream contains

$$D(x) = \dfrac{24x}{x^2 + 9}$$

hundred milligrams of the drug. When is the amount of the drug in the patient's bloodstream greatest?

23. **Optimal Production Levels** Each day, The Ice Cream Factory sells whatever amount of granita (an Italian sherbet) it produces. Thus, the daily profit

earned from selling the granita may be viewed as a function of the daily production level. Suppose the selling price is $15 per gallon, and the total daily cost of producing x gallons is

$$C(x) = \frac{1}{16}x^2 + 400$$

dollars.

a. Find the daily production level that maximizes the profit earned from selling the granita.

b. Find the daily production level that minimizes the average cost of producing each gallon of the granita.

24. **Optimal Production Levels** Mayport Motors builds a gasoline engine that delivers about 80 miles per gallon when used to power compact cars. Each month, the company sells all of the engines it produces. Thus, the monthly profit earned from selling the engine may be viewed as a function of the monthly production level. Suppose the selling price of each engine is $3000, and the total monthly cost of producing x engines is

$$C(x) = 0.04x^2 + 0.2x + 16$$

thousand dollars. In reality, the domain of this function is a set of integers. Nevertheless, assume the domain is the interval $[0, +\infty)$.

a. Find the monthly production level that maximizes the profit earned from selling the engine.

b. Find the monthly production level that minimizes the average cost of producing each engine.

25. **Optimal Processing Levels** The owners of Harbor Seafoods find that the total monthly cost of processing x tons of lump crab meat per month is

$$C(x) = 4x + 600$$

thousand dollars. The company can sell x tons of the crab meat monthly if the price is

$$D(x) = \frac{800}{x + 2}$$

thousand dollars per ton. Whenever the company processes x tons per month, the company sets the price at $800/(x + 2)$ thousand dollars per ton.

a. Show that the average cost of processing each ton decreases whenever the company increases the monthly processing level. Thus, there is no monthly processing level that minimizes the average cost of processing each ton.

b. Show that the monthly revenue obtained from selling the crab meat increases whenever the company increases the monthly processing level. Thus, there is no monthly processing level that maximizes the monthly revenue.

c. Find the monthly processing level that maximizes the profit earned from selling the crab meat. At this processing level, what is the selling price and what is the monthly profit?

d. If the cost of processing each ton increases $2250 (2.25 thousand dollars), what is the monthly processing level that maximizes the profit earned from selling the crab meat? At this processing level, what is the selling price and what is the monthly profit?

26. **Optimal Production Levels** Allobind is a super strong, lightweight alloy. The total monthly cost of producing x tons of the alloy per month is

$$C(x) = 0.25x + 500$$

thousand dollars. The company that produces Allobind can sell x tons of the alloy monthly if the price is

$$D(x) = \frac{2700}{x + 3}$$

thousand dollars per ton. Whenever the company produces x tons per month, the company sets the price at $2700/(x + 3)$ thousand dollars per ton.

a. Show that the average cost of producing each ton decreases whenever the company increases the monthly production level. Thus, there is no monthly production level that minimizes the average cost of producing each ton.

b. Show that the monthly revenue obtained from selling the alloy increases whenever the company increases the monthly production level. Thus, there is no monthly production level that maximizes the monthly revenue.

c. Find the monthly production level that maximizes the profit earned from selling the alloy. At this production level, what is the selling price and what is the monthly profit?

d. If the cost of producing each ton increases $750 (0.75 of a thousand dollars), what is the monthly production level that maximizes the profit earned from selling the alloy? At this production level, what is the selling price and what is the monthly profit?

27. **Optimal Order Size** During the coming year, a retailer will sell 1024 cases of Nino's Spaghetti at a constant rate. The retailer will place equal-size orders of at most 512 cases that arrive at the moment the supply is exhausted. Nino charges the retailer a delivery fee of $16 per order, regardless of the size of the order. It costs the retailer $0.125 to store each case for 1 year. How many cases should the retailer order each time to minimize the combined cost of delivery and storage?

Suggestion: See Example 2.23.

28. **Optimal Order Size** During the coming month, a restaurant will use 2304 pounds of ground beef at a constant rate. The restaurant will place equal-size orders of at most 1152 pounds that arrive at the moment the supply is exhausted. The wholesaler charges the restaurant a delivery fee of $9 per order, regardless of the size of the order. It costs the restaurant $0.50 to store each pound for 1 month. How many pounds should the restaurant order each time to minimize the combined cost of delivery and storage?

Suggestion: See Example 2.23.

29. **Maximum Volume** Suppose equal squares are cut from each of the four corners of a square cardboard whose sides are 72 centimeters long. See the figure for this problem. The resulting flaps are then folded up to form a box without a top. How long should the sides of the four squares be to maximize the volume of the box?

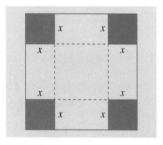

30. **Maximum Volume** Do Problem 29 again. This time, however, assume that each side of the cardboard is 80 centimeters long.

31. **Least Marginal Cost** Suppose the total weekly cost of producing x tons of a commodity per week is

$$C(x) = \tfrac{1}{15}x^3 - 4x^2 + 81x + 30$$

thousand dollars. Let $MC(x)$ represent the marginal cost at production level x. At low production levels, $MC(x)$ decreases if x is increased. However, at high production levels, $MC(x)$ increases if x is increased. Find the production level at which $MC(x)$ is least.

32. **Least Marginal Cost** Suppose the total monthly cost of producing x barrels of a product per month is

$$C(x) = \tfrac{1}{30}x^3 - 5x^2 + 251x + 75$$

thousand dollars. Let $MC(x)$ represent the marginal cost at production level x. At low production levels, $MC(x)$ decreases if x is increased. However, at high production levels, $MC(x)$ increases if x is increased. Find the production level at which $MC(x)$ is least.

33. **Maximum Profit** Suppose the total daily cost of producing x units of a product per day is

$$C(x) = \tfrac{1}{24}x^3 - 4x^2 + 129x + 60$$

dollars. Also, suppose x units of the product can be sold daily if the unit price is

$$D(x) = -0.5x + 159$$

dollars, where x is in the interval $[0, 318]$. Finally, suppose that whenever the daily production level is x units, the unit price is set at $-0.5x + 159$ dollars. Under these assumptions, the daily production level that maximizes the profit is the daily production level at which the marginal revenue is equal to the marginal cost. Use this fact to determine the daily production level that maximizes the profit.

34. **Minimum Average Cost** Suppose the total monthly cost of producing x tons of a product per month is

$$C(x) = 0.25x^2 + 16$$

thousand dollars. Under this assumption, the monthly production level that minimizes the average cost of producing each ton is the monthly production level at which the average cost of producing each ton is equal to the marginal cost. Use this fact to determine the monthly production level that minimizes the average cost of producing each ton.

35. **Minimum Average Cost** Suppose $A(x)$ is the average cost of producing each unit of a product when the daily production level is x units. See the graph of $A'(x)$ below. Explain how the graph of $A'(x)$ implies that there is a daily production level where the average cost of producing each unit is least.

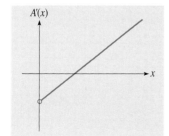

36. **Optimal Production Level** Suppose $P(x)$ is the weekly profit a firm earns when it produces x units of a product weekly. The figure for this problem shows the graph of $P'(x)$. Explain how the graph of $P'(x)$ implies that there is a weekly production level where the firm's weekly profit is greatest.

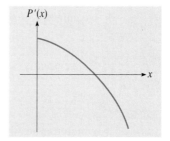

  a. Explain how the graphs are compatible with Theorem 2.13.

  b. Explain how the graphs are compatible with Theorem 2.15.

Graphing Calculator Problems

37. Set the viewing screen on your graphing calculator to $[0, 6] \times [-1, 3]$ with a scale of 1 on both axes. Then draw the graph of
$$f(x) = 0.025x^3 - 0.46x^2 + 2x$$
Also, draw the graph of the derivative of $f(x)$ so that it appears simultaneously with the graph of $f(x)$.

38. Do Problem 37 again. This time, however, let
$$f(x) = 0.3(\sqrt[3]{x})^5 - 1.2(\sqrt[3]{x})^2 + 1.3$$
and set the viewing screen to $[0.5, 3.5] \times [-1, 1]$ with a scale of 1 on both axes.

2.8 More on Approximating Change

In Section 2.4, we used calculus to approximate the amount $f(x)$ changes if x increases 1 unit. In this section, we generalize this use of calculus to include situations where the change in x is *not* a 1-unit increase.

Let $g(x)$ be the function whose graph is the curve in Figure 2.46. Also, let $T(x)$ be the linear function whose graph is the tangent to the graph of $g(x)$ at the point where $x = c$. In the figure, dy (read "dee y") represents the amount $T(x)$ changes if x changes from c to $c + \Delta x$. (Δx, read "delta x," represents the change in x.) The figure suggests that we can use dy to approximate $g(c + \Delta x) - g(c)$, which is the amount $g(x)$ changes if x changes from c to $c + \Delta x$. The quality of the approximation depends on how much the graph of $g(x)$ deviates from the tangent as x changes from c to $c + \Delta x$. Since the magnitude of Δx is relatively small, the graph of $g(x)$ does not deviate much from the tangent. So dy is a fairly good approximation for $g(c + \Delta x) - g(c)$.

Figure 2.46 The graph of $g(x)$ and its tangent at $(c, g(c))$.

The Differential of a Function

We can represent the slope of the line in Figure 2.46 by

$$\frac{dy}{\Delta x}$$

But, since the line is tangent to the graph of $g(x)$ at the point where $x = c$, we can represent its slope also by

$$g'(c)$$

Consequently,

$$\frac{dy}{\Delta x} = g'(c)$$

which implies that

$$dy = g'(c)\,\Delta x$$

This result motivates Definition 2.4.

Definition 2.4

The **differential** of a function $f(x)$ is

$$dy = f'(x)\,\Delta x$$

where Δx represents a real number.

So dy is a variable whose values depend on x and Δx according to the formula

$$dy = f'(x)\,\Delta x$$

In some books, the symbol dx is used in place of the symbol Δx. So, in such books, the formula for the differential is written as follows:

$$dy = f'(x)\,dx$$

Using the function $g(x)$, we illustrated the following fact:

> Suppose $f(x)$ is a function that is differentiable at $x = c$. If Δx is a number whose magnitude is small enough, then
>
> $$f'(c)\,\Delta x$$
>
> is a close approximation for the *amount* $f(x)$ changes if x changes from c to $c + \Delta x$.

The actual amount $f(x)$ changes if x changes from c to $c + \Delta x$ is

$$f(c + \Delta x) - f(c)$$

Of course, if x decreases, Δx will be a negative number.

In Example 2.24, we apply the differential concept to a medical problem.

Example 2.24

A physician believes that if one of his patients is administered x units of an experimental drug per day, the patient's systolic blood pressure will be maintained at

$$P(x) = \frac{115x + 254}{x + 2}$$

millimeters of mercury. Use the differential of $P(x)$ to approximate the amount the patient's systolic blood pressure changes if the daily dosage is decreased from 2 units to 1.1 units.

Solution

First note that if the daily dosage is decreased from 2 units to 1.1 units, it is decreased 0.9 of a unit. So $\Delta x = -0.9$.

Using the quotient rule, the derivative of $P(x)$ is

$$P'(x) = \frac{(x + 2)115 - (115x + 254)1}{(x + 2)^2}$$

$$= \frac{-24}{(x + 2)^2}$$

Thus, according to Definition 2.4, a formula for the differential of $P(x)$ is

$$dy = \frac{-24}{(x + 2)^2} \Delta x$$

Therefore, for $x = 2$ and $\Delta x = -0.9$, the value of the differential of $P(x)$ is

$$dy = \frac{-24}{(2 + 2)^2} (-0.9)$$

$$= 1.35$$

So $P(x)$ increases approximately 1.35 millimeters of mercury if x is decreased from 2 units to 1.1 units. In other words, the patient's systolic blood pressure increases about 1.35 millimeters of mercury if the patient's daily dosage of the drug is decreased from 2 units to 1.1 units.

Approximating Percentage Change

Sometimes we are interested in the percent the dependent variable of a function changes if the independent variable undergoes a change.

As we have seen, if the magnitude of Δx is small enough, then $f'(c)\,\Delta x$ is a good approximation for the amount $f(x)$ changes if x changes from c to $c + \Delta x$. Therefore, since $f'(c)\,\Delta x$ is

$$100\,\frac{f'(c)\,\Delta x}{f(c)}$$

percent of $f(c)$, we have the following result:

> Suppose $f(x)$ is differentiable at $x = c$ and $f(c)$ is positive. If the magnitude of Δx is small enough, then
>
> $$100\,\frac{f'(c)\,\Delta x}{f(c)}$$
>
> is a close approximation for the *percent* $f(x)$ changes if x changes from c to $c + \Delta x$.

The actual percent $f(x)$ changes if x changes from c to $c + \Delta x$ is

$$100\,\frac{f(c + \Delta x) - f(c)}{f(c)}$$

In Example 2.25, we approximate the change the independent variable of a function must undergo for the dependent variable to change a given percent.

Example 2.25

Presently, the concentration of EDB (ethylene dibromide) in Lake Blue Fly is 30 milligrams per kiloliter of water. Environmentalists believe the concentration will be

$$c(x) = \frac{60}{x + 2}$$

milligrams per kiloliter if \$$x$ million is spent in a cleanup project. Use calculus to approximate the amount of money that must be spent to reduce the concentration 20 percent.

Solution

The derivative of $c(x)$ is

$$c'(x) = \frac{-60}{(x + 2)^2}$$

Therefore,

$$100\,\frac{c'(x)\,\Delta x}{c(x)} = 100\,\frac{[-60/(x + 2)^2]\,\Delta x}{60/(x + 2)}$$

$$= \frac{-100\Delta x}{x + 2}$$

So the percent $c(x)$ changes as x increases from 0 to $0 + \Delta x$ is given approximately by

$$\frac{100\, c'(0)\, \Delta x}{c(0)} = \frac{-100 \Delta x}{0 + 2}$$
$$= -50 \Delta x$$

We want to find the value of Δx for which

$$-50 \Delta x = -20$$

(The -20 represents a 20 percent reduction in the concentration.) Dividing both sides of this equation by -50, we get

$$\Delta x = \frac{-20}{-50}$$
$$= 0.4$$

Therefore, since 0.4 of a million is 400,000, approximately \$400,000 must be spent to reduce the concentration of EDB 20 percent.

In Example 2.25, the *actual* amount that must be spent to reduce the concentration of EDB 20 percent is the value of Δx for which

$$100 \frac{c(0 + \Delta x) - c(0)}{c(0)} = -20$$

If we solve this equation for Δx, we would get

$$\Delta x = 0.5$$

Since 0.5 of a million is 500,000, the actual amount that must be spent is \$500,000.

Elasticity

Sometimes we are interested in the percent the dependent variable of a function changes if the independent variable *increases 1 percent*.

If we substitute $0.01c$ for Δx in the formula

$$100 \frac{f'(c)\, \Delta x}{f(c)}$$

we get

$$100 \frac{f'(c)(0.01c)}{f(c)}$$

which simplifies to

$$\frac{c f'(c)}{f(c)}$$

This result leads us to Definition 2.5.

Definition 2.5

Suppose c and $f(c)$ are positive. The **elasticity** of $f(x)$ with respect to x at $x = c$ is

$$E(c) = \frac{cf'(c)}{f(c)}$$

If there is no ambiguity regarding the independent variable, we often omit phrases such as the phrase "with respect to x" when we discuss elasticity.

Note that $0.01c$ is 1 percent of c. So, since the elasticity of $f(x)$ at $x = c$ is the special case of

$$100 \frac{f'(c) \Delta x}{f(c)}$$

in which $\Delta x = 0.01c$, we have the following result:

> If 1 percent of c is small enough, then the elasticity of $f(x)$ at $x = c$ is a close approximation for the percent $f(x)$ changes if x increases from c to an amount 1 percent greater.

In fact, if $f(x)$ is a *linear* function, the elasticity of $f(x)$ gives the exact percent change in $f(x)$ that results from a 1 percent increase in x. (You will be asked to prove this fact in a problem at the end of this section.)

The concept of elasticity is often applied to demand functions. Suppose $D(p)$ is the demand for a product during a specified period when its selling price is p. The elasticity of the demand is simply the elasticity of the function $D(p)$. In Example 2.26, we find the elasticity of a demand function at two price levels.

Example 2.26

Suppose the weekly demand for a ball-point pen is

$$D(p) = -0.4p + 28$$

thousand pens when its selling price is p dollars, where p is in the closed interval $[0, 70]$. Derive a formula for the elasticity of the demand. Then, for each of the following price increases, use the formula to approximate the corresponding percent decrease in demand.

a. The price increases from $30 to an amount 1 percent greater.
b. The price increases from $42 to an amount 1 percent greater.

Solution

Since $D'(p) = -0.4$,

$$\frac{pD'(p)}{D(p)} = \frac{p(-0.4)}{-0.4p + 28}$$

$$= \frac{4p}{4p - 280}$$

Thus, a formula for the elasticity of the demand is

$$E(p) = \frac{4p}{4p - 280}$$

where p is in the open interval $(0, 70)$.

Solution for (a)

The elasticity of the demand when $p = 30$ is

$$E(30) = \frac{4(30)}{4(30) - 280}$$

$$= -0.75$$

Therefore, the weekly demand decreases approximately 0.75 percent (less than 1 percent) if the price increases from \$30 to an amount 1 percent greater. (Actually, since the demand function is linear, this percent decrease in demand is exact.)

Solution for (b)

The elasticity of the demand when $p = 42$ is

$$E(42) = \frac{4(42)}{4(42) - 280}$$

$$= -1.5$$

Therefore, the weekly demand decreases approximately 1.5 percent (more than 1 percent) if the price increases from \$42 to an amount 1 percent greater.

In Example 2.26, the demand for the pen is more responsive to a price change when its price is \$42 than when its price is \$30. Note that the elasticity of the demand is negative at both prices. Why? Note also that in the formula for $E(p)$ we restricted the values of p to the open interval $(0, 70)$, even though the domain of the demand function $D(p)$ is the closed interval $[0, 70]$. We made this restriction to conform with the definition of elasticity (Definition 2.5), which requires p and $D(p)$ to be positive.

Inelastic and Elastic Demands

At prices where the elasticity of the demand for a product is between -1 and 0, the demand decreases less than 1 percent if the price increases 1 percent. Thus, since the revenue is the product of the demand and the price, at such prices the revenue increases as the price increases. At prices where the elasticity of the demand is less than -1, the demand decreases more than 1 percent if the price increases 1 percent. Thus, at such prices the revenue decreases as the price increases. Definition 2.6 focuses on the distinction between prices where the elasticity of the demand is greater than -1 and prices where the elasticity of the demand is less than -1.

Definition 2.6

> The demand for a product is **inelastic** at prices where the elasticity of the demand is *greater* than -1 and **elastic** at prices where the elasticity of the demand is *less* than -1.

We say the demand has **unit elasticity** at those prices where the elasticity is -1.

According to the above definition, the demand for the pen in Example 2.26 is inelastic when its price is $30 and elastic when its price is $42.

Example 2.27

For the pen in Example 2.26, find the prices where the demand is inelastic and the prices where the demand is elastic.

Solution

In Example 2.26, we found that the formula for the elasticity of the demand is

$$E(p) = \frac{4p}{4p - 280}$$

where p is in the interval $(0, 70)$. To find the prices where the demand is inelastic, we solve the inequality

$$\frac{4p}{4p - 280} > -1$$

Since $0 < p < 70$, it follows that $4p - 280$ is negative. Thus, multiplying both sides of the above inequality by $4p - 280$ reverses the direction of the inequality:

$$4p < (4p - 280)(-1) \quad \text{multiplying by } 4p - 280$$
$$4p < 280 - 4p \quad \text{since } (4p - 280)(-1) = 280 - 4p$$
$$8p < 280 \quad \text{adding } 4p$$
$$p < 35 \quad \text{dividing by 8}$$

Thus, the demand for the pen is inelastic at prices less than $35.
To find the prices where the demand is elastic, we solve the inequality

$$\frac{4p}{4p - 280} < -1$$

Since $4p - 280$ is negative, multiplying both sides by $4p - 280$ produces

$$4p > 280 - 4p$$
$$8p > 280$$
$$p > 35$$

Thus, the demand is elastic at prices greater than $35 (but less than $70).

In the problems at the end of this section, we will apply the idea of elasticity also in situations that do not involve demand functions.

2.8 Problems

In Problems 1 to 10, use the differential of $f(x)$ to approximate the amount $f(x)$ changes if x changes as indicated.

1. $f(x) = 0.25x^2 + x$ x increases from 8 to 8.4
2. $f(x) = 0.5x^2 + 5x$ x increases from 3 to 3.5
3. $f(x) = \dfrac{2x + 9}{3x + 1}$ x increases from 3 to 3.2
4. $f(x) = \dfrac{x + 8}{3x + 4}$ x increases from 2 to 2.3
5. $f(x) = \dfrac{48}{x}$ x decreases from 4 to 3.4
6. $f(x) = \dfrac{6}{x^2}$ x decreases from 2 to 1.75
7. $f(x) = 20\sqrt{2x + 7}$ x decreases from 9 to 8.25
8. $f(x) = 4\sqrt{3x + 1}$ x decreases from 5 to 4.7
9. $f(x) = x\sqrt{x^2 + 19}$ x increases from 9 to 9.6
10. $f(x) = x^2\sqrt{x + 9}$ x decreases from 7 to 6.8

In Problems 11 to 14, use calculus to approximate the percent $f(x)$ changes if x changes as indicated.

11. $f(x) = 0.04x^2 + 4$ x increases from 5 to 5.25
12. $f(x) = 0.05x^2 + 3$ x increases from 6 to 6.38
13. $f(x) = 5\sqrt{x}$ x decreases from 4 to 3.8
14. $f(x) = 8\sqrt{x}$ x decreases from 8 to 7.4

In Problems 15 to 18, find a formula for the elasticity of $f(x)$. Then use the formula to approximate the percent $f(x)$ changes if x increases from the given value to an amount 1 percent greater.

15. $f(x) = -0.2x + 50$ $x = 200$
16. $f(x) = -0.5x + 70$ $x = 40$
17. $f(x) = 480 - 0.3x^2$ $x = 20$
18. $f(x) = 1000 - 0.4x^2$ $x = 30$

In Problems 19 to 24, the given formula expresses the demand for a product as a function of its price. Determine whether the demand is inelastic, elastic, or has unit elasticity at the given price.

19. $D(p) = -p^2 + 400 \quad p = 15$

20. $D(p) = -p^2 + 400 \quad p = 10$

21. $D(p) = \dfrac{20p + 50}{p + 1} \quad p = 9$

22. $D(p) = \dfrac{400}{(p + 3)^2} \quad p = 5$

23. $D(p) = \dfrac{420}{p^2 + 144} \quad p = 12$

24. $D(p) = \dfrac{400}{(p + 3)^2} \quad p = 3$

25. **Unemployment Benefits** The government of a nation finds that when x million people are unemployed, the yearly expenditure for unemployment benefits is

$$B(x) = 4\sqrt{x}$$

billion dollars. Currently, 9 million people are unemployed. Use the differential of $B(x)$ to approximate how much the yearly expenditure for unemployment benefits drops if the number of unemployed people decreases by 750,000 (750,000 = 0.75 million).

26. **Factory Productivity** Suppose a shoe factory produces

$$P(x) = \sqrt{5x + 1}$$

thousand pairs of jogging shoes per week with a capital outlay of x million, where $x \geq 1$. The current capital outlay is $3 million. Use the differential of $P(x)$ to approximate how the weekly productivity responds to an increase of $500,000 in the capital outlay (500,000 = 0.5 million).

27. **Prison Population** In a country, 200,000 people are currently serving prison sentences. Taking into consideration a new probation policy, officials project that x years from now this number will be

$$P(x) = \dfrac{170x + 200}{x + 1}$$

thousand. Use the differential of $P(x)$ to approximate how much the prison population will change during the next 4 months.

A normal lung (left) and a smoker's lung.
(A. Glauberman/Science Source/Photo Researchers)

28. **Cigarette Smoking** The health officials of a nation are confident cigarette smoking will decline. They predict that x years from now

$$S(x) = \dfrac{20x + 30}{x + 1}$$

percent of the adults will smoke cigarettes. Use the differential of $S(x)$ to approximate the amount this percent will decrease during the 6-month period beginning 1 year from now.

29. **Demand for Protein Supplement** Protex is an inexpensive protein supplement. Officials in a Third World country claim that when the median weekly income per household is x hundred, the weekly demand for Protex is

$$D(x) = -\tfrac{1}{9}x^2 - 2x + 9$$

million tons, where $2 \leq x \leq 7$. The formula reflects the fact the demand decreases if the income increases. Use calculus to approximate the percent the weekly demand decreases if the median weekly income per household increases from $300 to $360.
Caution: The values of x represent hundred dollars.

30. **Consumption** Economists believe that Petroland, a smallish oil-producing country, consumes

$$c(x) = 0.75x + 0.2\sqrt{x}$$

billion dollars worth of goods per month when its monthly income is x billion, where $x \geq 1$. Use calculus to approximate the percent the monthly consumption would change if the monthly income increases $400 million from its current level of $16 billion (400 million = 0.4 billion).

31. **Peanut Yield** A Georgia farmer believes his land will produce

$$Y(x) = \frac{20x + 4}{x + 2}$$

tons of peanuts per year if he uses x thousand pounds of a special fertilizer. Use calculus to approximate the percent the yearly yield increases if he increases the fertilization level from 3000 pounds to 3800 pounds.

Caution: The values of x represent thousand pounds.

32. **Supply** Suppose producers supply the market with

$$S(x) = 2x\sqrt{x + 1}$$

hundred barrels of pear nectar per week when the nectar sells for $\$x$ per liter. Use calculus to approximate the percent the weekly supply decreases if the price decreases from $4.00 to $3.80 per liter.

33. **Marketing** The monthly demand for Sicilian Autumn, an expensive body soap, is

$$D(x) = -0.1x^2 + 4x + 5$$

million bars if the monthly advertising expenditure is $\$x$ million, where $0 \leq x \leq 20$. Currently, the monthly advertising expenditure is $10 million. The producer of the soap wants the monthly demand for the soap to increase 3 million bars. Use the differential of $D(x)$ to approximate the increase in the monthly advertising expenditure that generates this increase in demand.

34. **Production Cost** The producer of Sweetie, a sugar substitute, finds that the total monthly cost of producing x tons of the sweetener per month is

$$C(x) = 0.04x^2 + 6$$

thousand dollars. The current monthly production level is 40 tons. The producer wants the total monthly production cost to decrease $2 thousand. Use the differential of $C(x)$ to approximate the decrease in the monthly production level that results in this cost reduction.

35. **Marketing** Suppose the producer of the soap in Problem 33 wants the monthly demand for the soap to increase 2 percent, rather than 3 million bars. Use calculus to approximate the increase in the monthly advertising expenditure that generates this percent increase in demand.

36. **Production Cost** Suppose the producer of the sugar substitute in Problem 34 wants the total monthly production cost to decrease 4 percent, rather than $2 thousand. Use calculus to approximate the decrease in the monthly production level that results in this percent reduction in cost.

37. **Elasticity of Demand** The monthly demand for Block, a sunscreen lotion, is

$$D(p) = \frac{750}{p^2 + 81}$$

million liters when its selling price is $\$p$ per liter. Find the prices where the demand is inelastic and the prices where the demand is elastic.

38. **Elasticity of Demand** Angelo, the owner of Angelo's Pizzeria, believes he can sell

$$D(p) = \frac{2880}{(p + 12)^2}$$

hundred pizzas per week if he sells each pizza for $\$p$. Find the prices where the demand is inelastic and the prices where the demand is elastic.

39. **Elasticity of Demand** Suppose the daily demand for a product is

$$D(p) = \frac{576}{p + 62}$$

units if the selling price is $\$p$ per unit. Show that the demand is inelastic at every price.

40. **Elasticity of Demand** Suppose the weekly demand for a product is

$$D(p) = \frac{160}{\sqrt{p + 1}}$$

units when it sells for $\$p$ per unit. Show that the demand is inelastic at all price levels.

41. **Elasticity of Blood Sugar Level** A doctor assumes that if one of her patients is administered x

units of insulin per day, the patient's blood sugar will be maintained at

$$S(x) = \frac{60x + 1200}{x + 5}$$

milligrams per deciliter. Derive a formula for the elasticity of the blood sugar level with respect to the daily insulin dosage. Then use the formula to approximate the percent the blood sugar level changes if the daily insulin dosage is increased from 12 units to an amount 1 percent greater.

42. **Elasticity of Productivity** The director of a training program believes that if a new employee receives x hundred hours of training, the employee can produce an average of

$$Y(x) = \tfrac{1}{3}\sqrt{x^3 + 2}$$

units of a product per day. Find a formula for the elasticity of the daily productivity with respect to the training time. Then use the formula to estimate the percent the daily productivity increases if the training time is increased from 2 hundred hours to an amount 1 percent greater.

43. **Elasticity of Productivity** A firm produces

$$Q(x) = \frac{20x^2}{x^2 + 12}$$

thousand barrels of ketchup per day if it uses x units of labor. Find a formula for the elasticity of the daily productivity with respect to the labor force. Then use the formula to estimate the percent the daily productivity increases if the labor force is increased from 6 units to an amount 1 percent greater.

44. **Elasticity of Productivity** The management of a chemical firm believes that the firm can produce

$$Q(x) = 2\sqrt{x^3 + 200}$$

hundred barrels of an insecticide per day with a capital outlay of $\$x$ million, where $x \geq 1$. Find a formula for the elasticity of the daily productivity with respect to the capital. Then use the formula to estimate the percent the daily productivity increases if the capital outlay is increased from $10 million to an amount 1 percent greater.

45. **Percentage Error** The volume of a sphere is

$$V(r) = \tfrac{4}{3}\pi r^3$$

where r represents the radius. Suppose the radius of a spherical tank is measured with a possible error of 2 percent. The result is then substituted for r in the above formula. Use calculus to approximate the maximum percent the volume obtained through the formula could differ from the actual volume.

46. **Percentage Error** The top and bottom of the solid shown in the figure for this problem are equilateral triangles, while the sides are squares. Thus, all the edges are equal, and so the volume is given by the formula

$$V(x) = \frac{\sqrt{3}}{4} x^3$$

Suppose the value of x is obtained through a measurement with a possible error of 3 percent. Use calculus to approximate the maximum percent the volume obtained through the formula could differ from the actual volume.

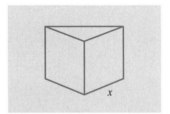

47. **Approximate Area** A woman who plans to paint the surface of a wooden deck that surrounds a circular pool needs to know (at least approximately) the area of the deck. See the figure for this problem. The radius of the pool is 90 feet and the deck is 2 feet wide. Use an appropriate differential to approximate the area of the deck.

Recall: The area of a circle is given by $A(r) = \pi r^2$, where r is the radius.

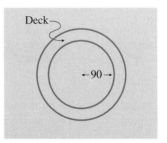

48. **Approximate Area** A 3-foot-wide strip at the margin of a circular field was paved to provide joggers with a place to run. See the figure for this problem. The field has a radius of 200 feet. Use an

appropriate differential to approximate the area of the paved strip. Note that the strip is inside the field.

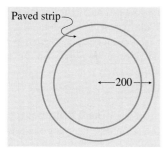

49. Suppose $f(x)$ is a *linear* function and Δx is *any* real number. Prove that
$$f'(c)\Delta x$$
is the *exact* amount $f(x)$ changes if x changes from c to $c + \Delta x$.

Hint: The derivative of a linear function is the slope of the line that represents the linear function.

50. Suppose $f(x)$ is a *linear* function and c is *any* positive real number for which $f(c)$ is also positive. Prove that the elasticity of $f(x)$ at $x = c$ is the *exact* percent $f(x)$ changes if x increases from c to an amount 1 percent greater.

Hint: The derivative of a linear function is the slope of the line that represents the linear function.

51. **Revenue and Elastic Demand** At prices where a product's demand is elastic, the revenue decreases as the price increases. Why?

52. **Revenue and Inelastic Demand** At prices where a product's demand is inelastic, the revenue increases as the price increases. Why?

 Graphing Calculator Problems

53. **Elasticity of Demand** Suppose the weekly demand for a product is
$$D(p) = 1800 - 0.25p^2$$
units if the selling price is $\$p$ per unit, where $20 \le p \le 60$. Set the viewing screen on your graphing calculator to $[20, 60] \times [-2, 0]$ with a scale of 5 on the horizontal axis and a scale of 1 on the vertical axis. Then draw the graph of $f(x) = -1$. Also, draw the graph of $E(p)$, the function that gives the elasticity of the product's demand, so that it appears simultaneously with the graph of $f(x)$. Finally, use the trace function to find the prices where the demand is inelastic and the prices where the demand is elastic.

54. **Elasticity of Demand** Do Problem 53 again. This time, however, suppose
$$D(p) = -0.3p + 42$$
where $40 \le p \le 90$, and set the viewing screen to $[40, 90] \times [-2, 0]$ with scales as in Problem 53.

Chapter 2 Important Terms

Section 2.1
tangent (67)

Section 2.2
derivative (76)
differentiable function (76)
difference quotient (76)
Leibniz notation (79)

Section 2.3
constant rule (84)
power rule (85)
sum and difference rule (86)

Section 2.4
marginal cost (93)
marginal revenue (93)
marginal productivity of labor (95)
marginal productivity of capital (95)

Section 2.5
power function (101)
generalized power rule (102)
product rule (104)
constant multiple rule (105)
quotient rule (106)
chain rule (108)

Section 2.7
interval on which a function
    increases (or decreases) (123)
sign chart (123)
high and low points on a graph (124)
critical point (124)

Section 2.8
the differential of a function (132)
amount change in f(x) (132)
percent change in f(x) (134)
elasticity (136)
inelastic and elastic demands (138)
unit elasticity (138)

Chapter 2 Review Problems

Section 2.2

1. Find the derivative by computing the limit of a difference quotient as h approaches zero.

   a. $f(x) = 5x^2 + 3$ b. $f(x) = x^3 - 2x$

   c. $f(x) = 7\sqrt{x}$ d. $f(x) = \dfrac{12}{x}$

Section 2.3

2. Find the derivative by applying the rules for computing derivatives.

   a. $f(x) = 7x^4$ b. $g(x) = -9x$

   c. $y = 13$ d. $f(x) = \dfrac{1}{x^4}$

   e. $f(x) = 14\sqrt{x}$ f. $y = \dfrac{15}{\sqrt[3]{x}}$

   g. $f(x) = 2x^3 - x + 5$ h. $g(x) = x^2 + \dfrac{3}{x}$

   i. $y = 6x - \dfrac{1}{\sqrt{x}}$ j. $f(x) = 8\sqrt{x} - \dfrac{1}{x^3} + 7$

3. Use the derivative to find the x coordinates of the points where the graph has horizontal tangents.

   a. $y = 2x^3 - 24x^2 + 72x - 13$

   b. $y = x^3 - 9x^2 + 30x$

   c. $y = 54x - 6(\sqrt{x})^3$

4. Find the slope of the tangent to the graph of

   $$f(x) = 7x - \dfrac{8}{x}$$

   at the point (2, 10).

Section 2.4

5. Use the derivative to approximate the amount $f(x)$ changes if x increases 1 unit as indicated. Then compute the actual change in $f(x)$.

   a. $f(x) = 5 + \dfrac{24}{\sqrt{x}}$ x increases from 4 to 5

   b. $f(x) = 0.02x^2 + 0.04x$ x increases from 6 to 7

6. **Marginal Revenue** Each week, a product is priced so that the amount produced during the week is sold that week. Thus, the weekly revenue obtained from selling the product can be viewed as a function of the weekly production level. Suppose the weekly revenue is given by

   $$R(x) = 250x - 0.125x^2$$

   when x units is produced per week. (This formula is valid up to $x = 2000$.) Compute the marginal revenue at each of the following production levels and interpret your answers.

   a. Production level 954

   b. Production level 1062

7. **Marginal Cost** The management of the company that produces Qwik Growth, a potent fertilizer, finds that the total weekly cost of producing x tons of the fertilizer per week is

   $$C(x) = 1.5x^3 - 189x^2 + 7940x + 5000$$

   dollars.

   a. Compute the marginal cost at production level 34 and interpret the result.

   b. The lowest point on the graph of the marginal cost function is the point where the tangent is horizontal. At what weekly production level is the marginal cost least? At this production level, what is the marginal cost and what is the total weekly production cost?

8. **Least Average Cost** Suppose the total monthly cost of producing x units of a product per month is given by

   $$C(x) = 0.04x^2 + 16$$

   Then the average cost of producing each unit is given by

   $$A(x) = \dfrac{C(x)}{x} = \dfrac{0.04x^2 + 16}{x} = 0.04x + \dfrac{16}{x}$$

   when x units is produced per month.

   a. Find the monthly production level at which the average cost of producing each unit is equal to the marginal cost.

b. The lowest point on the graph of the average cost function is the point where the tangent is horizontal. At what monthly production level is the average cost of producing each unit least? How does this production level compare with the production level found in part (a)?

Section 2.5

9. Find the derivative.

   a. $f(x) = (3x^2 + 5)^7$
   b. $y = 8\sqrt{3x + 4}$
   c. $g(x) = \dfrac{1}{x^2 + x - 8}$
   d. $f(x) = \dfrac{-6}{\sqrt[3]{x^2 + 9x}}$
   e. $y = \dfrac{2}{(4x^3 - x)^3}$
   f. $y = 5x(2x^3 - 4)^6$
   g. $f(x) = x\sqrt{x^3 - 2x}$
   h. $g(x) = (2x - 1)(7x + 3)^5$
   i. $y = (x + 7)^3(x^2 - 4x)^4$
   j. $f(x) = \dfrac{9x^2}{x^2 + 4}$
   k. $f(x) = \dfrac{3x + 4}{5x + 1}$
   l. $g(x) = \dfrac{4x}{x^3 - 8}$
   m. $f(x) = \dfrac{3x + 1}{(x + 2)^3}$
   n. $y = \dfrac{(2x - 8)^4}{7x + 1}$
   o. $g(x) = \left(\dfrac{5x - 1}{7x + 2}\right)^4$
   p. $f(x) = \dfrac{4x}{\sqrt{3x + 11}}$
   q. $f(x) = \sqrt{\dfrac{x + 3}{5x - 1}}$
   r. $f(x) = \dfrac{x^2 - 5x + 15}{x^3 + 8}$

10. **Marginal Productivity** Tina Rono, the owner of The Pasta Factory, believes her company can produce
    $$Q(w) = 12\sqrt{3w + 4}$$
    tons of pasta per month with w hundred hours of labor. Find the marginal productivity of labor when Tina uses 20 hundred hours of labor. Then interpret the result.

11. **Demand** The management of the company that distributes Green Springs Bottled Water finds that the monthly demand for the water is
    $$D(x) = \dfrac{6x + 5}{2x + 1}$$
    million gallons when its price is $\$x$ per gallon, provided x is in the interval $[1, 3]$. Use the derivative of $D(x)$ to approximate the decrease in the monthly demand if the price is increased from $\$1.50$ to $\$2.50$ per gallon.

Section 2.6

12. Each of the following functions is differentiable on its domain, which is the accompanying closed interval. Find the highest and lowest points on the graph.

    a. $f(x) = x^3 - 12x^2 + 36x + 5$ $[1, 10]$
    b. $f(x) = 2x^3 - 18x^2 + 30x - 10$ $[0, 4]$
    c. $f(x) = \dfrac{120}{x^2 - 8x + 20}$ $[2, 10]$
    d. $f(x) = \sqrt{x^2 - 14x + 64}$ $[0, 6]$

13. **Fossil Fuels Dependency** At the beginning of 1990, 76 percent of the energy used in a northern European country was derived from fossil fuels. At that time, a group concerned about global warming predicted that x years later this percent would be
    $$P(x) = 0.9x - 0.2(\sqrt{x})^3 + 76$$
    for x in the closed interval $[0, 30]$. According to the prediction, when during the three decades following the beginning of 1990 is the percent highest? How high is the percent at that time?

14. **Use of Steel-Producing Facilities** An Asian country presently uses 65 percent of its steel-producing facilities. Suppose x weeks from now it uses
    $$f(x) = 70 - \sqrt{x^2 - 8x + 25}$$
    percent of those facilities, where x is in the closed interval $[0, 10]$. When during the next 10 weeks will it make the most use of the facilities? What percent of the facilities will it use at that time?

15. **Optimal Work Force** A paint contractor has agreed to paint 500,000 square feet of walls for a client. The contractor will charge the client a fee of $\$50$ per painter plus $\$75$ per hour, regardless of how many painters the client decides to use. If no more than 20 painters are available and each painter paints at the rate of 1200 square feet per hour, how many painters should the client request?

Section 2.7

16. Find the intervals on which $f(x)$ increases and the intervals on which it decreases. Then identify the high and low points on the graph.

a. $f(x) = 2x^3 - 24x^2 + 72x + 6$

b. $f(x) = x^3 - 6x^2 + 12x$

c. $f(x) = -6x^4 + 8x^3 + 144x^2 + 5$

d. $f(x) = x^3 + 2x^2 + 7x$

e. $f(x) = -8x^3 + 33x^2 - 15x$

f. $f(x) = x^5 - \frac{40}{3}x^3 - 45x$

g. $f(x) = x^7 - 217x^4 - 875x + 20$

h. $f(x) = 0.25x^4 - 6x^3 + 40.5x^2 - 5$

i. $f(x) = \dfrac{5}{(x-7)^2}$

j. $f(x) = \dfrac{3x^2}{x-2}$

k. $f(x) = \dfrac{8x}{x^2 + 49}$

l. $f(x) = \dfrac{5}{x^2 - 12x + 37}$

m. $f(x) = \dfrac{x - 13}{x^2 - 25}$

17. **Maximum Infection Rate** A communicable disease was recently introduced to the inhabitants of an island in the South Atlantic. Presently, 5000 people have contracted the disease. Health officials predict that the number of people who contract the disease will never reach 120,000. The officials also predict that when x thousand people have contracted the disease, the people will be infected at the rate of

$$R(x) = 0.3x - 0.0025x^2$$

thousand per week, where x is in the half-open interval [5, 120). Use calculus to find how many people will have contracted the disease when the disease spreads most rapidly. Then find how rapidly it will spread when it spreads most rapidly.

18. **Maximum Carbon Emission Rate** At the beginning of 1992, a developing country was emitting carbon to the atmosphere (mostly through deforestation) at the rate of 300 million tons per year. At that time, experts projected that this rate would rise gradually before beginning a downward trend. When does the rate peak if x years after the beginning of 1992 the rate is

$$R(x) = \dfrac{25x + 100}{x^2 + 8x + 80} + 1.75$$

million tons per year? What is the rate when it peaks?

19. **Optimal Order Size** During the coming year, a nursery will use 3200 gallons of a biodegradable insecticide at a constant rate. The nursery will place equal-size orders of at most 1600 gallons that arrive at the moment the supply is exhausted. The chemical firm that produces the insecticide charges a delivery fee of $25, regardless of the size of the order. It costs the nursery $0.25 to store each gallon for 1 year. How many gallons should the nursery order each time to minimize the combined cost of delivery and storage?

20. **Optimal Production Level** The producers of Permoil, a motor oil that never requires changing, find that the total monthly cost of producing x cases of Permoil per month is

$$C(x) = 25x + 187{,}500$$

dollars. The producers also find that they can sell x cases of the oil monthly if the price is

$$D(x) = \dfrac{4{,}000{,}000}{x + 2500}$$

dollars per case. Whenever they produce x cases of the oil per month, they set the price at $4{,}000{,}000/(x + 2500)$ dollars per case.

a. Show that the average cost of producing each case decreases whenever the producers increase the monthly production level. Thus, there is no monthly production level that minimizes the average cost of producing each case.

b. Show that the monthly revenue obtained from selling the oil increases whenever the producers increase the monthly production level. Thus, there is no monthly production level that maximizes the monthly revenue.

c. Find the monthly production level that maximizes the profit earned from selling the oil. At this production level, what is the selling price and what is the monthly profit?

Section 2.8

21. Use the differential of $f(x)$ to approximate the amount $f(x)$ changes if x changes as indicated. Then compute the actual change in $f(x)$.

a. $f(x) = 0.24x^2 + 7x$ x increases from 9 to 9.6

b. $f(x) = \dfrac{x + 50}{2x + 4}$ x increases from 2 to 2.43

c. $f(x) = 32\sqrt{x^2 + 15}$ x decreases from 7 to 6.8

d. $f(x) = \dfrac{24}{\sqrt{x}}$ x decreases from 4 to 3.1

22. Use calculus to approximate the percent $f(x)$ changes if x changes as indicated. Then compute the actual percent change in $f(x)$.

a. $f(x) = 80 - 0.32x^2$ x decreases from 10 to 9.4

b. $f(x) = 0.04x^3$ x decreases from 50 to 49.2

c. $f(x) = 2x\sqrt{x + 1}$ x increases from 4 to 4.2

d. $f(x) = \dfrac{64}{x^2}$ x increases from 25 to 25.72

23. Find a formula for the elasticity of $f(x)$. Then use the formula to approximate the percent $f(x)$ changes if x increases from the given value to an amount 1 percent greater.

a. $f(x) = \dfrac{800}{(x + 7)^2}$ $x = 13$

b. $f(x) = \dfrac{5x}{x + 2}$ $x = 6$

24. **Elasticity of Demand** Suppose the monthly demand for a product is

$$D(p) = 720 - 0.6p^2$$

if its price is p. Determine whether the demand is inelastic, elastic, or has unit elasticity at each of the following prices.

a. $p = 15$

b. $p = 24$

c. $p = 20$

25. **Small-Company Employees** In a European country, 42 million workers are employed by small companies. (A small company is a company that employs fewer than 50 workers.) Economists expect this number to increase. In fact, they project that x years from now the number of small-company workers will be

$$W(x) = \dfrac{46x + 84}{x + 2}$$

million.

a. Use the differential of $W(x)$ to approximate how much the number of small-company workers will increase during the next 9 months.

b. Use calculus to approximate the percent the number of small-company workers will increase during the next 9 months.

26. **Medicine** A doctor believes that if a patient is administered x units of insulin per day, the patient's blood sugar will be maintained at

$$S(x) = \dfrac{65x + 700}{x + 2}$$

milligrams per deciliter. Currently, the patient receives 6 units of insulin per day. Thus, since $S(6) = 136.25$, the patient's blood sugar is currently maintained at 136.25 milligrams per deciliter.

a. Use the differential of $S(x)$ to approximate the insulin dosage increase that lowers the patient's blood sugar from its current level to 130 milligrams per deciliter. (Round your answer to the nearest tenth of a unit.)

b. Use calculus to approximate the insulin dosage decrease that raises the patient's blood sugar 5 percent from its current level. (Round your answer to the nearest tenth of a unit.)

27. **Elasticity of Demand** Zap is a new fuel that increases the efficiency of internal-combustion engines 20 percent. The producers of Zap assume its monthly demand is

$$D(p) = \dfrac{13{,}520}{p^2 + 2704}$$

million liters when its selling price is p cents per liter. Find the prices where the demand is inelastic and the prices where the demand is elastic.

28. **Elasticity of Productivity** The director of a training program believes that if a new employee receives x days of training, the employee can produce an average of

$$P(x) = 4\sqrt{x^3 + 2000}$$

units of a product per week. Derive a formula for the elasticity of the weekly productivity with respect to the training time. Then use the formula to estimate the percent increase in productivity if the training time is increased from 20 days to an amount 1 percent greater.

CHAPTER 3
More on Differentiation

- **3.1** Instantaneous Rates
- **3.2** The Second Derivative and Curve Sketching
- **3.3** More on the Second Derivative
- **3.4** Two Useful Rational Functions
- **3.5** More on Rational Functions
- **3.6** Percentage Rate of Change
- **3.7** Newton's Method for Approximating Zeros

Important Terms

Review Problems

In Chapter 2, we interpreted derivatives as slopes of tangents. In this chapter, we interpret derivatives as rates of change.

In Chapter 2, we used derivatives to find where functions increase and where they decrease. In this chapter, we use derivatives to find where functions change at an increasing rate and where they change at a decreasing rate. Geometrically, in Chapter 2, we used derivatives to find where graphs rise and where they fall. In this chapter, we use derivatives to find where graphs curve upward and where they curve downward.

In Section 3.7, we present a method for approximating where graphs cross the horizontal axis.

3.1 Instantaneous Rates

In Section 1.1, we saw that if $[a, b]$ is a closed interval contained in the domain of a function $f(x)$, the *average rate of change of $f(x)$* with respect to x as x increases from a to b is

$$\frac{f(b) - f(a)}{b - a}$$

In this section, we introduce the idea of rate of change of $f(x)$ with respect to x *at particular values of x*. We begin with a problem that involves the effect of gravity on an object.

From a point 15 feet above the ground, a calculus student throws a ball straight upward with an initial velocity of 92 feet per second. Physicists tell us that under these conditions, the height of the ball is

$$g(t) = -16t^2 + 92t + 15$$

feet t seconds after it is thrown. We want to determine when the ball reaches its maximum height.

Since $g(t)$ is a quadratic function, we can solve this problem by locating the vertex of a parabola. Or we can solve it by finding the point on the graph of $g(t)$ where the tangent is horizontal. However, to reveal another interpretation of derivatives, we use a new approach.

The student clocks the upward motion of the ball with a stopwatch. According to this measurement, the ball appears to reach its maximum height 3 seconds after it is thrown. But, according to the assumption that its height is $g(t) = -16t^2 + 92t + 15$ feet t seconds after it is thrown, the ball reaches its maximum height in slightly less time, as we will show.

Instantaneous Velocity

Under the assumption the height of the ball is given by the function $g(t)$, let $v(t)$ represent its *velocity*, in feet per second, at the *instant t* seconds after it is thrown. So, in particular, $v(3)$ represents its velocity at the instant 3 seconds after it is thrown. We will show that $v(3)$ is not zero. This result will allow us to conclude that the ball does not reach its maximum height 3 seconds after it is thrown, because at the instant it reaches its maximum height its velocity has to be zero. (Why?) Then we will find the value of t for which $v(t) = 0$. This value of t is the instant at which the ball reaches its maximum height.

Let h be a positive real number. When t increases from 3 to $3 + h$, the ball's height changes by

$$g(3 + h) - g(3)$$

feet. Therefore, the *average rate of change* of the ball's height with respect to time during those h seconds is

$$\frac{g(3 + h) - g(3)}{h}$$

feet per second. We call this average rate of change the **average velocity** of the ball during the interval of time from $t = 3$ to $t = 3 + h$.

Our intuition tells us that as h approaches zero, the corresponding average velocities get closer and closer to the velocity of the ball at the instant $t = 3$, which, as we agreed, is represented by $v(3)$. In fact, as h approaches zero (from either side and without reaching it), the corresponding values of the quotient

$$\frac{g(3 + h) - g(3)}{h}$$

eventually get and stay arbitrarily close to $v(3)$. We express this fact by writing

$$v(3) = \lim_{h \to 0} \frac{g(3 + h) - g(3)}{h}$$

Actually, the above result is valid for any permissible value of t, not just $t = 3$. In other words, for each permissible value of t,

$$v(t) = \lim_{h \to 0} \frac{g(t + h) - g(t)}{h}$$

Since the formula obtained by computing this limit is a formula for the derivative of $g(t) = -16t^2 + 92t + 15$ and since the derivative of $g(t)$ is

$$g'(t) = -32t + 92$$

we conclude that

$$v(t) = -32t + 92$$

Substituting 3 for t in the above formula, we get

$$v(3) = -32(3) + 92$$
$$= -4$$

Thus, the velocity of the ball is -4 feet per second at the instant $t = 3$. This result means that the height is *decreasing* (and so the ball is falling) at the rate of 4 feet per second at the instant $t = 3$. Therefore, the ball does not reach its maximum height 3 seconds after it is thrown, because at the instant it reaches its maximum height its velocity has to be zero.

We can use the formula $v(t) = -32t + 92$ to determine the velocity of the ball at any instant during its flight, not just at the instant $t = 3$. For example, at the instant $t = 2$, its velocity is

$$v(2) = -32(2) + 92$$
$$= 28$$

feet per second. Therefore, at that instant, the height of the ball is *increasing* (and so the ball is rising) at the rate of 28 feet per second.

We can also use the formula

$$v(t) = -32t + 92$$

to determine when the ball reaches its maximum height. Since the ball reaches its maximum height when its velocity is zero, we set the formula equal to zero and solve for x:

$$-32t + 92 = 0$$
$$-32t = -92$$
$$t = 2.875$$

Thus, $v(t) = 0$ at $t = 2.875$. This result tells us the ball reaches its maximum height 2.875 seconds after it is thrown.

Instantaneous Rate of Change

The velocity of the ball at an instant is an example of an instantaneous rate of change.

Definition 3.1

The **rate of change** of $f(x)$ with respect to x at $x = c$ is

$$f'(c)$$

Such a rate of change is often called an **instantaneous rate of change** (in contrast to an average rate of change), even if x does not represent time. If x does represent time, $f'(c)$ is a measure of how *fast $f(x)$* changes at the instant $x = c$.

According to the above definition, if

$$f(x) = 0.08x^2 + 2$$

then a formula for the instantaneous rate of change of $f(x)$ with respect to x is

$$f'(x) = 0.16x$$

In particular, since

$$f'(4) = 0.16(4)$$
$$= 0.64$$

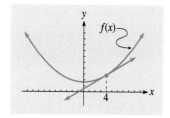

Figure 3.1 A line whose slope is an instantaneous rate of change.

the instantaneous rate of change at $x = 4$ is 0.64. Geometrically, this rate of change is the slope of the tangent to the graph of $f(x)$ at the point where $x = 4$. Figure 3.1 shows this tangent. Of course, we can also think of 0.64 as an approximation for the amount $f(x)$ changes if x increases from 4 to 5.

In Example 3.1, we find how fast a drug leaves a patient's bloodstream at a particular instant.

Example 3.1

A physician has just injected 200 units of a drug into a patient's bloodstream. Suppose that x hours from now the amount of the drug in the bloodstream is

$$D(x) = \frac{200}{\sqrt{x+1}}$$

units. How fast will the amount of the drug be decreasing 3 hours from now?

Solution

Since

$$\frac{200}{\sqrt{x+1}} = 200(x+1)^{-1/2}$$

we can write

$$D(x) = 200(x+1)^{-1/2}$$

Using the generalized power rule,

$$D'(x) = -100(x+1)^{-3/2}(1)$$
$$= \frac{-100}{(\sqrt{x+1})^3}$$

Thus, x hours from now the amount of the drug will be changing at the rate of

$$D'(x) = \frac{-100}{(\sqrt{x+1})^3}$$

units per hour. In particular, since

$$D'(3) = \frac{-100}{(\sqrt{3+1})^3}$$
$$= -12.5$$

the amount of the drug will be decreasing at the rate of 12.5 units per hour at the instant 3 hours from now.

In Example 3.1,

$$D(3) = \frac{200}{\sqrt{3+1}}$$
$$= 100$$

So there will be 100 units of the drug in the patient's bloodstream when its amount is decreasing at the rate of 12.5 units per hour.

Example 3.2 illustrates that the idea of instantaneous rate of change can be used with a function whose independent variable is not time.

Example 3.2

The amount of tomatoes a Central American country exports monthly to the United States depends on the price the United States agrees to pay for the tomatoes. To be more specific, the country exports

$$S(x) = 0.125x^2 + x$$

million cartons of tomatoes monthly when the United States pays x dollars for each carton. At what rate does the amount exported monthly increase with respect to price when the price is $4 per carton?

Solution

A formula for the rate at which the amount exported monthly increases with respect to price is

$$S'(x) = 0.25x + 1$$

Since $S'(4) = 2$, the amount exported monthly increases at the rate of 2 million cartons per dollar increase in price when the price is $4.

Related Rates

Suppose y is a differentiable function of x, and x, in turn, is a differentiable function of t, where t represents time. Then y is also a differentiable function of t. If we have a formula for y in terms of x, we can use an alternative version of the chain rule in Section 2.5 to derive a formula for dy/dt (the rate of change of y with respect to t). The derived formula involves dx/dt (the rate of change of x with respect to t). Thus, the two rates are *related* according to the formula. Before we present an example in which two rates are related, we restate the chain rule using the Leibniz notation for derivatives.

Chain Rule (Alternative Version)

Suppose y is a differentiable function of x, and x is a differentiable function of t. Then

$$\frac{dy}{dt} = \frac{dy}{dx} \cdot \frac{dx}{dt}$$

In Example 3.3, we use the above version of the chain rule to solve a problem about air pollution.

Example 3.3

Let y represent an index number that measures the air pollution level in a North American city. Environmentalists believe that y is a function of the city's population. In fact, they project that

$$y = \frac{7x + 8}{x + 4}$$

where x is the number of people, in millions, that will reside in the city t years from now.

a. How fast will the pollution level be changing when the population is 6 million, if at that time the population increases at the rate of 0.75 of a million per year? (Assume x is a differentiable function of t.)

b. How fast will the pollution level be changing 5 years from now, if the population is

$$x = 0.08t^2 + 2$$

million t years from now?

Solution for (a)

Using the quotient rule,

$$\frac{dy}{dx} = \frac{(x+4)7 - (7x+8)}{(x+4)^2}$$

$$= \frac{20}{(x+4)^2}$$

So, thinking of y as a function of t and applying the alternative version of the chain rule to the formula for y, we get

$$\frac{dy}{dt} = \frac{20}{(x+4)^2} \frac{dx}{dt}$$

Since the population increases at the rate of 0.75 of a million per year when it is 6 million,

$$\frac{dx}{dt} = 0.75$$

when

$$x = 6$$

Substituting these values into the formula for dy/dt, we obtain

$$\frac{dy}{dt} = \frac{20}{(6+4)^2}(0.75)$$
$$= 0.15$$

This result tells us that the pollution level will be increasing at the rate of 0.15 of a unit per year when the population is 6 million.

Solution for (b)

By substituting 5 for t into the formula

$$x = 0.08t^2 + 2$$

we obtain

$$x = 0.08(5^2) + 2$$
$$= 4$$

So 5 years from now, the population will be 4 million.
Note that

$$\frac{dx}{dt} = 0.16t$$

Thus, when $t = 5$,

$$\frac{dx}{dt} = 0.16(5)$$
$$= 0.8$$

So 5 years from now, the population will be increasing at the rate of 0.8 of a million per year.
From part (a), we know that

$$\frac{dy}{dt} = \frac{20}{(x+4)^2}\frac{dx}{dt}$$

By substituting 4 for x and 0.8 for dx/dt into this formula, we get

$$\frac{dy}{dt} = \frac{20}{(4+4)^2}(0.8)$$
$$= 0.25$$

Therefore, 5 years from now, the pollution level will be increasing at the rate of 0.25 of a unit per year.

3.1 Problems

In Problems 1 to 14, find a formula for the instantaneous rate of change of $f(x)$ with respect to x. Then compute the instantaneous rate of change at the given x value.

1. $f(x) = 0.5x^2 + 2x + 2$ $x = 5$
2. $f(x) = -x^2 + 36$ $x = 4$
3. $f(x) = 9 + \dfrac{16}{x^2}$ $x = 4$
4. $f(x) = 3x - \dfrac{98}{x} + 95$ $x = 7$
5. $f(x) = 6\sqrt{x}$ $x = 16$
6. $f(x) = 512 - 8(\sqrt{x})^3$ $x = 9$
7. $f(x) = \dfrac{1}{60}(x + 2)^3$ $x = 8$
8. $f(x) = 5\sqrt{2x + 4}$ $x = 6$
9. $f(x) = \dfrac{500}{(2x + 3)^2}$ $x = 1$
10. $f(x) = \dfrac{27}{4x + 1}$ $x = 2$
11. $f(x) = \dfrac{15x + 16}{x + 2}$ $x = 5$
12. $f(x) = \dfrac{7x + 45}{x + 3}$ $x = 1$
13. $f(x) = x(2x + 1)^3$ $x = 0.5$
14. $f(x) = 3x(5x + 2)^4$ $x = 0.2$

15. **Depreciation Rate** An accountant projects that x years from now a machine will be worth

$$V(x) = \dfrac{15x + 200}{x + 2}$$

thousand dollars. Find a formula for the rate at which the value of the machine will be changing x years from now. Then find the rate at which it will be depreciating 3 years from now. How much will the machine be worth at that time?

16. **Consumer Debt** Presently, the consumer debt in a country is $21 billion. However, the debt is declining gradually. Economists predict that t months from now the debt will be

$$d(t) = 12 + \dfrac{18}{\sqrt{t + 4}}$$

billion dollars. Find a formula for the rate at which the debt will be changing t months from now. Then compute the rate at which it will be declining 5 months from now. How large will the debt be at that time?

17. **Home Affordability** In a midwestern city, the number of families able to purchase homes costing at least $75,000 is decreasing. Suppose that t years from now the number of such families is

$$f(t) = \dfrac{25t + 60}{t + 2}$$

million. Find a formula that expresses how fast this number will be changing t years from now. Then determine how fast it will be decreasing 2 years from now.

18. **Prime Rate** In a country, the prime interest rate is rising. Suppose that x weeks from now the rate is

$$I(x) = \dfrac{22x + 54}{x + 3}$$

percent. Find a formula that expresses how rapidly the prime interest rate will be rising x weeks from now. Then determine how rapidly it will be rising 1 week from now.

19. **Population** A demographer believes that x years from now the population of Siliconia will be

$$P(x) = \dfrac{60x}{x^2 + 25} + 80$$

thousand.

a. At what rate will the population be changing 2 years from now? Will the population be growing at that time?

b. At what rate will the population be changing 7 years from now? Will the population be growing at that time?

20. **Unemployment Rate** Officials project that x months from now the unemployment rate in a southern city will be

$$U(x) = 12 - \dfrac{3}{x^2 - 16x + 65}$$

percent.

a. How fast will the unemployment rate be changing 5 months from now? Will it be increasing at that time?

b. How fast will the unemployment rate be changing 9 months from now? Will it be increasing at that time?

21. **Advertising** Suppose the weekly demand for Nutzy Peanut Butter is

$$D(x) = 20\sqrt{5x + 1600}$$

tons if $x thousand per week is spent on advertising. Find a formula for the rate of increase of the weekly demand with respect to the weekly advertising expenditure. Then compute the rate of increase when the weekly advertising expenditure is $85,000.

22. **Farmland Productivity** A farmer finds that each acre of his land produces

$$Y(x) = \frac{80x^2 + 100}{x^2 + 2}$$

bushels of corn if x thousand pounds of a fertilizer is applied to each acre. Find a formula for the rate of increase in the productivity of each acre with respect to the amount of fertilizer applied. Then compute the rate of increase when 2000 pounds of fertilizer is applied.

23. **Related Rates** A company has just started marketing a new product. Let x represent the number of units of the product that will be sold during the first t months. The management projects that the company will earn a profit of

$$y = 2\sqrt{3x + 4} - 6$$

thousand dollars from the sale of the first x units.

a. How rapidly will the company be earning the profit when it has sold 20 units of the product, if at that time the product is selling at the rate of 5 units per month? (Assume that x is a differentiable function of t.)

b. How rapidly will the company be earning the profit 16 months from now, if the company sells

$$x = 0.125t^2$$

units of the product during the first t months?

24. **Related Rates** So far, 0.125 of a million people in a South American country have contracted a communicable disease. Let x represent the total number of people (in millions) that will have contracted the disease by the time t more years elapse. Health officials believe that the total cost of the disease is

$$y = 0.12x^2$$

million dollars if x million people contract the disease.

a. How fast will the total cost be increasing when 5 million people have contracted the disease, if at that time the people are being infected at the rate of 0.75 of a million per year? (Assume that x is a differentiable function of t.)

b. How fast will the total cost be increasing 4 years from now, if a total of

$$x = \frac{109t^2 + 4}{8t^2 + 32}$$

million people contract the disease by the time t more years elapse.

25. **Sliding Plank Problem** The figure for this problem shows a 15-foot plank leaning against a wall. Suppose someone slides the plank's base toward the wall at the *constant rate* of 0.5 of a foot per second.

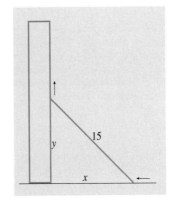

a. How fast will the plank's top be sliding up the wall when the plank's base is 12 feet from the wall?
Hint: Use the Pythagorean theorem to find a formula that expresses y as a function of x. Then use the fact that x is a differentiable function of t,

where t is the time that elapses after the sliding begins.

b. How fast will the plank's top be sliding up the wall when the plank's base is 9 feet from the wall?

c. Does the plank's top slide up the wall at a constant rate?

26. **Sphere Problem** The volume of a sphere decreases at the *constant rate* of 20 cubic meters per minute.

   a. How fast will the radius be decreasing when the radius is 5 meters? (The volume of the sphere is $v = \frac{4}{3}\pi r^3$, where r is the radius t minutes after the volume began decreasing.)

   b. How fast will the radius be decreasing when the radius is 3 meters?

   c. Does the radius decrease at a constant rate?

27. **Atmospheric Pollution** The concentration of N_2O (nitrous oxide) in the earth's atmosphere was about 302 parts per billion at the beginning of 1990. Suppose that x years after the beginning of 1990 the N_2O concentration is

   $$C(x) = 0.011x^2 + 302$$

   parts per billion. Find the average rate at which the concentration will increase during the first half of the 21st century. This average rate of increase will be equal to the concentration's instantaneous rate of increase at some instant in the first half of the 21st century. Find such an instant. Then interpret your answers geometrically.

28. **Atmospheric Pollution** The concentration of CO_2 (carbon dioxide) in the earth's atmosphere was about 345 parts per million at the beginning of 1990. (The preindustrial concentration was only 275 parts per million.) Suppose that x years after the beginning of 1990 the CO_2 concentration is

   $$C(x) = 0.0256x^2 + 345$$

   parts per million. Find the average rate at which the concentration will increase during the first quarter of the 21st century. This average rate of increase will be equal to the concentration's instantaneous rate of increase at some instant in the first quarter of the 21st century. Find such an instant. Then interpret your answers geometrically.

29. **Vertical Motion** From 6.5 feet above the ground, a ball is thrown straight upward with an initial velocity of 50 feet per second. Then, taking the upward direction as the positive direction of motion, its height is

   $$h(t) = -16t^2 + 50t + 6.5$$

   feet t seconds after it is thrown.

   a. Find the velocity of the ball at the instant 1.5 seconds after it is thrown. In which direction is it traveling at that instant?

   b. Find the velocity of the ball at the instant 3 seconds after it is thrown. In which direction is it traveling at that instant?

   c. When does the ball reach its maximum height? How high is it at that instant?

   d. The speed of the ball at an instant is the magnitude (absolute value) of its velocity at that instant. Find the speed at which the ball strikes the ground.

30. **Vertical Motion** Do Problem 29 again. This time assume the ball is thrown straight upward from 36 feet above the ground with an initial velocity of 64 feet per second. According to this assumption, its height is

   $$h(t) = -16t^2 + 64t + 36$$

   feet t seconds after it is thrown.

31. **Circle Problem** The radius of a circle is increasing. Thus, its area and its circumference likewise are increasing. Compute the instantaneous rate of increase of its area with respect to its radius when its radius is 10 meters. Then compare your answer with its circumference when its radius is 10 meters.

32. **Sphere Problem** The radius of a sphere is increasing. Thus, its volume and its surface area likewise are increasing. Compute the instantaneous rate of increase of its volume with respect to its radius when its radius is 10 meters. Then compare your answer with its surface area when its radius is 10 meters.

   Note: Formulas for the volume and surface area are, respectively,

$$V(r) = \tfrac{4}{3}\pi r^3 \quad \text{and} \quad A(r) = 4\pi r^2$$

where r represents the radius.

33. **Population Growth Rate** Suppose $P(x)$ is a population x years from now. The figure for this problem shows the graph of $P(x)$. Explain how the graph of $P(x)$ tells us that the population will increase at a decreasing rate.

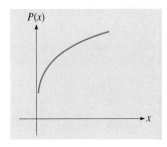

34. **Depreciation Rate** Suppose $V(x)$ is the value of an automobile when it is x years old. The figure for this problem shows the graph of $V(x)$. Explain how the graph of $V(x)$ tells us that the automobile's value at first decreases at an increasing rate and eventually decreases at a decreasing rate.

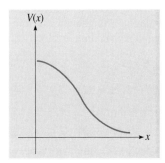

Graphing Calculator Problems

35. **Population Growth Rate** A demographer believes that x years from now the population of an island in the South Pacific will be

$$P(x) = \frac{5x^2 + 730}{0.1x^2 + 24.3}$$

million. Set the viewing screen on your graphing calculator to $[0, 25] \times [0, 1]$ with a scale of 1 on both axes and draw the graph of $P'(x)$. Then explain how the graph of $P'(x)$ tells us that during the next 25 years the population will at first increase at an increasing rate and eventually increase at a decreasing rate. Finally, use the trace function to estimate when the population will be increasing fastest and how fast it will be increasing at that time.

36. **Depletion Rate of a Mineral** Geologists project that x years from now the reserve amount of a useful mineral will be

$$Q(x) = \frac{864}{x^2 + 108}$$

billion tons. Set the viewing screen on your graphing calculator to $[0, 20] \times [-1, 0]$ with a scale of 1 on both axes and draw the graph of $Q'(x)$. Then explain how the graph of $Q'(x)$ tells us that during the next 20 years the reserve amount will at first decrease at an increasing rate and eventually decrease at a decreasing rate. Finally, use the trace function to estimate when the reserve amount will be decreasing most rapidly and how rapidly it will be decreasing at that time.

3.2 The Second Derivative and Curve Sketching

The derivative of a function is also a function. Therefore, if the derivative is a differentiable function, we can compute its derivative and obtain the **second derivative** of the original function. If the original function is represented by $f(x)$, we represent the second derivative by

$$f''(x)$$

(read "f double prime of x").

Recall that if the dependent variable of a function is y, we represent the derivative by dy/dx. In this case, we represent the second derivative by

$$\frac{d^2y}{dx^2}$$

From now on, we will sometimes call the derivative of a function the **first derivative**.

In Example 3.4, we compute the second derivative of a rational function.

Example 3.4

Find the second derivative of

$$f(x) = \frac{4x + 6}{x + 5}$$

Solution

Using the quotient rule, the first derivative of $f(x)$ is

$$f'(x) = \frac{(x + 5)4 - (4x + 6)1}{(x + 5)^2}$$

$$= \frac{14}{(x + 5)^2}$$

$$= 14(x + 5)^{-2}$$

Using the generalized power rule, the derivative of $f'(x)$, which is the second derivative of $f(x)$, is

$$f''(x) = -28(x + 5)^{-3}$$

$$= \frac{-28}{(x + 5)^3}$$

In Chapter 2 we saw that the derivative of a function $f(x)$ can be used to determine where $f(x)$ decreases and where it increases. We saw that if $f'(x)$ is *negative* at each x in an interval, then $f(x)$ *decreases* on the interval. We also saw that if $f'(x)$ is *positive* for each x in an interval, then $f(x)$ *increases* on the interval. In this section we use the second derivative of $f(x)$ to determine the *concavity* of the graph of $f(x)$.

Concavity

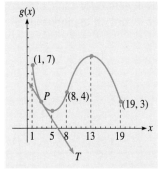

Figure 3.2 A function whose graph is concave upward on the interval from 1 to 8 and concave downward on the interval from 8 to 19.

Let $g(x)$ represent the function whose graph is the curve in Figure 3.2. Imagine point P moving along the path of the curve in the direction whereby its x coordinate increases. Line T is tangent to the curve at point P. As point P moves from point $(1, 7)$ toward point $(8, 4)$, line T rotates counterclockwise, and so its slope increases. Therefore, since $g'(x)$ is the slope of line T, $g'(x)$ increases on the interval from 1 to 8. After point P passes point (8, 4), line T rotates clockwise, and thus its slope decreases. Therefore, $g'(x)$ decreases on the interval from 8 to 19.

The above observations lead us to Definition 3.2.

Definition 3.2

> The graph of a function $f(x)$ is **concave upward** on an interval, if $f'(x)$ increases on the interval. The graph of $f(x)$ is **concave downward** on an interval, if $f'(x)$ decreases on the interval.

Thus, the curve in Figure 3.2 is *concave upward* on the interval from 1 to 8 and *concave downward* on the interval from 8 to 19.

Figure 3.2 shows that the dependent variable of a function can *decrease* either on an interval where the graph is concave upward (look at the interval from 1 to 5), or on an interval where the graph is concave downward (look at the interval from 13 to 19). Figure 3.2 also shows that the dependent variable of a function can *increase* either on an interval where the graph is concave upward (look at the interval from 5 to 8), or on an interval where the graph is concave downward (look at the interval from 8 to 13).

A Test for Concavity

Since $f''(x)$ is the derivative of $f'(x)$, it follows that if $f''(x)$ is positive for each x in an interval, then $f'(x)$ increases on the interval. Therefore, according to Definition 3.2, if $f''(x)$ is positive for each x in an interval, then the graph of $f(x)$ is concave upward on the interval. Similarly, if $f''(x)$ is negative for each x in an interval, then $f'(x)$ decreases on the interval. Therefore, according to Definition 3.2, if $f''(x)$ is negative for each x in an interval, then the graph of $f(x)$ is concave downward on the interval. Theorem 3.1 summarizes these results.

Theorem 3.1 (Test for Concavity)

> If $f''(x)$ is *positive* for each x in an interval, then the graph of $f(x)$ is *concave upward* on the interval. If $f''(x)$ is *negative* for each x in an interval, then the graph of $f(x)$ is *concave downward* on the interval.

In Example 3.5, we use the above theorem to determine the concavity of a polynomial function.

Example 3.5

Determine where the graph of

$$g(x) = \tfrac{1}{3}x^3 - x^2 - 3x + 2$$

is concave upward and where it is concave downward.

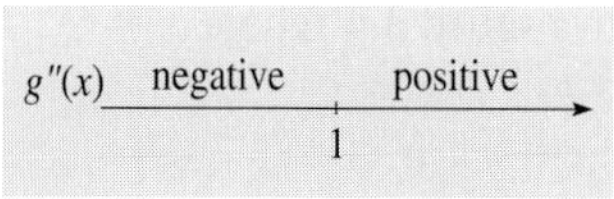

Solution

The derivative of $g(x)$ is
$$g'(x) = x^2 - 2x - 3$$
and so the second derivative of $g(x)$ is
$$g''(x) = 2x - 2$$
Thus, $g''(x) = 0$ when $x = 1$. As the sign chart in Figure 3.3 shows, $g''(x)$ is negative when $x < 1$ and positive when $x > 1$. Therefore, the test for concavity tells us the graph of $g(x)$ is concave downward on the interval $(-\infty, 1)$ and concave upward on the interval $(1, +\infty)$.

Figure 3.3 The sign chart for $g''(x) = 2x - 2$.

Inflection Points

In Figure 3.2, the concavity of the curve changes at the point (8, 4). Points on the graph of a function where the concavity changes are special points.

Definition 3.3

Suppose $(c, f(c))$ is a point on the graph of a function $f(x)$ where the concavity changes. Then $(c, f(c))$ is an **inflection point**, if one of the following conditions is satisfied: (1) $f''(c) = 0$, or (2) $f''(c)$ does not exist, but the graph of $f(x)$ has a vertical tangent at the point $(c, f(c))$.

In this book, we deal mostly with inflection points of the type where the second derivative is zero.

In Example 3.5, we saw that the concavity of the graph of
$$g(x) = \tfrac{1}{3}x^3 - x^2 - 3x + 2$$
changes at the point where $x = 1$. We also saw that $g''(x) = 0$ at $x = 1$. Thus, since
$$g(1) = \tfrac{1}{3} \cdot 1^3 - 1^2 - 3 \cdot 1 + 2$$
$$= -\tfrac{5}{3}$$
we can conclude that the point $(1, -\tfrac{5}{3})$ is an inflection point on the graph of $g(x)$.

The concavity of a curve may change at a point which is not an inflection point. For example, the curve in Figure 3.4 is concave downward on the interval from 2 to 9 and concave upward on the interval from 9 to 14. But the point on the curve where $x = 9$ is not an inflection point. (In Problem 42 at the end of this section, you are asked to explain why this point is not an inflection point.)

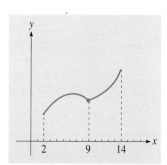

Figure 3.4 A curve whose concavity changes at a point which is not an inflection point.

Some authors allow the point on the curve in Figure 3.4 where $x = 9$ to be an inflection point.

Second-Order Critical Points

The numbers in the domain of a function where the second derivative is either zero or fails to exist are sufficiently important to be labeled.

Definition 3.4

The **second-order critical points** of a function $f(x)$ are the values of x in the domain where $f''(x) = 0$ or $f''(x)$ fails to exist.

If c is a second-order critical point of $f(x)$, we refer to $(c, f(c))$ as a second-order critical point on the graph of $f(x)$.

In Chapter 2, we saw that the critical points of a function are the points where the first derivative is either zero or fails to exist. Henceforth, we call such points **first-order critical points**.

According to Definition 3.3, the inflection points on the graph of a function are among the second-order critical points. However, it is possible for a second-order critical point to not be an inflection point. For example, the second derivative of

$$f(x) = x^4$$
$$f''(x) = 12x^2$$

is

Since $f''(0) = 0$, the point where $x = 0$ is a second-order critical point. But $f''(x)$ is positive when $x < 0$ and when $x > 0$. Therefore, the concavity of the graph of $f(x)$ does not change at the point where $x = 0$, and so this second-order critical point is not an inflection point.

The above observation resembles our observation in Chapter 2 regarding the location of high and low points on the graph of a function. If you recall, excluding endpoints, the high and low points on the graph of a function are among the first-order critical points. But it is possible for a first-order critical point to be neither a high point nor a low point.

Curve Sketching

We now return to the function

$$g(x) = \tfrac{1}{3}x^3 - x^2 - 3x + 2$$

In Example 3.5, we showed that

$$g'(x) = x^2 - 2x - 3$$

Factoring, we get

$$g'(x) = (x + 1)(x - 3)$$

Thus, we see that $g'(x) = 0$ when $x = -1$ and when $x = 3$. Since

$$g(-1) = \tfrac{1}{3}(-1)^3 - (-1)^2 - 3(-1) + 2$$
$$= \tfrac{11}{3}$$

and

$$g(3) = \tfrac{1}{3} \cdot 3^3 - 3^2 - 3 \cdot 3 + 2$$
$$= -7$$

$x+1$	negative	positive	positive
$x-3$	negative	negative	positive
$g'(x)$	positive	negative	positive

           -1 3

Figure 3.5 The sign chart for $g'(x) = (x+1)(x-3)$.

the points $(-1, \tfrac{11}{3})$ and $(3, -7)$ are first-order critical points on the graph of $g(x)$.

The sign chart in Figure 3.5 analyzes the sign of $g'(x)$ in terms of the factors of $g'(x)$. The chart shows that $g'(x)$ is positive when $x < -1$ and when $x > 3$. The chart also shows that $g'(x)$ is negative when $-1 < x < 3$.

Using Theorem 2.13 in Section 2.7, we can conclude that $g(x)$ increases on the interval $(-\infty, -1)$ and on the interval $(3, +\infty)$. We can also conclude that $g(x)$ decreases on the interval $(-1, 3)$.

In Example 3.5, we showed that $g''(x) = 0$ when $x = 1$. Thus, since $g(1) = -\tfrac{5}{3}$, the point $(1, -\tfrac{5}{3})$ is a second-order critical point on the graph of $g(x)$. We also showed that the graph of $g(x)$ is concave downward on the interval $(-\infty, 1)$ and concave upward on the interval $(1, +\infty)$.

Now we are ready to sketch the graph of

$$g(x) = \tfrac{1}{3}x^3 - x^2 - 3x + 2$$

We begin by plotting the first- and second-order critical points, which are $(-1, \tfrac{11}{3})$, $(3, -7)$, and $(1, -\tfrac{5}{3})$. Then we sketch a curve that passes through these points and is compatible with the rest of the information about $g(x)$ that we gathered. Our curve should look like the curve in Figure 3.6.

In Figure 3.6, we also plotted the y intercept, since it is obtainable by simply substituting zero for x in the formula for $g(x)$.

The following procedure is useful in sketching the graphs of the functions in this section.

A Procedure for Curve Sketching

Step 1. Use the first derivative to find the first-order critical points.

Step 2. Use the first derivative to find where the dependent variable increases and where it decreases.

Step 3. Use the second derivative to find the second-order critical points.

Step 4. Use the second derivative to determine where the graph is concave upward and where it is concave downward.

Step 5. If zero is in the domain, find the y intercept by substituting zero for x in the formula.

Step 6. If the domain has endpoints, find the corresponding endpoints of the graph.

Step 7. Plot the points found in steps 1, 3, 5, and 6.

Step 8. Sketch a curve that passes through the points plotted in step 7 and is compatible with the information gathered in steps 2 and 4.

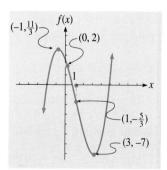

Figure 3.6
$g(x) = \tfrac{1}{3}x^3 - x^2 - 3x + 2$

If steps 1, 3, 5, and 6 do not give rise to enough points, calculate and plot several points of your own choosing before sketching the curve.

The graph of the function in Example 3.6 has two endpoints and no first-order critical points.

Example 3.6

Suppose the total weekly cost of producing x tons of a product per week is

$$C(x) = \tfrac{1}{72}x^3 - \tfrac{1}{4}x^2 + 2x + 3$$

thousand dollars, where $0 \leq x \leq 12$. Sketch the graph of this total cost function.

Solution

The first derivative of $C(x)$ is

$$C'(x) = \tfrac{1}{24}x^2 - \tfrac{1}{2}x + 2$$

The discriminant of $\tfrac{1}{24}x^2 - \tfrac{1}{2}x + 2$ is negative. In fact, the discriminant is

$$(-\tfrac{1}{2})^2 - 4(\tfrac{1}{24})(2) = -\tfrac{1}{12}$$

Thus, $C'(x)$ is never zero. So $C'(x)$ must be either always negative or always positive. Since, for example, $C'(0) = 2$, we conclude that $C'(x)$ is always positive. Therefore, $C(x)$ has no first-order critical points and, as expected, increases if x is increased.

We now examine the second derivative of $C(x)$, which is

$$C''(x) = \tfrac{1}{12}x - \tfrac{1}{2}$$

Note that $C''(x) = 0$ when $x = 6$. Thus, since $C(6) = 9$, the point $(6, 9)$ is a second-order critical point on the graph of $C(x)$. Note also that $C''(x)$ is negative when $x < 6$ and positive when $x > 6$. So, according to Theorem 3.1, the graph of $C(x)$ is concave downward on the interval $[0, 6)$ and concave upward on the interval $(6, 12]$.

Since the domain of $C(x)$ has endpoints, namely, zero and 12, the graph of $C(x)$ has endpoints. Since $C(0) = 3$ and $C(12) = 15$, the endpoints of the graph are $(0, 3)$ and $(12, 15)$. The point $(0, 3)$ also happens to be the y intercept of the graph.

We are now ready to sketch the graph of $C(x)$. We begin by plotting $(6, 9)$, which, as we saw, is the only critical point. Then we plot the two endpoints. Finally, we sketch a curve that passes through these points and is compatible with the rest of the information we gathered using the first and second derivatives. Our curve should look like the curve in Figure 3.7.

The point $(6, 9)$ is an inflection point of the curve in Figure 3.7.

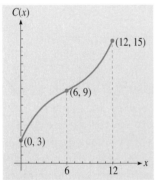

Figure 3.7
$C(x) = \tfrac{1}{72}x^3 - \tfrac{1}{4}x^2 + 2x + 3$ for x in $[0, 12]$.

3.2 Problems

In Problems 1 to 12, find where the graph is concave downward and where it is concave upward. Also, give the coordinates of the inflection points.

1. $f(x) = 0.5x^3 - 4x^2 + 5$
2. $f(x) = -x^3 + 9x^2 - 7x$
3. $f(x) = -\frac{1}{12}x^4 + \frac{3}{2}x^3 - 9x^2$
4. $f(x) = \frac{1}{12}x^4 - \frac{1}{2}x^3 - 5x^2$
5. $y = 12\sqrt{x}$
6. $y = 16(\sqrt[4]{x})$
7. $f(x) = \sqrt{x^2 + 2}$
8. $f(x) = \sqrt{3x^2 + 1}$
9. $f(x) = (x^2 - 9)^5$
10. $f(x) = (x^2 - 25)^3$
11. $f(x) = \dfrac{3x^2 + 145}{x^2 + 48}$
12. $f(x) = \dfrac{2x^2}{x^2 + 27}$

In Problems 13 to 26, sketch the graph of a function $f(x)$ that satisfies the listed conditions.

13. $f'(x)$ is negative for $x < 2$ and for $x > 7$.
    $f'(x)$ is positive for $2 < x < 7$.
    $f(2) = 1$ and $f(7) = 8$.
    $f''(x)$ is positive for $x < 4$ and negative for $x > 4$.
    $f(4) = 3$ and $f(0) = 4$.

14. $f'(x)$ is positive for $x < 4$ and for $x > 8$.
    $f'(x)$ is negative for $4 < x < 8$.
    $f(4) = 6$ and $f(8) = 1$.
    $f''(x)$ is negative for $x < 6$ and positive for $x > 6$.
    $f(6) = 3$ and $f(0) = 0$.

15. $f'(x)$ is positive for $x < 0$ and for $0 < x < 9$.
    $f'(x)$ is negative for $x > 9$.
    $f(9) = 6$.
    $f''(x)$ is negative for $x < 0$ and for $x > 5$.
    $f''(x)$ is positive for $0 < x < 5$.
    $f(0) = 0$ and $f(5) = 3$.

16. $f'(x)$ is negative for $x < 2$.
    $f'(x)$ is positive for $2 < x < 10$ and for $x > 10$.
    $f(2) = 2$.
    $f''(x)$ is positive for $x < 6$ and for $x > 10$.
    $f''(x)$ is negative for $6 < x < 10$.
    $f(6) = 4$, $f(10) = 7$, and $f(0) = 4$.

17. The domain is the interval $[0, +\infty)$.
    $f'(x)$ is negative for $x < 2$ and positive for $x > 2$.
    $f(2) = 1$.
    $f''(x)$ is positive for $x < 7$ and negative for $x > 7$.
    $f(7) = 4$ and $f(0) = 5$.

18. The domain is the interval $[0, +\infty)$.
    $f'(x)$ is positive for $x > 0$.
    $f''(x)$ is positive for $x < 3$ and negative for $x > 3$.
    $f(3) = 2$ and $f(0) = 0$.

19. The domain is the interval $[0, +\infty)$.
    $f'(x)$ is negative for $x > 0$.
    $f''(x)$ is negative for $x < 4$ and positive for $x > 4$.
    $f(4) = 2$ and $f(0) = 3$.

20. The domain is the interval $[0, +\infty)$.
    $f'(x)$ is positive for $x < 3$ and negative for $x > 3$.
    $f(3) = 7$.
    $f''(x)$ is negative for $x < 7$ and positive for $x > 7$.
    $f(7) = 4$ and $f(0) = 6$.

21. The domain is the interval $[1, 9]$.
    $f'(x)$ is negative for $x < 6$ and positive for $x > 6$.
    $f(6) = 4$.
    $f''(x)$ is always positive.
    $f(1) = 8$ and $f(9) = 5$.

22. The domain is the interval $[2, 8]$.
    $f'(x)$ is positive for $x < 4$ and negative for $x > 4$.
    $f(4) = 5$.
    $f''(x)$ is always negative.
    $f(2) = 3$ and $f(8) = 0$.

23. The domain is the interval $[0, 10]$.
    $f'(x)$ is always positive.
    $f''(x)$ is always negative.
    $f(0) = 1$ and $f(10) = 7$.

24. The domain is the interval $[3, 12]$.
    $f'(x)$ is always negative.
    $f''(x)$ is always positive.
    $f(3) = 5$ and $f(12) = 1$.

25. The domain is the interval $[1, 9]$.
    $f'(x)$ and $f''(x)$ are always negative.
    $f(1) = 4$ and $f(9) = 0$.

26. The domain is the interval $[0, 7]$.
    $f'(x)$ and $f''(x)$ are always positive.
    $f(0) = 2$ and $f(7) = 6$.

In Problems 27 to 30, sketch the graph of the given function.

27. $f(x) = x^3 - 6x^2 + 9x$

28. $f(x) = \frac{1}{3}x^3 - 3x^2 + 9x - 2$

29. $f(x) = -\frac{1}{9}x^4 + \frac{4}{9}x^3 + 1$

30. $f(x) = \frac{1}{48}x^4 - \frac{1}{4}x^3 + x^2$

31. **Consumption Rate** Officials believe that the citizens of Oldfoundland consume

$$c(x) = \frac{4}{5}x + \frac{1}{7}\sqrt[3]{x} + \frac{37}{35}$$

billion dollars worth of goods per month when their monthly income is $x billion. Sketch the graph of this consumption function for x in the interval $[1, 8]$.

32. **Worker Productivity** The foreman of La Luna Cigar Factory finds that an average worker produces

$$P(x) = \frac{14x^2 + 216}{x^2 + 108}$$

hundred cigars per day after receiving x hours of training. Sketch the graph of this productivity function for x in the interval $[0, 10]$.

33. **Profit** The demand for a product is so great that whatever amount is produced is immediately sold. So we may view the weekly profit earned from selling the product as a function of its weekly production level. Suppose the total weekly cost of producing x units of the product per week is

$$C(x) = \frac{1}{30}x^3 - \frac{1}{2}x^2 + \frac{13}{5}x + 10$$

thousand dollars, and the product is sold for $5000 per unit. Sketch the graph of the corresponding profit function for x in the interval $[0, +\infty)$.

34. **Average Production Cost** The total monthly cost of producing x units of a product per month is

$$C(x) = 0.5x^2 + 32$$

thousand dollars. Sketch the graph of the corresponding average cost function for x in the interval $[4, +\infty)$.

35. Suppose $f(x)$ is a function whose derivative is represented by the curve in the figure for this problem.

   a. Where does $f(x)$ increase and where does it decrease?

   b. Where is the graph of $f(x)$ concave upward and where is it concave downward?

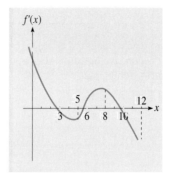

36. Do Problem 35 again. This time suppose the derivative of $f(x)$ is represented by the curve in the figure for this problem.

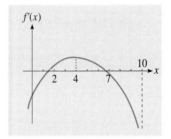

37. **Public Transportation** Transportation officials predict that x years from now

$$T(x) = \frac{12x + 8}{x + 4}$$

million people will make frequent use of the public transportation system in a New England metropolitan area. Sketch the graph of $T'(x)$ for x in the interval $[0, 6]$. Then explain how the graph of $T'(x)$ tells us that during the next 6 years the number of frequent users will increase at a slower and slower rate.

   Note: To determine the concavity of the graph of $T'(x)$, you will have to examine the sign of the second derivative of $T'(x)$, which is the *third* derivative of $T(x)$.

38. **Depreciation Rate** An accountant assumes that x years from now the value of a machine will be

$$V(x) = \frac{3x + 45}{x + 3}$$

thousand dollars. Sketch the graph of $V'(x)$ for x in the interval $[0, 7]$. Then explain how the graph of

$V'(x)$ tells us that during the next 7 years the machine's value will decrease at a slower and slower rate.

   Note: To determine the concavity of the graph of $V'(x)$, you will have to examine the sign of the second derivative of $V'(x)$, which is the *third* derivative of $V(x)$.

39. Let
$$f(x) = \frac{2x^2}{x^2 + 243}$$
for x in the interval $[0, +\infty)$. Find the point on the graph of $f(x)$ where the tangent has maximum slope.

40. Let
$$f(x) = \frac{3x^2}{x^2 + 75}$$
for x in the interval $[0, +\infty)$. Find the point on the graph of $f(x)$ where the tangent has maximum slope.

41. Sketch a curve that has an inflection point where the corresponding function's second derivative fails to exist.

42. We saw that the concavity of the curve in Figure 3.4 changes at the point where $x = 9$. But we concluded that this point is not an inflection point. Why did we make this conclusion?

Graphing Calculator Problems

43. Set the viewing screen on your graphing calculator to $[0, 6] \times [-1.5, 2.5]$ with a scale of 1 on both axes. Then draw the graph of
$$f(x) = 0.08x^3 - 0.72x^2 + 2.16x - 1.66$$
Also, draw the graph of the second derivative of $f(x)$ so that it appears simultaneously with the graph of $f(x)$. Finally, explain how the graphs are compatible with Theorem 3.1.

44. Do Problem 43 again. This time let
$$f(x) = -0.07x^3 + 0.63x^2 - 1.89x + 2.39$$

3.3 More on the Second Derivative

In Section 3.2, we saw that the second derivative of a function can be used to find where the graph is concave upward and where it is concave downward. In this section, we examine the relationship between the concavity of the graph and the manner in which the dependent variable changes. We begin with a demographic situation.

Describing How a Dependent Variable Increases

In a European country, the number of centenarians (persons at least 100 years old) is presently 3000. Let $c(t)$ represent the number of centenarians (in thousands) t years from now. Demographers predict that $c(t)$ will increase during the next 10 years. In fact, they believe that for t in the interval $[0, 10]$, $c(t)$ is the function whose graph is the curve in Figure 3.8.

The graph of $c(t)$ is concave downward on the interval $[0, 4)$ and concave upward on the interval $(4, 10]$. So, according to the definition of concavity (Definition 3.2), $c'(t)$ decreases on the interval $[0, 4)$ and increases on the interval $(4, 10]$. But $c'(t)$ is the (instantaneous) rate of increase of $c(t)$. Thus, $c(t)$ increases at a *decreasing rate* on the interval $[0, 4)$ and at

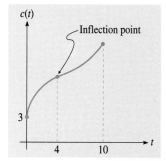

Figure 3.8 A function whose dependent variable increases at a decreasing rate on the interval $[0, 4)$ and at an increasing rate on the interval $(4, 10]$.

an *increasing rate* on the interval (4, 10]. This result tells us that the number of centenarians will increase at a slower and slower rate during the next 4 years. After that time, the number of centenarians will increase faster and faster until 10 years from now.

Figure 3.8 illustrates Theorem 3.2.

Theorem 3.2

> Suppose $f(x)$ is a function that *increases* on an interval. If the graph of $f(x)$ is concave upward on the interval, then $f(x)$ increases at an increasing rate. If the graph of $f(x)$ is concave downward on the interval, then $f(x)$ increases at a decreasing rate.

In Example 3.7, we use Theorem 3.2 to describe how a disease spreads through a herd of cows.

Example 3.7

Recently, a veterinarian discovered that some of the cows in a large herd are afflicted with a communicable disease. She predicts that x months from now

$$D(x) = \frac{12x^2 + 9}{x^2 + 27}$$

thousand cows will have contracted the disease. Verify that the number of cows that contract the disease will increase with the passage of time. Then describe how it will increase.

Solution

Using the quotient rule, the first derivative of $D(x)$ is

$$D'(x) = \frac{(x^2 + 27)24x - (12x^2 + 9)2x}{(x^2 + 27)^2}$$

$$= \frac{630x}{(x^2 + 27)^2}$$

The value of $(x^2 + 27)^2$ is always positive. So, since $630x$ is positive when $x > 0$, it follows that $D'(x)$ is positive when $x > 0$. Therefore, although $D'(x) = 0$ at $x = 0$, $D(x)$ increases on the interval $[0, +\infty)$. This result means that the number of cows that contract the disease indeed will increase with the passage of time.

Now we describe how the disease will spread. Applying the quotient rule and the generalized power rule, the second derivative of $D(x)$ is

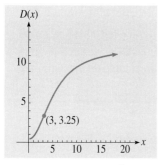

Figure 3.9 $D(x) = \dfrac{12x^2 + 9}{x^2 + 27}$ for $x \geq 0$.

$$D''(x) = \frac{(x^2 + 27)^2 630 - 630x \cdot 2(x^2 + 27)2x}{(x^2 + 27)^4}$$

$$= \frac{(x^2 + 27)[(x^2 + 27)630 - 2520x^2]}{(x^2 + 27)^4} \quad \text{factoring out } x^2 + 27$$

$$= \frac{(x^2 + 27)630 - 2520x^2}{(x^2 + 27)^3} \quad \text{reducing}$$

$$= \frac{17{,}010 - 1890x^2}{(x^2 + 27)^3}$$

$$= \frac{1890(9 - x^2)}{(x^2 + 27)^3}$$

Note that $D''(x) = 0$ at $x = 3$. Note also that for nonnegative values of x, $D''(x)$ is positive when $x < 3$ and negative when $x > 3$. [For example, $D''(1)$ is approximately 0.69, while $D''(4)$ is approximately -0.17.] Thus, according to the test for concavity (Theorem 3.1), the graph of $D(x)$ is concave upward on the interval $[0, 3)$ and concave downward on the interval $(3, +\infty)$. So, according to Theorem 3.2, the disease will spread at a faster and faster rate during the next 3 months. But after that time, it will spread at a slower and slower rate. Evidently, the disease will be spreading most rapidly at the instant 3 months from now. Figure 3.9 shows the graph of $D(x)$. The point $(3, 3.25)$ is an inflection point on the graph of $D(x)$.

Next we examine *decreasing* functions.

Describing How a Dependent Variable Decreases

Let $h(x)$ be a function whose derivative is negative for each x in the interval $[0, 12]$. Then $h(x)$ decreases on the interval $[0, 12]$. In fact, let $h(x)$ be the function whose graph is the curve in Figure 3.10.

The graph of $h(x)$ is concave upward on the interval $[0, 5)$. So, according to the definition of concavity, $h'(x)$ increases on the interval $[0, 5)$. But, since $h'(x)$ is negative for each x in $[0, 5)$, the magnitude (absolute value) of $h'(x)$ decreases on $[0, 5)$. Therefore, $h(x)$ decreases at a *decreasing rate* as x increases toward 5, because the magnitude of $h'(x)$ is the rate at which $h(x)$ decreases.

The graph of $h(x)$ is concave downward on the interval $(5, 12]$. So $h'(x)$ decreases on this interval, which implies that the magnitude of $h'(x)$ increases. Therefore, $h(x)$ decreases at an *increasing rate* after x passes 5.

Figure 3.10 illustrates Theorem 3.3.

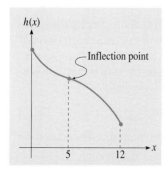

Figure 3.10 A function whose dependent variable decreases at a decreasing rate on the interval $[0, 5)$ and at an increasing rate on the interval $(5, 12]$.

Theorem 3.3

Suppose $f(x)$ is a function that *decreases* on an interval. If the graph of $f(x)$ is concave upward on the interval, then $f(x)$ decreases at a decreasing rate. If the graph of $f(x)$ is concave downward on the interval, then $f(x)$ decreases at an increasing rate.

We use Theorem 3.3 in Example 3.8.

Example 3.8

The *index of income concentration* of a community is a statistic used by the Census Bureau. It is a number between zero and 1 that measures how evenly the income of the community is distributed. The closer the index is to zero, the more evenly the income is distributed. The index of income concentration of a particular community is presently 0.5. Suppose that x years from now the index is

$$i(x) = \frac{x+8}{4x+16}$$

Show that the index will decrease with the passage of time. Then describe how it will decrease.

Solution

Using the quotient rule, the first derivative of $i(x)$ is

$$i'(x) = \frac{(4x+16)1 - (x+8)4}{(4x+16)^2}$$

$$= \frac{-16}{(4x+16)^2}$$

The value of $i'(x)$ is negative for $x \geq 0$. Thus, $i(x)$ decreases on the interval $[0, +\infty)$. In other words, the index indeed will decrease with the passage of time.

Now we describe how the index will decrease. The first derivative of $i(x)$ can be expressed as

$$i'(x) = -16(4x+16)^{-2}$$

Thus, we can use the generalized power rule to obtain the second derivative of $i(x)$:

$$i''(x) = 32(4x+16)^{-3} 4$$

$$= \frac{128}{(4x+16)^3}$$

The value of $i''(x)$ is positive for $x \geq 0$. So, according to the test for con-

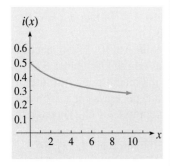

Figure 3.11 $i(x) = \dfrac{x+8}{4x+16}$ for $x \geq 0$.

cavity, the graph of $i(x)$ is concave upward on the interval $[0, +\infty)$. Therefore, according to Theorem 3.3, the index of income concentration of the community will decrease at a slower and slower rate with the passage of time. Figure 3.11 shows the graph of $i(x)$.

3.3 Problems

In Problems 1 to 8, use the first derivative to show that $f(x)$ increases on the interval $[0, +\infty)$. Then use Theorem 3.2 to determine how $f(x)$ increases on this interval.

1. $f(x) = \dfrac{9x + 1}{x + 2}$
2. $f(x) = \dfrac{15x + 6}{2x + 3}$
3. $f(x) = (x + 2)^3$
4. $f(x) = \frac{1}{6}(x + 1)^3$
5. $f(x) = \dfrac{6x^2 + 1}{x^2 + 75}$
6. $f(x) = \dfrac{9x^2 + 44}{x^2 + 12}$
7. $f(x) = \frac{1}{3}x^3 - 3x^2 + 10x$
8. $f(x) = \frac{1}{6}x^3 - 2x^2 + 9x$

In Problems 9 to 16, use the first derivative to show that $f(x)$ decreases on the interval $[0, +\infty)$. Then use Theorem 3.3 to determine how $f(x)$ decreases on this interval.

9. $f(x) = \dfrac{x + 9}{2x + 1}$
10. $f(x) = \dfrac{3x + 28}{x + 4}$
11. $f(x) = -x^2 - 4x + 32$
12. $f(x) = -x^2 - 2x + 24$
13. $f(x) = -\frac{1}{3}x^3 + 3x^2 - 10x + 20$
14. $f(x) = -\frac{1}{3}x^3 + 7x^2 - 50x + 30$
15. $f(x) = \dfrac{72}{x^2 + 48}$
16. $f(x) = \dfrac{x^2 + 60}{x^2 + 12}$

In Problems 17 to 24, sketch the graph of a function $f(x)$ that satisfies the listed conditions.

17. The domain is $[0, +\infty)$.
    $f(0) = 2$.
    $f(x)$ increases at a decreasing rate.

18. The domain is $[0, 8]$.
    $f(0) = 2$ and $f(8) = 6$.
    $f(x)$ increases at an increasing rate.

19. The domain is $[0, 7]$.
    $f(0) = 5$ and $f(7) = 0$.
    $f(x)$ decreases at an increasing rate.

20. The domain is $[0, +\infty)$.
    $f(0) = 6$.
    $f(x)$ decreases at a decreasing rate.

21. The domain is $[0, +\infty)$.
    $f(0) = 0$ and $f(5) = 6$.
    $f(x)$ increases at a decreasing rate as x increases toward 5 and at an increasing rate after x passes 5.

22. The domain is $[0, +\infty)$.
    $f(0) = 1$ and $f(4) = 6$.
    $f(x)$ increases at an increasing rate as x increases toward 4 and at a decreasing rate after x passes 4.

23. The domain is $[0, 10]$.
    $f(0) = 8, f(4) = 4$, and $f(10) = 1$.
    $f(x)$ decreases at an increasing rate on $[0, 4)$ and at a decreasing rate on $(4, 10]$.

24. The domain is $[0, 10]$.
    $f(0) = 9, f(5) = 6$, and $f(10) = 0$.
    $f(x)$ decreases at a decreasing rate on $[0, 5)$ and at an increasing rate on $(5, 10]$.

25. **Family Income** In a city, the annual income needed by a family of four to maintain a moderate standard of living is presently \$27,000. Suppose that x years from now the necessary income is

$$I(x) = \dfrac{32x + 27}{x + 1}$$

thousand dollars. Show that the necessary income will increase. Then determine how it will increase.

26. **Sperm Count** Due to environmental contaminants, the average sperm count of the male population of an island has been declining. Suppose that x years from now the average sperm count is

$$s(x) = \dfrac{124x + 225}{2x + 3}$$

million per cubic centimeter. Show that the sperm count will continue to decline. Then describe how it will decline.

27. **Weight Gain** A man who is concerned about being underweight has just started a special diet high in caloric intake. Suppose that x weeks from now his weight is

$$W(x) = \frac{175x^2 + 13{,}500}{x^2 + 108}$$

pounds. Verify that he will gain weight. Then describe how his weight will increase.

28. **Acidic Rain** Suppose that x years from now the pH of rainwater in a North American forest is

$$A(x) = \frac{4x^2 + 450}{x^2 + 75}$$

Verify that the pH will decrease. (Therefore, the rainwater will become more acidic.) Then describe how it will decrease.

29. **Average Production Cost** Silkon, a fabric made of synthetic fibers, has many of the qualities of silk. Suppose the total monthly cost of producing x bolts of silkon per month is

$$C(x) = 120x + 9000$$

dollars. Let $A(x)$ represent the average cost of producing each bolt when the monthly production level is x bolts. Show that $A(x)$ decreases if x is increased. Then determine how it decreases.
Recall: For $x > 0$, $A(x) = C(x)/x$.

30. **Average Production Cost** Suppose the total weekly cost of producing x units of a commodity per week is

$$C(x) = 5x + 87$$

thousand dollars. Let $A(x)$ represent the average cost of producing each unit when the weekly production level is x units. Verify that $A(x)$ decreases if x is increased. Then describe how it decreases.
Recall: For $x > 0$, $A(x) = C(x)/x$.

31. **Spread Rate of Information** Fluffy, a new ultrasoft bathroom tissue, has just been introduced in Ebbton. Suppose that x weeks from now

$$P(x) = \frac{100x^2}{x^2 + 27}$$

percent of the people are aware of the product. When will the news about the product be spreading most rapidly? How rapidly will it be spreading at that time?

32. **Spread Rate of a Disease** A communicable disease has just been introduced into a community. Suppose that during the first x weeks

$$D(x) = \frac{90x^2}{x^2 + 48}$$

percent of the people contract the disease. When will the disease be spreading fastest? How fast will it be spreading at that time?

33. **Least Marginal Cost** Suppose the total daily cost of producing x units of a commodity per day is

$$C(x) = \tfrac{1}{3}x^3 - 20x^2 + 403x + 9$$

dollars. Find the production level at which the marginal cost is least.

34. **Least Marginal Cost** Redo Problem 33 under the new assumption that

$$C(x) = \tfrac{1}{6}x^3 - 15x^2 + 452x + 13$$

35. **Farmland Productivity** A farmer finds that if he uses x tons of a special fertilizer, his land produces

$$Y(x) = \frac{200x^2 + 22{,}500}{x^2 + 300}$$

tons of string beans. If x is increased, $Y(x)$ at first increases at an increasing rate and eventually increases at a decreasing rate. Consequently, there exists a point after which the returns due to further increases in fertilization begin to diminish. Find this *point of diminishing returns*.

36. **Decontamination** Suppose that with an expenditure of $\$x$ million, the percent of a pollutant that can be removed from a lake is

$$P(x) = \frac{95x^2}{x^2 + 192}$$

If x is increased, $P(x)$ at first increases at an increasing rate and eventually increases at a decreasing rate. Therefore, there exists a point after which the returns due to further increases in expenditure begin to diminish. Find this *point of diminishing returns*.

37. **Aging of a Population** Demographers project that the median age of a population will increase. Suppose $A(x)$ represents the population's median age x

years from now. The figure for this problem shows the graph of $A''(x)$. Explain how the graph of $A''(x)$ implies that the increase rate of the median age will at first increase and eventually decrease.

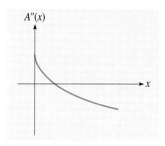

38. **Dieting** As long as a man remains on a diet, his weight will gradually decrease. Suppose $W(x)$ represents his weight after x weeks of dieting. The figure for this problem shows the graph of $W'''(x)$. Explain how the graph of $W'''(x)$ implies that his weight's decrease rate will at first increase and eventually decrease.

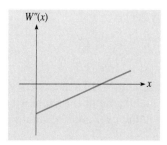

Graphing Calculator Problems

39. **Public Transportation** An authority on public transportation projects that x years from now

$$T(x) = \frac{7x + 5}{x + 2}$$

million people will make frequent use of the public transportation system in a large metropolitan area. Set the viewing screen on your graphing calculator to $[0, 12] \times [0, 8]$ with a scale of 1 on both axes. Then draw the graph of $T(x)$. The graph will show that the number of people that frequently use the system will increase during the next 12 years. Use the graph to determine how this number will increase.

40. **Population** A major manufacturing firm has just moved away from a community. Officials project that x years from now the population of the community will be

$$P(x) = \frac{3x + 11}{2x + 2}$$

million. Set the viewing screen on your graphing calculator to $[0, 9] \times [0, 6]$ with a scale of 1 on both axes. Then draw the graph of $P(x)$. The graph will show that the population will decrease during the next 9 years. Use the graph to determine how the population will decrease.

3.4 Two Useful Rational Functions

In this section, we deal with quantities that approach a *fixed level* either at a rate that always decreases or at a rate that at first increases and eventually decreases. With certain restrictions on the coefficients, rational functions of the type

$$f(x) = \frac{ax + b}{cx + d}$$

and rational functions of the type

$$f(x) = \frac{ax^2 + b}{cx^2 + d}$$

have properties compatible with the behavior of such quantities. We introduce the first of these two types of functions through a situation in the cable television industry.

Functions of the Type $f(x) = (ax + b)/(cx + d)$

Presently, Teleshow, a cable television company, has 500,000 (0.5 million) subscribers in a large metropolitan area. Officials project that x years from now this number will be $T(x) = (7x + 1)/(2x + 2)$ million. We will examine how the number of subscribers will behave as time passes.

The function $T(x)$ is a rational function for which the degree of the numerator is equal to the degree of the denominator. The leading coefficient of the numerator is 7 and the leading coefficient of the denominator is 2. Therefore, using Theorem 1.4 in Section 1.5, we conclude that the limit of $T(x)$, as x approaches positive infinity, is $\frac{7}{2}$. This statement is expressed by writing

$$\lim_{x \to +\infty} T(x) = \frac{7}{2}$$

The above statement means that as x increases without bound, the values of $T(x)$ eventually get and stay arbitrarily close to $\frac{7}{2}$. Consequently, since $\frac{7}{2} = 3.5$, the number of cable television subscribers will approach 3.5 million as time passes. We will now determine how the number of subscribers will approach this level.

Using the quotient rule, the derivative of $T(x)$ is

$$T'(x) = \frac{(2x + 2)7 - (7x + 1)2}{(2x + 2)^2} = \frac{12}{(2x + 2)^2}$$

Note that $T'(x)$ is positive for $x \geq 0$. Thus, the number of subscribers will increase as time passes. This result implies that the number of subscribers will approach 3.5 million from *below* and without ever quite reaching 3.5 million.

The derivative of $T(x)$ can be expressed as

$$T'(x) = 12(2x + 2)^{-2}$$

Thus, we can use the generalized power rule to obtain the second derivative of $T(x)$:

$$T''(x) = -24(2x + 2)^{-3}2 = \frac{-48}{(2x + 2)^3}$$

The value of $T''(x)$ is negative for $x \geq 0$. So the graph of $T(x)$ is concave downward throughout its domain, which we assume to be the interval $[0, +\infty)$. Therefore, according to Theorem 3.2 in Section 3.3, the number of subscribers will approach 3.5 million from below while increasing at a slower and slower rate.

If we sketch a curve that is compatible with the information we gathered about the function $T(x)$, our curve should look like the curve in Figure 3.12.

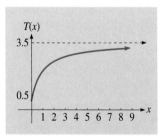

Figure 3.12 $T(x) = \dfrac{7x + 1}{2x + 2}$ for $x \geq 0$.

The dashed horizontal line in Figure 3.12 is the graph of

$$y = 3.5$$

The figure suggests that as x approaches positive infinity, the graph of $T(x)$ eventually gets and stays arbitrarily close to this line.

Definition 3.5

The horizontal line

$$y = k$$

is a **horizontal asymptote** of the graph of a function $f(x)$ if k is a number such that

$$\lim_{x \to +\infty} f(x) = k \text{ or } \lim_{x \to -\infty} f(x) = k$$

According to Definition 3.5, the line $y = 3.5$ is a horizontal asymptote of the graph of $T(x)$.

A proof of Theorem 3.4 would parallel our analysis of the function $T(x) = (7x + 1)/(2x + 2)$.

Theorem 3.4

Suppose

$$f(x) = \frac{ax + b}{cx + d}$$

where a, b, c, and d are positive constants, and $x \geq 0$. (Actually, a or b can be zero, but not both.) Then the line

$$y = \frac{a}{c}$$

is a horizontal asymptote of the graph of $f(x)$.

1. If

$$\frac{a}{c} > \frac{b}{d}$$

the graph of $f(x)$ is generally like the curve in Figure 3.13.

2. If

$$\frac{a}{c} < \frac{b}{d}$$

the graph of $f(x)$ is generally like the curve in Figure 3.14.

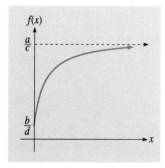

Figure 3.13 $f(x) = \dfrac{ax + b}{cx + d}$ if $\dfrac{a}{c} > \dfrac{b}{d}$.

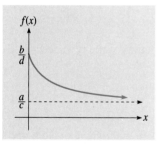

Figure 3.14 $f(x) = \dfrac{ax + b}{cx + d}$ if $\dfrac{a}{c} < \dfrac{b}{d}$.

The function $T(x)$ is a function of the type described in Theorem 3.4. For the function $T(x)$,

$$\frac{a}{c} = \frac{7}{2} = 3.5$$

and

$$\frac{b}{d} = \frac{1}{2} = 0.5$$

Thus, the function $T(x)$ illustrates the case in Theorem 3.4 where $a/c > b/d$. We use Theorem 3.4 in Example 3.9.

Example 3.9

Sketch the graph of

$$g(x) = \frac{3x + 31}{2x + 5}$$

for $x \geq 0$. Then describe how $g(x)$ behaves as x increases.

Solution

The function $g(x)$ is of the type described in Theorem 3.4 with

$$\frac{a}{c} = \frac{3}{2}$$

and

$$\frac{b}{d} = \frac{31}{5}$$

Since $a/c = \frac{3}{2}$, the horizontal asymptote is the line

$$y = \frac{3}{2}$$

Since $\frac{3}{2} < \frac{31}{5}$, the graph of $g(x)$ descends from the point $(0, \frac{31}{5})$ toward the horizontal asymptote and is concave upward. Figure 3.15 shows the graph of $g(x)$. From the graph we see that as x increases, $g(x)$ approaches $\frac{3}{2}$ from above while decreasing at a decreasing rate.

Example 3.10 also deals with a rational function of the type described in Theorem 3.4.

Example 3.10

Five years ago a company paid $70,000 for a new machine. One year ago the value of the machine was $50,000. An accountant assumes that as the machine ages, its value approaches zero while it decreases at a slower and slower rate. The accountant also assumes that when the machine is x years

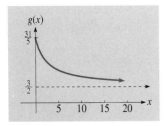

Figure 3.15 $g(x) = \dfrac{3x + 31}{2x + 5}$ for $x \geq 0$.

old, its value is given in thousands of dollars by a formula of the form

$$V(x) = \frac{ax + b}{cx + d}$$

If the company sells the machine when its value is $25,000, how old is the machine when the company sells it?

Solution

First we will find a formula for $V(x)$. Since the machine was worth $70,000 when it was new,

$$V(0) = 70$$

Since 1 year ago the machine was 4 years old and worth $50,000,

$$V(4) = 50$$

So we are looking for a function of the type

$$V(x) = \frac{ax + b}{cx + d}$$

that satisfies the following conditions:

$$V(0) = 70$$
$V(x)$ approaches zero as x increases
$$V(4) = 50$$

Theorem 3.4 tells us that $V(0) = b/d$. Therefore, since $V(0) = 70$, the values of b and d must be such that

$$\frac{b}{d} = 70$$

which implies that $\qquad b = 70d \qquad (1)$

Theorem 3.4 also tells us that the line $y = a/c$ is a horizontal asymptote of the graph of $V(x)$, which means that $V(x)$ approaches a/c as x increases. Therefore, if we set $a = 0$ and $c = 1$, then $V(x)$ approaches zero as x increases, no matter what values we assign to b and d. Moreover,

$$V(x) = \frac{0 \cdot x + b}{1 \cdot x + d}$$

$$= \frac{b}{x + d} \qquad (2)$$

Since $V(4) = 50$, the values of b and d also must be such that

$$\frac{b}{4 + d} = 50$$

Multiplying both sides by $4 + d$, we get
$$b = 200 + 50d \tag{3}$$

From formula (1) we see that we can substitute $70d$ for b in formula (3) and obtain
$$70d = 200 + 50d$$

Solving for d, we get
$$d = 10$$

Substituting 10 for d in formula (1), we get
$$b = 70(10) = 700$$

Substituting 700 for b and 10 for d in formula (2), we finally have a formula for $V(x)$, namely,
$$V(x) = \frac{700}{x + 10}$$

To find the machine's age when the company sells it, we solve the following equation:
$$\frac{700}{x + 10} = 25$$
$$700 = 25x + 250$$
$$450 = 25x$$
$$18 = x$$

So the company sells the machine when it is 18 years old. Since the machine was new when the company purchased it 5 years ago, the company will keep the machine 13 more years. Figure 3.16 shows the graph of $V(x)$.

We now introduce functions of the type
$$f(x) = \frac{ax^2 + b}{cx^2 + d}$$
through a situation in the life sciences.

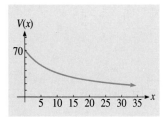

Figure 3.16 $V(x) = \dfrac{700}{x + 10}$ for $x \geq 0$.

Functions of the Type $f(x) = (ax^2 + b)/(cx^2 + d)$

A wildlife scientist predicts that x years from now the iguana population of an island in the Caribbean will be
$$I(x) = \frac{5x^2 + 192}{2x^2 + 24}$$

thousand. We will examine how the population will behave as time passes.

The function $I(x)$ is a rational function for which the degree of the numerator is equal to the degree of the denominator. The leading coefficients of the numerator and denominator are, respectively, 5 and 2. Thus,

$$\lim_{x \to +\infty} I(x) = \frac{5}{2}$$
$$= 2.5$$

This result tells us that the iguana population will approach 2500 (2.5 thousand) as time passes.

Using the quotient rule, the derivative of $I(x)$ is

$$I'(x) = \frac{(2x^2 + 24)10x - (5x^2 + 192)4x}{(2x^2 + 24)^2}$$
$$= \frac{-528x}{(2x^2 + 24)^2}$$

Since $(2x^2 + 24)^2$ is always positive and $-528x$ is negative when $x > 0$, the value of $I'(x)$ is negative when $x > 0$. Therefore, the iguana population will decrease as time passes and, thus, will approach 2500 from *above*.

Applying the quotient rule and the generalized power rule, the second derivative of $I(x)$ is

$$I''(x) = \frac{(2x^2 + 24)^2(-528) - (-528x)2(2x^2 + 24)4x}{(2x^2 + 24)^4}$$
$$= \frac{(2x^2 + 24)[(2x^2 + 24)(-528) + 4224x^2]}{(2x^2 + 24)^4} \quad \text{factoring out } 2x^2 + 24$$
$$= \frac{3168(x^2 - 4)}{(2x^2 + 24)^3}$$

Note that $I''(x) = 0$ at $x = 2$. For nonnegative values of x, $I''(x)$ is negative when $x < 2$ and positive when $x > 2$. So the graph of $I(x)$ is concave downward on the interval $[0, 2)$ and concave upward on the interval $(2, +\infty)$. Therefore, according to Theorem 3.3 in Section 3.3, the iguana population will approach 2500 from above while decreasing at a faster and faster rate during the next 2 years and at a slower and slower rate thereafter.

Figure 3.17 shows the graph of the function $I(x)$.
The formula for $I(x)$ is of the form

$$\frac{ax^2 + b}{cx^2 + d}$$

with $a = 5$, $b = 192$, $c = 2$, and $d = 24$. Note that if we substitute 24 for d and 2 for c in the formula

$$\sqrt{\frac{d}{3c}}$$

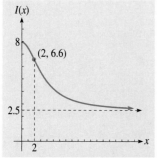

Figure 3.17 $I(x) = \dfrac{5x^2 + 192}{2x^2 + 24}$ for $x \geq 0$.

we obtain

$$\sqrt{\frac{24}{3 \cdot 2}} = 2$$

which is the x coordinate of the inflection point on the graph of $I(x)$. This result is not a coincidence!

A proof of Theorem 3.5 would follow steps similar to those we used to analyze the function $I(x) = (5x^2 + 192)/(2x^2 + 24)$.

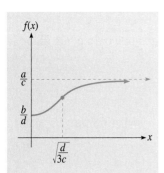

Figure 3.18 $f(x) = \dfrac{ax^2 + b}{cx^2 + d}$ if $\dfrac{a}{c} > \dfrac{b}{d}$.

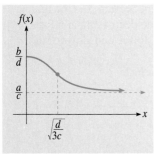

Figure 3.19 $f(x) = \dfrac{ax^2 + b}{cx^2 + d}$ if $\dfrac{a}{c} < \dfrac{b}{d}$.

Theorem 3.5

Suppose

$$f(x) = \frac{ax^2 + b}{cx^2 + d}$$

where a, b, c, and d are positive constants, and $x \geq 0$. (Actually, a or b can be zero, but not both.) Then the line

$$y = \frac{a}{c}$$

is a horizontal asymptote of the graph of $f(x)$.

1. If

$$\frac{a}{c} > \frac{b}{d}$$

the graph of $f(x)$ is generally like the curve in Figure 3.18.

2. If

$$\frac{a}{c} < \frac{b}{d}$$

the graph of $f(x)$ is generally like the curve in Figure 3.19.

In either case, the graph has an inflection point where

$$x = \sqrt{\frac{d}{3c}}$$

If $f(x)$ is a function of the type described in Theorem 3.5, the magnitude of the rate of change of $f(x)$ is greatest when

$$x = \sqrt{\frac{d}{3c}}$$

Therefore, if x represents time, $f(x)$ changes most rapidly when $x = \sqrt{d/(3c)}$.

The function $I(x)$ in the iguana problem is a function of the type described in Theorem 3.5. For the function $I(x)$,

$$\frac{a}{c} = \frac{5}{2} = 2.5$$

and

$$\frac{b}{d} = \frac{192}{24} = 8$$

So the function $I(x)$ illustrates the case in Theorem 3.5 where $a/c < b/d$. Example 3.11 uses Theorem 3.5.

Example 3.11

The management of Dukes School Supplies finds that the firm produces

$$P(x) = \frac{31x^2}{4x^2 + 192}$$

thousand loose-leaf notebooks per day with a capital outlay of $\$x$ million. Sketch the graph of $P(x)$ for $x \geq 0$. Then describe how the daily production level behaves as the capital outlay is increased.

Solution

The function $P(x)$ is of the type described in Theorem 3.5 with $a = 31$, $b = 0$, $c = 4$, and $d = 192$. Thus,

$$\frac{a}{c} = \frac{31}{4} = 7.75$$

$$\frac{b}{d} = \frac{0}{192} = 0$$

and

$$\sqrt{\frac{d}{3c}} = \sqrt{\frac{192}{3 \cdot 4}} = 4$$

Since $a/c = 7.75$, the horizontal asymptote is the line

$$y = 7.75$$

Since $a/c > b/d$, the graph of $P(x)$ rises from the point $(0, 0)$ toward the horizontal asymptote. It is concave upward on the interval $[0, 4)$ and concave downward on the interval $(4, +\infty)$. Figure 3.20 shows the graph of $P(x)$.

The graph of $P(x)$ shows that the daily notebook production level approaches 7.75 thousand as the capital outlay is increased. At first, the production level increases at an increasing rate. After the capital outlay passes $\$4$ million, the production level increases at a decreasing rate.

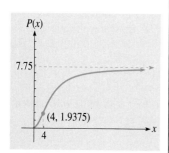

Figure 3.20 $P(x) = \dfrac{31x^2}{4x^2 + 192}$ for $x \geq 0$.

3.4 Problems

In Problems 1 to 16, sketch the graph of $f(x)$ for $x \geq 0$. Then describe how $f(x)$ behaves as x increases.

1. $f(x) = \dfrac{3x + 12}{2x + 3}$
2. $f(x) = \dfrac{10x + 7}{2x + 3}$
3. $f(x) = \dfrac{15x + 3}{3x + 2}$
4. $f(x) = \dfrac{x + 12}{3x + 3}$
5. $f(x) = \dfrac{4x}{x + 2}$
6. $f(x) = \dfrac{3}{2x + 1}$
7. $f(x) = \dfrac{12}{3x + 2}$
8. $f(x) = \dfrac{18x}{4x + 3}$
9. $f(x) = \dfrac{17x^2 + 54}{2x^2 + 54}$
10. $f(x) = \dfrac{3x^2 + 144}{4x^2 + 48}$
11. $f(x) = \dfrac{x^2 + 30}{2x^2 + 6}$
12. $f(x) = \dfrac{7x^2 + 24}{x^2 + 48}$
13. $f(x) = \dfrac{78}{x^2 + 12}$
14. $f(x) = \dfrac{13x^2}{3x^2 + 81}$
15. $f(x) = \dfrac{7x^2}{x^2 + 75}$
16. $f(x) = \dfrac{15}{x^2 + 3}$

In Problems 17 to 24, find the function of the type

$$f(x) = \dfrac{ax + b}{cx + d}$$

that satisfies the given conditions as x increases on the interval $[0, +\infty)$.

17. $f(0) = 8$
    $f(x)$ approaches 3
    $f(5) = 4$

18. $f(0) = 2$
    $f(x)$ approaches 6
    $f(3) = 4$

19. $f(0) = 1$
    $f(x)$ approaches 7
    $f(4) = 5$

20. $f(0) = 9$
    $f(x)$ approaches 4
    $f(7) = 5$

21. $f(0) = 0$
    $f(x)$ approaches 4
    $f(9) = 3$

22. $f(0) = 5$
    $f(x)$ approaches zero
    $f(1) = 2$

23. $f(0) = 8$
    $f(x)$ approaches zero
    $f(5) = 1$

24. $f(0) = 0$
    $f(x)$ approaches 10
    $f(8) = 6$

In Problems 25 to 28, find the function of the type

$$f(x) = \dfrac{ax^2 + b}{cx^2 + d}$$

that satisfies the given conditions as x increases on the interval $[0, +\infty)$.

25. $f(0) = 4$
    $f(x)$ approaches 2
    $f(1) = 3$

26. $f(0) = 1$
    $f(x)$ approaches 5
    $f(2) = 4$

27. $f(0) = 7$
    $f(x)$ approaches zero
    $f(4) = 2$

28. $f(0) = 4$
    $f(x)$ approaches zero
    $f(5) = 3$

In Problems 29 to 32, sketch the graph of the function of the type

$$f(x) = \dfrac{ax^2 + b}{cx^2 + d}$$

that satisfies the given conditions as x increases on the interval $[0, +\infty)$.

29. $f(0) = 3$
    $f(x)$ approaches 8
    $f(12) = 6.75$

30. $f(0) = 7$
    $f(x)$ approaches 4
    $f(6) = 4.75$

31. $f(0) = 0$
    $f(x)$ approaches 6
    $f(5) = 1.5$

32. $f(0) = 0$
    $f(x)$ approaches 9
    $f(9) = 6.75$

33. **Factory Utilization** A country presently uses 68 percent of its manufacturing facilities. Suppose that x weeks from now it uses

$$M(x) = \dfrac{225x + 136}{3x + 2}$$

percent of its facilities. Sketch the graph of $M(x)$ for $x \geq 0$. Then describe how the utilization will behave as time passes.

34. **Puberty Age** In a Southern European country, experts project that t years from now the average puberty age will be

$$f(t) = \dfrac{12t + 720}{t + 48}$$

Sketch the graph of $f(t)$ for $t \geq 0$. What will happen to the average puberty age as time passes?

35. **Hyacinth Removal** A new process for removing water hyacinths from lakes is applied to a lake in Georgia. Let $P(x)$ represent the percent of hyacinths that remain in the lake if $\$x$ million is spent applying the process. Suppose

$$P(x) = \frac{30x + 100}{x + 1}$$

for $x \geq 0$. Describe how $P(x)$ behaves as x is increased.

36. **Perspiration Rate** A man has just moved from Sweden to Venezuela. Let $f(t)$ represent the maximum rate (in liters per hour) at which he can perspire t months from now. If

$$f(t) = \frac{19t + 7}{5t + 5}$$

how will his capacity for producing perspiration behave as time passes?

37. **Demand** Suppose consumers purchase

$$D(x) = \frac{375}{x^2 + 75}$$

million barrels of a product per month when its selling price is x cents per quart. Sketch the graph of $D(x)$ for $x \geq 0$. How does the monthly demand behave as the price increases?

38. **Service Jobs** There are 3 million service related jobs in a northeastern city. Officials project that x years from now there will be

$$S(x) = \frac{13x^2 + 450}{2x^2 + 150}$$

million such jobs. Sketch the graph of $S(x)$ for $x \geq 0$. Then describe how the number of service related jobs will behave as time passes.

39. **TV Watching** In a country, 20 percent of the population watches more than 2 hours of television per day. If x years from now this percent is

$$P(x) = \frac{40x^2 + 2160}{x^2 + 108}$$

how will the percent behave as time passes?

40. **Females to Males Ratio** In a community, the ratio of females to males is $\frac{4}{3}$. Suppose that x years from now this ratio is

$$R(x) = \frac{5x^2 + 96}{6x^2 + 72}$$

Describe how the ratio will behave as time passes.

41. **Dieting** A woman started a special diet 6 weeks ago. At that time she weighed 170 pounds, but 2 weeks ago she weighed 160 pounds. Her doctor believes that as long as she remains on the diet, her weight will approach 125 pounds while it decreases at a slower and slower rate. If her weight after x weeks of dieting is given by a formula of the form

$$W(x) = \frac{ax + b}{cx + d}$$

what will her weight be 3 weeks from now?
Suggestion: See Example 3.10.

42. **Population** When a major manufacturing firm relocated in a community 4 years ago, the population of the community was only 120,000. Three years ago the population was 125,000. A demographer believes that as long as the firm remains in the community, the population will approach 185,000 while it increases at a slower and slower rate. If population x years after the firm's relocation is given in thousands by a formula of the form

$$P(x) = \frac{ax + b}{cx + d}$$

what will the population be 6 years from now?
Suggestion: See Example 3.10.

43. **Advertising** Without spending any money on advertising, Crown Imports sells 200 barrels of Spanish olive oil weekly. Whenever the weekly advertising expenditure is $3000, the company sells 350 barrels of the oil weekly. As the weekly advertising expenditure increases, the weekly demand for the oil approaches 500 barrels while increasing at a decreasing rate. Suppose the weekly demand is given by a formula of the form

$$D(x) = \frac{ax + b}{cx + d}$$

when the weekly advertising expenditure is $$x$ thousand. How much must Crown Imports spend on advertising to sell 400 barrels of the oil weekly?
Suggestion: See Example 3.10.

44. **Labor Productivity** If a laundry uses 300 hours of labor daily, it launders 8 tons of laundry daily. Understandably, no laundry can be laundered without any labor. As the daily labor force increases, the amount laundered daily approaches 20 tons while increasing at a decreasing rate. Suppose the amount laundered daily is given by a formula of the form

$$L(x) = \frac{ax + b}{cx + d}$$

when the laundry uses x hundred hours of labor daily. How much labor must the laundry use to launder 15 tons of laundry daily?
Suggestion: See Example 3.10.

45. **Full-size Cars** Transportation officials project that the number of full-size automobiles will decrease. If x years from now this number is

$$A(x) = \frac{19x^2 + 1440}{x^2 + 48}$$

million, when will it be decreasing most rapidly?

46. **Public Transportation** Transportation experts predict that the demand for public transportation will increase in a large metropolitan area. If x years from now

$$T(x) = \frac{19x^2 + 270}{2x^2 + 54}$$

million people use public transportation, when will the demand be growing fastest?

47. **Median Age** When a housing development for retirees became available two years ago, the median age of the community was 29. Today the median age is 30. Suppose the median age approaches 37 as time passes. Also, suppose that x years after the development became available, the median age is given by a formula of the form

$$M(x) = \frac{ax^2 + b}{cx^2 + d}$$

When will the median age increase at the fastest rate? (Round your answer to the nearest year.)

48. **A Product's Popularity** Three months ago a new brand of soap was introduced in a community. Presently, 5 thousand people use the soap. Suppose the number of users approaches 33 thousand as time passes. Also, suppose that t months after the soap's introduction, this number is given in thousands by a formula of the form

$$S(t) = \frac{at^2 + b}{ct^2 + d}$$

When does the number of people that use the soap increase most rapidly? (Round your answer to the nearest month.)

49. **Gold Price** In a country, the price of gold is presently $360 per troy ounce. Officials predict that the price will approach $330 while declining at an increasing rate during the next 5 weeks and at a decreasing rate thereafter. Suppose that x weeks from now the price (in dollars per troy ounce) is given by a formula of the form

$$G(x) = \frac{ax^2 + b}{cx^2 + d}$$

What will the price be 20 weeks from now?

50. **Decontamination** A process has been developed for extracting a pollutant from a bay. Its effectiveness depends on how much money is spent applying it. As the expenditure is increased, the remaining percent of the pollutant approaches 20. It decreases at an increasing rate prior to the expenditure reaching $7 million and at a decreasing rate thereafter. Suppose the remaining percent of the pollutant is given by a formula of the form

$$P(x) = \frac{ax^2 + b}{cx^2 + d}$$

when the expenditure is x million. What percent of the pollutant remains if the expenditure is $10 million? (If no money is spent, 100 percent of the pollutant remains.)

51. **Retraining** The government of a city has decided to support a program that retrains the unemployed. Experts project that if the government spends x million on the program,

$$E(x) = \frac{90x}{x + 8}$$

percent of the unemployed become employed. This formula is of the form $(ax + b)/(cx + d)$ with $a/c > b/d$. Therefore, $E(x)$ increases at a decreasing rate as the government channels money into the program. The government plans to channel money into the program as long as the rate of increase of $E(x)$ is large enough. Specifically, the government will increase x (the total expenditure) until $E'(x)$ is reduced to 5. How much money will the government spend on the program by the time it stops supporting the program?

52. **Employee Productivity** The director of a training program in a toy factory believes that a new employee can assemble an average of

$$Y(x) = \frac{1000x}{2x + 5}$$

dolls per week after receiving x hours of training. This formula is of the form $(ax + b)/(cx + d)$ with $a/c > b/d$. Thus, the weekly productivity of a new employee increases at a decreasing rate as the training time is increased. The management plans to train a new employee as long as the rate of increase of the employee's weekly productivity is sufficiently great. Specifically, the management will increase x (the training time) until $Y'(x)$ is reduced to 8. How many hours of training will the management give a new employee?

53. **Learning** A student believes that the grade she will make on an examination is

$$G(x) = \frac{93x^2 + 1920}{x^2 + 192}$$

if she studies x hours. This formula is of the form $(ax^2 + b)/(cx^2 + d)$ with $a/c > b/d$. Therefore, as she increases her study time, her grade at first increases at an increasing rate and eventually increases at a decreasing rate. Consequently, there exists a point after which the returns due to further studying begin to diminish. Find this **point of diminishing returns**.

54. **Factory Productivity** The management of a manufacturing firm believes that

$$P(x) = \frac{53x^2}{2x^2 + 486}$$

metric tons of a product can be produced daily with a capital outlay of $\$x$ million. This formula is of the form $(ax^2 + b)/(cx^2 + d)$ with $a/c > b/d$. Therefore, as the capital outlay is increased, the daily productivity at first increases at an increasing rate and eventually increases at a decreasing rate. Thus, there exists a point after which the returns due to further increases in the capital outlay begin to diminish. Find this point of diminishing returns.

Graphing Calculator Problems

55. The function

$$f(x) = \frac{7x^2 + 294}{x^2 + 147}$$

is a function of the type described in Theorem 3.5. Set the viewing screen on your graphing calculator to $[0, 12] \times [0, 0.5]$ with a scale of 1 on both axes and draw the graph of $f'(x)$. Then explain how the graph of $f'(x)$ is compatible with the information we can obtain through Theorem 3.5 regarding the concavity of the graph of $f(x)$.

56. Do Problem 55 again. This time, however, let

$$f(x) = \frac{x^2 + 600}{x^2 + 75}$$

and set the viewing screen to $[0, 12] \times [0, -1]$ with a scale of 1 on both axes.

3.5 More on Rational Functions

In Section 3.4, we analyzed rational functions of the type

$$f(x) = \frac{ax + b}{cx + d}$$

and rational functions of the type

$$f(x) = \frac{ax^2 + b}{cx^2 + d}$$

In this section, we analyze other types of rational functions. We begin with the rational function

$$g(x) = \frac{x^2}{2x - 4}$$

Since the denominator of $x^2/(2x - 4)$ is zero only when $x = 2$, the domain of $g(x)$ is the set of all real numbers except 2. Figure 3.21 and

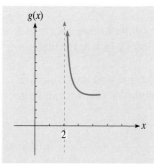

Figure 3.21 $g(x) = \dfrac{x^2}{2x - 4}$ for values of x close to and greater than 2.

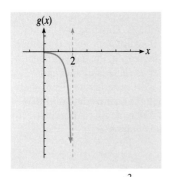

Figure 3.22 $g(x) = \dfrac{x^2}{2x - 4}$ for values of x close to and less than 2.

Table 3.1 both suggest that the values of $g(x)$ increase without bound as x approaches 2 from the right. The dashed vertical line in the figure is the graph of

$$x = 2$$

Table 3.1

x	4	3	2.5	2.1	2.01	2.001	2.0001
$g(x)$	4	4.5	6.25	22.05	202.005	2002.0005	20,002.00005

Figure 3.22 and Table 3.2 both suggest that the values of $g(x)$ decrease without bound as x approaches 2 from the left.

Table 3.2

x	0	1	1.5	1.9	1.99	1.999	1.9999
$g(x)$	0	−0.5	−2.25	−18.05	−198.005	−1998.0005	−19,998.00005

Since

$$g(x) = \frac{x^2}{2x - 4}$$

is a rational function for which the degree of the numerator is greater than the degree of the denominator, as we saw in Section 1.5, there is no number equal to $\lim\limits_{x \to +\infty} g(x)$. Nor is there a number equal to $\lim\limits_{x \to -\infty} g(x)$. Therefore, the graph of $g(x)$ has no horizontal asymptote.

However, the graph of

$$g(x) = \frac{x^2}{2x - 4}$$

does eventually get and stay arbitrarily close to a *slant line* as x increases without bound and also as x decreases without bound. In what follows, we find this line.

If we divide x^2 by $2x - 4$, the quotient is $0.5x + 1$ and the remainder is 4. (See Example A.12 in Section A.3 of Appendix A.) So we can write

$$\frac{x^2}{2x - 4} = 0.5x + 1 + \frac{4}{2x - 4}$$

which implies that

$$\frac{x^2}{2x - 4} - (0.5x + 1) = \frac{4}{2x - 4}$$

Therefore,
$$\lim_{x \to +\infty}\left(\frac{x^2}{2x-4} - (0.5x+1)\right) = \lim_{x \to +\infty} \frac{4}{2x-4}$$
$$= 0 \quad \text{Theorem 1.5, Section 1.5}$$

Similarly,
$$\lim_{x \to -\infty}\left(\frac{x^2}{2x-4} - (0.5x+1)\right) = 0$$

These two results tell us that the *difference* between the graph of $g(x)$ and the line
$$y = 0.5x + 1$$
eventually gets and stays arbitrarily small as x either increases or decreases without bound. Our awareness of this fact will help us when we sketch the graph of $g(x)$.

Next, we determine where $g(x)$ increases and where it decreases. Using the quotient rule, the first derivative of $g(x) = x^2/(2x-4)$ is
$$g'(x) = \frac{(2x-4)2x - x^2 \cdot 2}{(2x-4)^2}$$
$$= \frac{2x(x-4)}{(2x-4)^2}$$

Thus, we see that $g'(x) = 0$ when $x = 0$ and when $x = 4$. Since
$$g(0) = \frac{0^2}{2(0)-4} = 0$$
and
$$g(4) = \frac{4^2}{2(4)-4} = 4$$

the points $(0, 0)$ and $(4, 4)$ are first-order critical points on the graph of $g(x)$.

We use the point 2 along with the critical points 0 and 4 to subdivide the number line into intervals on which we examine the sign of $g'(x)$. The sign chart in Figure 3.23 analyzes the sign of $g'(x)$.

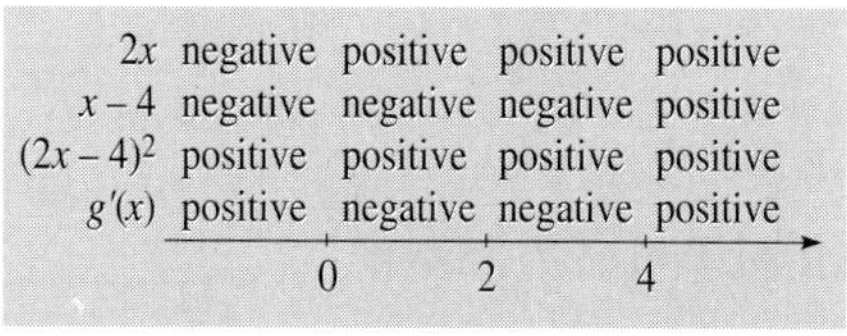

$2x$	negative	positive	positive	positive
$x-4$	negative	negative	negative	positive
$(2x-4)^2$	positive	positive	positive	positive
$g'(x)$	positive	negative	negative	positive
	0	2	4	

Figure 3.23 The sign chart for $g'(x) = 2x(x-4)/(2x-4)^2$.

Using Theorem 2.13 in Section 2.7, we conclude that $g(x)$ increases on the interval $(-\infty, 0)$ and on the interval $(4, +\infty)$. We also conclude that $g(x)$ decreases on the interval $(0, 2)$ and on the interval $(2, 4)$.

To determine the concavity of the graph of $g(x)$, we use the second derivative of $g(x)$. But first note that we can express the first derivative of $g(x)$ as follows:

$$g'(x) = \frac{2x^2 - 8x}{(2x - 4)^2}$$

Using the quotient rule and the generalized power rule, the second derivative of $g(x)$ is

$$g''(x) = \frac{(2x - 4)^2(4x - 8) - (2x^2 - 8x)2(2x - 4)2}{(2x - 4)^4}$$

$$= \frac{32}{(2x - 4)^3}$$

There is no number in the *domain* of $g(x)$ where $g''(x) = 0$ or $g''(x)$ fails to exist. So $g(x)$ has no second-order critical points. Figure 3.24 shows the sign chart for $g''(x)$.

Using the test for concavity (Theorem 3.1), we conclude that the graph of $g(x)$ is concave downward on the interval $(-\infty, 2)$ and concave upward on the interval $(2, +\infty)$. Since 2 is not in the domain, the graph does not have an inflection point.

We now have enough data to sketch the graph of $g(x) = x^2/(2x - 4)$. We begin by plotting the vertical line

$$x = 2$$

and the slant line

$$y = 0.5x + 1$$

Then we plot the first-order critical points, which are $(0, 0)$ and $(4, 4)$. [As we saw, $g(x)$ has no second-order critical points.] Finally, we sketch a curve that passes through these points and is compatible with the rest of the information we gathered. Our curve should look like the curve in Figure 3.25. The dashed slant line is the graph of $y = 0.5x + 1$.

Figure 3.24 The sign chart for $g''(x) = 32/(2x - 4)^3$.

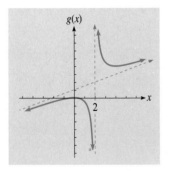

Figure 3.25 $g(x) = \dfrac{x^2}{2x - 4}$

Vertical Asymptotes

In our discussion of the function $g(x) = x^2/(2x - 4)$, we illustrated the idea of a vertical asymptote.

Definition 3.6

The vertical line

$$x = k$$

is a **vertical asymptote** of the graph of a function $f(x)$ if k is a number for which *at least one* of the following conditions holds:

1. As x approaches k from the left, the values of $f(x)$ either increase without bound or decrease without bound.
2. As x approaches k from the right, the values of $f(x)$ either increase without bound or decrease without bound.

According to Definition 3.6, the line $x = 2$ is a vertical asymptote of the graph of $g(x) = x^2/(2x - 4)$.

The function $g(x)$ suggests Theorem 3.6, which provides an easy method to determine the vertical asymptotes of a rational function.

Theorem 3.6

Suppose $f(x)$ is a *rational function* and k is a number. If at $x = k$ the denominator is zero but the numerator is not zero, then the line

$$x = k$$

is a vertical asymptote of the graph of $f(x)$.

The graph of a rational function whose denominator is never zero has no vertical asymptotes.

Slant Asymptotes

In our discussion of the function $g(x) = x^2/(2x - 4)$, we also illustrated the idea of a slant asymptote.

Definition 3.7

The line

$$y = mx + b$$

is a **slant asymptote** of the graph of a function $f(x)$ if $m \neq 0$ and

$$\lim_{x \to +\infty} (f(x) - (mx + b)) = 0$$

or

$$\lim_{x \to -\infty} (f(x) - (mx + b)) = 0$$

According to Definition 3.7, the line $y = 0.5x + 1$ is a slant asymptote of the graph of $g(x) = x^2/(2x - 4)$.

The function $g(x)$ suggests Theorem 3.7, which provides an easy method to determine slant asymptotes of rational functions.

Theorem 3.7

Suppose $f(x)$ is a rational function for which the degree of the numerator exceeds the degree of the denominator by 1. Then the quotient that results when the numerator is divided by the denominator has the form

$$mx + b$$

where $m \neq 0$. Moreover, the line

$$y = mx + b$$

is the slant asymptote of the graph of $f(x)$.

In Theorem 3.7, we assume the domain of $f(x)$ is large enough to include an interval of the type $(-\infty, a)$ and/or an interval of the type $(b, +\infty)$. Why?

The graph of a rational function has *no* slant asymptote if the degree of the numerator exceeds the degree of the denominator by more than 1.

Example 3.12

Suppose the total monthly cost of producing x tons of a product is

$$C(x) = 0.6x^2 + 15$$

thousand dollars for $x \geq 0$. Let $A(x)$ represent the average cost of producing each ton when the monthly production level is x tons. Sketch the graph of $A(x)$ for $x > 0$.

Solution

The formula for the average cost function in question is

$$A(x) = \frac{C(x)}{x}$$

$$= \frac{0.6x^2 + 15}{x}$$

At $x = 0$ the denominator is zero but the numerator is not zero. So, according to Theorem 3.6, the line

$$x = 0$$

(the vertical axis) is a vertical asymptote of the graph of $A(x)$.

The degree of the numerator exceeds the degree of the denominator by 1. Therefore, according to Theorem 3.7, the graph of $A(x)$ has a slant asymptote. When we divide $0.6x^2 + 15$ by x, we obtain $0.6x$ as the quotient. (See Example A.12 in Section A.3 of Appendix A.) So, using Theorem 3.7, we conclude that the slant asymptote of the graph of $A(x)$ is the line

$$y = 0.6x$$

Now we determine where $A(x)$ decreases and where it increases. Using the quotient rule,

$$A'(x) = \frac{x(1.2x) - (0.6x^2 + 15)}{x^2}$$

$$= \frac{0.6(x^2 - 25)}{x^2}$$

For $x > 0$, $A'(x) = 0$ only when $x = 5$. Thus, since

$$A(5) = \frac{0.6(5^2) + 15}{5}$$

$$= 6$$

the point $(5, 6)$ is a first-order critical point on the graph of $A(x)$. Note that for $x > 0$, $A'(x)$ is negative when $x < 5$ and positive when $x > 5$. So, using Theorem 2.13 in Section 2.7, we conclude that $A(x)$ decreases on the interval $(0, 5)$ and increases on the interval $(5, +\infty)$.

Next we determine the concavity of the graph of $A(x)$. Note that we can express the first derivative of $A(x)$ as follows:

$$A'(x) = \frac{0.6x^2 - 15}{x^2}$$

So the second derivative of $A(x)$ is

$$A''(x) = \frac{x^2(1.2x) - (0.6x^2 - 15)2x}{x^4}$$

$$= \frac{30}{x^3}$$

Therefore, $A''(x)$ is positive for all $x > 0$. Using the test for concavity, we conclude that the graph of $A(x)$ is concave upward throughout its domain, which is the interval $(0, +\infty)$.

Now we can sketch the graph of $A(x) = (0.6x^2 + 15)/x$ for $x > 0$. We begin by plotting the slant asymptote, which is the line

$$y = 0.6x$$

(Since the vertical asymptote is the vertical axis, we do not need to plot it.) Then we plot the only first-order critical point, which is $(5, 6)$. [The

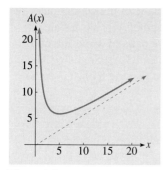

Figure 3.26
$A(x) = \dfrac{0.6x^2 + 15}{x}$ for $x > 0$.

graph of $A(x)$ has no second-order critical points.] Finally, we sketch a curve that passes through the point (5, 6) and is compatible with the rest of the information we gathered. Our curve should resemble the curve in Figure 3.26. The dashed line is the slant asymptote.

In Example 3.13, we analyze a rational function whose graph has more than one vertical asymptote but no slant asymptote.

Example 3.13

Sketch the graph of

$$h(x) = \frac{x^2 + 8}{x^2 - 4}$$

Solution

At $x = -2$ and at $x = 2$ the denominator of $h(x)$ is zero but the numerator is not zero. Therefore, according to Theorem 3.6, the lines

$$x = -2$$

and

$$x = 2$$

are vertical asymptotes of the graph of $h(x)$.

The degree of the numerator is equal to the degree of the denominator. Therefore, since 1 is the leading coefficient of both the numerator and denominator, using Theorem 1.4 in Section 1.5, we conclude that

$$\lim_{x \to +\infty} h(x) = \lim_{x \to -\infty} h(x) = 1$$

So, according to Definition 3.5 in Section 3.4, the line

$$y = 1$$

is a horizontal asymptote of the graph of $h(x)$.
The derivative of $h(x)$ is

$$h'(x) = \frac{-24x}{(x^2 - 4)^2}$$

Note that $h'(x) = 0$ at $x = 0$. Thus, since $h(0) = -2$, the point $(0, -2)$ is a first-order critical point on the graph of $h(x)$.

We use the points -2 and 2 along with the critical point 0 to subdivide the number line into intervals on which we examine the sign of $h'(x)$. Figure 3.27 shows the sign chart for $h'(x)$.

$-24x$	positive	positive	negative	negative
$(x^2 - 4)^2$	positive	positive	positive	positive
$h'(x)$	positive	positive	negative	negative
	-2	0	2	

Figure 3.27 The sign chart for $h'(x) = -24x/(x^2 - 4)^2$.

Using Theorem 2.13 in Section 2.7, we conclude that $h(x)$ increases on the interval $(-\infty, -2)$ and on the interval $(-2, 0)$. We also conclude that $h(x)$ decreases on the interval $(0, 2)$ and on the interval $(2, +\infty)$.

The second derivative of $h(x)$ is

$$h''(x) = \frac{24(3x^2 + 4)}{(x^2 - 4)^3}$$

There is no number in the *domain* of $h(x)$ where $h''(x) = 0$ or $h''(x)$ fails to exist. So $h(x)$ has no second-order critical points. Figure 3.28 shows the sign chart for $h''(x)$.

$24(3x^2 + 4)$	positive	positive	positive
$(x^2 - 4)^3$	positive	negative	positive
$h''(x)$	positive	negative	positive
	-2		2

Figure 3.28 The sign chart for $h''(x) = 24(3x^2 + 4)/(x^2 - 4)^3$.

Using the test for concavity, we conclude that the graph of $h(x)$ is concave upward on the interval $(-\infty, -2)$ and on the interval $(2, +\infty)$. We also conclude that the graph is concave downward on the interval $(-2, 2)$. Since -2 and 2 are not in the domain of $h(x)$, the graph does not have any inflection points.

Now we are ready to sketch the graph of $h(x) = (x^2 + 8)/(x^2 - 4)$. We begin by plotting the vertical and horizontal asymptotes. As we saw, the vertical asymptotes are $x = -2$ and $x = 2$. The horizontal asymptote is $y = 1$. Then we plot $(0, -2)$, the only first-order critical point. [As we saw, $h(x)$ has no second-order critical points.] Finally, we sketch a curve that passes through the point $(0, -2)$ and is compatible with the rest of the information we gathered. Our curve should look like the curve in Figure 3.29. The vertical dashed lines are the vertical asymptotes.

In Example 3.14, we analyze a rational function whose graph passes through its horizontal asymptote.

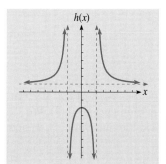

Figure 3.29 $h(x) = \dfrac{x^2 + 8}{x^2 - 4}$

Example 3.14

Sketch the graph of

$$i(x) = \frac{6x}{x^2 + 3}$$

Solution

The denominator is never zero. Therefore, the graph of $i(x)$ has no vertical asymptotes. The degree of the numerator is less than the degree of the denominator. Therefore, using Theorem 1.5 in Section 1.5, we conclude that

$$\lim_{x \to +\infty} i(x) = \lim_{x \to -\infty} i(x) = 0$$

So, according to Definition 3.5 in Section 3.4, the line

$$y = 0$$

(the x axis) is a horizontal asymptote of the graph of $i(x)$.

The derivative of $i(x)$ is

$$i'(x) = \frac{6(3 - x^2)}{(x^2 + 3)^2}$$

Note that $i'(x) = 0$ at $x = -\sqrt{3}$ and $x = \sqrt{3}$. Thus, since $i(-\sqrt{3}) = -\sqrt{3}$ and $i(\sqrt{3}) = \sqrt{3}$, the points $(-\sqrt{3}, -\sqrt{3})$ and $(\sqrt{3}, \sqrt{3})$ are first-order critical points on the graph of $i(x)$.

Figure 3.30 shows the sign chart for $i'(x)$.

$6(3 - x^2)$	negative	positive	negative
$(x^2 + 3)^2$	positive	positive	positive
$i'(x)$	negative	positive	negative
	$-\sqrt{3}$		$\sqrt{3}$

Figure 3.30 The sign chart for $i'(x) = 6(3 - x^2)/(x^2 + 3)^2$.

Using Theorem 2.13 in Section 2.7, we conclude that $i(x)$ decreases on the interval $(-\infty, -\sqrt{3})$ and on the interval $(\sqrt{3}, +\infty)$. We also conclude that $i(x)$ increases on the interval $(-\sqrt{3}, \sqrt{3})$.

The second derivative of $i(x)$ is

$$i''(x) = \frac{12x(x^2 - 9)}{(x^2 + 3)^3}$$

Note that $i''(x) = 0$ at $x = -3$, $x = 0$, and $x = 3$. Thus, since $i(-3) = -1.5$, $i(0) = 0$, and $i(3) = 1.5$, the points $(-3, -1.5)$, $(0, 0)$, and $(3, 1.5)$ are second-order critical points on the graph of $i(x)$.

Figure 3.31 shows the sign chart for $i''(x)$.

$12x$	negative	negative	positive	positive
$x^2 - 9$	positive	negative	negative	positive
$(x^2 + 3)^3$	positive	positive	positive	positive
$i''(x)$	negative	positive	negative	positive
	-3	0	3	

Figure 3.31 The sign chart for $i''(x) = 12x(x^2 - 9)/(x^2 + 3)^3$.

Using the test for concavity, we conclude that the graph of $i(x)$ is concave downward on the interval $(-\infty, -3)$ and on the interval $(0, 3)$. We also conclude that the graph is concave upward on the interval $(-3, 0)$ and on the interval $(3, +\infty)$.

We now have enough information to sketch the graph of $i(x) = 6x/(x^2 + 3)$. We begin by plotting the first-order critical points, which are $(-\sqrt{3}, -\sqrt{3})$ and $(\sqrt{3}, \sqrt{3})$. We also plot the second-order

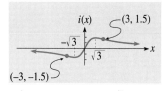

Figure 3.32 $i(x) = \dfrac{6x}{x^2 + 3}$

critical points, namely, $(-3, -1.5)$, $(0, 0)$, and $(3, 1.5)$. Then we sketch a curve that passes through these five points and is compatible with the other information we gathered. Our curve should look like the curve in Figure 3.32. Note that all three second-order critical points are inflection points. Note also that the curve passes through its horizontal asymptote, which is the x axis.

3.5 Problems

In Problems 1 to 26, find any existing horizontal, slant, and vertical asymptotes of the graph.

1. $f(x) = \dfrac{x + 1}{x^3 - 8}$

2. $f(x) = \dfrac{x^2 - 2x + 5}{3x^2 - 27}$

3. $f(x) = \dfrac{5x^2 + 1}{2x - 8}$

4. $f(x) = \dfrac{3}{x^2 + x + 5}$

5. $f(x) = \dfrac{3x + 2}{4x + 24}$

6. $f(x) = \dfrac{8x^3}{x^2 + 3}$

7. $f(x) = \dfrac{7}{x^2 + 2x + 3}$

8. $f(x) = \dfrac{9x - 2}{x^3 - 64}$

9. $f(x) = \dfrac{3x^2 + 2x + 1}{x^2 - 25}$

10. $f(x) = \dfrac{x + 2}{4x + 12}$

11. $f(x) = \dfrac{x^3 + 1}{x^2 + 7x}$

12. $f(x) = \dfrac{x^2 - 4}{3x - 27}$

13. $f(x) = \dfrac{x^2}{x^2 - 2x - 3}$

14. $f(x) = \dfrac{3x^2 + 7}{2x^2 - 20x + 32}$

15. $f(x) = \dfrac{x^2}{5x^2 + 1}$

16. $f(x) = \dfrac{x^3 - 7}{5x}$

17. $f(x) = \dfrac{2x^4}{5x^2 + 1}$

18. $f(x) = \dfrac{x^2 + 3}{x^2 - 14x + 49}$

19. $f(x) = \dfrac{5x^2}{2x^2 - 12x + 18}$

20. $f(x) = \dfrac{x^5 - 8}{x^2 + 2}$

21. $f(x) = \dfrac{12x^2 - 5x + 1}{3x^2}$

22. $f(x) = \dfrac{x - 6}{x^2}$

23. $f(x) = \dfrac{x - 6}{x^2 - 12x + 36}$

24. $f(x) = \dfrac{x^2 - x}{2x^2 - 12x + 10}$

25. $f(x) = \dfrac{x^2 - x}{4x^2 + 8x - 12}$

26. $f(x) = \dfrac{3x^2 - 3x - 6}{x^2 - 4x + 4}$

In Problems 27 to 36, sketch the graph of a function $f(x)$ that satisfies the listed conditions.

27. The horizontal asymptote is the x axis.
The graph has no vertical asymptotes.
$f'(x)$ is positive for $x < 0$ and negative for $x > 0$.
$f(0) = 5$.
$f''(x)$ is positive for $x < -2$ and for $x > 2$.
$f''(x)$ is negative for $-2 < x < 2$.
$f(-2) = 2.5$ and $f(2) = 2.5$.

28. The horizontal asymptote is the line $y = 2$.
The graph has no vertical asymptotes.
$f'(x)$ is positive for $x < 3$ and negative for $x > 3$.
$f(3) = 6$.
$f''(x)$ is positive for $x < 1$ and for $x > 5$.
$f''(x)$ is negative for $1 < x < 5$.
$f(1) = 4$ and $f(5) = 4$.

29. The line $y = 1$ is the horizontal asymptote.
The only vertical asymptote is the line $x = 4$.
$f'(x)$ is positive for $x < 4$ and for $x > 4$.
$f''(x)$ is positive for $x < 4$ and negative for $x > 4$.

30. The x axis is the horizontal asymptote.
The only vertical asymptote is the line $x = 3$.
$f'(x)$ is negative for $x < 3$ and for $x > 3$.
$f''(x)$ is negative for $x < 3$ and positive for $x > 3$.

31. The slant asymptote is the line $y = 0.75x - 1$.
The only vertical asymptote is the line $x = 3$.
$f'(x)$ is positive for $x < 0$ and for $x > 6$.
$f'(x)$ is negative for $0 < x < 3$ and for $3 < x < 6$.
$f(0) = -2$ and $f(6) = 5$.
$f''(x)$ is negative for $x < 3$ and positive for $x > 3$.

32. The slant asymptote is the line $y = -0.5x$.
    The only vertical asymptote is the y axis.
    $f'(x)$ is negative for $x < -4$ and for $x > 4$.
    $f'(x)$ is positive for $-4 < x < 0$ and for $0 < x < 4$.
    $f(-4) = 3$ and $f(4) = -3$.
    $f''(x)$ is positive for $x < 0$ and negative for $x > 0$.

33. The horizontal asymptote is the x axis.
    The only vertical asymptote is the y axis.
    $f'(x)$ is positive for $x < 0$ and for $x > 2$.
    $f'(x)$ is negative for $0 < x < 2$.
    $f(2) = -4$.
    $f''(x)$ is positive for $x < 0$ and for $0 < x < 7$.
    $f''(x)$ is negative for $x > 7$.
    $f(7) = -2$.

34. The horizontal asymptote is the line $y = 2$.
    The only vertical asymptote is the y axis.
    $f'(x)$ is positive for $x < 0$ and for $0 < x < 3$.
    $f'(x)$ is negative for $x > 3$.
    $f(3) = 4$.
    $f''(x)$ is positive for $x < 0$ and for $x > 6$.
    $f''(x)$ is negative for $0 < x < 6$.
    $f(6) = 3$.

35. The horizontal asymptote is the x axis.
    The lines $x = -3$ and $x = 3$ are the vertical asymptotes.
    $f'(x)$ is negative for $x < -3$, for $-3 < x < 0$, for $0 < x < 3$, and for $x > 3$.
    $f''(x)$ is negative for $x < -3$ and for $0 < x < 3$.
    $f''(x)$ is positive for $-3 < x < 0$ and for $x > 3$.
    $f(0) = 0$.

36. The horizontal asymptote is the x axis.
    The lines $x = -2$ and $x = 2$ are the vertical asymptotes.
    $f'(x)$ is negative for $x < -2$ and for $x > 2$.
    $f'(x)$ is positive for $-2 < x < 0$ and for $0 < x < 2$.
    $f''(x)$ is negative for $x < -2$ and for $-2 < x < 0$.
    $f''(x)$ is positive for $0 < x < 2$ and for $x > 2$.
    $f(0) = 0$.

In Problems 37 to 44, sketch the graph of the given function.

37. $f(x) = \dfrac{12}{x^2 - 14x + 52}$

38. $f(x) = \dfrac{x^2 + 4x + 1}{x^2 + 1}$

39. $f(x) = \dfrac{x^2 + 16x - 32}{x^2}$

40. $f(x) = \dfrac{2x + 3}{x + 1}$

41. $f(x) = \dfrac{4x^3}{243(x - 3)}$

42. $f(x) = \dfrac{x^2 - 15}{x - 4}$

43. $f(x) = \dfrac{x^2 - 9}{2x}$

44. $f(x) = \dfrac{32x - 64}{x^2}$

45. **House Affordability** In a region of the United States, approximately 18 percent of the families can afford houses costing $80,000. A team of financial experts predicts that x months from now this percent will be

$$H(x) = \dfrac{20x^2 - 120x + 372}{x^2 - 6x + 21}$$

Sketch the graph of $H(x)$ for $x \geq 0$. Then describe how the percent will behave as time passes.

46. **Defense Spending** Officials predict that x years from now a nation will spend

$$D(x) = \dfrac{8x^2 - 64x + 425}{x^2 - 8x + 43}$$

percent of its gross domestic product on defense. Sketch the graph of $D(x)$ for $x \geq 0$. Then describe how the percent will behave as time passes.

Fighters like this F-22 prototype will go into production in December 1998. (Courtesy Boeing Defense and Space Group)

47. **Demand** Market researchers believe that x months from now a product's daily demand will be

$$D(x) = \dfrac{3x^2 + 20x + 12}{x^2 + 4}$$

million barrels. Sketch the graph of $D(x)$ for x in the interval $[0, +\infty)$. Then describe how the daily demand will behave as time passes.

48. **Insect Infestation** An orange grove has just been sprayed with a biodegradable insecticide. Suppose that x days from now each acre is infested with

$$I(x) = \frac{5x^2 - 30x + 125}{x^2 + 25}$$

hundred thousand insects. Sketch the graph of $I(x)$ for $x \geq 0$. Then describe how the insect population will behave as time passes.

49. **Advertising** The amount a company spends on advertising a product depends on the percent of the market the company wants to capture. Company officials believe that to capture x percent of the market the monthly advertising expenditure should be

$$A(x) = \frac{5x}{300 - 3x}$$

thousand dollars, where $0 \leq x < 100$. Sketch the graph of $A(x)$. Then explain how the graph shows that no matter how much the company spends on advertising, it cannot capture 100 percent of the market.

50. **Fertilization** A farmer can produce up to 50 tons of a crop without using fertilizer. The farmer finds that to produce x tons of the crop he must use

$$F(x) = \frac{50 - x}{x - 122}$$

sacks of fertilizer, where $50 \leq x < 122$. Sketch the graph of $F(x)$. Then explain how the graph shows that no matter how much fertilizer he uses, he cannot produce 122 tons of the crop.

51. Explain why the graph of a rational function can have at most one horizontal asymptote.

52. Write a theorem that provides an easy method for finding horizontal asymptotes of rational functions. The theorem should cover three cases.

3.6 Percentage Rate of Change

Suppose a building appreciates at the rate of $9000 per year. If the value of the building is $80,000, we are inclined to say the building appreciates rapidly. If the value of the building is $400,000, we are inclined to say it appreciates slowly. This situation suggests that we need a measure of how fast a quantity changes that takes the quantity into consideration. In this section, we introduce such a measure. But first we examine an environmental situation that also suggests the need for the measure.

The concentration of EDB (ethylene dibromide) in Lake Blue Rock has been increasing. Let $c(t)$ be the concentration (in milligrams of EDB per kiloliter of water) t months from now. Environmentalists believe that

$$c(t) = 0.01t^2 + t + 25$$

for values of t in the interval $[0, 20]$.

The derivative of the function $c(t)$ is

$$c'(t) = 0.02t + 1$$

Note that $c'(t)$ is positive for values of t in the interval $[0, 20]$. Thus, the concentration will continue to increase during the next 20 months. Note also that $c'(t)$ increases as t increases. So, since $c'(t)$ is the rate at which the concentration will increase t months from now, the concentration, in fact, will increase at an increasing rate. To illustrate,

$$c'(5) = 0.02(5) + 1$$
$$= 1.1$$

while

$$c'(15) = 0.02(15) + 1$$
$$= 1.3$$

Therefore, 5 months from now the concentration will be increasing at the rate of 1.1 milligrams per month, while 15 months from now the concentration will be increasing at the rate of 1.3 milligrams per month.

Substituting 5 and then 15 in the formula for $c(t)$, we get

$$c(5) = 0.01(5^2) + 5 + 25$$
$$= 30.25$$

and

$$c(15) = 0.01(15^2) + 15 + 25$$
$$= 42.25$$

Thus, 5 months from now the concentration will be 30.25 milligrams, while 15 months from now the concentration will be 42.25 milligrams. Since 1.1 is about 3.6 percent of 30.25, five months from now the concentration's increase rate will be about 3.6 percent of the concentration. Since 1.3 is about 3.1 percent of 42.25, fifteen months from now the concentration's increase rate will be about 3.1 percent of the concentration. So relative to the concentration's level, the concentration will be increasing less rapidly 15 months from now than 5 months from now. The environmentalists are somewhat consoled by this fact.

As with the building situation, the environmental situation suggests that we need a measure of how fast a quantity changes that takes the quantity into consideration. Definition 3.8 provides such a measure.

Definition 3.8

Suppose $f(x)$ is a function and c is a number such that $f(c)$ is positive. The **percentage rate of change** of $f(x)$ with respect to x at $x = c$ is

$$100 \cdot \frac{f'(c)}{f(c)}$$

The reasonableness of the above definition is based on the fact that $f'(c)$ is

$$100 \cdot \frac{f'(c)}{f(c)}$$

percent of $f(c)$.

Using Definition 3.8, we can derive a formula that gives the percentage rate of increase of the EDB concentration at any instant during the next 20 months. Let $P(t)$ represent the concentration's percentage rate of increase t months from now. According to Definition 3.8,

$$P(t) = 100 \cdot \frac{c'(t)}{c(t)}$$
$$= 100 \cdot \frac{0.02t + 1}{0.01t^2 + t + 25}$$

$$= \frac{2t + 100}{0.01t^2 + t + 25}$$

$$= \frac{200t + 10,000}{t^2 + 100t + 2500} \quad \text{multiplying by } \frac{100}{100}$$

If we substitute 5 for t in the formula for $P(t)$, we get

$$P(5) = 3\tfrac{7}{11}$$

which is approximately 3.6. Thus, 5 months from now the EDB concentration will be increasing at a percentage rate close to 3.6 percent per month. If we substitute 15 for t in the formula for $P(t)$, we get

$$P(15) = 3\tfrac{1}{13}$$

which is approximately 3.1. Thus, 15 months from now the EDB concentration will be increasing at a percentage rate close to 3.1 percent per month. So, although the concentration will increase at an increasing rate, the percentage rate of increase will be less 15 months from now than 5 months from now. Actually, the concentration's percentage rate of increase will decrease throughout the next 20 months. We can verify this claim by examining the sign of the derivative of $P(x)$.

Using the quotient rule, the derivative of

$$P(t) = \frac{200t + 10,000}{t^2 + 100t + 2500}$$

is

$$P'(t) = \frac{(t^2 + 100t + 2500)200 - (200t + 10,000)(2t + 100)}{(t^2 + 100t + 2500)^2}$$

$$= \frac{-200}{t^2 + 100t + 2500}$$

$$= \frac{-200}{(t + 50)^2}$$

Since $(t + 50)^2$ is positive for all values of t except $t = -50$, the value of $P'(t)$ is *negative* for all values of t except $t = -50$. In particular, $P'(t)$ is negative when $0 \le t \le 20$. So $P(t)$ decreases on the interval $[0, 20]$. Thus, as we claimed, the EDB concentration will increase at a decreasing percentage rate during the next 20 months. In other words, in terms of percentages, the concentration will increase at a slower and slower rate during the next 20 months.

Figure 3.33 shows the graph of $P(t)$. The graph is compatible with our finding, namely, that the concentration's percentage rate of increase will decrease during the next 20 months.

Example 3.15 involves a function whose dependent variable decreases at an increasing percentage rate.

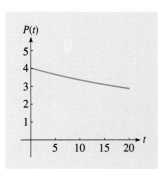

Figure 3.33
$P(t) = \dfrac{200t + 10,000}{t^2 + 100t + 2500}$
for $0 \le t \le 20$.

Example 3.15

A machine used in manufacturing suitcases will depreciate linearly during the next 10 years. To be more specific, an accountant projects that t years from now the value of the machine will be

$$V(t) = -2t + 30$$

thousand dollars for $0 \leq t \leq 10$. Since the machine will depreciate linearly, its value will decrease at a constant rate. In fact, it will decrease at the constant rate of $2000 per year.

a. Find a formula for the percentage rate of change of the value t years from now.
b. At what percentage rate will the value be decreasing 5 years from now? 7 years from now?
c. We already know that the value will decrease during the next 10 years. Show that it will decrease at an increasing percentage rate.

Solution for (a)

Let $P(t)$ represent the percentage rate of change of the machine's value t years from now. According to Definition 3.8,

$$P(t) = 100 \cdot \frac{V'(t)}{V(t)}$$

$$= 100 \cdot \frac{-2}{-2t + 30}$$

$$= \frac{100}{t - 15}$$

Note that $P(t)$ is negative for $t < 15$. In particular, $P(t)$ is negative for $0 \leq t \leq 10$, which is compatible with the fact the machine will depreciate during the next 10 years.

Solution for (b)

Substituting 5 in the formula for $P(t)$, we get

$$P(5) = \frac{100}{5 - 15}$$

$$= -10$$

So 5 years from now the value will be decreasing at the rate of 10 percent per year. Substituting 7 in the formula for $P(t)$, we get

$$P(7) = \frac{100}{7 - 15}$$

$$= -12.5$$

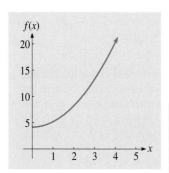

Figure 3.34 $P(t) = \dfrac{100}{t-15}$ for $0 \leq t \leq 10$.

So 7 years from now the value will be decreasing at the rate of 12.5 percent per year.

Solution for (c)

The formula for the percentage rate of change can be expressed as

$$P(t) = 100(t-15)^{-1}$$

Using the generalized power rule, the derivative of $P(t)$ is

$$P'(t) = -100(t-15)^{-2}$$
$$= \dfrac{-100}{(t-15)^2}$$

Since $(t-15)^2$ is positive for all values of t except $t = 15$, the value of $P'(t)$ is *negative* for all values of t except $t = 15$. In particular, $P'(t)$ is negative when $0 \leq t \leq 10$. Thus, $P(t)$ decreases on the interval $[0, 10]$. Therefore, since $P(t)$ is negative for each t in $[0, 10]$, the magnitude (absolute value) of $P(t)$ increases on $[0, 10]$. This result means that the value of the machine will decrease at an increasing percentage rate during the next 10 years.

Figure 3.34 shows the graph of the percentage rate of change function we derived in Example 3.15. The graph is consistent with what we concluded, namely, that the machine's percentage rate of depreciation will increase during the next 10 years.

Example 3.16 involves a function whose dependent variable at first increases at an increasing percentage rate, but eventually increases at a decreasing percentage rate.

Example 3.16

Let

$$f(x) = x^2 + 4$$

for $x \geq 0$. Show that $f(x)$ increases on the interval $[0, +\infty)$. Then find the value of x where the percentage rate of increase of $f(x)$ is greatest.

Solution

The derivative of $f(x)$ is

$$f'(x) = 2x$$

Since $f'(x)$ is positive when $x > 0$, $f(x)$ indeed increases on the interval $[0, +\infty)$. In fact, as Figure 3.35 shows, $f(x)$ increases at an increasing rate.

Figure 3.35 $f(x) = x^2 + 4$ for $x \geq 0$.

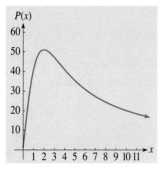

Figure 3.36 $P(x) = \dfrac{200x}{x^2 + 4}$ for $x \geq 0$.

According to Definition 3.8, a formula for the percentage rate of increase of $f(x)$ is

$$P(x) = 100 \cdot \frac{f'(x)}{f(x)}$$

$$= 100 \cdot \frac{2x}{x^2 + 4}$$

$$= \frac{200x}{x^2 + 4}$$

Using the quotient rule, the derivative of $P(x)$ is

$$P'(x) = \frac{(x^2 + 4)200 - (200x)2x}{(x^2 + 4)^2}$$

$$= \frac{200(4 - x^2)}{(x^2 + 4)^2}$$

For x in $[0, +\infty)$, the value of $P'(x)$ is positive when $x < 2$ and negative when $x > 2$. Thus, $P(x)$ increases as x increases toward 2 and decreases after x passes 2. This result means that $f(x)$ increases at an increasing percentage rate as x increases toward 2 and at a decreasing percentage rate after x passes 2. Therefore, the percentage rate of increase of $f(x)$ is greatest at $x = 2$. Figure 3.36 shows the graph of $P(x)$.

In Chapter 4, we will deal with functions whose dependent variables change at a *constant* percentage rate.

3.6 Problems

In Problems 1 to 12, find a formula for the percentage rate of change of $f(x)$ with respect to x. Then compute the percentage rate of change at the given x value.

1. $f(x) = -0.5x + 12$ $x = 6$
2. $f(x) = 3x + 15$ $x = 120$
3. $f(x) = \frac{1}{30}x^2 + 3$ $x = 10$
4. $f(x) = -0.05x^2 - 0.1x + 130$ $x = 30$
5. $f(x) = \sqrt{x^2 + 100}$ $x = 20$
6. $f(x) = \frac{1}{15}(x^3 + 92)^5$ $x = 2$
7. $f(x) = \dfrac{35x + 90}{x + 2}$ $x = 5$
8. $f(x) = \dfrac{40x + 65}{x + 1}$ $x = 9$
9. $f(x) = \dfrac{500}{(x + 1)^2}$ $x = 99$
10. $f(x) = \dfrac{600}{\sqrt{x + 5}}$ $x = 25$
11. $f(x) = x\sqrt{x^2 + 10}$ $x = 10$
12. $f(x) = x^2\sqrt{x + 9}$ $x = 10$

13. **Gasoline Price** Experts predict that the price of gasoline will increase linearly during the next few weeks. In fact, they predict that t weeks from now the price will be

$$f(t) = 0.05t + 1.1$$

dollars per gallon, where $0 \leq t \leq 10$.

    a. Find a formula for the percentage rate of increase of the price t weeks from now.

b. At what percentage rate will the price be increasing 3 weeks from now? 8 weeks from now?

c. Since the price will increase linearly, it will increase at a constant rate. The percentage rate of increase, however, will decrease. Verify this fact.

14. **Population** Demographers predict that the population of an island will be

$$f(x) = 0.04x^2 + 2x + 8$$

million x years from now.

a. Use the first and second derivatives of $f(x)$ to show that the population will increase at an increasing rate.

b. Find a formula for the percentage rate of increase of the population x years from now.

c. At what percentage rate will the population be increasing 5 years from now? 10 years from now?

d. Although the population will increase at an increasing rate, the percentage rate of increase will decrease. Verify this fact.

15. **Phosphate Depletion** Geologists project that the supply of mineable phosphate in a region of the United States will decrease. To be more specific, they project that t years from now the supply will be

$$f(t) = \frac{70{,}000}{t + 10}$$

billion tons. Find a formula for the percentage rate at which the supply will be changing t years from now. Then show that the supply will decrease at a decreasing percentage rate.

16. **A Spice's Value** The value of the dollar relative to a rare spice is decreasing. Let $V(t)$ represent the number of ounces of the spice that one dollar can buy t months from now. Suppose that during the next 30 months $V(t)$ decreases linearly according to the formula

$$V(t) = -0.25t + 25$$

Find a formula for the percentage rate at which the value of the dollar will be changing t years from now. Then show that the value of the dollar will decrease at an increasing percentage rate.

17. **Appreciation** A real estate appraiser believes that a certain lot will be worth

$$V(x) = 0.5x^2 + 32$$

thousand dollars x years after the beginning of 1993. According to her belief, the value will increase always at an increasing rate. But the percentage rate of increase of the value will increase only for a while. Eventually, it will decrease. When will the percentage rate of increase be greatest? What will it be at that time?

18. **Toucan Population** A conservationist projects that x years after the beginning of 1994,

$$T(x) = 0.4x^2 + 10$$

thousand toucans will inhabit a certain region in South America. According to his projection, the number of toucans will increase always at an increasing rate. But the percentage rate of increase will increase only for a while. Eventually, it will decrease. When will the percentage rate of increase be greatest? What will it be at that time?

19. **Investing $1 Million** Suppose you want to invest $1 million for a period of 10 years, and only two investment plans are available: plan A and plan B. In plan A, the balance will increase at the constant *rate* of $0.2 million per year. In plan B, the balance will increase at the constant *percentage rate* of 8 percent per year. How should you invest the $1 million?
Hint: First find a formula for the balance after t years if $1 million is invested in plan A.

20. **Investing $1 Million** Do Problem 19 under the new assumption that in plan A the balance will increase at the constant *rate* of $0.1 million per year, and in plan B the balance will increase at the constant *percentage rate* of 6 percent per year.

Graphing Calculator Problems

21. **Consumer Debt** Economists predict that x months from now the consumer debt of a country will be

$$d(x) = \frac{20x^2 + 4320}{x^2 + 432}$$

billion dollars. According to this prediction, the debt will increase in a manner whereby both the rate of increase and the percentage rate of increase eventually peak, but not at the same time.

a. Set the viewing screen on your graphing calculator to [0, 27] × [0, 0.5] with a scale of 1 on both axes and draw the graph of the debt's rate of increase function. Then use the trace function to estimate when the rate of increase will peak. Approximately what will the rate of increase be when it peaks?

b. Set the viewing screen on your graphing calculator to [0, 27] × [0, 3] with a scale of 1 on both axes and draw the graph of the debt's percentage rate of increase function. Then use the trace function to estimate when the percentage rate of increase will peak. Approximately what will the percentage rate of increase be when it peaks?

22. **Insect Population** An entomologist projects that x weeks from now an insect population will be

$$E(x) = 0.005x^3 + 0.015x^2 + 0.015x + 8$$

million, where $0 \leq x \leq 36$. According to this projection, the population will increase at an increasing rate. The percentage rate of increase, however, will increase for a while and then decrease.

a. Set the viewing screen on your graphing calculator to [0, 36] × [0, 24] with a scale of 1 on both axes and draw the graph of $E'(x)$. Then explain how the graph shows that the population will increase at an increasing rate.

b. With the viewing screen set as in part (a), draw the graph of the population's percentage rate of increase function. Then use the trace function to estimate when the percentage rate of increase will be greatest. Approximately what will the percentage rate of increase be when it is greatest?

c. The population's percentage rate of increase will be 10 percent on two occasions. Use the trace function to estimate when this percentage rate of increase occurs.

3.7 Newton's Method for Approximating Zeros

A **zero** of a function $f(x)$ is a number c for which

$$f(c) = 0$$

So far, we have found the zeros of functions (mostly derivatives) either by inspection, by factoring, or by using the quadratic formula. For some functions, however, these methods are either not practical or not applicable. The following problem involves such a function.

A shrimp tycoon believes that the monthly demand for shrimp is x million pounds when the price is

$$g(x) = -\tfrac{3}{4}x + 12$$

dollars per pound. (This formula is valid up to $x = 16$.) The tycoon also believes that shrimpers supply the market with x million pounds of shrimp per month when the price is

$$h(x) = \tfrac{1}{64}x^3$$

dollars per pound. We want to determine how much shrimp is supplied monthly when the market is in **equilibrium**. In other words, we want to determine how much shrimp is supplied monthly at the price for which the monthly demand equals the monthly supply. Thus, we want to find the value of x for which

$$g(x) = h(x)$$

If we let

$$E(x) = g(x) - h(x)$$
$$= (-\tfrac{3}{4}x + 12) - \tfrac{1}{64}x^3$$
$$= -\tfrac{1}{64}x^3 - \tfrac{3}{4}x + 12$$

the problem amounts to finding a zero of the function $E(x)$. However, none of the methods we have used so far for finding zeros of functions can be used to find the zeros of this function.

In this section, we introduce a method for approximating zeros of functions. The method was developed by Isaac Newton and, therefore, is called **Newton's method**. As we will see, Theorem 3.8 is a basis for the method.

Theorem 3.8

Suppose $f(x)$ is differentiable at $x = n$ and $f'(n) \neq 0$. Then the tangent to the graph of $f(x)$ at the point $(n, f(n))$ crosses the x axis at

$$n - \frac{f(n)}{f'(n)}$$

Figure 3.37 shows a graphical representation of Theorem 3.8.

Proof of Theorem 3.8

The assumption that $f'(n) \neq 0$ implies that the tangent in question is not horizontal. Thus, the tangent indeed crosses the x axis, say at $x = x_0$. We want to show that $x_0 = n - f(n)/f'(n)$.

Since $(n, f(n))$ and $(x_0, 0)$ are points on the tangent, the slope of the tangent is

$$\frac{0 - f(n)}{x_0 - n} = \frac{-f(n)}{x_0 - n}$$

But the slope of the tangent is also $f'(n)$. Therefore,

$$f'(n) = \frac{-f(n)}{x_0 - n}$$

We now solve this equation for x_0:

$$(x_0 - n)f'(n) = -f(n)$$
$$x_0 - n = \frac{-f(n)}{f'(n)}$$
$$x_0 = n - \frac{f(n)}{f'(n)}$$

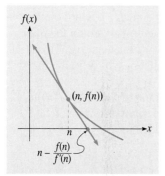

Figure 3.37 A graphical representation of Theorem 3.8.

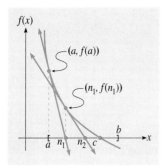

Figure 3.38 The idea behind Newton's method.

Since x_0 represents the point where the tangent crosses the x axis, the proof is complete.

Before we show how Newton's method for approximating zeros uses Theorem 3.8, we will illustrate the idea on which the method is based.

The Idea Behind Newton's Method

Let $f(x)$ be the function whose graph is the curve in Figure 3.38. The figure shows that the graph crosses the x axis at $x = c$. Thus, c is a zero of $f(x)$. The figure also shows that the tangent at the point $(a, f(a))$ crosses the x axis at $x = n_1$, and the tangent at the point $(n_1, f(n_1))$ crosses the x axis at $x = n_2$. Note that n_2 approximates c better than n_1 approximates c.

The value of n_1 is the approximation for c obtained through one application of Newton's method starting at $x = a$. Similarly, n_2 is the approximation for c obtained through two applications of Newton's method starting at $x = a$.

Figure 3.38 does not show the tangent to the graph of $f(x)$ at the point $(n_2, f(n_2))$. Nevertheless, if we let n_3 represent the x coordinate of the point where this tangent crosses the x axis, then n_3 would approximate c better than n_2 approximates c. Moreover, n_3 would be the approximation for c obtained through three applications of Newton's method starting at $x = a$.

For each positive integer k, let n_k represent the approximation for c obtained through k applications of Newton's method starting at $x = a$. The above discussion correctly suggests that for any positive integer k, the value of n_{k+1} approximates c better than n_k approximates c. In fact, as k increases without bound, the values of n_k eventually get and stay arbitrarily close to c. Thus, c is the limit of n_k as k approaches positive infinity. This statement is expressed by writing

$$\lim_{k \to +\infty} n_k = c$$

We started applying Newton's method at $x = a$, rather than at $x = b$, because $f(a)$ is positive and the graph of $f(x)$ is concave upward on the interval $[a, b]$. If the graph were instead concave downward on the interval $[a, b]$, we would have started at $x = b$ since $f(b)$ is negative. This strategy for determining where to start guarantees that n_1 will be in the interval $[a, b]$.

Figure 3.38 shows that the tangent to the graph of $f(x)$ at $(a, f(a))$ is neither vertical nor horizontal. Thus, $f(x)$ is differentiable at $x = a$ and $f'(a) \neq 0$. So we can use Theorem 3.8 to conclude that

$$n_1 = a - \frac{f(a)}{f'(a)}$$

Similarly, we can use Theorem 3.8 to conclude that

$$n_2 = n_1 - \frac{f(n_1)}{f'(n_1)}$$
$$n_3 = n_2 - \frac{f(n_2)}{f'(n_2)}$$

and so on. In other words, we can determine n_1 by substituting the number a for n in the formula

$$n - \frac{f(n)}{f'(n)}$$

Once we determine n_1, we can determine n_2 by substituting n_1 for n in the above formula, and so on. By repeating this process enough times, we can approximate c within any degree of closeness specified.

Zero Intervals

Figure 3.38 also shows that the following statements are true:

1. $f(a)$ and $f(b)$ have opposite signs.
2. $f(x)$ is continuous on the interval $[a, b]$.
3. $f(x)$ decreases on $[a, b]$.
4. The graph of $f(x)$ is concave upward on $[a, b]$.

So, according to Definition 3.9, the closed interval $[a, b]$ is a zero interval of the function $f(x)$.

Definition 3.9

A closed interval $[a, b]$ is a **zero interval** of a function $f(x)$ if the following conditions hold:

1. $f(a)$ and $f(b)$ have opposite signs.
2. $f(x)$ is continuous on $[a, b]$.
3. $f(x)$ either increases on $[a, b]$ or decreases on $[a, b]$.
4. The graph of $f(x)$ is either concave upward on $[a, b]$ or concave downward on $[a, b]$.

Conditions 1, 2, and 3 imply that $f(x)$ has exactly one zero between a and b. This fact justifies calling $[a, b]$ a zero interval of $f(x)$.

There are four basic configurations in which a closed interval is a zero interval of a function. Figure 3.38 illustrates one of these configurations, while Figure 3.39 illustrates the other three configurations.

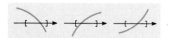

Figure 3.39 Three zero intervals.

With the function $f(x)$ whose graph is the curve in Figure 3.38, we illustrated Theorem 3.9.

Theorem 3.9

Suppose c is a zero of a function $f(x)$ and $[a, b]$ is a zero interval of $f(x)$ that contains c. If the sign of $f(a)$ is the same as the sign of $f''(x)$ on the interval $[a, b]$, let

$$n_1 = a - \frac{f(a)}{f'(a)}$$

Otherwise, let

$$n_1 = b - \frac{f(b)}{f'(b)}$$

Then, if

$$n_2 = n_1 - \frac{f(n_1)}{f'(n_1)}$$

$$n_3 = n_2 - \frac{f(n_2)}{f'(n_2)}$$

and so on, it follows that

$$\lim_{k \to +\infty} n_k = c$$

Now we can state Newton's method.

Newton's Method

Suppose c is a zero of a function $f(x)$. To approximate c within a specified degree of closeness, first find a zero interval of $f(x)$ that contains c. Then, using the procedure described in Theorem 3.9, compute n_1, n_2, n_3, and so on, until an n_k is obtained that approximates c within the specified degree of closeness.

Accuracy

We will use the following assumption to determine when an n_k approximates c within the specified degree of closeness. (The assumption is not always valid. But situations where it fails are rare.)

Accuracy Assumption

If the last two n_k's are identical through $i + 1$ decimal places, the last n_k approximates the actual zero with an error less than $(1/10)^i$.

For example, suppose we want to approximate a zero with an error less than 0.00001. Since $0.00001 = (1/10)^5$, we compute each n_k until two consecutive n_k's are identical through six decimal places. Then, according to the above assumption, the last n_k approximates the zero with an error less than 0.00001.

The Shrimp Problem Revisited

Now we can solve the supply and demand problem presented at the beginning of this section. As we said, the problem amounts to finding a zero of the function

$$E(x) = -\frac{1}{64}x^3 - \frac{3}{4}x + 12$$

The derivative of this function is

$$E'(x) = -\frac{3}{64}x^2 - \frac{3}{4}$$

Note that $E'(x)$ is negative for each value of x. Thus, $E(x)$ always decreases as x increases. This result implies $E(x)$ can have no more than one zero. In fact, it has exactly one zero. In Example 3.17, we will approximate this zero with an error less than 0.001.

Example 3.17

Use Newton's method to approximate the zero of

$$E(x) = -\frac{1}{64}x^3 - \frac{3}{4}x + 12$$

with an error less than 0.001.

Solution

We begin by finding a zero interval of $E(x)$. A rough sketch of the graph of $E(x)$ suggests that the closed interval $[7, 8]$ is such an interval. But we must be certain. So we will verify that this interval satisfies the four conditions listed in Definition 3.9.

Note that $E(7) = \frac{89}{64}$ and $E(8) = -2$. Thus, $E(7)$ and $E(8)$ have opposite signs. (So condition 1 is satisfied.) Since $E(x)$ is a polynomial function, it is continuous everywhere. In particular, $E(x)$ is continuous on $[7, 8]$. (So condition 2 is satisfied.) We already noted that $E'(x)$ is negative for each value of x. In particular, $E'(x)$ is negative for each x in $[7, 8]$. Therefore, $E(x)$ decreases on $[7, 8]$. (So condition 3 is satisfied.) The second derivative of $E(x)$ is

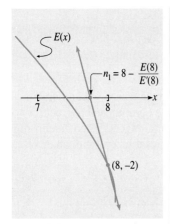

Figure 3.40 Newton's method applied once to $E(x)$.

$$E''(x) = -\tfrac{3}{32}x$$

Thus, we see that $E''(x)$ is negative for each x in $[7, 8]$. Therefore, the graph of $E(x)$ is concave downward on $[7, 8]$. (So condition 4 is satisfied.) Since $[7, 8]$ satisfies the four conditions listed in Definition 3.9, the interval $[7, 8]$ is indeed a zero interval of $E(x)$.

To compute the n_k's, we need a formula for $n - E(n)/E'(n)$:

$$n - \frac{E(n)}{E'(n)} = n - \frac{-\tfrac{1}{64}n^3 - \tfrac{3}{4}n + 12}{-\tfrac{3}{64}n^2 - \tfrac{3}{4}}$$

$$= \frac{2n^3 + 768}{3n^2 + 48}$$

Since the sign of $E(8)$ is the same as the sign of $E''(x)$ on the interval $[7, 8]$ (both are negative), we begin computing the n_k's by substituting 8 (not 7) for n in the above formula:

$$n_1 = 8 - \frac{E(8)}{E'(8)} \approx 7.4666$$

$$n_2 \approx 7.4666 - \frac{E(7.4666)}{E'(7.4666)} \approx 7.4356$$

$$n_3 \approx 7.4356 - \frac{E(7.4356)}{E'(7.4356)} \approx 7.4355$$

$$n_4 \approx 7.4355 - \frac{E(7.4355)}{E'(7.4355)} \approx 7.4355$$

Since n_3 and n_4 are identical through four decimal places, according to the assumption we made about accuracy, 7.4355 approximates the zero of $E(x)$ with an error less than 0.001. Figure 3.40 shows one application of Newton's method to the function $E(x)$.

The result we obtained in Example 3.17 is impressive, because

$$E(7.4355) = -\tfrac{1}{64}(7.4355)^3 - \tfrac{3}{4}(7.4355) + 12$$
$$= 0.0001881$$

which is very close to zero. The result tells us that approximately 7.4355 million pounds of shrimp is supplied when the market is in equilibrium.

3.7 Problems

In Problems 1 to 8, the given function has one zero. Use Newton's method to approximate the zero with an error less than 0.001.

1. $f(x) = x^3 - 17$
2. $f(x) = -x^3 + 3$
3. $f(x) = -x^3 - 7x + 50$
4. $f(x) = x^3 + 3x - 5$
5. $f(x) = x^5 - 40$
6. $f(x) = x^5 - 500$

7. $f(x) = x^3 - 27x^2 + 243x - 700$

8. $f(x) = -x^3 + 15x^2 - 75x + 115$

In Problems 9 and 10, the given function has three zeros, a negative zero and two positive zeros. Use Newton's method to approximate the smaller of the positive zeros with an error less than 0.001.

9. $f(x) = x^3 - 27x + 30$

10. $f(x) = x^3 - 12x + 12$

In Problems 11 and 12, the given function has two zeros, a negative zero and a positive zero. Use Newton's method to approximate the positive zero with an error less than 0.001.

11. $f(x) = x^4 - 300$

12. $f(x) = 5 - x^4$

13. **Optimal Selling Level** The monthly profit earned from selling x units of a commodity per month is

$$P(x) = -x^4 + 400x^2$$

The maximum profit occurs at the positive x value where the derivative of the profit function is zero. Use Newton's method to approximate this x value with an error less than 0.001.

14. **Electricity Production Cost** Let A and B represent two sources of electrical energy. It now costs more to produce each megawatt hour from source B than from source A. However, the cost of producing the energy from source B will decrease gradually and will eventually be less than the cost of producing it from source A. To be more specific, energy experts predict that t years from now (up to $t = 20$) the cost of producing each megawatt hour from source B will be

$$B(t) = -0.5t + 75$$

dollars. They also predict that t years from now the cost of producing each megawatt hour from source A will be

$$A(t) = 0.02t^3 + 30$$

dollars. Use Newton's method to approximate (with an error less than 0.001) when the cost of producing electrical energy from source B will be the same as the cost of producing it from source A.

Computer Problems

The Basic program NEWTON listed below uses Newton's method to approximate a zero of a function $f(x)$. Execution of the program prompts you to enter an endpoint of a zero interval. After you enter the endpoint, the program prints the approximation obtained through one application of Newton's method. Striking the space bar prints subsequent approximations. You can exit the program by striking E.

As listed, NEWTON applies to the function

$$f(x) = -0.015625x^3 - 0.75x + 12$$

But you can make the program applicable to other functions by editing lines 70 and 80. (Line 80 contains the derivative of the function.)

The Program NEWTON and a Sample Run

```
LIST
10   REM                          NEWTON
20   REM
30   REM A PROGRAM THAT USES NEWTON'S METHOD TO APPROXIMATE ZEROS
40   REM
50   REM LINE 70 CONTAINS THE FUNCTION, AND LINE 80 CONTAINS ITS DERIVATIVE.
60   REM
70   DEF FNF(X)=-.015625*X^3-.75*X+12
80   DEF FND(X)=-.046875*X^2-.75
90   PRINT
100  PRINT "FIND A ZERO INTERVAL. IF THE SIGN OF THE FUNCTION AT THE LEFT"
110  PRINT "ENDPOINT OF THE ZERO INTERVAL IS THE SAME AS THE SIGN OF THE"
120  PRINT "2ND DERIVATIVE AT THAT POINT, ENTER THE LEFT ENDPOINT. OTHERWISE,"
130  PRINT "ENTER THE RIGHT ENDPOINT. THE PROGRAM WILL PRINT THE APPROXIMATION"
140  PRINT "OBTAINED THROUGH ONE APPLICATION OF NEWTON'S METHOD. AFTERWARDS,"
```

```
150   PRINT "EACH STROKE OF THE SPACE BAR PRINTS THE NEXT APPROXIMATION. WHEN"
160   PRINT "2 APPROXIMATIONS ARE EQUAL, EXIT THE PROGRAM BY STRIKING E."
170   INPUT N
180   PRINT
190   LET M=N-FNF(N)/FND(N)
200   PRINT M
210   LET N=M
220   LET B$=INKEY$
230   IF B$="E" THEN 260
240   IF B$=" " THEN 190
250   GOTO 220
260   END
Ok
Run
FIND A ZERO INTERVAL. IF THE SIGN OF THE FUNCTION AT THE LEFT
ENDPOINT OF THE ZERO INTERVAL IS THE SAME AS THE SIGN OF THE
2ND DERIVATIVE AT THAT POINT, ENTER THE LEFT ENDPOINT. OTHERWISE,
ENTER THE RIGHT ENDPOINT. THE PROGRAM WILL PRINT THE APPROXIMATION
OBTAINED THROUGH ONE APPLICATION OF NEWTON'S METHOD. AFTERWARDS,
EACH STROKE OF THE SPACE BAR PRINTS THE NEXT APPROXIMATION. WHEN
2 APPROXIMATIONS ARE EQUAL, EXIT THE PROGRAM BY STRIKING E.
? 8

7.466667
7.435657
7.435557
7.435557
Ok
```

The next Basic program, PROBE, finds the general location of the zeros of a continuous function on a closed interval whose endpoints are integers. Execution of the program prompts you to enter the endpoints of the interval. After you enter the endpoints, the program identifies subintervals of length one that contain zeros. (These subintervals are not necessarily zero intervals.) On rare instances, it actually identifies zeros. But it can also fail to detect zeros.

As listed, PROBE applies to the function

$$f(x) = x^3 - 19x^2 + 108x - 175$$

You can, however, render it applicable to other functions by just editing line 60.

The Program PROBE and Two Sample Runs

```
LIST
10    REM                          PROBE
20    REM
30    REM A PROGRAM THAT SEARCHES FOR THE ZEROS OF A CONTINUOUS FUNCTION ON
40    REM A CLOSED INTERVAL WHOSE ENDPOINTS ARE INTEGERS.
50    REM
60    DEF FNF(X)=X^3-19*X^2+108*X-175
70    PRINT "ENTER THE ENDPOINTS OF THE INTERVAL."
80    INPUT A,B
```

```
90   PRINT
100  FLAG=0
110  FOR N=A TO B-1 STEP 1
120  IF FNF(N)<0 AND FNF(N+1)>0 THEN 170
130  IF FNF(N)>0 AND FNF(N+1)<0 THEN 170
140  IF FNF(N)<>0 THEN 190
150  PRINT N;"IS A ZERO OF F(X)."
160  GOTO 180
170  PRINT "F(X) HAS A ZERO BETWEEN";N;"AND";N+1;"."
180  FLAG=1
190  NEXT N
200  IF FLAG=1 THEN 220
210  PRINT "THE PROGRAM DID NOT FIND ANY ZEROS IN THE INTERVAL FROM";A;"TO";B;"."
220  PRINT
230  PRINT "WARNING: THE PROGRAM DOES NOT NECESSARILY FIND ALL OF THE ZEROS"
240  PRINT "IN AN INTERVAL."
250  END
Ok
RUN
ENTER THE ENDPOINTS OF THE INTERVAL.
? 0,15

F(X) HAS A ZERO BETWEEN 2 AND 3.
F(X) HAS A ZERO BETWEEN 6 AND 7.
F(X) HAS A ZERO BETWEEN 9 AND 10.

WARNING: THE PROGRAM DOES NOT NECESSARILY FIND ALL OF THE ZEROS IN AN INTERVAL.
Ok

RUN
ENTER, THE ENDPOINTS OF THE INTERVAL.
? 15,25

THE PROGRAM DID NOT FIND ANY ZEROS IN THE INTERVAL FROM 15 TO 25.

WARNING: THE PROGRAM DOES NOT NECESSARILY FIND ALL OF THE ZEROS IN AN INTERVAL.
Ok
```

In Problems 15 and 16, the given function has one zero. Use the program PROBE to establish the general location of the zero. Then use the program NEWTON to approximate the zero.

15. $f(x) = 0.02x^3 + 5x - 75$

16. $f(x) = -0.04x^3 - 7x + 80$

In Problems 17 and 18, the given function has two positive zeros. Use the program PROBE to establish the general location of the positive zeros. Then use the program NEWTON to approximate these zeros.

17. $f(x) = -x^3 + 100x - 40$

18. $f(x) = x^3 - 64x + 50$

The program PROBE concludes that a function $f(x)$ has a zero between N and $N + 1$ if it finds that $f(N)$ and $f(N + 1)$ have opposite signs. Use this fact in Problems 19 and 20.

19. Explain how the program PROBE can fail to detect zeros.

20. What could happen if we apply the program PROBE to a function that fails to be continuous at some points in its domain?

Chapter 3 Important Terms

Section 3.1
average rate of change (150)
instantaneous rate of change (152)
related rates (154)

Section 3.2
second derivative (160)
concavity (161)
test for concavity (162)
inflection point (163)
first- and second-order critical points (164)

Section 3.3
increasing rate (170 and 172)
decreasing rate (170 and 172)

Section 3.4
horizontal asymptote (177)

Section 3.5
vertical asymptote (191)
slant asymptote (191)

Section 3.6
percentage rate of change (200)

Section 3.7
zero interval (209)
Newton's method (210)

Chapter 3 Review Problems

Section 3.1

1. Find a formula for the instantaneous rate of change of $f(x)$ with respect to x. Then compute the instantaneous rate of change at the given x value.

   a. $f(x) = 224 - 6(\sqrt{x})^3$ $x = 4$

   b. $f(x) = \dfrac{2x + 18}{x + 3}$ $x = 2$

   c. $f(x) = 4\sqrt{3x + 1}$ $x = 5$

   d. $f(x) = 5x - \dfrac{72}{x}$ $x = 3$

   e. $f(x) = x(0.25x + 1)^3$ $x = 4$

2. **Price Increase Rate** The price of each liter of a dairy product is expected to increase. Experts believe that x weeks from now the price will be
$$p(x) = -\tfrac{1}{32}x^2 + \tfrac{4}{3}x + 40$$
cents, where $0 \le x \le 20$. Find a formula for the rate at which the price will be increasing x weeks from now. Then find the rate at which it will be increasing 4 weeks from now. What will be the price at that time?

3. **Demand** Suppose the weekly demand for Honey Bunch Candy Bars is
$$D(x) = \dfrac{192}{x + 2}$$
million bars when they sell for x cents per bar. Find a formula for the rate of change of the weekly demand with respect to the price. Then compute the rate of change (decrease) when the price is 30 cents.

4. **Debt Increase Rate** At the beginning of 1990, Third World countries owed American banks about $400 billion. Suppose this debt is
$$D(x) = 0.5\sqrt{x} + 400$$
billion dollars x years after the beginning of 1990. Find the average rate at which the debt will increase during the 21-year period between the beginning of 1994 and the beginning of 2015. This average rate of increase will be equal to the debt's instantaneous rate of increase at some instant in the 21-year period under consideration. Find such an instant. Then interpret your answers geometrically.

5. **Related Populations** Let y represent the bobcat population (in thousands) in a state park. Environmentalists believe that y is a function of the park's mice population. In fact, they project that
$$y = \dfrac{12x}{x + 4}$$
where x is the park's mice population (in millions) t years from now.

   a. How fast will the bobcat population be changing when the mice population is 2 million, if at that time the mice population decreases at the rate of 1.2 million per year? (Assume x is a differentiable function of t.)

b. How fast will the bobcat population be changing 3 years from now, if the mice population is

$$x = \frac{24}{t+3}$$

million t years from now?

Section 3.2

6. Find where the graph is concave downward and where it is concave upward. Also, give the coordinates of the inflection points.

   a. $f(x) = x^4 - 6x^3 - 108x^2 + 5$
   b. $f(x) = \sqrt{4x^2 + 5}$
   c. $f(x) = \dfrac{3x^2 + 20}{x^2 + 75}$
   d. $f(x) = x^3 - 6x^2 + 12x + 4$

7. Sketch the graph of a function $f(x)$ that satisfies the listed conditions.

   a. $f'(x)$ is positive for $x < 1$ and for $x > 8$.
      $f'(x)$ is negative for $1 < x < 8$.
      $f(1) = 7$ and $f(8) = 2$.
      $f''(x)$ is negative for $x < 5$ and positive for $x > 5$.
      $f(5) = 4$ and $f(0) = 1$.

   b. The domain is the interval $[0, +\infty)$.
      $f'(x)$ is negative for $x < 2$ and for $2 < x < 9$.
      $f'(x)$ is positive for $x > 9$.
      $f(9) = 2$.
      $f''(x)$ is positive for $x < 2$ and for $x > 6$.
      $f''(x)$ is negative for $2 < x < 6$.
      $f(2) = 8, f(6) = 5$, and $f(0) = 10$.

   c. The domain is the interval $[3, 8]$.
      $f'(x)$ is always negative.
      $f''(x)$ is always positive.
      $f(3) = 9$ and $f(8) = 0$.

   d. The domain is the interval $[0, 12]$.
      $f'(x)$ is positive for $x < 7$ and negative for $x > 7$.
      $f(7) = 5$.
      $f''(x)$ is always negative.
      $f(0) = -6$ and $f(12) = -3$.

   e. The domain is the interval $[0, +\infty)$.
      $f'(x)$ is always positive.

   $f''(x)$ is positive for $x < 6$ and negative for $x > 6$.
   $f(6) = 3$ and $f(0) = 1$.

8. Sketch the graph of the given function.

   a. $f(x) = 0.1x^3 - 1.5x^2 + 4.8x + 3$
   b. $f(x) = \dfrac{1}{120}x^4 - \dfrac{3}{200}x^3 + \dfrac{9}{10}x^2$

9. **Disease Transmission** A communicable disease was recently introduced to the inhabitants of an island near the coast of Alaska. Let $D(x)$ represent the number of inhabitants (in thousands) who will have contracted the disease x months from now. Health officials project that

   $$D(x) = 0.225x^2 - 0.04(\sqrt{x})^5 + 2$$

   for x in the interval $[0, 16]$. Sketch the graph of the function $D(x)$ for x in $[0, 16]$.

Section 3.3

10. Let

    $$f(x) = x^3 - 15x^2 + 78x$$

    for x in the interval $[0, +\infty)$. Use the first derivative to show that $f(x)$ increases on the interval $[0, +\infty)$. Then use Theorem 3.2 to determine how $f(x)$ increases on this interval.

11. Let

    $$f(x) = \frac{8}{\sqrt{3x+5}}$$

    for x in the interval $[0, +\infty)$. Use the first derivative to show that $f(x)$ decreases on the interval $[0, +\infty)$. Then use Theorem 3.3 to determine how $f(x)$ decreases on this interval.

12. Sketch the graph of a function $f(x)$ that satisfies the listed conditions.

    a. The domain is $[0, 9]$.
       $f(0) = 7, f(3) = 5$, and $f(9) = 1$.
       $f(x)$ decreases at a decreasing rate on $[0, 3)$ and at an increasing rate on $(3, 9]$.

    b. The domain is $[0, +\infty)$.
       $f(0) = 0$.
       $f(x)$ increases at an increasing rate.

    c. The domain is $[0, 8]$.
       $f(0) = 6$ and $f(8) = 0$.
       $f(x)$ decreases at an increasing rate.

d. The domain is $[0, +\infty)$.
$f(0) = 1$ and $f(5) = 3$.
$f(x)$ increases at an increasing rate on $[0, 5)$ and at a decreasing rate on $(5, +\infty)$.

13. **Acidity of Rainwater** Let $A(x)$ represent the pH of rainwater in a South American rain forest x years from now. Environmentalists believe that

$$A(x) = \sqrt{2.5x + 9}$$

for x in the interval $[0, 10]$. Verify that the pH of the rainwater will increase during the next 10 years. (Thus, the rainwater will become less acidic.) Then describe how it will increase.

14. **Rabbit Population** Due to deforestation, the rabbit population of an island near the coast of Chile has been declining. Let $P(x)$ represent the rabbit population (in thousands) x years from now. Environmentalists predict that

$$P(x) = (\sqrt{x})^5 - 3.75x^2 + 70$$

for x in the interval $[0, 8]$. Show that the rabbit population will continue to decline during the next 8 years. Then describe how it will decline.

15. **Worker Productivity** In a factory, a work shift consists of five continuous hours of work. Suppose a worker produces

$$P(x) = 8x^3 - x^4$$

units of a product during the first x hours on a work shift. When does the worker produce the product most rapidly? When the worker produces the product most rapidly, at what rate does the worker produce the product?

Section 3.4

16. Sketch the graph of $f(x)$ for $x \geq 0$. Then describe how $f(x)$ behaves as x increases.

    a. $f(x) = \dfrac{28x + 5}{7x + 2}$ b. $f(x) = \dfrac{8}{3x + 1}$

    c. $f(x) = \dfrac{5x^2}{x^2 + 27}$ d. $f(x) = \dfrac{x^2 + 336}{x^2 + 48}$

17. Find the function of the type

$$f(x) = \frac{ax + b}{cx + d}$$

that satisfies the listed conditions as x increases on the interval $[0, +\infty)$.

a. $f(0) = 6$
$f(x)$ approaches 2
$f(4) = 2.8$

b. $f(0) = 0$
$f(x)$ approaches 3
$f(10) = 2.4$

c. $f(0) = 2$
$f(x)$ approaches 5.2
$f(8) = 4.8$

d. $f(0) = 13$
$f(x)$ approaches zero
$f(11) = 2$

18. Find the function of the type

$$f(x) = \frac{ax^2 + b}{cx^2 + d}$$

that satisfies the listed conditions as x increases on the interval $[0, +\infty)$.

a. $f(0) = 3$
$f(x)$ approaches 9
$f(1) = 6.6$

b. $f(0) = 6.25$
$f(x)$ approaches zero
$f(4) = 1.25$

19. Sketch the graph of the function of the type

$$f(x) = \frac{ax^2 + b}{cx^2 + d}$$

that satisfies the listed conditions as x increases on the interval $[0, +\infty)$.

a. $f(0) = 0$
$f(x)$ approaches 7.5
$f(12) = 5.625$

b. $f(0) = 6$
$f(x)$ approaches 1
$f(9) = 2.25$

20. **Depreciation** An accountant projects that x years from now a paper recycling machine will be worth

$$V(x) = \frac{3x^2 + 4500}{x^2 + 75}$$

thousand dollars. Sketch the graph of $V(x)$ for $x \geq 0$. Then describe how the machine's value will behave as time passes.

21. **Body Building** Nine months ago Tipo enrolled in a physical fitness program. At that time he weighed 130 pounds, but 4 months ago he weighed 150 pounds. As long as Tipo participates in the program, his weight will approach 175 pounds while it increases at a slower and slower rate. Suppose his weight after x months of participating in the program is given by a formula of the form

$$W(x) = \frac{ax + b}{cx + d}$$

What will be his weight 3 months from now? (Round your answer to the nearest tenth of a pound.)

22. **Advertising** Suppose a company sells

$$D(x) = \frac{9x^2 + 1350}{x^2 + 675}$$

million units of a product per month when its monthly advertising expenditure is $x thousand. This formula is of the form $(ax^2 + b)/(cx^2 + d)$ with $a/c > b/d$. Therefore, as the monthly advertising expenditure is increased, the monthly demand at first increases at an increasing rate and eventually increases at a decreasing rate. Thus, there exists a point after which the returns due to further increases in the monthly advertising expenditure begin to diminish. Find this point of diminishing returns.

23. **Immunization** The government of a South American country has decided to support a program for immunizing the people against a new communicable disease. Health officials project that if the government spends $x million on the program,

$$I(x) = \frac{100x + 40}{x + 2}$$

percent of the people will become immune to the disease. This formula is of the form $(ax + b)/(cx + d)$ with $a/c > b/d$. Therefore, $I(x)$ increases at a decreasing rate as the government channels money into the program. The government plans to channel money into the program as long as the rate of increase of $I(x)$ is large enough. Specifically, the government will increase x (the total expenditure) until $I'(x)$ is reduced to 2.5. How much money will the government spend on the program by the time it stops supporting the program?

Section 3.5

24. Find any existing horizontal, slant, and vertical asymptotes of the graph.

   a. $f(x) = \dfrac{6x + 1}{3x - 15}$ b. $f(x) = \dfrac{5x^3}{x^2 - 49}$

   c. $f(x) = \dfrac{4x}{x^2 + 1}$ d. $f(x) = \dfrac{x + 2}{x^2 - 5x + 4}$

25. Sketch the graph of a function $f(x)$ that satisfies the listed conditions.

   a. The horizontal asymptote is the x axis. The graph has no vertical asymptotes.
   $f'(x)$ is positive for $x < 6$ and negative for $x > 6$.
   $f(6) = 4$.
   $f''(x)$ is positive for $x < 3$ and for $x > 9$.
   $f''(x)$ is negative for $3 < x < 9$.
   $f(3) = 1$ and $f(9) = 1$.

   b. The line $y = 2$ is the horizontal asymptote. The y axis and the line $x = 4$ are the vertical asymptotes.
   $f'(x)$ is positive for $x < 0$ and for $x > 4$.
   $f'(x)$ is negative for $0 < x < 4$.
   $f''(x)$ is positive for $x < 0$ and for $0 < x < 2$.
   $f''(x)$ is negative for $2 < x < 4$ and for $x > 4$.
   $f(2) = 2$.

   c. The slant asymptote is the line $y = 0.625x - 2$. The only vertical asymptote is the line $x = 5$.
   $f'(x)$ is positive for $x < 2$ and for $x > 8$.
   $f'(x)$ is negative for $2 < x < 5$ and for $5 < x < 8$.
   $f(2) = -2$ and $f(8) = 5$.
   $f''(x)$ is negative for $x < 5$ and positive for $x > 5$.

   d. The domain is the interval $[0, +\infty)$.
   The horizontal asymptote is the line $y = 7$.
   The graph has no vertical asymptotes.
   $f'(x)$ is negative for $0 \leq x < 3$ and positive for $x > 3$.
   $f(3) = 2$.
   $f''(x)$ is positive for $x < 8$ and negative for $x > 8$.
   $f(8) = 5$.

26. Sketch the graph of the given function.

   a. $f(x) = \dfrac{x^2}{2x - 4}$ b. $f(x) = \dfrac{8x - 16}{x^2 - 4x + 5}$

Section 3.6

27. Find a formula for the percentage rate of change of $f(x)$ with respect to x. Then compute the percentage rate of change at the given x value.

   a. $f(x) = 3x^2 + 50$ $x = 20$

   b. $f(x) = \sqrt{x^2 + 11}$ $x = 33$

   c. $f(x) = \dfrac{2x + 20}{x + 5}$ $x = 10$

   d. $f(x) = \dfrac{250}{(x + 4)^2}$ $x = 36$

28. **Electric Cars** Experts project that x years after the beginning of 1994

$$E(x) = 0.3x^2 + 4.8$$

million electric cars will be in use in a northern European country. According to this projection, the number of electric cars will increase at an increasing rate. But the percentage rate of increase will increase for a while only. Eventually, it will decrease. When will the percentage rate of increase be greatest? What will it be at that time?

Section 3.7

29. The given function has one zero. Use Newton's method to approximate the zero with an error less than 0.001.

   a. $f(x) = x^3 + 2x - 30$

   b. $f(x) = -x^3 - 8x + 100$

30. The given function has three zeros, a negative zero and two positive zeros. Use Newton's method to approximate the smaller of the positive zeros with an error less than 0.001.

   a. $f(x) = x^3 - 48x + 120$

   b. $f(x) = -x^3 + 75x - 150$

CHAPTER 4

Exponential and Logarithmic Functions

4.1 Compound Interest

4.2 Exponential Functions

4.3 The Natural Logarithmic Function

4.4 Differentiation of $\ln u(x)$

4.5 Differentiation of $e^{u(x)}$

4.6 Two Models for Bounded Growth

Important Terms

Review Problems

In Section 1.3, we studied functions that are used to model quantities that change at a *constant rate*. Such functions, recall, are linear functions. In this chapter, we will introduce functions that are used to model quantities that change at a *constant percentage rate*. Such functions are called *exponential functions*.

We will also introduce a function that is closely related to the exponential functions. Known as the *natural logarithmic function*, we will use this function primarily to help solve problems that involve exponential functions.

In Section 4.6 at the end of this chapter, we will examine two types of functions constructed with exponential functions. These functions are used to model quantities that experience bounded growth.

4.1 Compound Interest

Before we examine a situation that involves compound interest, we introduce a number which has an important role in this chapter.

The Number e

Consider

$$\left(1 + \frac{1}{n}\right)^n$$

where n is a positive integer. The values of $1 + 1/n$ get closer and closer to 1 as n increases. Thus, since 1 raised to any power is 1, we are tempted to conclude that the values of $(1 + 1/n)^n$ likewise get closer and closer to 1 as n increases. But this conclusion is false. The truth is that as n increases without bound, the values of $(1 + 1/n)^n$ eventually get and stay arbitrarily close to an irrational number. The value of the irrational number correct through 10 decimal places is 2.7182818284. So, using e to represent this irrational number, we can write

$$\lim_{n \to +\infty} \left(1 + \frac{1}{n}\right)^n = e$$

where $e \approx 2.7182818284$

The plausibility of this claim is supported by Table 4.1.

Table 4.1 Several Approximations of e

n	$(1 + 1/n)^n$
10	2.5937424
100	2.7048138
1000	2.7169239
10,000	2.7181459
100,000	2.7182682
1,000,000	2.7182804

The above discussion deals with the special case of Theorem 4.1, where $r = 1$.

Theorem 4.1

If r is *any* positive rational number,

$$\lim_{n \to +\infty} \left(1 + \frac{r}{n}\right)^{n/r} = e$$

Note that r/n and n/r are reciprocals of each other.

The limit in Theorem 4.1 is represented by the letter e to honor the prolific Swiss mathematician Leonhard Euler (1707–1783).

As we said, e is an irrational number, which means that its decimal representation neither terminates nor repeats. In fact, e is a **transcendental number**, which means that e is not a zero of a polynomial function with integer coefficients. Thus, e is a rather exotic number.

We now present a situation in which e arises.

Compound Interest

We begin by finding the balance at the end of 5 years for $2500 invested at an annual interest rate of 8 percent compounded semiannually.

Since the interest is compounded semiannually, the balance increases 4 percent at the end of each half-year. So the balance at the end of the first half-year (and throughout the second half-year) is 104 percent of $2500. This balance is expressed by

$$2500(1.04)$$

The balance at the end of the second half-year (and throughout the third half-year) is 104 percent of 2500(1.04). This balance is expressed by

$$2500(1.04)^2$$

Similarly, the balance at the end of the third half-year (and throughout the fourth half-year) is

$$2500(1.04)^3$$

Continuing in this manner, the balance at the end of the 10th half-year is

$$2500(1.04)^{10}$$

Since 5 years is equal to 10 half-years and since

$$2500(1.04)^{10} = 3700.61$$

we conclude that the $2500 grows to $3700.61 in 5 years.

Actually, 3700.61 is not really equal to $2500(1.04)^{10}$. It is the value of $2500(1.04)^{10}$ rounded to the nearest hundredth. We took the liberty of claiming equality because we are dealing with money. From now on we will take this kind of liberty whenever it is convenient.

The above discussion suggests that a formula for the balance at the end of t years is

$$B(t) = 2500(1.04)^{2t}$$

This formula gives correct balances only for values of t that are integral multiples of 0.5. According to the formula, the balance increases continuously. In reality, however, the balance increases only at the end of each half-year. After each increase, the balance remains unchanged for a half-year period.

Note that

$$1.04 = 1 + 0.04$$
$$= 1 + \frac{0.08}{2}$$

So

$$2500(1.04)^{2t} = 2500\left(1 + \frac{0.08}{2}\right)^{2t}$$

Thus, if the interest is compounded semiannually (2 times per year), we can alternatively express the balance at the end of t years by the formula

$$B(t) = 2500\left(1 + \frac{0.08}{2}\right)^{2t}$$

This result suggests (correctly) that if the interest is compounded n times per year, the balance at the end of t years is

$$B(t) = 2500\left(1 + \frac{0.08}{n}\right)^{nt}$$

provided t is an integral multiple of $1/n$.

Therefore, if the interest on the $2500 investment is compounded quarterly rather than semiannually, the balance at the end of t years is given by the formula

$$B(t) = 2500\left(1 + \frac{0.08}{4}\right)^{4t}$$
$$= 2500(1.02)^{4t}$$

provided t is an integral multiple of 0.25. For example, the balance at the end of 5 years is

$$B(5) = 2500(1.02)^{4 \cdot 5}$$
$$= 2500(1.02)^{20}$$
$$= 3714.87$$

dollars if the interest is compounded quarterly. This balance is larger than the balance we obtained for interest compounded semiannually.

If the interest is compounded monthly, the balance at the end of t years is given by

$$B(t) = 2500\left(1 + \frac{0.08}{12}\right)^{12t}$$

provided t is an integral multiple of $\frac{1}{12}$. For any specified amount of time, this deal produces a larger balance than interest compounded quarterly. If the interest is compounded daily (assuming 360 days in a year), the balance at the end of t years is given by

$$B(t) = 2500\left(1 + \frac{0.08}{360}\right)^{360t}$$

provided t is an integral multiple of $\frac{1}{360}$. This deal produces even larger balances.

The best deal for the investor is the deal in which the interest is compounded continuously.

Definition 4.1

Suppose P units of money is invested at an annual interest rate of $100r$ percent. The interest is **compounded continuously** if the balance at the end of t years is

$$\lim_{n \to +\infty} P\left(1 + \frac{r}{n}\right)^{nt}$$

According to Definition 4.1, if the interest on the $2500 investment is compounded continuously, the balance at the end of t years is

$$\lim_{n \to +\infty} 2500\left(1 + \frac{0.08}{n}\right)^{nt}$$

To compute this limit, we first note that

$$nt = \left(\frac{n}{0.08}\right)0.08t$$

Then, using the rule $b^{xy} = (b^x)^y$, we can say that

$$2500\left(1 + \frac{0.08}{n}\right)^{nt} = 2500\left(1 + \frac{0.08}{n}\right)^{(n/0.08)0.08t}$$

$$= 2500\left[\left(1 + \frac{0.08}{n}\right)^{n/0.08}\right]^{0.08t}$$

From this result and the fact that

$$\lim_{n \to +\infty} \left(1 + \frac{0.08}{n}\right)^{n/0.08} = e \qquad \text{Theorem 4.1}$$

we get

$$\lim_{n \to +\infty} 2500\left(1 + \frac{0.08}{n}\right)^{nt} = \lim_{n \to +\infty} 2500\left[\left(1 + \frac{0.08}{n}\right)^{n/0.08}\right]^{0.08t}$$

$$= 2500 e^{0.08t}$$

Therefore, if the interest on the $2500 investment is compounded continuously, the balance at the end of t years is given by the formula

$$B(t) = 2500 e^{0.08t}$$

In particular, the balance at the end of 5 years is

$$B(5) = 2500 e^{0.08(5)}$$
$$= 2500 e^{0.4}$$
$$= 3729.56$$

dollars. So the $2500 grows to $3729.56 in 5 years if the interest is compounded continuously at 8 percent per year. Are you disappointed that this balance is only about $15 more than the balance for interest compounded quarterly?

We can prove Theorem 4.2 by going through steps that parallel the steps in our $2500 investment example.

Theorem 4.2

Suppose P units of money is invested at an annual interest rate of $100r$ percent.

1. If the interest is compounded n times per year, the balance at the end of t years is

$$B(t) = P\left(1 + \frac{r}{n}\right)^{nt}$$

provided t is an integral multiple of $1/n$.

2. If the interest is compounded continuously, the balance at the end of t years is

$$B(t) = P e^{rt}$$

If the interest is compounded n times per year, the balance increases at the end of each $1/n$ of a year. If the interest is compounded continuously, the balance increases continuously.

We use Theorem 4.2 in Example 4.1.

Example 4.1

Suppose $1000 is invested at an annual interest rate of 6 percent.

a. Find a formula for the balance at the end of t years if the interest is compounded monthly. Then compute the balance at the end of 8.5 years.
b. Find a formula for the balance at the end of t years if the interest is compounded continuously. Then compute the balance at the end of 8.5 years.

Solution for (a)

If the interest is compounded monthly, a formula for the balance at the end of t years is

$$B(t) = 1000\left(1 + \frac{0.06}{12}\right)^{12t}$$
$$= 1000(1.005)^{12t}$$

provided t is an integral multiple of $\frac{1}{12}$. Since

$$B(8.5) = 1000(1.005)^{12(8.5)}$$
$$= 1000(1.005)^{102}$$
$$= 1663.18$$

the balance at the end of 8.5 years is $1663.18.

Solution for (b)

If the interest is compounded continuously, a formula for the balance at the end of t years is

$$B(t) = 1000e^{0.06t}$$

Since

$$B(8.5) = 1000e^{0.51}$$
$$= 1665.29$$

the balance at the end of 8.5 years is $1665.29.

In Example 4.1, the balance if the interest is compounded continuously is only about $2 more than the balance if the interest is compounded monthly.

Irrational Exponents

In our introductory investment example, we found that the formula

$$B(t) = 2500e^{0.08t}$$

gives the balance at the end of t years for \$2500 invested at an annual interest rate of 8 percent compounded continuously. This formula is valid for any nonnegative real number t.

We already know the meaning of $e^{0.08t}$ if $0.08t$ is an integer. For instance, if $t = 100$, then

$$e^{0.08t} = e^8$$
$$= e \cdot e \cdot e \cdot e \cdot e \cdot e \cdot e \cdot e$$

We also already know the meaning of $e^{0.08t}$ if $0.08t$ is a nonintegral rational number. For example, if $t = 10$, then

$$e^{0.08t} = e^{0.8}$$
$$= e^{4/5}$$
$$= (\sqrt[5]{e})^4$$

But there are many values of t for which $0.08t$ is an irrational number. For example, $0.08t$ is an irrational number when $t = \sqrt{2}$. For such values of t, what does $e^{0.08t}$ mean?

Considering our objectives, we do not need a precise answer to the above question. We do need to know, however, that the definition of irrational exponents allows

$$B(t) = 2500e^{0.08t}$$

to be a continuous function whose graph (for $t \geq 0$) is the curve in Figure 4.1. The curve shows that the balance grows at an increasing rate. The percentage rate of growth, however, is constantly 8 percent. Later we will be able to prove this claim.

The basic properties of exponents remain valid if we permit irrational exponents. For convenience, we list these properties using the number e as base.

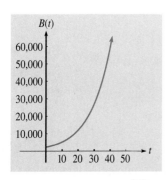

Figure 4.1 $B(t) = 2500e^{0.08t}$ for $t \geq 0$.

Properties of Exponents

E1 $e^m \cdot e^n = e^{m+n}$

E2 $\dfrac{e^m}{e^n} = e^{m-n}$

E3 $(e^m)^n = e^{mn}$

E4 $e^0 = 1$

E5 $e^{-m} = \dfrac{1}{e^m}$

E6 If $e^m = e^n$, then $m = n$.

We now consider the problem of determining how much must be invested to obtain a specified balance at the end of a specified time interval.

Present Value

As we have seen, the formula

$$B = Pe^{rt}$$

gives the balance at the end of t years if P units of money is invested at an annual interest rate of $100r$ percent compounded continuously. If we divide both sides of the above formula by e^{rt}, we obtain

$$\frac{B}{e^{rt}} = P$$

which is equivalent to

$$P = Be^{-rt}$$

This formula gives the amount that must be invested at an annual interest rate of $100r$ percent compounded continuously to attain a balance of B units of money in t years. In this context, P is called the **present value** of B units of money attainable t years from now. If the interest is compounded n times per year rather than continuously, the formula for the present value is

$$P = B\left(1 + \frac{r}{n}\right)^{-nt}$$

The derivation of this formula is similar to the derivation of the formula $P = Be^{-rt}$.

Example 4.2

How much must an investor invest now at an annual interest rate of 9 percent to obtain a balance of $80,000 in 20 years if the interest is compounded (a) monthly? (b) continuously?

Solution for (a)

If the interest is compounded monthly at the annual rate of 9 percent, a formula for the present value of $80,000 attainable t years from now is

$$P = 80{,}000\left(1 + \frac{0.09}{12}\right)^{-12t}$$
$$= 80{,}000(1.0075)^{-12t}$$

Substituting 20 for t, we get

$$P = 80{,}000(1.0075)^{-240}$$
$$= 13{,}313.02$$

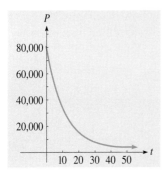

Figure 4.2 $P = 80{,}000e^{-0.09t}$ for $t \geq 0$.

So, if the interest is compounded monthly, the investor must invest $13,313.02 now to obtain a balance of $80,000 in 20 years.

Solution for (b)

If the interest is compounded continuously at the annual rate of 9 percent, a formula for the present value of $80,000 attainable t years from now is

$$P = 80{,}000e^{-0.09t}$$

(Figure 4.2 shows the graph of this function for $t \geq 0$.) Substituting 20 for t, we get

$$P = 80{,}000e^{-1.8}$$
$$= 13{,}223.91$$

So, if the interest is compounded continuously, the investor must invest $13,223.91 now to obtain a balance of $80,000 in 20 years. This amount is only $89.11 less than the amount the investor must invest if the interest is compounded monthly.

Effective Interest Rate

If the interest is compounded more frequently than one time per year, the percent a balance increases each year is *greater* than the annual interest rate.

Definition 4.2

The **effective interest rate** is the percent the balance actually increases each year.

From now on, we will sometimes call the annual interest rate the **nominal interest rate**.

If the interest is compounded *yearly*, the effective interest rate is equal to the nominal interest rate. If the interest is compounded other than yearly, the effective interest rate can be obtained by substituting the nominal interest rate in a formula. In what follows, we derive such a formula.

Let B represent the balance of an investment at an arbitrary time after the investment is made. If the nominal interest rate is $100r$ percent and the interest is compounded continuously, the B units of money increase to Be^r units of money in 1 year. Thus, the balance increases $Be^r - B$ units of money in 1 year. Since $Be^r - B$ is

$$100 \cdot \frac{Be^r - B}{B}$$

percent of B and since

$$100 \cdot \frac{Be^r - B}{B} = 100(e^r - 1)$$

we just proved the second part of Theorem 4.3. (The proof of the first part is similar.)

Theorem 4.3

Suppose money is invested at a nominal interest rate of $100r$ percent.

1. If the interest is compounded n times per year, the effective interest rate is

$$100\left[\left(1 + \frac{r}{n}\right)^n - 1\right]$$

percent.

2. If the interest is compounded continuously, the effective interest rate is

$$100(e^r - 1)$$

percent.

We use Theorem 4.3 in Example 4.3.

Example 4.3

Find the effective interest rate that corresponds to a nominal interest rate of 12 percent if the interest is compounded (*a*) quarterly, and (*b*) continuously.

Solution for (*a*)

If the interest is compounded quarterly, the effective interest rate (rounded to the nearest one-hundredth of 1 percent) is

$$100\left[\left(1 + \frac{0.12}{4}\right)^4 - 1\right] = 12.55$$

percent. So, if money is invested at a nominal interest rate of 12 percent compounded quarterly, the balance increases about 12.55 percent each year.

Solution for (*b*)

If the interest is compounded continuously, the effective interest rate (rounded to the nearest one-hundredth of 1 percent) is

$$100(e^{0.12} - 1) = 12.75$$

percent. So, if money is invested at a nominal interest rate of 12 percent compounded continuously, the balance increases about 12.75 percent each year.

4.1 Problems

1. **Future Value** Suppose $300 is invested at an annual interest rate of 9 percent.

   a. Find a formula for the balance at the end of t years if the interest is compounded quarterly. Then compute the balance at the end of 2.75 years.

   b. Find a formula for the balance at the end of t years if the interest is compounded daily. (Assume a year consists of 360 days.) Then compute the balance at the end of 2.75 years.

   c. Find a formula for the balance at the end of t years if the interest is compounded continuously. Then compute the balance at the end of 2.75 years. As you will see, this balance is extremely close to the balance obtained in part (*b*).

2. **Future Value** Suppose $100 is invested at an annual interest rate of 15 percent.

   a. Find a formula for the balance at the end of t years if the interest is compounded yearly. Then compute the balance at the end of 4 years.

   b. Find a formula for the balance at the end of t years if the interest is compounded monthly. Then compute the balance at the end of 4 years.

   c. Find a formula for the balance at the end of t years if the interest is compounded continuously. Then compute the balance at the end of 4 years.

3. **Present Value** Determine the amount that must be invested now at an annual interest rate of 7 percent to obtain a balance of $50,000 in 18 years if the following conditions hold:

   a. The interest is compounded annually.

   b. The interest is compounded continuously.

4. **Present Value** Determine the amount that must be invested now at an annual interest rate of 5 percent to obtain a balance of $75,000 in 10 years if the following conditions hold:

   a. The interest is compounded semiannually.

   b. The interest is compounded continuously.

5. **Effective Rate** Find the effective interest rate (rounded to the nearest one-hundredth of 1 percent) that corresponds to a nominal interest rate of 8 percent if the interest is compounded (*a*) semiannually, and (*b*) continuously.

6. **Effective Rate** Find the effective interest rate (rounded to the nearest one-hundredth of 1 percent) that corresponds to a nominal interest rate of 6 percent if the interest is compounded (*a*) every 4 months (three times per year), and (*b*) continuously.

7. **Comparing Investment Plans** Investment plan A offers an annual interest rate of 18 percent compounded monthly. Investment plan B offers an annual interest rate of 18.2 percent compounded semiannually. Use the concept of effective interest rate to decide which plan is better.

8. **Comparing Investment Plans** Investment plan C offers an annual interest rate of 14 percent compounded continuously. Investment plan D offers an annual interest rate of 14.4 percent compounded quarterly. Use the concept of effective interest rate to decide which plan is better.

9. **Optimal Transferring Time** Suppose you invest $3000 in plan G, where t years from now your balance will be

   $$A(t) = 0.25t + 3$$

   thousand dollars. In another investment plan, plan H, interest is compounded continuously at the annual rate of 5 percent. Thus, in plan H, a balance will increase at the constant rate of 5 percent per year. In plan G, at first the percentage rate of increase of your balance will be greater than 5 percent. However, this percentage rate of increase will gradually decrease and eventually be less than 5 percent. When should you transfer your balance to plan H?

   Recall: From Section 3.6, we know that in plan G your balance's percentage rate of increase t years from now is given by $100 \cdot A'(t)/A(t)$.

10. **Optimal Transferring Time** Do Problem 9 again. This time suppose

    $$A(t) = 0.2t + 3$$

Graphing Calculator Problems

11. **Future Value** Set the viewing screen on your graphing calculator to $[0, 40] \times [0, 80{,}000]$ with a scale of 10 on the x axis and a scale of 10,000 on the y axis. Then draw the graph of the function that

gives the balance at the end of x years for $5000 invested at an annual interest rate of 7 percent compounded continuously.

a. Use the trace function to estimate the balance at the end of 17 years.

b. Use the trace function to estimate when the balance will be $37,900. (Round your answer to the nearest year.)

12. **Present Value** Set the viewing screen on your graphing calculator to $[0, 40] \times [0, 70{,}000]$ with a scale of 10 on the x axis and a scale of 10,000 on the y axis. Then draw the graph of the function that gives the present value of $60,000 attainable x years from now if the annual interest rate is 8 percent compounded continuously.

a. Use the trace function to estimate how much an investor must invest now to obtain a balance of $60,000 in 14 years.

b. Use the trace function to estimate how long it takes a $24,760 investment to yield a balance of $60,000. (Round your answer to the nearest year.)

13. **Effective Rate** Let $E(r)$ be the effective interest rate that corresponds to a nominal interest rate of $100r$ percent if the interest is compounded continuously. According to Theorem 4.3,

$$E(r) = 100(e^r - 1)$$

Set the viewing screen on your graphing calculator to $[0, 1] \times [0, 200]$ with a scale of 0.1 on the x axis and a scale of 20 on the y axis. Then draw the graph of $E(r)$.

a. Use the trace function to estimate the effective interest rate that corresponds to a nominal interest rate of 11.7 percent. (Round your answer to the nearest tenth of 1 percent.)

b. Use the trace function to estimate the nominal interest rate that corresponds to an effective interest rate of 16 percent. (Round your answer to the nearest tenth of 1 percent.)

14. **Optimal Transferring Time** Suppose you invest $2 million in plan J, where your balance will increase at the constant rate of $0.4 million per year. In another investment plan, plan K, interest is compounded continuously at the annual rate of 7 percent. Thus, in plan K, a balance will increase at the constant rate of 7 percent per year.

a. Set the viewing screen on your graphing calculator to $[0, 15] \times [0, 24]$ with a scale of 2 on both axes. Then draw the graph of the function that gives the percentage rate of increase of your balance in plan J.

b. The graph you drew in part (*a*) should show that in plan J your balance's percentage rate of increase at first will be greater than 7 percent. However, the graph should also show that this percentage rate of increase will gradually decrease and eventually be less than 7 percent. Use the trace function to estimate when you should transfer your balance to plan K. (Round your answer to the nearest tenth of a year.)

4.2 Exponential Functions

In Section 4.1, we saw that if $2500 is invested at an annual interest rate of 8 percent compounded continuously, the balance at the end of t years is given by the formula

$$B(t) = 2500e^{0.08t}$$

This function is an example of an exponential function.

Definition 4.3

An **exponential function** is a function that can be expressed in the form

$$f(x) = Ae^{rx}$$

where $A > 0$ and $r \neq 0$.

The function

$$f(x) = e^x$$

is another example of an exponential function. Here, A and r both are 1. This exponential function is sometimes referred to as *the* exponential function.

If we substitute zero for x in an exponential function expressed in the form

$$f(x) = Ae^{rx}$$

we get

$$\begin{aligned} f(0) &= Ae^{r(0)} \\ &= Ae^0 \\ &= A \quad \text{since } e^0 = 1 \end{aligned}$$

So A has the same value as $f(0)$, which means that the graph intersects the y axis at $y = A$.

Exponential Functions in which $r > 0$

Suppose $f(x)$ is an exponential function in which $r > 0$. Then the graph of $f(x)$ is generally like the curve in Figure 4.3. The curve shows that as x increases, $f(x)$ increases at an increasing rate. However, as we will show in Section 4.5, the percentage rate at which $f(x)$ increases is constantly $100r$ percent.

Note that the curve in Figure 4.3 is asymptotic to the x axis.

If $f(x)$ is an exponential function in which $r > 0$, we can say that $f(x)$ **increases exponentially**. For example, suppose that t months from now a product's monthly demand is

$$D(t) = 48e^{0.03t}$$

million barrels. Then the product's monthly demand will increase exponentially at the rate of 3 percent per month.

Under ideal conditions, populations tend to grow exponentially. Example 4.4 deals with such a population.

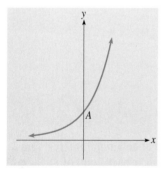

Figure 4.3 $f(x) = Ae^{rx}$ if $A > 0$ and $r > 0$.

The world population, which was less than 5.5 billion at the beginning of the 1990's, is expected to reach 10 billion by the middle of the 21st century. (Will McIntyre/Photo Researchers)

Example 4.4

The population of a country is presently 30 million. Demographers project that the population will grow exponentially at the rate of 2 percent per year. Find a formula for the population x years from now. Then determine what the population will be 7 years from now.

Solution

Let $P(x)$ represent the population (in millions) x years from now. Since the population will grow exponentially at the rate of 2 percent per year,

$$P(x) = Ae^{0.02x}$$

where $A > 0$. Since the population is presently 30 million, $P(0) = 30$, which implies that $A = 30$. So

$$P(x) = 30e^{0.02x}$$

In particular,

$$P(7) = 30e^{0.02(7)}$$
$$= 30e^{0.14}$$
$$\approx 34.5$$

which tells us that 7 years from now the population will be approximately 34.5 million.

Exponential Functions in which $r < 0$

Suppose $f(x)$ is an exponential function in which $r < 0$. Then the graph of $f(x)$ is generally like the curve in Figure 4.4. The curve shows that as x increases, $f(x)$ decreases at a decreasing rate. However, as we will show in Section 4.5, the percentage rate at which $f(x)$ decreases is constantly $100|r|$ percent.

Note that the curve in Figure 4.4 is asymptotic to the x axis.

If $f(x)$ is an exponential function in which $r < 0$, we can say that $f(x)$ **decreases exponentially**. For example, suppose that x years from now a machine is worth

$$V(x) = 75e^{-0.05x}$$

thousand dollars. Then the machine will depreciate exponentially at the rate of 5 percent per year.

Radioactive substances such as plutonium, uranium, and carbon-14 decay exponentially. Example 4.5 deals with such a substance.

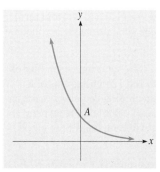

Figure 4.4 $f(x) = Ae^{rx}$ if $A > 0$ and $r < 0$.

Example 4.5

Five hours ago, a chemist acquired a 100-gram sample of actinium-228, a radioactive substance that decays exponentially at the rate of 11 percent per hour. How much of the sample will remain 3 hours from now?

Solution

Let $Q(t)$ represent the amount that remains t hours after the chemist acquired the sample. Since the substance decays exponentially at the rate of 11 percent per hour,

$$Q(t) = Ae^{-0.11t}$$

where $A > 0$. Since the sample weighed 100 grams when the chemist acquired it, $Q(0) = 100$, which implies that $A = 100$. So

$$Q(t) = 100e^{-0.11t}$$

Three hours from now is 8 hours after the chemist acquired the sample. Substituting 8 for t in the above formula, we get

$$\begin{aligned} Q(8) &= 100e^{-0.11(8)} \\ &= 100e^{-0.88} \\ &\approx 41.5 \end{aligned}$$

Therefore, 3 hours from now about 41.5 grams of the actinium-228 sample will remain. Figure 4.5 shows the graph of $Q(t) = 100e^{-0.11t}$ for $t \geq 0$.

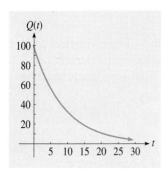

Figure 4.5 $Q(t) = 100e^{-0.11t}$ for $t \geq 0$.

4.2 Problems

1. **Bacteria** A colony of bacteria consists of 8000 bacteria. During the next 200 minutes, the colony will grow exponentially at the rate of 3.5 percent per minute. Find a formula for the number of bacteria in the colony t minutes from now. (The formula will be valid up to $t = 200$.) Then determine how large the colony will be 40 minutes from now.

2. **Demand** The monthly demand for an insecticide is presently 12 million barrels. However, it will decrease exponentially at the rate of 5 percent per month. Find a formula for the monthly demand t months from now. Then determine what the monthly demand will be 9 months from now.

3. **Depreciation** A company purchased a new machine for $75,000. Suppose the machine depreciates exponentially at the rate of 12 percent per year. Find a formula that gives the value of the machine when it is x years old. Then determine how much it will be worth when it is 10 years old.

4. **Necessary Income** In a city, the necessary annual income for a family of four to maintain a moderate standard of living is presently $25,000. Suppose that during the next 15 years the necessary income increases exponentially at the rate of 3 percent per year. Find a formula that gives the necessary income x years from now. (The formula will be valid up to $x = 15$.) Then determine what the necessary income will be 5 years from now.

5. **Drug Concentration** Six hours ago, each liter of a woman's blood contained 125 milligrams of a drug. Suppose the concentration decreases exponentially at the rate of 8 percent per hour. How many milligrams of the drug will each liter of her blood contain 4 hours from now?

6. **Excess Weight** A man started a special diet 2 weeks ago when he was 40 pounds heavier than recommended for his body type. As long as he stays on the diet, his excess weight will decrease exponen-

tially at the rate of 12 percent per week. If he stays on the diet, how many pounds overweight will he be 7 weeks from now?

7. **Dilution** A tank contains 500,000 gallons of a mixture consisting of water and a poisonous chemical. Pure water is pumped into the tank at the rate of 2000 gallons per hour, while a valve allows the mixture to pour out of the tank at the same rate. If the chemical is kept evenly mixed, the quantity of the chemical in the tank decreases exponentially at the rate of 0.4 percent per hour. Determine how many pounds of the chemical will still be in the tank 30 hours from now if presently there is 90 pounds of the chemical in the tank.

8. **Cooling Law** A ceramist places a vase whose temperature is 120 degrees in a room where the temperature is kept at 70 degrees. According to what is called Newton's law of cooling, the difference between the temperature of the vase and the temperature of the room will decrease exponentially. Suppose that for this vase this difference decreases exponentially at the rate of 25 percent per hour. Determine what the temperature of the vase will be 8 hours after the ceramist places it in the room.

9. **Advertising** Without spending any money on advertising, a firm sells 75 tons of a commodity per week. As the firm increases the weekly advertising expenditure, the weekly demand increases exponentially at the rate of 0.5 percent per thousand dollars. Find the weekly demand if the firm spends $20,000 per week on advertising.

10. **Atmospheric Pressure** At sea level, the atmospheric pressure on an object is 14.7 pounds per square inch. As the object gains altitude, the atmospheric pressure decreases exponentially at the rate of 21 percent per mile. What is the pressure when the object is 4 miles above sea level?

11. **Population** When a manufacturing firm moved to a city 6 years ago, the population of the city was 65,000. If the population has been growing exponentially at the rate of 1.5 percent per year since the firm moved to the city, what is the present population?

12. **Appreciation** The value of a plot of land is presently $58,000. If the value increases exponentially at the rate of 9 percent per year, what will it be 5 years from now?

13. Express the function

$$f(x) = e^{0.03x + 5}$$

in the form

$$f(x) = Ae^{rx}$$

where $A > 0$ and $r \neq 0$. [So $f(x) = e^{0.03x + 5}$ is an exponential function. Actually, any function of the form $f(x) = e^{mx + b}$ is an exponential function, provided $m \neq 0$.]

14. Let $h(x)$ represent the product of the following two exponential functions:

$$f(x) = 15e^{-0.25x}$$
$$g(x) = 76e^{0.08x}$$

Express $h(x)$ in the form

$$h(x) = Ae^{rx}$$

where $A > 0$ and $r \neq 0$. [So $h(x)$ is an exponential function. Actually, the product of two exponential functions is always an exponential function, unless the sum of the exponents is zero.]

Graphing Calculator Problems

15. Set the viewing screen on your graphing calculator to $[-18, 18] \times [0, 24]$ with a scale of 2 on both axes. Then draw the graphs of the following exponential functions so that they appear simultaneously on the screen:

$$f(x) = 6e^{-0.07x}$$
$$f(x) = 6e^{-0.12x}$$
$$f(x) = 6e^{-0.19x}$$

What seems to happen to the graph of an exponential function if we increase the absolute value of r?

16. Set the viewing screen on your graphing calculator to $[-36, 36] \times [0, 48]$ with a scale of 6 on both axes. Then draw the graphs of the following exponential functions so that they appear simultaneously on the screen:

$$f(x) = 6e^{0.06x}$$
$$f(x) = 10e^{0.06x}$$
$$f(x) = 16e^{0.06x}$$

If we increase the value of A in an exponential function in which $r > 0$, what do you think happens to the graph? (What do you think would happen if $r < 0$?)

4.3 The Natural Logarithmic Function

In this section, we use what is called the natural logarithmic function to help solve problems that involve exponential functions.

Definition 4.4

The **natural logarithmic function** is the function that associates with each *positive* real number x the power to which e must be raised to produce x. This function is represented by

$$\ln x$$

(read "the natural logarithm of x").

So, if we write

$$y = \ln x$$

we mean that y is such that

$$e^y = x$$

According to Definition 4.4,

$$\ln 1 = 0 \text{ because } e^0 = 1$$
$$\ln e = 1 \text{ because } e^1 = e$$
$$\ln 9 \approx 2.1972246 \text{ because } e^{2.1972246} \approx 9$$

We can express the meaning of $\ln x$ by writing

$$e^{\ln x} = x \text{ for each } x > 0$$

The values of x must be positive because any power of e is positive.

Figure 4.6 shows the graph of $\ln x$. The graph shows that $\ln x$ is *continuous* on its domain, which is the set of positive real numbers. Note that the graph crosses the horizontal axis at $x = 1$ and is asymptotic to the vertical axis.

The following properties of $\ln x$ are implied by the definition of $\ln x$ and the basic properties of exponents listed in Section 4.1.

Properties of ln x

1. $\ln ab = \ln a + \ln b$
2. $\ln \dfrac{a}{b} = \ln a - \ln b$
3. $\ln a^k = k \ln a$
4. $\ln \dfrac{1}{b} = -\ln b$
5. $\ln e^k = k$

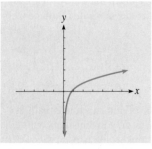

Figure 4.6 The graph of $\ln x$.

Here is the proof of property 1:

$$e^{\ln ab} = ab \qquad \text{definition of } \ln x$$
$$= (e^{\ln a})(e^{\ln b}) \qquad \text{definition of } \ln x$$
$$= e^{\ln a + \ln b} \qquad \text{property E1 in Section 4.1}$$

So

$$e^{\ln ab} = e^{\ln a + \ln b}$$

Thus,

$$\ln ab = \ln a + \ln b \qquad \text{property E6 in Section 4.1}$$

The proofs of properties 2 and 3 are similar to the proof of property 1. Since $\ln 1 = 0$, property 4 is a special case of property 2 with $a = 1$. Since $\ln e = 1$, property 5 is a special case of property 3 with $a = e$.

We use $\ln x$ in Example 4.6.

Example 4.6

The owners of a firm find that they can sell

$$D(x) = 50 - 30e^{-0.07x}$$

tons of their main product daily if they spend \$$x$ thousand per day on advertising. If the owners want to sell 40 tons of the product daily, how much must they spend on advertising? [Figure 4.7 shows the graph of $D(x)$.]

Solution

To sell 40 tons of the product daily, the amount the owners must spend per day on advertising is the value of x that satisfies the equation

$$40 = 50 - 30e^{-0.07x}$$
$$-10 = -30e^{-0.07x}$$
$$\tfrac{1}{3} = e^{-0.07x}$$
$$-0.07x = \ln \tfrac{1}{3} \qquad \text{definition of } \ln x$$
$$-0.07x = -\ln 3 \qquad \text{property 4 of } \ln x$$
$$x = \frac{\ln 3}{0.07}$$

Using a calculator, we obtain

$$\frac{\ln 3}{0.07} \approx 15.7$$

Therefore, the owners of the firm must spend about \$15,700 per day on advertising to sell 40 tons of the product daily.

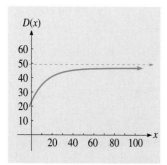

Figure 4.7
$D(x) = 50 - 30e^{-0.07x}$ for $x \geq 0$.

Doubling Time and Half-Life

Suppose a quantity increases exponentially at the rate of $100r$ percent per unit of time. Let Q_0 represent the quantity at an *arbitrary* instant. Then

$$Q(t) = Q_0 e^{rt}$$

gives the quantity t units of time after the arbitrary instant. We want to find how much time must elapse for $Q(t)$ to become *twice* as large as Q_0. In other words, we want to find the value of t that satisfies the equation

$$2Q_0 = Q_0 e^{rt}$$
$$2 = e^{rt}$$
$$rt = \ln 2 \qquad \text{definition of } \ln x$$
$$t = \frac{\ln 2}{r}$$

Thus, $(\ln 2)/r$ units of time must elapse for $Q(t)$ to become twice as large as Q_0.

The value of $(\ln 2)/r$ depends only on the value of r, and, as we said, Q_0 represents the quantity at an arbitrary instant. Therefore, we have shown that if a quantity increases exponentially, the time that must elapse for the quantity to double remains constant as the quantity increases. This time, called the **doubling time**, is $(\ln 2)/r$ units of time, where $100r$ is the quantity's percentage rate of increase.

Similarly, if a quantity decreases exponentially, the time that must elapse for the quantity to halve remains constant as the quantity decreases. This time, called the **half-life**, is $(\ln 2)/|r|$, where $100|r|$ is the quantity's percentage rate of decrease.

We summarize the above results in Theorem 4.4.

Theorem 4.4

If a quantity changes exponentially at the rate of $100|r|$ percent per unit of time, the doubling time or the half-life (depending on whether the quantity increases or decreases) is

$$\frac{\ln 2}{|r|}$$

units of time.

We use Theorem 4.4 in Examples 4.7 and 4.8.

Example 4.7

A woman deposited a sum of money in a bank where the interest is compounded continuously at the annual rate of 6 percent. How frequently will the balance double?

Solution

According to Theorem 4.2 in Section 4.1, the balance at the end of t years is given by the formula

$$B(t) = Pe^{0.06t}$$

where P represents the amount the woman deposited. So the balance increases exponentially, and, therefore, it has a doubling time. According to Theorem 4.4, the doubling time is

$$\frac{\ln 2}{0.06} \approx 11.6$$

years, which means that the balance doubles approximately every 11.6 years. This doubling time is independent of the amount the woman deposited.

Example 4.8

Find the half-life of a radioactive substance that decays exponentially at the rate of 4 percent per month.

Solution

According to Theorem 4.4, the half-life is

$$\frac{\ln 2}{0.04} \approx 17.3$$

months. Thus, the amount of the substance halves approximately every 17.3 months.

In Theorem 4.4, let

$$n = \frac{\ln 2}{|r|}$$

Then, solving for $|r|$, we get

$$|r| = \frac{\ln 2}{n}$$

So

$$100|r| = 100 \frac{\ln 2}{n}$$

This result shows that Theorem 4.5 is a direct consequence of Theorem 4.4.

Theorem 4.5

If a quantity changes exponentially, the percentage rate of change is

$$100 \frac{\ln 2}{n}$$

where n represents either the doubling time or the half-life, depending on whether the quantity increases or decreases.

We use Theorem 4.5 in Example 4.9.

Example 4.9

Thanks to a new purifying process, the amount of a pollutant in a pond is decreasing. In fact, it is decreasing exponentially with a half-life of 7 weeks. Find the percentage rate of decrease.

Solution

Since

$$100 \frac{\ln 2}{7} \approx 9.9$$

the amount of the pollutant is decreasing at the rate of about 9.9 percent per week.

In Example 4.10, we find the percentage rate of increase of a quantity that increases exponentially even though we are not given the doubling time.

Example 4.10

A substance whose weight increases exponentially weighed 24 grams 20 days ago and 36 grams 2 days ago.

a. At what percentage rate does the weight increase?
b. How much will the substance weigh 5 days from now?

Solution for (a)

Let $W(x)$ represent the weight of the substance x days after it weighed 24 grams. Since the weight increases exponentially,

$$W(x) = 24e^{rx}$$

where $100r$ is the percentage rate of increase. Two days ago was 18 days after the substance weighed 24 grams. Thus, since 2 days ago the weight was 36 grams, $W(18) = 36$. So we want to find the value of r that satisfies

$$36 = 24e^{r(18)}$$
$$1.5 = e^{18r}$$
$$18r = \ln 1.5 \quad \text{definition of } \ln x$$
$$r = \frac{\ln 1.5}{18}$$
$$\approx 0.023$$

Therefore, the weight increases at the rate of approximately 2.3 percent per day.

Solution for (b)

We can use the formula

$$W(x) = 24e^{0.023x}$$

to approximate the weight of the substance x days after it weighed 24 grams. Five days from now is 25 days after the substance weighed 24 grams. Therefore, since

$$W(25) = 24e^{0.023(25)}$$
$$= 24e^{0.575}$$
$$\approx 42.7$$

the substance will weigh approximately 42.7 grams 5 days from now.

Functions of the Form $f(x) = Ab^{kx}$, where $b \neq e$

According to the definition of the natural logarithmic function (Definition 4.4), if $b > 0$,

$$b = e^{\ln b}$$

Therefore, if $b > 0$,

$$Ab^{kx} = A(e^{\ln b})^{kx}$$
$$= Ae^{(\ln b)kx} \quad \text{property E3 in Section 4.1}$$

This result is a basis for Theorem 4.6.

Theorem 4.6

Functions of the form

$$f(x) = Ab^{kx}$$

are exponential functions (even if $b \neq e$), provided A and b are positive, $b \neq 1$, and $k \neq 0$.

An advantage in having $b = e$ is that the percentage rate of change is visible in the formula.

Example 4.11

An accountant believes that a machine depreciates exponentially. In fact, the accountant believes that the machine is worth

$$V(x) = 80{,}000\left(\frac{1}{5}\right)^{0.04x}$$

dollars when it is x years old. Find the percentage rate of depreciation.

Solution

We want to express the formula for $V(x)$ in the form

$$V(x) = 80{,}000 e^{rx}$$

because then we can conclude that the percentage rate of depreciation is $100|r|$.

According to the definition of $\ln x$,

$$\frac{1}{5} = e^{\ln(1/5)}$$

So,

$$\begin{aligned}
80{,}000\left(\frac{1}{5}\right)^{0.04x} &= 80{,}000(e^{\ln(1/5)})^{0.04x} \\
&= 80{,}000(e^{-\ln 5})^{0.04x} \quad \text{property 4 of } \ln x \\
&= 80{,}000 e^{-(\ln 5)0.04x} \quad \text{property E3 in Section 4.1}
\end{aligned}$$

Thus,

$$V(x) = 80{,}000 e^{-(\ln 5)0.04x}$$

Since

$$(\ln 5)0.04 \approx 0.064$$

the machine depreciates at the rate of about 6.4 percent per year.

Since

$$\frac{\ln 2}{(\ln 5)0.04} \approx 10.8$$

the half-life of the value of the machine in Example 4.11 is about 10.8 years.

Carbon Dating

After an organism dies, the amount of carbon-14 in the remains of the organism decreases exponentially with a half-life of 5730 years. This phenomenon is the basis of a method for determining the age of an organism's remains. The method was discovered by Willard F. Libby in the latter part of the 1940s at the Institute for Nuclear Studies of the University of Chicago. Libby eventually received a Nobel prize for his discovery. We illustrate the method in Example 4.12.

Example 4.12

An archaeologist found an individual's sternum at an excavation site outside the city of Messina, Sicily. If the sternum contains 0.375 the amount of carbon-14 it contained at the time of death, when did the individual die?

Solution

Let A represent the amount of carbon-14 the sternum contained when the individual died. Then the amount of carbon-14 in the sternum t years later is

$$Q(t) = Ae^{rt}$$

where

$$r = -\frac{\ln 2}{5730}$$

We want to find the value of t that satisfies

$$0.375A = Ae^{rt}$$
$$0.375 = e^{rt}$$
$$rt = \ln 0.375 \qquad \text{definition of } \ln x$$
$$t = \frac{\ln 0.375}{r}$$
$$t = \frac{\ln 0.375}{-(\ln 2)/5730} \qquad \text{since } r = -\frac{\ln 2}{5730}$$
$$\approx 8108$$

Thus, the individual died approximately 8100 years ago.

In Example 4.12, $\ln 0.375$ is a negative number. This fact explains why the value of t turns out to be a positive number.

Composite Functions of the Form ln f(x)

Suppose the composite function $\ln f(x)$ is a *nonconstant* linear function. In other words, suppose

$$\ln f(x) = rx + b$$

where $r \neq 0$. Then

$$f(x) = e^{rx+b}$$ Definition 4.4
$$= e^{rx}e^{b}$$ property E1 in Section 4.1
$$= Ae^{rx}$$ letting $A = e^b$

Thus, we have established the validity of Theorem 4.7.

Theorem 4.7

Suppose
$$\ln f(x) = rx + b$$
where $r \neq 0$. Then $f(x)$ is an exponential function that can be expressed in the form
$$f(x) = Ae^{rx}$$
where $A = e^b$.

We use Theorem 4.7 in Example 4.13.

Example 4.13

Let $P(t)$ represent a country's population (in millions) t years from now. A demographer believes that
$$\ln P(t) = 0.03t + 4$$
for $t \geq 0$. Find a formula for $P(t)$.

Solution

Using Theorem 4.7,
$$P(t) = e^4 e^{0.03t}$$
Therefore, since
$$e^4 \approx 54.6$$
we can use the formula
$$P(t) = 54.6 e^{0.03t}$$
to estimate what the population will be t years from now.

We can perceive that several points lie along the path of a line more easily than we can perceive that several points lie along the path of an exponential curve. Therefore, we often use a combination of linear regression and Theorem 4.7 to derive exponential functions from empirical data. (Linear regression is discussed in Section 1.3.) Two graphing calculator problems at the end of this section (Problems 35 and 36) illustrate this approach for deriving exponential functions.

4.3 Problems

1. **Future Value** If $1500 is deposited in a bank where the interest is compounded continuously at the annual rate of 8 percent, when will the balance be $4500?

2. **Population** A demographer expects the population of a country to grow exponentially at the rate of 2.5 percent per year. If the population is now 80 million, when will it be 400 million?

3. **Demand** The weekly demand for a fertilizer is presently 20 million tons. However, it will decrease exponentially at the rate of 15 percent per week. When will the weekly demand be 5 million tons?

4. **Depreciation** A company purchased a new machine for $60,000. If the machine depreciates exponentially at the rate of 18 percent per year, when will it be worth $10,000?

5. **Decontamination** The effectiveness of a process for extracting a pollutant from a lake depends on how much is spent applying the process. Experts claim that the percent of the pollutant that remains after spending x thousand is given by

$$P(x) = 30 + 70e^{-0.09x}$$

How much must be spent to reduce the amount of the pollutant to 40 percent of the original amount?

6. **Labor Productivity** The management of a manufacturing firm finds that the firm produces

$$Q(x) = 160 - 160e^{-0.08x}$$

barrels of its principal product per day using x units of labor. How much labor should the firm use to produce 140 barrels of the product per day?

7. **Weight Reduction** A woman who now weighs 160 pounds plans to start a special diet. Her doctor predicts that she will weigh

$$W(t) = 125 + 35e^{-0.05t}$$

pounds after t weeks of dieting. How long must she diet to bring her weight down to 140 pounds?

8. **Lead Concentration** The average lead concentration in the blood of the residents of a European country is decreasing. Suppose that t years after the beginning of 1995 the concentration is

$$L(t) = 4 + 6e^{-0.2t}$$

micrograms of lead per deciliter of blood. When is the concentration 5.5 micrograms per deciliter?

9. **Half-Life** Find the half-life of a radioactive substance that decays exponentially at the rate of 5 percent per week.

10. **Doubling Time** A man deposited a sum of money in a bank where the interest is compounded continuously at the annual rate of 9 percent. Find the doubling time of the balance. That is, find how frequently the balance doubles.

11. **Doubling Time** Find the doubling time of a population that grows exponentially at the rate of 3 percent per year.

12. **Half-Life** Find the half-life of the value of a machine that depreciates exponentially at the rate of 20 percent per year.

13. **Medicine** A medical doctor believes that the concentration of a drug in the blood of a man (in milligrams per deciliter) decreases exponentially at the rate of 30 percent per hour. How frequently does the concentration halve? In other words, find the half-life of the concentration.

14. **Appreciation** If the value of a plot of land increases exponentially at the rate of 4 percent per year, how often does the value double?

15. **Pollution** During the past 8 years, the concentration of a pollutant in a river (in milligrams per liter) has been decreasing exponentially with a half-life of 5 months. Find the percentage rate at which the concentration has been decreasing.

16. **Demand** Marketing experts project that the weekly demand for a new product will grow exponentially with a doubling time of 16 months. At what percentage rate will the weekly demand grow?

17. **Bacteria** If the doubling time of a colony of bacteria that grows exponentially is 27 minutes, what is its percentage rate of growth?

18. **Exponential Decay** If the half-life of a substance that decays exponentially is 2300 years, what is its percentage rate of decay?

19. **Radioactivity** Fifteen weeks ago a radioactive substance weighed 40 pounds. Three weeks ago the substance weighed 25 pounds. If the weight

decreases exponentially, find what it will be 7 weeks from now.

20. **Bacteria** When a colony of bacteria began growing 4 hours ago, it consisted of 5000 bacteria. Now it consists of 9000 bacteria. If the colony grows exponentially, find what the size of the colony will be 6 hours from now.

21. **Decontamination** The effectiveness of a process for extracting a pollutant from a pond depends on how much is spent applying the process. Let $P(x)$ represent the percent of the pollutant that remains in the pond after spending $\$x$ thousand. The engineer who designed the process believes that $P(x)$ decreases exponentially as x is increased. If 60 percent of the pollutant remains after spending $3000, what percent remains after spending $6000?
    Note: If no money is spent, all of the pollutant remains. Thus, $P(0) = 100$.

22. **Advertising** Without spending any money on advertising, a company sells 56 million barrels of its main product daily. If the company spends $40,000 per day on advertising, it sells 72 million barrels daily. Suppose the daily demand increases exponentially as the company increases the daily advertising expenditure. What is the daily demand if the company spends $120,000 per day on advertising?

23. **Weight Gain** Suppose that t hours from now a substance weighs
    $$W(t) = 15 \cdot 7^{0.03t}$$
    kilograms. So its weight increases exponentially. Find the percentage rate of increase.
    Warning: The answer is not 3 percent.

24. **Weight Loss** Suppose that t days from now a substance weighs
    $$W(t) = 25\left(\frac{1}{9}\right)^{0.08t}$$
    pounds. So its weight decreases exponentially. Find the percentage rate of decrease.
    Warning: The answer is not 8 percent.

25. **Book Sales** So far, 5000 copies of a recently published book have been sold. The management of the firm that published the book believes that x months from now the number of copies sold will be
    $$B(x) = \frac{80}{1 + 15e^{-0.12x}}$$
    thousand.
    a. Will the number of copies sold eventually reach 70,000? If so, when?
    b. Will the number of copies sold eventually reach 90,000? If so, when?

26. **Mortgage Rates** In an area of the United States, the interest rate for conventional mortgages is presently 11 percent. However, experts expect this rate to increase. They predict that x months from now the interest rate will be
    $$R(x) = \frac{165}{11 + 4e^{-0.25x}}$$
    percent.
    a. Will the rate eventually reach 14 percent? If so, when?
    b. Will the rate eventually reach 18 percent? If so, when?

27. **Carbon Dating** If a sample of an organism's remains contains $\frac{1}{6}$ the amount of carbon-14 it contained when the organism died, when did it die?

28. **Carbon Dating** If a sample of an organism's remains contains $\frac{2}{7}$ the amount of carbon-14 it contained when the organism died, when did it die?

29. **Pollution** Presently, the concentration of a pollutant in a lake is 8 times the maximum concentration that can be tolerated by swimmers. However, the source of the pollutant has been removed, and so the concentration of the pollutant in the lake will gradually decrease. Environmentalists predict that the concentration will decrease exponentially at the rate of 35 percent per month. When will it be safe to swim in the lake?

30. **Radioactive Contamination** Officials closed a recreational facility after it was accidentally contaminated by a radioactive substance that decays exponentially with a half-life of 9 years. The level of contamination is presently 15 times the maximum level that humans can tolerate. When can the officials reopen the facility?

31. If $e^b = c$, we can use the definition of the natural logarithmic function (Definition 4.4) to conclude that $\ln c = b$. Elaborate.

32. In Theorem 4.6, we require b to be different from 1 and k to be different from zero. Why?

Graphing Calculator Problems

33. **Depreciation** An accountant projects that a machine will be worth
$$V(x) = 15(0.25)^{0.03x}$$
million dollars when it is x years old. Thus, the machine depreciates exponentially. Set the viewing screen on your graphing calculator to $[0, 24] \times [0, 16]$ with a scale of 2 on both axes and draw the graph of $V(x)$. Then use the trace function to estimate the half-life of the machine's value.

34. **Population** Demographers believe that a population will be
$$P(x) = 4 \cdot 5^{0.02x}$$
million x years from now. So the population will grow exponentially. Set the viewing screen on your graphing calculator to $[0, 24] \times [0, 16]$ with a scale of 2 on both axes and draw the graph of $P(x)$. Then use the trace function to estimate the population's doubling time.

35. **Depreciation** Twenty years ago, a manufacturing firm purchased a new machine. Let $V(x)$ represent the machine's value (in millions of dollars) when it is x years old. The following table shows the machine's value at six different ages:

x	3	5	7	10	14	19
$V(x)$	32	29	26	23	18	15

a. Set the viewing screen on your graphing calculator to $[0, 20] \times [0, 4]$ with a scale of 1 on both axes. Then plot the data points of the form
$$(x, \ln V(x))$$
for $x = 3, 5, 7, 10, 14,$ and 19.

b. Draw the regression line for the data points you plotted in part (a).

c. Find a formula for the least-squares linear function that corresponds to the regression line you drew in part (b).

d. The formula you found in part (c) approximates the functional relationship between $\ln V(x)$ and x. Use this fact and Theorem 4.7 to find a formula that approximates the functional relationship between $V(x)$ and x.

e. Use the formula you found in part (d) to estimate what the value of the machine will be 8 years from now. (Eight years from now the machine will be 28 years old.)

36. **Population** The following table shows the population of a Third World country at six different times:

Date, Jan. 1	1981	1983	1987	1989	1992	1994
Population, millions	25	26	27	28.5	30.5	31

Let $P(x)$ represent the country's population x years after January 1, 1980.

a. Set the viewing screen on your graphing calculator to $[0, 15] \times [0, 4]$ with a scale of 1 on both axes. Then plot the data points of the form
$$(x, \ln P(x))$$
for $x = 1, 3, 7, 9, 12,$ and 14.

b. Draw the regression line for the data points you plotted in part (a).

c. Find a formula for the least-squares linear function that corresponds to the regression line you drew in part (b).

d. The formula you found in part (c) approximates the functional relationship between $\ln P(x)$ and x. Use this fact and Theorem 4.7 to find a formula that approximates the functional relationship between $P(x)$ and x.

e. Use the formula you found in part (d) to estimate what the population of the Third World country will be on January 1, 2000.

4.4 Differentiation of ln u(x)

In this section, we derive a rule for computing derivatives of functions that can be expressed in the form

$$f(x) = \ln u(x)$$

where $u(x)$ is a differentiable function. We begin with the special case in which $u(x) = x$.

Theorem 4.8 (Logarithm Rule)

The derivative of

$$f(x) = \ln x$$

is

$$f'(x) = \frac{1}{x}$$

According to Definition 2.1 in Section 2.2, the derivative of

$$f(x) = \ln x$$

is

$$f'(x) = \lim_{h \to 0} \frac{\ln(x+h) - \ln x}{h}$$

So, if we show that the graph of

$$s(x) = \frac{\ln(x+h) - \ln x}{h}$$

is close to the graph of

$$f'(x) = \frac{1}{x}$$

when h is a small constant, we would have evidence supporting Theorem 4.8. We provide such evidence in Figure 4.8. The figure shows that when $h = 0.6$, the graph of $s(x)$ is close to the graph of $f'(x) = 1/x$. [The graph of $s(x)$ lies below the graph of $f'(x)$.] For a value of h less than 0.6, the graph of $s(x)$ would be closer yet to the graph of $f'(x) = 1/x$.

We use Theorem 4.8 in Examples 4.14, 4.15, and 4.16.

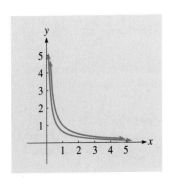

Figure 4.8 $f'(x) = \dfrac{1}{x}$ and $s(x) = \dfrac{\ln(x+0.6) - \ln x}{0.6}$.

Example 4.14

Find the derivative of $f(x) = x^4 \ln x$.

Solution

Using the product rule and the logarithm rule, we get

$$f'(x) = (x^4)\frac{1}{x} + (\ln x)4x^3$$
$$= x^3 + 4x^3 \ln x$$

Example 4.15

Find the derivative of $f(x) = (\ln x)^3$.

Solution

Using the generalized power rule and the logarithm rule, we get

$$f'(x) = 3(\ln x)^2 \left(\frac{1}{x}\right)$$
$$= \frac{3(\ln x)^2}{x}$$

Example 4.16

Find the derivative of $f(x) = \ln x^3$.

Solution

According to property 3 of $\ln x$ in Section 4.3,

$$\ln x^3 = 3 \ln x$$

Thus,

$$f(x) = 3 \ln x$$

Then, using the constant multiple rule and the logarithm rule, we get

$$f'(x) = 3 \cdot \frac{1}{x}$$
$$= \frac{3}{x}$$

A **logarithmic function** is a function that can be expressed in the form

$$f(x) = \ln u(x)$$

Thus,

$$f(x) = \ln(x^2 + 7)$$

is a logarithmic function. In this case, $u(x) = x^2 + 7$. The natural logarithmic function is the special logarithmic function in which $u(x) = x$.

Theorem 4.8 provides us with a rule for finding the derivative of this special logarithmic function. In what follows we derive a rule for finding the derivative of *any* logarithmic function in which $u(x)$ is a differentiable function.

The Derivative of a Logarithmic Function

Suppose
$$f(x) = \ln u(x)$$
where $u(x)$ is a differentiable function. If we let
$$g(x) = \ln x$$
we can think of $f(x)$ as the composite of $g(x)$ and $u(x)$ and thus write
$$f(x) = g[u(x)]$$
Using the chain rule (Section 2.5),
$$f'(x) = g'[u(x)]u'(x) \qquad (1)$$
According to Theorem 4.8,
$$g'(x) = \frac{1}{x}$$
So
$$g'[u(x)] = \frac{1}{u(x)}$$
Thus, substituting $1/u(x)$ for $g'[u(x)]$ in formula (1), we get
$$f'(x) = \frac{1}{u(x)} \cdot u'(x)$$
$$= \frac{u'(x)}{u(x)}$$
We have just derived Theorem 4.9.

Theorem 4.9 (Generalized Logarithm Rule)

Suppose $u(x)$ is a differentiable function. Then the derivative of
$$f(x) = \ln u(x)$$
is
$$f'(x) = \frac{u'(x)}{u(x)}$$

The rule in Theorem 4.8 is a special case of Theorem 4.9. [In Theorem 4.8, $u(x) = x$.] Theorem 4.9, in turn, is a special case of the chain rule.

Example 4.17

Find the derivative of $f(x) = \ln(x^2 + 7)$.

Solution

The derivative of $x^2 + 7$ is $2x$. So, according to the generalized logarithm rule,

$$f'(x) = \frac{2x}{x^2 + 7}$$

In Example 4.18, we use property 3 of $\ln x$ to make the computation easier.

Example 4.18

If $y = \ln(x^3 + 9)^5$, find a formula for dy/dx.

Solution

According to property 3 of $\ln x$,

$$\ln(x^3 + 9)^5 = 5 \ln(x^3 + 9)$$

So

$$y = 5 \ln(x^3 + 9)$$

Using the constant multiple rule and the generalized logarithm rule, we get

$$\frac{dy}{dx} = 5 \cdot \frac{3x^2}{x^3 + 9}$$
$$= \frac{15x^2}{x^3 + 9}$$

In Example 4.19, we use property 2 of $\ln x$ to simplify the computation.

Example 4.19

Find the slope of the tangent to the graph of

$$f(x) = \ln\left(\frac{9x - 50}{6x - 11}\right)$$

at the point where $x = 6$.

Solution

According to property 2 of ln x,

$$\ln\left(\frac{9x - 50}{6x - 11}\right) = \ln(9x - 50) - \ln(6x - 11)$$

So

$$f(x) = \ln(9x - 50) - \ln(6x - 11)$$

Using the generalized logarithm rule, we get

$$f'(x) = \frac{9}{9x - 50} - \frac{6}{6x - 11}$$

In particular,

$$f'(6) = \frac{9}{4} - \frac{6}{25}$$
$$= 2.01$$

Thus, the slope of the tangent in question is 2.01.

4.4 Problems

In Problems 1 to 30, find the derivative of the given function.

1. $f(x) = 3x^2 \ln x$
2. $f(x) = 5x \ln x$
3. $f(x) = 9 \ln x$
4. $f(x) = 13 \ln x$
5. $f(x) = \dfrac{\ln x}{8x}$
6. $f(x) = \dfrac{\ln x}{x^3}$
7. $f(x) = (\ln x)^2$
8. $f(x) = (\ln x)^5$
9. $y = \ln(3x - 4)$
10. $y = \ln(2x + 1)$
11. $y = \ln(3x^2 + 7)$
12. $y = \ln(2x^4 - 9)$
13. $y = 10 \ln(x^2 + 3)$
14. $y = 9 \ln(x^3 + 1)$
15. $f(x) = \ln(x^3 + 5x)$
16. $f(x) = \ln(2x^4 + x)$
17. $f(x) = \ln 17x$
18. $f(x) = \ln 5x$
19. $f(x) = 8 \ln x^2$
20. $f(x) = 5 \ln x^3$
21. $y = \ln(3x + 1)^4$
22. $y = \ln(x + 9)^6$
23. $f(x) = \ln \sqrt{x + 3}$
24. $f(x) = \ln \sqrt{6x + 5}$
25. $y = \ln\left(\dfrac{x + 3}{2x + 1}\right)$
26. $y = \ln\left(\dfrac{4x + 3}{x + 1}\right)$
27. $f(x) = \ln[3x(x + 5)^4]$
28. $f(x) = \ln[x\sqrt{2x + 8}]$
29. $f(x) = \ln\sqrt{\dfrac{4x - 1}{3x + 2}}$
30. $f(x) = \ln\left(\dfrac{2x + 7}{3x + 1}\right)^3$

31. **Sales** The manager of The Stereo Place finds that her store sells

$$R(x) = 120 - 25 \ln(4x + 1)$$

thousand dollars worth of stereo equipment monthly when the unemployment rate is x percent, provided $5 \le x \le 25$. Use the derivative of $R(x)$ to approximate the amount the monthly sales level decreases if the unemployment rate increases from 9 percent to 10 percent.

32. **Labor Productivity** The owner of Rapido Courier Service believes that his company can deliver

$$D(x) = 80 \ln(9x + 1)$$

tons of packages per month using x units of labor. Use the derivative of $D(x)$ to approximate the amount the monthly delivery capacity increases if the labor force is increased from 27 units to 28 units.

33. **Population** Officials predict that t years from now the population of a city will be

$$P(t) = 50 \ln(3t + 8)$$

thousand. How fast will the population be growing 4 years from now?

34. **Depreciation** The management of a manufacturing firm believes that t years from now a machine will be worth

$$V(t) = 170 - 34 \ln(5t + 18)$$

thousand dollars, provided $0 \leq t \leq 15$. At what rate will the machine be depreciating 6.5 years from now?

35. **Consumer Debt** Economists predict that x months from now the consumer debt of a country will be

$$D(x) = \ln(x^2 - 18x + 120)$$

billion dollars. When will the debt be least? What will be the debt when it is least?

36. **Optimal Fertilization Level** Let $Y(x)$ represent the number of tons of a crop each acre produces if x sacks of fertilizer is applied to each acre. A professor of agriculture assumes that

$$Y(x) = 8 - \ln(x^2 - 12x + 40)$$

for $0 \leq x \leq 20$. If at most 20 sacks of fertilizer can be applied to each acre, how much fertilizer should be applied to maximize the yield? What is the maximum possible per-acre yield?

Graphing Calculator Problems

37. Set the viewing screen on your graphing calculator to $[0, 6] \times [0, 4]$ with a scale of 1 on both axes. Then draw the graphs of the following functions so that they appear simultaneously:

$$y = \frac{1}{x}$$

$$y = \frac{\ln(x + 0.4) - \ln x}{0.4}$$

$$y = \frac{\ln(x + 0.2) - \ln x}{0.2}$$

Explain how the graphs support Theorem 4.8.

38. **Population** Set the viewing screen on your graphing calculator to $[0, 30] \times [0, 20]$ with a scale of 1 on both axes. Then draw the graph of the function that gives the rate at which the population in Problem 33 will be growing t years from now. Use the trace function to estimate when the population will be growing at the rate of 5.5 thousand per year.

4.5 Differentiation of $e^{u(x)}$

A few inhabitants of an island in the South Pacific were recently exposed to a virus that causes nausea lasting about a day. Health officials predict that x weeks from now

$$S(x) = e^{-x^2 + 6x - 9}$$

million inhabitants will be experiencing the nausea. Using the derivative of this function, we can determine when the number of inhabitants experiencing the nausea will be greatest.
 The function

$$S(x) = e^{-x^2 + 6x - 9}$$

is a function of the form

$$f(x) = e^{u(x)}$$

with $u(x) = -x^2 + 6x - 9$. In what follows, we use the generalized logarithm rule to derive a rule for computing derivatives of such functions. Then we will use the rule to obtain the derivative of $S(x)$.

The Derivative of a Function of the Form $e^{u(x)}$

Suppose
$$f(x) = e^{u(x)}$$
where $u(x)$ is a differentiable function. Then
$$\ln f(x) = \ln e^{u(x)}$$
$$= u(x) \quad \text{property 5 of } \ln x \text{ in Section 4.3}$$

Thus,
$$D_x(\ln f(x)) = u'(x)$$

According to the generalized logarithm rule,
$$D_x(\ln f(x)) = \frac{f'(x)}{f(x)}$$

So
$$\frac{f'(x)}{f(x)} = u'(x)$$
$$f'(x) = u'(x) \cdot f(x) \quad \text{multiplying by } f(x)$$
$$= u'(x)e^{u(x)} \quad \text{since } f(x) = e^{u(x)}$$

We have just derived Theorem 4.10.

Theorem 4.10 (Generalized Exponential Rule)

Suppose $u(x)$ is a differentiable function. Then the derivative of
$$f(x) = e^{u(x)}$$
is
$$f'(x) = u'(x)e^{u(x)}$$

We name the rule in Theorem 4.10 the generalized exponential rule even though $e^{u(x)}$ is not necessarily an exponential function. (See Definition 4.3 in Section 4.2.)

Now we can determine when the number of islanders experiencing the nausea will be greatest. We begin by finding the derivative of $S(x) = e^{-x^2 + 6x - 9}$.

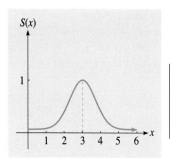

Figure 4.9 $S(x) = e^{-x^2 + 6x - 9}$ for $x \geq 0$.

Example 4.20

Find the derivative of $S(x) = e^{-x^2 + 6x - 9}$.

Solution

The derivative of $-x^2 + 6x - 9$ is $-2x + 6$. Thus, according to the generalized exponential rule,

$$S'(x) = (-2x + 6)e^{-x^2 + 6x - 9}$$

Let us examine the sign of $S'(x)$. First, note that since e raised to any power is positive, $e^{-x^2 + 6x - 9}$ is always positive. Then note that $-2x + 6 = 0$ when $x = 3$. Consequently, $S'(x) = 0$ when $x = 3$. Finally, note that $-2x + 6$ is positive when $x < 3$ and negative when $x > 3$. Therefore, $S'(x)$ likewise is positive when $x < 3$ and negative when $x > 3$.

Thus, according to Theorem 2.15 in Section 2.7, $S(x)$ achieves its maximum value at $x = 3$. So the number of islanders experiencing the nausea will be greatest 3 weeks from now. Figure 4.9 shows the graph of $S(x)$.

In Example 4.20, we used the generalized exponential rule to find the derivative of $f(x) = e^{-x^2 + 6x - 9}$ even though this function is not an exponential function according to Definition 4.3 in Section 4.2.

In Example 4.21, we find the derivative of an exponential function using a combination of the constant multiple rule and the generalized exponential rule.

Example 4.21

Find the derivative of $y = 70e^{-0.02t}$.

Solution

Using the generalized exponential rule, the derivative of $e^{-0.02t}$ is $-0.02e^{-0.02t}$. So, according to the constant multiple rule,

$$\frac{dy}{dt} = 70(-0.02e^{-0.02t})$$
$$= -1.4e^{-0.02t}$$

The function

$$f(x) = e^x$$

is a function of the form $f(x) = e^{u(x)}$. In this case, $u(x) = x$. Since the derivative of x is 1, using the generalized exponential rule, we get

$$f'(x) = 1 \cdot e^x$$
$$= e^x$$

Thus, we have reached the surprising conclusion that *the derivative of e^x is e^x itself*! Now we can see that the generalized exponential rule (Theorem 4.10) is the special case of the chain rule where $g(x) = e^x$.

In Example 4.22, we combine the above result with the quotient rule to find a derivative.

Example 4.22

Suppose
$$f(x) = \frac{e^x}{x^2 + 3}$$

Find a formula for $f'(x)$.

Solution

Using the quotient rule,
$$f'(x) = \frac{(x^2 + 3)e^x - e^x(2x)}{(x^2 + 3)^2}$$
$$= \frac{(x^2 - 2x + 3)e^x}{(x^2 + 3)^2}$$

In Example 4.23, we find the first and second derivatives of a function using the product rule and the generalized exponential rule.

Example 4.23

Sketch the graph of $f(x) = 12xe^{-0.02x}$ for $x \geq 0$.

Solution

According to the product rule and the generalized exponential rule,
$$f'(x) = 12x(-0.02e^{-0.02x}) + (e^{-0.02x})12$$
$$= 12e^{-0.02x} - (0.02x)12e^{-0.02x}$$
$$= (1 - 0.02x)12e^{-0.02x}$$

The sign of $f'(x)$ is determined by the factor $1 - 0.02x$, because $12e^{-0.02x}$ is always positive. Note that $f'(x) = 0$ when $x = 50$. Thus, since $f(50) = 600e^{-1}$, the point $(50, 600e^{-1})$ is a first-order critical point on the graph ($600e^{-1} \approx 221$).

As the sign chart in Figure 4.10 shows, $f'(x)$ is positive when $x < 50$ and negative when $x > 50$. So, according to Theorem 2.13 in Section 2.7, $f(x)$ increases on the interval $[0, 50)$ and decreases on the interval $(50, +\infty)$.

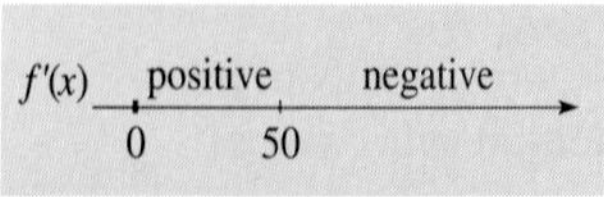

Figure 4.10 The sign chart for $f'(x) = (1 - 0.02x)12e^{-0.02x}$.

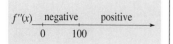

Figure 4.11 The sign chart for $f''(x) = (0.0048x - 0.48)e^{-0.02x}$.

We now examine the second derivative of $f(x)$. Applying the product rule and the generalized exponential rule to

$$f'(x) = (1 - 0.02x)12e^{-0.02x}$$

we get

$$f''(x) = (1 - 0.02x)(-0.24e^{-0.02x}) + 12e^{-0.02x}(-0.02)$$
$$= (0.0048x - 0.48)e^{-0.02x}$$

Note that $f''(x) = 0$ when $x = 100$. Thus, since $f(100) = 1200e^{-2}$, the point $(100, 1200e^{-2})$ is a second-order critical point on the graph $(1200e^{-2} \approx 162)$.

As the sign chart in Figure 4.11 shows, $f''(x)$ is negative when $x < 100$ and positive when $x > 100$. Therefore, according to Theorem 3.1 in Section 3.2, the graph of $f(x)$ is concave downward on the interval $[0, 100)$ and concave upward on the interval $(100, +\infty)$.

Since 0 is in the domain of $f(x)$ and $f(0) = 0$, the graph of $f(x)$ intersects the vertical axis at the point $(0, 0)$, which is also an endpoint of the graph.

Now we are ready to sketch the graph of $f(x)$. We begin by plotting the first- and second-order critical points. We also plot the vertical axis intercept. Then we sketch a curve that passes through these points and is compatible with the information about $f(x)$ that we gathered. Our curve should look like the curve in Figure 4.12.

The highest point on the curve in Figure 4.12 is the point where $x = 50$. Thus, $f(x)$ achieves its maximum value at $x = 50$. The point on the curve where $x = 100$ is an inflection point.

Figure 4.12 suggests that

$$\lim_{x \to +\infty} 12xe^{-0.02x} = 0$$

which is indeed true. This result is a special case of the following result:

If $g(x)$ is a polynomial function and $r < 0$,

$$\lim_{x \to +\infty} g(x)e^{rx} = 0$$

So, if $g(x)$ is a polynomial function and $r < 0$, the graph of

$$f(x) = g(x)e^{rx}$$

is *asymptotic* to the x axis.

The above result is a consequence of a theorem known as **L'Hôpital's rule**. This theorem, which usually appears in more advanced calculus books, is named after Guillaume François de L'Hôpital (1661–1704), a Frenchman who in 1696 published the first calculus textbook.

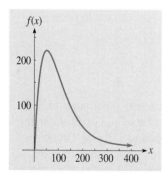

Figure 4.12 $f(x) = 12xe^{-0.02x}$ for $x \geq 0$.

Percentage Rate of Change of Exponential Functions

We have been assuming that if a quantity, say $f(x)$, changes exponentially as x increases, then the percentage rate of change of $f(x)$ remains constant. For example, suppose a population grows exponentially so that t years from now it is

$$P(t) = 53e^{0.02t}$$

million. Then, according to our assumption, the population grows at the constant rate of 2 percent per year. We now are able to prove this assumption.

Theorem 4.11

Suppose

$$f(x) = Ae^{rx}$$

where $A > 0$ and $r \neq 0$.

1. If $r > 0$, then $f(x)$ increases at an increasing rate. In terms of percentages, however, $f(x)$ increases at the constant rate of $100r$ percent per unit increase in x.
2. If $r < 0$, then $f(x)$ decreases at a decreasing rate. In terms of percentages, however, $f(x)$ decreases at the constant rate of $100|r|$ percent per unit increase in x.

We will only prove part 1 of Theorem 4.11. The proof of part 2 is similar to the proof of part 1.

Proof of Part 1 of Theorem 4.11

According to the constant multiple rule and the generalized exponential rule,

$$f'(x) = Are^{rx}$$

and

$$f''(x) = Ar^2 e^{rx}$$

Since A and r are positive and e raised to any power is positive, it follows that $f'(x)$ and $f''(x)$ are both positive for every value of x. Therefore, using Theorem 2.13 in Section 2.7, Theorem 3.1 in Section 3.2, and Theorem 3.2 in Section 3.3, we conclude that $f(x)$ increases at an increasing rate.

Now we will show that in terms of percentages $f(x)$ increases at the constant rate of $100r$ percent per unit increase in x. According to Definition 3.8 in Section 3.6, for *any* value of x, the percentage rate of increase of $f(x)$ is

$$100\frac{f'(x)}{f(x)} = 100\frac{Are^{rx}}{Ae^{rx}}$$
$$= 100r$$

Thus, in terms of percentages, $f(x)$ increases as we claimed, namely, at the constant rate of $100r$ percent per unit increase in x. So part 1 of Theorem 4.11 is indeed true.

In Example 4.24, we use Theorem 4.11 to show that a function is not an exponential function.

Example 4.24

Show that
$$f(x) = e^{\sqrt{x}}$$
is not an exponential function.

Solution

The derivative of $\sqrt{x}$ is $1/(2\sqrt{x})$. Thus, according to the generalized exponential rule,
$$f'(x) = \frac{1}{2\sqrt{x}} e^{\sqrt{x}}$$

Therefore, a formula for the percentage rate of change of $f(x)$ is
$$100\frac{f'(x)}{f(x)} = 100\frac{[1/(2\sqrt{x})]e^{\sqrt{x}}}{e^{\sqrt{x}}}$$
$$= \frac{50}{\sqrt{x}}$$

So the percentage rate of change of $f(x)$ does not remain constant as x increases. In fact, it decreases. Using Theorem 4.11, we conclude that $f(x) = e^{\sqrt{x}}$ is not an exponential function.

The converse of Theorem 4.11 is also true. But we are not yet able to verify it.

Converse of Theorem 4.11

If a quantity, say $f(x)$, changes so that its percentage rate of change remains constant as x increases, then the quantity changes exponentially.

The converse of Theorem 4.11 implies Theorem 4.12, which can help us solve two of the graphing calculator problems at the end of this section (Problems 45 and 46).

Theorem 4.12

Suppose
$$f'(x) = rf(x)$$
where $r \neq 0$ and $f(x) > 0$. Then $f(x)$ is an exponential function that can be expressed in the form
$$f(x) = Ae^{rx}$$

4.5 Problems

In Problems 1 to 28, find the derivative of the given function.

1. $f(x) = e^{x^3 + 5x}$
2. $f(x) = e^{x^2 - 9x}$
3. $f(t) = 55e^{-0.04t}$
4. $f(t) = 72e^{0.05t}$
5. $y = 0.5e^x + 0.5e^{-x}$
6. $y = 0.5e^x - 0.5e^{-x}$
7. $f(x) = x^3 e^x$
8. $f(x) = (3x + 1)e^x$
9. $y = \dfrac{5x}{1 + e^x}$
10. $y = \dfrac{3 + e^x}{x^2 - 1}$
11. $f(t) = 40(1 - 0.2e^{-0.08t})^3$
12. $f(t) = 25(1 - 0.4e^{-0.06t})^3$
13. $f(x) = \dfrac{1}{\sqrt{2 + 4e^x}}$
14. $f(x) = \dfrac{8}{(1 + 5e^x)^2}$
15. $f(x) = (x - 5)e^{0.03x}$
16. $f(x) = (x - 6)e^{-0.06x}$
17. $f(t) = \dfrac{25}{1 + 8e^{-0.3t}}$
18. $f(t) = \dfrac{32}{1 + 5e^{-0.4t}}$
19. $f(t) = \dfrac{9 + 12e^{-0.5t}}{1 + 4e^{-0.5t}}$
20. $f(t) = \dfrac{3 + 40e^{-0.2t}}{1 + 5e^{-0.2t}}$
21. $f(t) = 14 - 5e^{-0.3t}$
22. $f(t) = 20 - 8e^{-0.2t}$
23. $f(x) = e^x \ln x$
24. $f(x) = \dfrac{\ln x}{e^x}$
25. $f(x) = \dfrac{\ln(x^2 + 9)}{e^x}$
26. $f(x) = (e^{5x + 3}) \ln x$
27. $f(x) = \ln(e^{0.05x} + 1)$
28. $f(x) = \ln(e^{0.3x} - 4)$

In Problems 29 to 32, find the value of x where $f(x)$ is maximum.

29. $f(x) = 3e^{-x^2 + 10x - 25}$
30. $f(x) = 5e^{-x^2 + 4x - 4}$
31. $f(x) = 3e^{-4x} - 4e^{-6x}$
32. $f(x) = e^{-6x} - 3e^{-8x}$

In Problems 33 and 34, find the value of x where $f(x)$ is minimum if the domain is the interval $[0, +\infty)$.

33. $f(x) = 10 - x^2 e^{-0.25x}$
34. $f(x) = 12 - x^3 e^{-0.5x}$

In Problems 35 and 36, sketch the graph of the given function for $x \geq 0$.

35. $f(x) = (x + 2)e^{-0.1x}$
36. $f(x) = 3xe^{-0.2x}$

In Problems 37 and 38, verify that the given function is not an exponential function by showing that the percentage rate of change of $f(x)$ does not remain constant as x increases. Let the domain of each function be the interval $[0, +\infty)$.

37. $f(x) = 10e^{0.04x^2}$
38. $f(x) = 15 - 8e^{-0.03x}$

39. **Elderly Population** In an Asian country, the number of people at least 80 years old is increasing. Suppose that x years after the beginning of 1994 the number of such people is
$$N(x) = 30 - 24e^{-0.5x}$$
million. Use the derivative of $N(t)$ to approximate the amount this number increases during 1998.

40. **Lead Concentration** The average lead concentration in the blood of the residents of a European country is decreasing. Suppose that t years after the beginning of 1994 the concentration is
$$L(t) = 4 + 6e^{-0.2t}$$
micrograms of lead per deciliter of blood. Use the derivative of $L(t)$ to approximate the amount the concentration decreases during 1999.

This machine was used to tunnel under the English Channel. (AP/Wide World.)

41. **Rodent Population** A biologist believes that the population of a colony of rodents will increase exponentially with a doubling time of 5 months. If the population is presently 8000, at what rate (not percentage rate) will it be increasing 10 months from now?

42. **Depreciation** An accountant assumes that a machine's value decreases exponentially with a half-life of 4 years. If the machine is presently worth $144,000, at what rate (not percentage rate) will it be depreciating 12 years from now?

43. **Health Care Spending** In a South American country, health care spending was occurring at the rate of $222 billion per year at the beginning of 1994. Suppose that t years after the beginning of 1994 health care spending occurs at the rate of

$$H(t) = 100 + (6t + 122)e^{-0.04t}$$

billion dollars per year. When does the rate peak?

44. **Utilization of Industrial Capacity** At the beginning of 1994, the industries of an African nation were operating at 70 percent of their capacity. If t months after the beginning of 1994 the industries operate at

$$I(t) = 70 + 2te^{-0.05t}$$

percent of their capacity, when is the utilization greatest?

Graphing Calculator Problems

45. **Population** The following table shows the population of a South American country at seven different times. The table also shows the rate at which the population was growing at each of the seven times.

Date, Jan. 1	1980	1982	1986	1989	1991	1992	1994
Population, millions	20	21.2	24	26	27.8	28.7	30
Growth rate, millions per year	0.6	0.64	0.72	0.79	0.83	0.86	0.91

Let $P(t)$ represent the country's population t years after January 1, 1980.

a. Set the viewing screen on your graphing calculator to $[20, 30] \times [0, 1]$ with a scale of 1 on the x axis and a scale of 0.1 on the y axis. Then plot the data points of the form

$$(P(t), P'(t))$$

for $t = 0, 2, 6, 9, 11, 12,$ and 14.

b. Draw the regression line for the data points you plotted in part (a).

c. Find a formula for the least-squares linear function that corresponds to the regression line you drew in part (b).

d. The formula you found in part (c) has the form

$$y = rx + b$$

where $r \neq 0$ and b is nearly zero. Therefore,

$$P'(t) \approx rP(t)$$

Use this fact and Theorem 4.12 to find a formula that we can use to approximate the value of $P(t)$ for *any* nonnegative value of t that is not too large. (Why should the value of t be not too large?)

e. Use the formula you found in part (d) to estimate what the population of the South American country will be on January 1, 2005. How fast will the population be growing at that time?

46. **Depreciation** Eighteen years ago, a manufacturing firm purchased a new machine that depreciates with age. The following table shows the machine's value at seven different ages. The table also shows the depreciation rate at each of the seven ages.

Age, years	0	1	3	5	8	12	17
Value, $ millions	60	59	56.5	54.3	51	47	42.7
Depreciation rate, $ millions per year	1.25	1.18	1.13	1.09	1.02	0.94	0.85

Let $V(t)$ represent the machine's value when it is t years old.

a. Set the viewing screen of your graphing calculator to $[40, 60] \times [-1.3, -0.7]$ with a scale of 1 on the x axis and a scale of 0.1 on the y axis. Then plot the data points of the form

$$(V(t), V'(t))$$

for $t = 0, 1, 3, 5, 8, 12,$ and 17. [Since the machine depreciates with age, $V'(t)$ is *negative* for each value of t.]

b. Draw the regression line for the data points you plotted in part (a).

c. Find a formula for the least-squares linear function that corresponds to the regression line you drew in part (b).

d. The formula you found in part (c) has the form

$$y = rx + b$$

where $r \neq 0$ and b is nearly zero. Therefore,

$$V'(t) \approx rV(t)$$

Use this fact and Theorem 4.12 to find a formula that we can use to approximate the value of $V(t)$ for *any* nonnegative value of t that is not too large. (Why should the value of t be not too large?)

e. Use the formula you found in part (d) to estimate what the machine's value will be 2 years from now. (Two years from now the machine will be 20 years old.) How rapidly will the machine be depreciating at that time?

4.6 Two Models for Bounded Growth

In Section 3.4, we used rational functions to model quantities that approach a fixed level. In this section, we examine two types of nonrational functions that often can be used to model such quantities also. We begin with a situation in veterinary medicine.

Functions of the Type
$f(x) = b - ce^{-rx}$

A veterinarian predicts that a dog, who recently recovered from a serious disease, will weigh

$$S(t) = 12 - 7e^{-0.17t}$$

pounds t months from now. We want to examine how the dog's weight will behave as time passes.

Since
$$\lim_{t\to+\infty} 7e^{-0.17t} = 0$$
it follows that
$$\lim_{t\to+\infty} S(t) = 12$$

So the dog's weight will approach 12 pounds as time passes.

The derivative of $S(t)$ is
$$\begin{aligned} S'(t) &= 0 - 7(-0.17)e^{-0.17t} \\ &= 1.19e^{-0.17t} \end{aligned}$$

Since e raised to any power is positive, $S'(t)$ is positive for every value of t. Thus, the dog's weight will increase as time passes and, hence, will approach 12 pounds from below.

The second derivative of $S(x)$ is
$$S''(t) = -0.2023e^{-0.17t}$$

The value of $S''(t)$ is negative for every value of t. So the graph of $S(t)$ is concave downward. Therefore, according to Theorem 3.2 in Section 3.3, the dog's weight will approach 12 pounds from below while increasing at a slower and slower rate.

Substituting zero for t, we get
$$\begin{aligned} S(0) &= 12 - 7e^{-0.17(0)} \\ &= 12 - 7 \quad \text{since } e^{-0.17(0)} = 1 \\ &= 5 \end{aligned}$$

Thus, the dog's weight is presently 5 pounds.

Figure 4.13 shows the graph of $S(t)$ for $t \geq 0$. Since $\lim_{t\to+\infty} S(t) = 12$, the line $y = 12$ is a horizontal asymptote of the graph.

We can prove Theorem 4.13 with steps that parallel those we used to analyze $S(t) = 12 - 7e^{-0.17t}$.

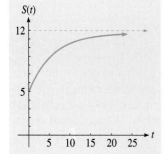

Figure 4.13
$S(t) = 12 - 7e^{-0.17t}$ for $t \geq 0$.

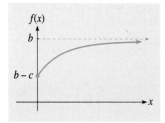

Figure 4.14
$f(x) = b - ce^{-rx}$ for $x \geq 0$.

Theorem 4.13

Suppose
$$f(x) = b - ce^{-rx}$$
where c and r are positive constants and $x \geq 0$. Then the graph of $f(x)$ is generally like the curve in Figure 4.14. The line
$$y = b$$
is a horizontal asymptote of the graph.

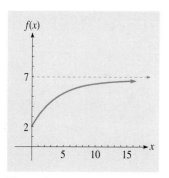

Figure 4.15
$f(x) = 7 - 5e^{-0.25x}$ for $x \geq 0$.

Functions of the type described in Theorem 4.13 are frequently used to model the relationship between a person's job performance and the person's training. Thus, the graphs of such functions are sometimes called **learning curves**. (See Problems 17 and 22 at the end of this section.)

We use Theorem 4.13 in Example 4.25.

Example 4.25

Sketch the graph of

$$f(x) = 7 - 5e^{-0.25x}$$

for $x \geq 0$. Then describe how $f(x)$ behaves as x increases.

Solution

The function $f(x)$ is of the type described in Theorem 4.13 with $b = 7$, $c = 5$, and $r = 0.25$. Since $b = 7$, the horizontal asymptote is the line

$$y = 7$$

Since $b - c = 2$, the graph rises from the point $(0, 2)$ toward the horizontal asymptote and is concave downward. Figure 4.15 shows the graph of $f(x)$. The graph shows that as x increases, $f(x)$ approaches 7 from below while increasing at a decreasing rate.

In Example 4.26, we use a function of the type $f(x) = b - ce^{-rx}$ to model a substance's temperature increase.

Example 4.26

Two hours ago a chemist placed a substance in a room where the temperature is always 35 degrees Celsius. At that time, the temperature of the substance was 5 degrees Celsius. Obviously, the temperature of the substance approaches the temperature of the room from below. What is not as obvious is that the temperature of the substance increases at a decreasing rate. Thus, it is reasonable to use a function of the type

$$T(t) = b - ce^{-rt}$$

where c and r are positive, to find the temperature of the substance t hours after the chemist placed it in the room. Use such a function to predict what the temperature of the substance will be 4 hours from now, if now its temperature is 11 degrees Celsius.

Solution

Since the temperature of the room is 35 degrees Celsius, $T(t)$ approaches 35. Thus, using Theorem 4.13, we can conclude that $b = 35$. So

$$T(t) = 35 - ce^{-rt}$$

Since the temperature of the substance was 5 degrees Celsius when the chemist placed it in the room, $T(0) = 5$. Consequently,

$$5 = 35 - ce^{-r(0)}$$

which implies that $c = 30$. Therefore,

$$T(t) = 35 - 30e^{-rt}$$

Since the temperature of the substance now is 11 degrees Celsius, and the chemist placed it in the room 2 hours ago, $T(2) = 11$. Thus,

$$11 = 35 - 30e^{-r(2)}$$

Subtracting 35 from both sides and dividing both sides by -30, we get

$$e^{-2r} = 0.8$$
$$-2r = \ln 0.8 \quad \text{definition of } \ln x$$
$$r = \frac{\ln 0.8}{-2}$$

Using a calculator, we get that

$$r \approx 0.11$$

So we can use the function

$$T(t) = 35 - 30e^{-0.11t}$$

to approximate the temperature of the substance t hours after the chemist placed it in the room.

Four hours from now is 6 hours after the chemist placed the substance in the room. Thus, since

$$T(6) = 35 - 30e^{-0.11(6)}$$
$$\approx 19.5$$

we conclude that 4 hours from now the temperature of the substance will be about 19.5 degrees Celsius.

Theorem 4.14, whose proof we are not yet able to produce, can help us solve two of the graphing calculator problems at the end of this section (Problems 25 and 26).

Theorem 4.14

Suppose

$$f'(x) = r[b - f(x)]$$

where $r > 0$ and $f(0) < b$. Then

$$f(x) = b - ce^{-rx}$$

where $c > 0$.

We now introduce functions of the type

$$f(x) = \frac{b}{1 + ce^{-rx}}$$

through a situation involving the growth of a population of rats.

Functions of the Type $f(x) = b/(1 + ce^{-rx})$

Recently, an island in the North Atlantic that had no rats became infested with rats as a result of a shipwreck. A retired biology professor, who lives on the island, predicts that the rat population of the island will be

$$P(x) = \frac{10}{1 + 4e^{-0.2x}}$$

thousand x months from now. We want to examine how the rat population will behave as time passes.

Since

$$\lim_{x \to +\infty} 4e^{-0.2x} = 0$$

it follows that

$$\lim_{x \to +\infty} P(x) = 10$$

Therefore, the rat population will approach 10,000 as time passes.

The derivative of $P(x)$ is

$$P'(x) = \frac{(1 + 4e^{-0.2x})0 - 10(-0.8e^{-0.2x})}{(1 + 4e^{-0.2x})^2}$$

$$= \frac{8e^{-0.2x}}{(1 + 4e^{-0.2x})^2}$$

Since e raised to any power is positive, $P'(x)$ is positive for every value of x. Thus, the rat population will increase as time passes and, therefore, will approach 10,000 from below.

The second derivative of $P(x)$ is

$$P''(x) = \frac{(1 + 4e^{-0.2x})^2(-1.6e^{-0.2x}) - (8e^{-0.2x})2(1 + 4e^{-0.2x})(-0.8e^{-0.2x})}{(1 + 4e^{-0.2x})^4}$$

$$= \frac{(4e^{-0.2x} - 1)1.6e^{-0.2x}}{(1 + 4e^{-0.2x})^3}$$

The denominator and the factor $1.6e^{-0.2x}$ are positive for every value of x. So $P''(x) = 0$ only when the factor

$$4e^{-0.2x} - 1 = 0$$

Adding 1 to both sides and then dividing both sides by 4 produces

$$e^{-0.2x} = \tfrac{1}{4}$$
$$-0.2x = \ln \tfrac{1}{4} \qquad \text{definition of } \ln x$$
$$-0.2x = -\ln 4 \qquad \text{property 4 of } \ln x$$
$$x = \frac{\ln 4}{0.2} \approx 7$$

So $P''(x) = 0$ when $x = (\ln 4)/0.2$.

The factor $4e^{-0.2x} - 1$ [in the formula for $P''(x)$] is positive when $x < (\ln 4)/0.2$ and negative when $x > (\ln 4)/0.2$. Therefore, $P''(x)$ is positive when $x < (\ln 4)/0.2$ and negative when $x > (\ln 4)/0.2$. Thus, by Theorem 3.1 in Section 3.2 and Theorem 3.2 in Section 3.3, the rat population will approach 10,000 from below while increasing at an increasing rate for about 7 months and at a decreasing rate thereafter. [Keep in mind that $(\ln 4)/0.2 \approx 7$.]

The information we gathered about the sign of $P''(x)$ implies that the graph of $P(x)$ has an inflection point where $x = (\ln 4)/0.2$. Note that

$$P\left(\frac{\ln 4}{0.2}\right) = \frac{10}{1 + 4e^{-0.2[(\ln 4)/0.2]}}$$
$$= \frac{10}{1 + 4e^{-\ln 4}}$$
$$= 5 \qquad \text{since } e^{-\ln 4} = \tfrac{1}{4}$$

So the ordinate of the inflection point is 5.

Note also that

$$P(0) = \frac{10}{1 + 4e^{-0.2(0)}}$$
$$= 2 \qquad \text{since } e^{-0.2(0)} = 1$$

Thus, the graph of $P(x)$ intersects the vertical axis at the point $(0, 2)$. (This result means that the rat population is presently 2000.)

Finally, since $\lim_{x \to +\infty} P(x) = 10$, the line $y = 10$ is a horizontal asymptote of the graph of $P(x)$.

If the domain is the interval $[0, +\infty)$, then the graph of $P(x)$ is the curve in Figure 4.16.

The function the professor selected to model the growth of the rat population has properties compatible with the growth pattern of populations in a limited environment. While a population is relatively small, the increasing number of propagators causes it to grow at an increasing rate. Eventually, however, when the population is relatively large, environmental limitations overpower the effect of the increasing number of propagators and cause the growth rate to decrease. Environmental limitations also impose an upper bound on the population. In the case of the rat population, the professor believes this upper bound is 10,000.

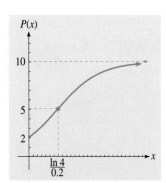

Figure 4.16
$P(x) = \dfrac{10}{1 + 4e^{-0.2x}}$ for $x \geq 0$.

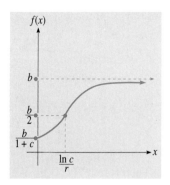

Figure 4.17 $f(x) = \dfrac{b}{1 + ce^{-rx}}$ for $x \geq 0$.

The formula for $P(x)$ is of the form

$$\frac{b}{1 + ce^{-rx}}$$

with $b = 10$, $c = 4$, and $r = 0.2$. As we saw, the ordinate of the inflection point of the graph of $P(x)$ is 5, which is half the value of b. According to Theorem 4.15, this result is not a coincidence. We can establish the validity of the theorem with steps similar to those we used to analyze $P(x) = 10/(1 + 4e^{-0.2x})$.

Theorem 4.15

Suppose

$$f(x) = \frac{b}{1 + ce^{-rx}}$$

where b and r are positive constants, c is a constant greater than 1, and $x \geq 0$. Then the graph of $f(x)$ is generally like the curve in Figure 4.17. The line

$$y = b$$

is a horizontal asymptote of the graph, and the point where

$$x = \frac{\ln c}{r}$$

is an inflection point of the graph.

If $f(x)$ is a function of the type described in Theorem 4.15, the rate of increase of $f(x)$ is greatest when

$$x = \frac{\ln c}{r}$$

Therefore, if x represents time, $f(x)$ increases most rapidly when $x = (\ln c)/r$.

The use of functions of the type described in Theorem 4.15 to model the growth of populations was introduced in the late 1830s by Pierre-François Verhulst, a Belgian.

Example 4.27 uses Theorem 4.15.

Example 4.27

Sketch the graph of

$$f(x) = \frac{52}{5 + 45e^{-0.27x}}$$

for $x \geq 0$. Then describe how $f(x)$ behaves as x increases.

Sec. 4.6 Two Models for Bounded Growth   271

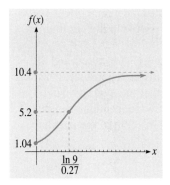

Figure 4.18
$f(x) = \dfrac{52}{5 + 45e^{-0.27x}}$ for $x \geq 0$.

Solution

By dividing the numerator and denominator by 5, we get

$$f(x) = \dfrac{10.4}{1 + 9e^{-0.27x}}$$

Thus, $f(x)$ is a function of the type described in Theorem 4.15. In this case, $b = 10.4$, $c = 9$, and $r = 0.27$. Since $b = 10.4$, the horizontal asymptote is the line

$$y = 10.4$$

Since

$$\dfrac{b}{1+c} = \dfrac{10.4}{1+9}$$
$$= 1.04$$

the graph rises from the point $(0, 1.04)$ toward the horizontal asymptote. It is concave upward on the interval $[0, (\ln 9)/0.27)$ and concave downward on the interval $((\ln 9)/0.27, +\infty)$. [*Note*: $(\ln 9)/0.27 \approx 8$.] Since

$$\dfrac{b}{2} = \dfrac{10.4}{2}$$
$$= 5.2$$

the inflection point is $((\ln 9)/0.27, 5.2)$. Figure 4.18 shows the graph of $f(x)$.

From the graph we see that $f(x)$ approaches 10.4 from below while increasing at an increasing rate as x increases toward $(\ln 9)/0.27$ and at a decreasing rate thereafter.

As we saw, functions of the type

$$f(x) = \dfrac{b}{1 + ce^{-rx}}$$

where b, c, and r are restricted as in Theorem 4.15, have properties compatible with the growth pattern of populations in a limited environment. The properties of these functions are also compatible with the way information spreads when it spreads by word of mouth.

For example, suppose *no advertising* accompanies the opening of a department store in a community. While the number of uninformed people is relatively large, the increasing number of informers causes information about the store to spread at an increasing rate. Eventually, however, the effect of the increasing number of informers is overpowered by the fact there are relatively few uninformed people left, and so the information spreads at a decreasing rate. In this situation, b represents the population of the community.

Communicable diseases also tend to spread in the manner described above. In this case, b represents the number of people in the community not immune to the disease.

Functions of the type described in Theorem 4.15 are called **logistic functions** and their graphs are called **logistic curves**. Therefore, if we say that a quantity *increases logistically*, we mean that the quantity can be expressed by a logistic function in terms of some independent variable, which is usually time. We use this terminology in some of the problems at the end of this section.

4.6 Problems

In Problems 1 to 8, sketch the graph of $f(x)$ for $x \geq 0$. Then describe how $f(x)$ behaves as x increases.

1. $f(x) = 9 - 6e^{-0.08x}$
2. $f(x) = 12 - 10e^{-0.32x}$
3. $f(x) = \dfrac{5}{1 + 4e^{-0.24x}}$
4. $f(x) = \dfrac{7}{1 + 3e^{-0.3x}}$
5. $f(x) = 6 - 6e^{-0.1x}$
6. $f(x) = 4 - 4e^{-0.25x}$
7. $f(x) = \dfrac{9}{2 + 4e^{-0.3x}}$
8. $f(x) = \dfrac{16}{3 + 9e^{-0.28x}}$

In Problems 9 and 10, find the function of the type $f(x) = b - ce^{-rx}$ that satisfies the given conditions as x increases on the interval $[0, +\infty)$. Express r as a decimal rounded to two decimal places.

9. $f(x)$ approaches 8
   $f(0) = 2$
   $f(3) = 5$

10. $f(x)$ approaches 12
    $f(0) = 0$
    $f(6) = 9$

In Problems 11 and 12, find the function of the type $f(x) = b/(1 + ce^{-rx})$ that satisfies the given conditions as x increases on the interval $[0, +\infty)$. Express r as a decimal rounded to two decimal places.

11. $f(x)$ approaches 8
    $f(0) = 1$
    $f(12) = \frac{64}{15}$

12. $f(x)$ approaches 9
    $f(0) = 3$
    $f(6) = 5$

13. **Demography** The census bureau of a country predicts that
$$S(t) = 15 - 8e^{-0.16t}$$
million people will be living alone t years from now. Sketch the graph of $S(t)$ for $t \geq 0$. Then describe how the number of people living alone will behave as time passes.

14. **Dairy Products Consumption** In a developing country, the per-capita consumption rate of dairy products is presently 150 pounds per year. Officials project that this consumption rate will be
$$D(t) = 260 - 110e^{-0.14t}$$
pounds per year t years from now. Sketch the graph of $D(t)$ for $t \geq 0$. Then describe how the consumption rate will behave as time passes.

15. **Disease Transmission** A communicable disease was recently introduced to the inhabitants of an island in the South Pacific. Health officials believe that x months from now
$$D(x) = \dfrac{54}{1 + 17e^{-0.5x}}$$
thousand inhabitants will have contracted the disease. Sketch the graph of $D(x)$ for $x \geq 0$. Then describe how the number of inhabitants that contract the disease will behave as time passes.

16. **Chemistry** A chemical reaction is causing substance A to become substance B. Suppose that x hours from now
$$A(x) = \dfrac{96}{1 + 23e^{-0.35x}}$$
milliliters of substance B will have formed. Sketch the graph of $A(x)$ for $x \geq 0$. Then describe how the amount of substance B will behave as time passes.

17. **Employee Productivity** The director of the training program in an electronics firm believes that a new employee can assemble
$$P(x) = 5 - 5e^{-0.2x}$$

dozen pulse monitors per day after receiving x hours of training. Thus, an employee's daily productivity increases at a decreasing rate as the training time is increased. The firm continues training an employee as long as the rate of increase of the employee's daily productivity is sufficiently large. To be more specific, the firm increases x (the number of hours of training) until $P'(x)$ is reduced to $\frac{1}{6}$. How much training does the firm give to an employee?

18. **Aerodynamics** An aerodynamicist plans to drop an object from a point high above the ground. If the air in its path is somehow eliminated, the velocity of the object (in feet per second) t seconds after it is dropped would be given by the formula $v(t) = 32t$. Thus, the velocity would increase at a constant rate, namely, 32 feet per second per second. In other words, the acceleration would be constantly 32 feet per second per second. In reality, however, due to the resistance of air, the velocity will approach a "terminal velocity" while increasing at a decreasing rate. Therefore, it is reasonable to use a formula of the type $v(t) = b - ce^{-rt}$ for the actual velocity t seconds after the object is dropped. In fact, the aerodynamicist, after examining the object, predicts that the actual velocity will be

$$v(t) = 400 - 400e^{-0.08t}$$

feet per second t seconds after it is dropped. This prediction implies that the acceleration will approach zero, starting at 32 feet per second per second. When will the acceleration be 4 feet per second per second?

Hint: The acceleration is the derivative of the velocity.

19. **Sales** Astro Walkers is the name given to trendy shoes recently put on the market by a famous manufacturer of designer clothes. The manufacturer has decided not to use a major advertising campaign to market the shoes. Therefore, it is reasonable to assume that the sale of the shoes will follow a logistic curve. In fact, the manufacturer predicts that x weeks from now

$$S(x) = \frac{70}{1 + 34e^{-0.25x}}$$

thousand pairs will have been sold. When will the shoes be selling fastest?

20. **Population** Demographers believe that the population of a Third World country will grow logistically. More specifically, they believe that x years from now the population will be

$$P(x) = \frac{198}{1 + 1.2e^{-0.03x}}$$

million. When will the population be growing most rapidly?

21. **Pulse Rate** Chester started pedaling a stationary bicycle 8 minutes ago. At that time his pulse rate was 70. Now it is 125. Suppose that t minutes after he started pedaling, his pulse rate is given by a function of the type

$$R(t) = b - ce^{-kt}$$

Also, suppose his pulse rate approaches 180 as he pedals. What will his pulse rate be 16 minutes from now?

Hint: Sixteen minutes from now is 24 minutes after he started pedaling.

22. **Learning** Without studying, Jane always makes a 45 on Professor Perez's tests. Whenever she studies 5 hours, she manages to make an 85. Suppose her grade on any of Professor Perez's tests is given by a function of the type

$$G(t) = b - ce^{-rt}$$

where t represents the number of hours she studies. Also, suppose her grade approaches 95 as she increases her studying time. What will her grade be on a test for which she studies 10 hours?

23. **Book Sales** *University* is the title of a novel written by a young, obscure English professor. Although several months have passed since the novel was published, only 5000 copies have been sold so far. But the editor in charge of the project is optimistic about future sales. She predicts that 2 months from now a total of 20,000 copies will have been sold. Suppose the volume of sales increases logistically while approaching 500,000 copies. How many copies will have been sold 6 months from now?

24. **Spread of a Rumor** A rumor about a county official's willingness to accept bribes is circulating. So far, 25,000 of the 300,000 citizens of the county have heard the rumor. Suppose the rumor spreads logistically through the county, and during the next 8 days, 10,000 *more* citizens will hear the rumor. How many of the county's citizens will have heard the rumor 15 days from now?

Graphing Calculator Problems

25. **Intravenous Feeding** A nurse has been feeding glucose intravenously to a patient since midnight. The following table shows the glucose level in the patient's bloodstream at eight different times. The table also shows the rate at which the glucose level was increasing at each of the eight times.

Time	Midnight	1:00 a.m.	2:00 a.m.	3:00 a.m.	4:00 a.m.	5:00 a.m.	6:00 a.m.	7:00 a.m.
Level, grams	2	2.7	3.3	3.8	4.2	4.5	4.8	5
Rate, grams/hour	0.8	0.7	0.5	0.4	0.4	0.3	0.2	0.2

Let $G(t)$ represent the glucose level after t hours of feeding.

a. Set the viewing screen on your graphing calculator to $[0, 6] \times [0, 1]$ with a scale of 1 on the x axis and a scale of 0.1 on the y axis. Then plot the data points of the form

$$(G(t), G'(t))$$

for $t = 0, 1, 2, 3, 4, 5, 6,$ and 7.

b. Draw the regression line for the data points you plotted in part (a).

c. Find a formula for the least-squares linear function that corresponds to the regression line you drew in part (b).

d. The formula you found in part (c) has the form

$$y = -rx + A$$

where $r > 0$. Rewrite the formula in the form

$$y = r(b - x)$$

where $b = A/r$. Therefore,

$$G'(t) \approx r[b - G(t)]$$

Use this fact and Theorem 4.14 to find a formula that we can use to approximate the value of $G(t)$ for *any* nonnegative value of t that is not too large. (Why should the value of t be not too large?)

e. Use the formula you found in part (d) to estimate what the glucose level will be at 10:00 a.m. if the nurse continues the intravenous feeding. How rapidly will the glucose level be increasing at that time?

26. **Consumer Debt** The following table shows the consumer debt in an African country at seven different times. The table also shows the rate at which the debt was increasing at each of the seven times.

Date, Jan. 1	1988	1989	1990	1991	1992	1993	1994
Debt, $ billions	5	6.7	8.1	9.2	9.9	10.4	10.8
Rate, $ billions per year	2.1	1.5	1.1	0.9	0.6	0.5	0.3

Let $D(t)$ represent the consumer debt t years after January 1, 1988.

a. Set the viewing screen of your graphing calculator to $[0, 12] \times [0, 3]$ with a scale of 1 on both axes. Then plot the data points of the form

$$(D(t), D'(t))$$

for $t = 0, 1, 2, 3, 4, 5,$ and 6.

b. Draw the regression line for the data points you plotted in part (a).

c. Find a formula for the least-squares linear function that corresponds to the regression line you drew in part (b).

d. The formula you found in part (c) has the form

$$y = -rx + A$$

where $r > 0$. Rewrite the formula in the form

$$y = r(b - x)$$

where $b = A/r$. Therefore,

$$D'(t) \approx r[b - D(t)]$$

Use this fact and Theorem 4.14 to find a formula that we can use to approximate the value of $D(t)$ for *any* nonnegative value of t that is not too large. (Why should the value of t be not too large?)

e. Use the formula you found in part (d) to estimate what the consumer debt will be on January 1, 2000. How fast will the debt be increasing at that time?

Chapter 4 Important Terms

Section 4.1
the number e (222)
interest compounded periodically (226)
interest compounded continuously (225 and 226)
present value (229)
effective interest rate (230)
nominal interest rate (230)

Section 4.2
exponential function (234)

Section 4.3
natural logarithmic function (238)
doubling time and half-life (240)
carbon dating (244)

Section 4.4
logarithm rule (250)
logarithmic function (251)
generalized logarithm rule (252)

Section 4.5
generalized exponential rule (256)

Section 4.6
learning curve (266)
logistic function (272)
logistic curve (272)

Chapter 4 Review Problems

Section 4.1

1. **Future Value** Suppose $8000 is invested at an annual interest rate of 12 percent.
   a. Find a formula for the balance at the end of t years if the interest is compounded semiannually. Then compute the balance at the end of 7.5 years.
   b. Find a formula for the balance at the end of t years if the interest is compounded continuously. Then compute the balance at the end of 7.5 years.

2. **Present Value** Determine how much must be invested now at an annual interest rate of 8 percent to obtain a balance of $90,000 in 15 years (*a*) if the interest is compounded quarterly, and (*b*) if the interest is compounded continuously.

3. **Effective Rate** Find the effective interest rate (rounded to the nearest one-hundredth of 1 percent) that corresponds to a nominal interest rate of 6 percent if the interest is compounded (*a*) monthly, and (*b*) continuously.

4. **Comparing Investment Plans** Investment A offers an annual interest rate of 16.8 percent compounded quarterly. Investment B offers an annual interest rate of 16.5 percent compounded continuously. Use the concept of effective interest rate to decide which investment is better.

Section 4.2

5. **Microbiology** A biologist expects a colony of microorganisms to grow exponentially at the rate of 4.2 percent per hour. Presently, the colony consists of 6000 microorganisms. Find a formula that gives the size of the colony x hours from now. Then determine how large the colony will be 5 hours from now.

6. **Drug Concentration** Four days ago, each liter of a man's blood contained 120 milligrams of a drug. Suppose the concentration decreases exponentially at the rate of 12 percent per day. What will the concentration be 2 days from now?

Section 4.3

7. **Demography** In an African country, the number of people at least 60 years old is increasing. Demographers project that x years from now the number of such people will be

$$N(x) = 32 - 14e^{-0.08x}$$

million. When will this number reach 25 million?

8. **Lake Pollution** The concentration of a pollutant in a lake (in milligrams per liter) decreases exponentially at the rate of 15 percent per year. Find the half-life of the concentration.

9. **Appreciation** If the value of a plot of land increases exponentially with a doubling time of 12 years, at what percentage rate does the value increase?

10. **Depreciation** A machine was worth $750,000 at the beginning of 1992 and $690,000 at the beginning of 1994. If the machine depreciates exponentially, what will its value be at the beginning of 2004?

11. **Rain Forest Shrinkage** Environmentalists predict that t years from now a South American country will contain

$$R(t) = 80\left(\frac{1}{4}\right)^{0.06t}$$

million hectares of rain forest. Thus, the size of the rain forest will decrease exponentially. Find the percentage rate of decrease.

12. **Carbon Dating** If a sample of the remains of an organism contains 0.18 the amount of carbon-14 it contained when it died, when did the organism die?

Section 4.4

13. Find the derivative.

   a. $f(x) = 5x^3 \ln x$ b. $f(x) = \dfrac{\ln x}{8x}$

   c. $f(x) = (\ln x)^4$ d. $f(x) = \ln (7x + 12)$

   e. $y = \ln x^7$ f. $y = \ln (x^5 + 3x)$

   g. $y = \ln (x^2 + 6x)^8$ h. $y = \ln \left(\dfrac{4x + 1}{x^3 + 5}\right)$

   i. $f(x) = \ln (x\sqrt{3x + 1})$

   j. $f(x) = \ln \left(\dfrac{2x + 5}{6x + 7}\right)^4$

14. **Capital Productivity** A company can produce

$$P(x) = 98 \ln (4x + 1)$$

tons of a product per month with a capital outlay of $\$x$ million. Use the derivative of $P(x)$ to approximate the amount the monthly production level increases if management increases the capital outlay from $12 million to $13 million.

15. **Depreciation** An accountant projects that t years from now a tractor will be worth

$$V(t) = 230 - 35 \ln (9t + 16)$$

thousand dollars, provided $0 \le t \le 10$. At what rate will the tractor be depreciating 6 years from now?

Section 4.5

16. Find the derivative.

   a. $y = e^{x^2 + 7x}$ b. $y = 26e^{-0.08x}$

   c. $f(x) = x^5 e^x$ d. $f(x) = \dfrac{2x}{3 + e^x}$

   e. $g(x) = (2 - 0.5e^{-x})^4$ f. $g(x) = \dfrac{4}{\sqrt{1 + e^{-0.2x}}}$

   g. $f(x) = 18 - 6e^{-0.5x}$ h. $f(x) = \dfrac{20}{1 + 6e^{-0.4x}}$

17. **Optimal Fertilization Level** Let $Y(x)$ represent the number of bushels of peaches each hectare produces if x sacks of a special fertilizer is applied to each hectare. A farmer believes that

$$Y(x) = e^{-x^2 + 36x - 318}$$

How much fertilizer should the farmer apply to each hectare to maximize the yield?

18. Sketch the graph of $f(x) = 8xe^{-0.5x}$ for $x \ge 0$.

19. Verify that the function

$$f(x) = 21 - 12e^{-0.02x}$$

where $x \ge 0$, is not an exponential function by showing that the percentage rate of change of $f(x)$ does not remain constant as x increases.

Section 4.6

20. Sketch the graph of $f(x)$ for $x \ge 0$. Then describe how $f(x)$ behaves as x increases.

   a. $f(x) = 10 - 7e^{-0.18x}$ b. $f(x) = 5 - 5e^{-0.2x}$

   c. $f(x) = \dfrac{8}{1 + 3e^{-0.26x}}$ d. $f(x) = \dfrac{13}{2 + 14e^{-0.3x}}$

21. **Intravenous Feeding** A doctor decided to feed glucose intravenously to a patient for a period of 10 hours. A nurse began the feeding 3 hours ago. At that time, the patient's bloodstream contained 3 grams of glucose. One hour ago, it contained 5 grams of glucose. The doctor believes that after t hours of feeding, the amount of glucose in the bloodstream is given by a function of the type

$$G(t) = b - ce^{-rt}$$

The doctor also believes that the glucose level in the bloodstream approaches 8 grams while the feeding is in progress. How much glucose will the patient's bloodstream contain 5 hours from now?

22. **Population** The population of a Third World country is presently 50 million. Demographers predict that the population will increase to 60 million during the next 5 years. They also predict that it will increase logistically while approaching 210 million. When will the population be increasing most rapidly? (Round your answer to the nearest year.)

CHAPTER 5

Integration

5.1 Antidifferentiation
5.2 The Definite Integral
5.3 Area and Definite Integrals
5.4 Integration by Substitution
5.5 Integration by Parts
5.6 Riemann Sums

Important Terms

Review Problems

Geometrically, there are two basic problems in calculus: the problem of finding the slope of a tangent, which we introduced in Chapter 2, and the problem of finding the area of a region, which we will introduce in this chapter. One of the most profound results in calculus is that these two apparently unrelated problems are linked by the derivative concept.

5.1 Antidifferentiation

We have seen that there are many problems that can be solved by finding the derivative of a function. In this section, we will see that some problems can be solved by the inverse process: finding a function whose derivative is a given function.

For example, suppose the management of a manufacturing firm has a formula for the marginal cost function and wants to find a formula for the corresponding total cost function. Since the marginal cost function is the derivative of the total cost function, the management, in fact, wants to find a function whose derivative is the marginal cost function. Definition 5.1 assigns a label to such a function.

Definition 5.1

A function $F(x)$ is an **antiderivative** of a function $f(x)$ if
$$F'(x) = f(x)$$
for each x in the domain of $f(x)$.

Thus, since $4x^3$ is a function whose derivative is $12x^2$, we can say that $4x^3$ is an antiderivative of $12x^2$. Note that $4x^3 + 8$ is also a function whose derivative is $12x^2$. Therefore, $4x^3 + 8$ is *another* antiderivative of $12x^2$.

Indefinite Integrals

Suppose $F(x)$ is an antiderivative of $f(x)$. Then, for *any* constant C,
$$F(x) + C$$
is also an antiderivative of $f(x)$. This statement is true because the derivative of a constant function is zero. It is also true that any antiderivative of $f(x)$ can be expressed in the form $F(x) + C$. Thus, we have a basis for Definition 5.2.

Definition 5.2

Suppose $F(x)$ is an antiderivative of $f(x)$, and C represents an arbitrary constant. Then the **indefinite integral** of $f(x)$ is

$$F(x) + C$$

Sometimes we refer to $F(x) + C$ as the **general antiderivative** of $f(x)$. Either way, we call $f(x)$ the **integrand** and C the **constant of integration** or the **constant of antidifferentiation**.

We use the symbol

$$\int f(x)\, dx$$

to represent the indefinite integral (general antiderivative) of $f(x)$. So, if $F(x)$ is an antiderivative of $f(x)$ and C is an arbitrary constant, we can write

$$\int f(x)\, dx = F(x) + C$$

We emphasize that finding an indefinite integral is the inverse of finding a derivative.

Example 5.1

Find the indefinite integral of

$$2x + 5$$

That is, find

$$\int (2x + 5)\, dx$$

Solution

The derivative of $x^2 + 5x$ is $2x + 5$. Therefore, an antiderivative of $2x + 5$ is $x^2 + 5x$, and so

$$\int (2x + 5)\, dx = x^2 + 5x + C$$

This result tells us that the antiderivatives of

$$2x + 5$$

are the functions of the form

$$x^2 + 5x + C$$

Particular Antiderivatives

If we assign a particular value to the constant of antidifferentiation, we obtain a **particular antiderivative**. Thus, $x^2 + 5x$ and $x^2 + 5x - 6$ are particular antiderivatives of $2x + 5$. In the first case, $C = 0$. In the second case, $C = -6$.

Example 5.2

Suppose $F(x)$ represents the particular antiderivative of $2x + 5$ that satisfies the condition $F(3) = 28$. Find a formula for $F(x)$.

Solution

From Example 5.1 we know that

$$\int (2x + 5)\, dx = x^2 + 5x + C$$

Therefore,
$$F(x) = x^2 + 5x + C$$

for a certain value of C. Since $F(3) = 28$, the value of C must satisfy the equation

$$3^2 + 5(3) + C = 28$$

Solving this equation for C, we get

$$C = 4$$

Consequently,
$$F(x) = x^2 + 5x + 4$$

is the particular antiderivative of $2x + 5$ that satisfies the condition $F(3) = 28$.

We now present rules which will help us find antiderivatives more systematically. We begin with a rule for constant functions.

The Integral of a Constant Function

If k is a constant, the derivative of kx is k. Thus, an antiderivative of k is kx, and so we have the rule shown in Theorem 5.1.

Theorem 5.1 (Constant Rule for Integrals)

If k is a real number,

$$\int k\, dx = kx + C$$

According to the above rule,

$$\int 13\, dx = 13x + C$$

We now turn to functions that can be expressed in the form kx^n.

The Integral of a Function of the Form kx^n

The rule shown in Theorem 5.2 follows from the fact the derivative of

$$\frac{k}{n+1} x^{n+1}$$

is kx^n. (See Theorem 2.4 in Section 2.3.)

Theorem 5.2 (Power Rule for Integrals)

If k is a real number and n is a rational number *different* from -1,

$$\int kx^n \, dx = \frac{k}{n+1} x^{n+1} + C$$

We require n to be different from -1 to prevent k from being divided by zero.

In Example 5.3, we use the power rule for integrals to find four indefinite integrals.

Example 5.3

Find the following indefinite integrals:

a. $\int 3x^4 \, dx$ b. $\int 8x \, dx$

c. $\int \frac{1}{4x^3} \, dx$ d. $\int \frac{1}{\sqrt[3]{x}} \, dx$

Solution for (a)

$$\int 3x^4 \, dx = \frac{3}{4+1} x^{4+1} + C$$

$$= \frac{3}{5} x^5 + C$$

Solution for (b)

$$\int 8x \, dx = \frac{8}{2} x^2 + C$$

$$= 4x^2 + C$$

Solution for (c)

Note that

$$\frac{1}{4x^3} = \frac{1}{4} x^{-3}$$

So we can use the power rule for integrals:

$$\int \frac{1}{4x^3} dx = \int \frac{1}{4} x^{-3} dx$$

$$= \frac{1/4}{-2} x^{-2} + C$$

$$= \frac{-1}{8x^2} + C$$

Solution for (d)

Note that

$$\frac{1}{\sqrt[3]{x}} = \frac{1}{x^{1/3}}$$

$$= x^{-1/3}$$

So we can use the power rule for integrals:

$$\int \frac{1}{\sqrt[3]{x}} dx = \int x^{-1/3} dx$$

$$= \frac{1}{2/3} x^{2/3} + C$$

$$= \frac{3}{2} (\sqrt[3]{x})^2 + C$$

Next we consider functions formed by adding or subtracting other functions.

The Integral of a Sum or Difference

Recall that the derivative of the sum (or difference) of two functions is the sum (or difference) of their respective derivatives. (See Theorem 2.5 in Section 2.3.) This result implies the rule shown in Theorem 5.3.

Theorem 5.3 (Sum and Difference Rule for Integrals)

Suppose $G(x)$ and $H(x)$ are antiderivatives of $g(x)$ and $h(x)$, respectively. Then

$$\int (g(x) + h(x)) \, dx = G(x) + H(x) + C$$

and

$$\int (g(x) - h(x)) \, dx = G(x) - H(x) + C$$

In Example 5.1, we concluded that

$$\int (2x + 5) \, dx = x^2 + 5x + C$$

Note that x^2 is an antiderivative of $2x$ and $5x$ is an antiderivative of 5. Therefore, our conclusion in that example illustrates Theorem 5.3.

The sum and difference rule for integrals extends to sums and differences of *more than two* functions. In Example 5.4, we apply the rule to the difference of two functions.

Example 5.4

The fish population of a lake is presently 250,000. Due to pollutants, environmentalists expect this population to decrease. In fact, they predict that x months from now the population will be changing at the rate of

$$0.2x - 10$$

thousand per month, where $0 \leq x \leq 36$. Find a formula that gives the fish population x months from now. (The formula will be valid for $0 \leq x \leq 36$.) Then determine what the population will be 8 months from now.

Solution

Let $P(x)$ represent the fish population (in thousands) x months from now. Since $0.2x - 10$ is the rate of change of the population x months from now,

$$P'(x) = 0.2x - 10$$

Thus, since the population is presently 250,000, $P(x)$ is the particular antiderivative of $0.2x - 10$ that satisfies the condition $P(0) = 250$.

According to the power rule for integrals, an antiderivative of $0.2x$ is $0.1x^2$. According to the constant rule for integrals, an antiderivative of 10 is $10x$. Thus, according to the sum and difference rule for integrals,

$$\int (0.2x - 10)\, dx = 0.1x^2 - 10x + C$$

So

$$P(x) = 0.1x^2 - 10x + C$$

for a certain value of C. Since $P(0) = 250$, the value of C must satisfy the equation

$$0.1(0^2) - 10(0) + C = 250$$

Thus,

$$C = 250$$

Consequently,

$$P(x) = 0.1x^2 - 10x + 250$$

is a formula that gives the fish population x months from now. So, since

$$P(8) = 0.1(8^2) - 10(8) + 250$$
$$= 176.4$$

we conclude that 8 months from now there will be 176,400 fish in the lake.

Recall that the function
$$f(x) = e^x$$
is sometimes referred to as **the exponential function**.

The Integral of The Exponential Function

Theorem 5.4 is valid because the derivative of e^x is e^x.

Theorem 5.4 (Exponential Rule for Integrals)

$$\int e^x \, dx = e^x + C$$

The Integral of a Constant Times a Function

In Section 2.5, we saw that the derivative of a constant times a function is the constant times the derivative of the function (Theorem 2.9). From this result we get Theorem 5.5.

Theorem 5.5 (Constant Multiple Rule for Integrals)

If k is a real number and $F(x)$ is an antiderivative of $f(x)$, then
$$\int k \cdot f(x) \, dx = k \cdot F(x) + C$$

We use Theorems 5.4 and 5.5 in Example 5.5.

Example 5.5

Find the general antiderivative (indefinite integral) of
$$f(x) = 7e^x$$
Then find a formula for $F(x)$, where $F(x)$ represents the particular antiderivative of $f(x)$ that satisfies the condition $F(0) = 4$.

Solution

The exponential rule for integrals tells us that an antiderivative of e^x is e^x. Thus, according to the constant multiple rule for integrals, an antiderivative of $7e^x$ is $7e^x$. Consequently, the general antiderivative of $7e^x$ is

$$\int 7e^x \, dx = 7e^x + C$$

So
$$F(x) = 7e^x + C$$

for a certain value of C. Since $F(0) = 4$, the value of C must satisfy the equation

$$7e^0 + C = 4$$

Since $e^0 = 1$, solving this equation for C, we get

$$C = -3$$

Thus,
$$F(x) = 7e^x - 3$$

is the particular antiderivative of $f(x) = 7e^x$ that satisfies the condition $F(0) = 4$.

Note that

$$\frac{1}{x} = 1x^{-1}$$

So $1/x$ is a function that can be expressed in the form kx^n. Nevertheless, we cannot use the power rule for integrals (Theorem 5.2) to find the indefinite integral of $1/x$, because that rule requires that $n \neq -1$.

The Integral of 1/x

If $x > 0$, $\ln|x| = \ln x$. So, if $x > 0$,
$$D_x(\ln|x|) = D_x(\ln x)$$
$$= \frac{1}{x} \qquad \text{logarithm rule}$$

If $x < 0$, $\ln|x| = \ln(-x)$. So, if $x < 0$,
$$D_x(\ln|x|) = D_x(\ln(-x))$$
$$= \frac{-1}{-x} \qquad \text{generalized logarithm rule}$$
$$= \frac{1}{x}$$

So, for $x \neq 0$,
$$D_x(\ln|x|) = \frac{1}{x}$$

We have just proven Theorem 5.6.

Theorem 5.6 (Logarithm Rule for Integrals)

For $x \neq 0$,

$$\int \frac{1}{x} dx = \ln|x| + C$$

In Example 5.6, we use all the integral rules presented in this section, except the exponential rule.

Example 5.6

Find

$$\int \frac{5x^3 + 7x^2 + 6}{x^3} dx$$

Solution

We begin by expressing the integrand in a form that facilitates the application of our integral rules:

$$\frac{5x^3 + 7x^2 + 6}{x^3} = \frac{5x^3}{x^3} + \frac{7x^2}{x^3} + \frac{6}{x^3}$$

$$= 5 + 7\left(\frac{1}{x}\right) + 6x^{-3}$$

So

$$\int \frac{5x^3 + 7x^2 + 6}{x^3} dx = \int \left(5 + 7\left(\frac{1}{x}\right) + 6x^{-3}\right) dx$$

$$= 5x + 7 \ln|x| + \left(\frac{6}{-2}\right)x^{-2} + C$$

$$= 5x + 7 \ln|x| - \frac{3}{x^2} + C$$

Therefore, the antiderivatives of

$$\frac{5x^3 + 7x^2 + 6}{x^3}$$

are the functions of the form

$$5x + 7 \ln|x| - \frac{3}{x^2} + C$$

We can verify the correctness of an indefinite integral by showing that its derivative is the integrand. For instance, in Example 5.6 we concluded that

$$\int \frac{5x^3 + 7x^2 + 6}{x^3}\, dx = 5x + 7\ln|x| - \frac{3}{x^2} + C$$

Note that

$$D_x\left(5x + 7\ln|x| - \frac{3}{x^2} + C\right) = D_x(5x + 7\ln|x| - 3x^{-2} + C)$$

$$= 5 + 7\left(\frac{1}{x}\right) + 6x^{-3} + 0$$

$$= 5 + \frac{7}{x} + \frac{6}{x^3}$$

$$= \frac{5x^3 + 7x^2 + 6}{x^3}$$

So the result we obtained in Example 5.6 is correct.

In Example 5.6, we began the process of finding the indefinite integral by rewriting the integrand in a form compatible with our integral rules. Another example in which rewriting the integrand helps is

$$\int 12x^2(x^3 - 5x + 1)\, dx$$

Here we use the distributive property to rewrite the integrand:

$$12x^2(x^3 - 5x + 1) = 12x^5 - 60x^3 + 12x^2$$

Then we can use the integral versions of the power rule and the sum and difference rule to find the indefinite integral:

$$\int 12x^2(x^3 - 5x + 1)\, dx = \int (12x^5 - 60x^3 + 12x^2)\, dx$$
$$= 2x^6 - 15x^4 + 4x^3 + C$$

5.1 Problems

In Problems 1 to 22, find the indefinite integral. Then check your answer by showing that the derivative of the indefinite integral is the integrand.

1. $\int 3x^2\, dx$
2. $\int 5x\, dx$
3. $\int x^4\, dx$
4. $\int 7\, dx$
5. $\int \frac{6}{x^4}\, dx$
6. $\int -\frac{1}{x^2}\, dx$
7. $\int 6\sqrt{x}\, dx$
8. $\int 3\sqrt{x}\, dx$
9. $\int (2x^2 + 4)\, dx$
10. $\int (4x - 9)\, dx$
11. $\int x(8x^2 - 3x)\, dx$
12. $\int x(x + 2)\, dx$
13. $\int \left(x - \frac{1}{x^2} - 1\right) dx$
14. $\int \left(4x^3 + \frac{10}{x^3}\right) dx$
15. $\int (6x + \sqrt[3]{x})\, dx$
16. $\int (3x^3 - \sqrt[3]{x} + 8)\, dx$
17. $\int \frac{x^3 + 10}{2x}\, dx$
18. $\int \frac{x - 1}{x}\, dx$
19. $\int \left(e^x + \frac{3}{x}\right) dx$
20. $\int 6e^x\, dx$
21. $\int \left(6x + \frac{1}{\sqrt{x}} + 3\right) dx$
22. $\int \left(\frac{2}{\sqrt[3]{x}} + 7\right) dx$

In Problems 23 to 40, for the function $f(x)$, find the particular antiderivative $F(x)$ that satisfies the given condition.

23. $f(x) = 8x^3$ $F(2) = 35$
24. $f(x) = x^2$ $F(0) = 7$
25. $f(x) = 3\sqrt{x}$ $F(0) = 4$
26. $f(x) = 12\sqrt[3]{x}$ $F(8) = 145$
27. $f(x) = -\dfrac{4}{x^5}$ $F(2) = \dfrac{1}{16}$
28. $f(x) = \dfrac{1}{x^2}$ $F(3) = 0$
29. $f(x) = \dfrac{2}{x}$ $F(1) = -3$
30. $f(x) = \dfrac{1}{x}$ $F(e) = 5$
31. $f(x) = e^x$ $F(0) = -6$
32. $f(x) = 9e^x$ $F(0) = 15$
33. $f(x) = (x+3)^2$ $F(0) = 5$
34. $f(x) = (6x-1)^2$ $F(1) = 10$
35. $f(x) = 4x - \dfrac{3}{x^2}$ $F(3) = 21$
36. $f(x) = 7 + \dfrac{8}{x^3}$ $F(2) = 13$
37. $f(x) = 3x^2 + 9\sqrt{x}$ $F(4) = 100$
38. $f(x) = 6\sqrt{x} - \dfrac{9}{x^2}$ $F(9) = 112$
39. $f(x) = 2 - \dfrac{1}{x}$ $F(1) = 7$
40. $f(x) = e^x + 1$ $F(0) = 0$

41. **Chipmunk Population** In Crown Island, the chipmunk population is presently 175,000. A conservationist predicts that t months from now the population will be increasing at the rate of

$$0.05t + 4$$

thousand per month. Let $P(t)$ represent the population (in thousands) t months from now. Find a formula for $P(t)$. Then determine what the population will be 20 months from now.

Hint: $P'(t) = 0.05t + 4$ and $P(0) = 175$. Thus, $P(t)$ is the particular antiderivative of $0.05t + 4$ that satisfies the condition $P(0) = 175$.

42. **Depreciation** Presently, a machine is worth $48,600 ($48.6 thousand). An accountant believes that x years from now its worth will be changing at the rate of

$$-\sqrt{x}$$

thousand dollars per year, where $0 \leq x \leq 10$. For such values of x, let $V(x)$ represent the machine's worth x years from now. Find a formula for $V(x)$. Then determine what the machine will be worth 9 years from now.

Hint: $V'(x) = -\sqrt{x}$ and $V(0) = 48.6$. Thus, $V(x)$ is the particular antiderivative of $-\sqrt{x}$ that satisfies the condition $V(0) = 48.6$.

43. **Production Cost** The total monthly cost of producing 9 tons of a product per month is $2010. The management of the firm that produces the product believes that the marginal cost is

$$x^2 - 36x + 325$$

when the monthly production level is x tons. Let $C(x)$ represent the total monthly cost of producing x tons per month. Find a formula for $C(x)$. Then determine the total monthly cost of producing 15 tons per month.

Hint: $C'(x) = x^2 - 36x + 325$ and $C(9) = 2010$. Thus, $C(x)$ is the particular antiderivative of $x^2 - 36x + 325$ for which $C(9) = 2010$.

44. **Productivity of Capital** Let $Q(x)$ represent the number of gallons of Laxatol a firm can produce daily with a capital outlay of x thousand dollars. Suppose the marginal productivity of capital is

$$\dfrac{20}{x^2} + 30$$

for $x \geq 5$. The firm can produce 798 gallons of the product daily with a capital outlay of $10,000. Find a formula for $Q(x)$. Then determine how many gallons the firm can produce daily with a capital outlay of $20,000.

Hint: $Q'(x) = 20/x^2 + 30$ and $Q(10) = 798$. Thus, $Q(x)$ is the particular antiderivative of $20/x^2 + 30$ that satisfies the condition $Q(10) = 798$.

45. **Agriculture** Let $F(x)$ represent the number of tons of cabbage a farmer can produce if she uses x barrels of Growmore Fertilizer. Suppose

$$F'(x) = \frac{1}{\sqrt{x}}$$

for $4 \leq x \leq 25$. Also, suppose the farmer can produce 36 tons of cabbage if she uses 9 barrels of the fertilizer. Find a formula for $F(x)$. How much of the fertilizer must the farmer use to produce 38 tons of cabbage?

46. **Land Use** Presently, a South American country has 30 hectares of cropland per capita. Let $H(t)$ represent the amount of cropland per capita t years from now. Suppose that t years from now $H(t)$ will be changing at the rate of

$$0.005t - 0.5$$

hectares per year, where $0 \leq t \leq 100$. Find a formula for $H(t)$. When during the next hundred years will the country have only 14 hectares of cropland per capita?

47. **Affordability of Homes** Let $A(t)$ represent the number of families (in millions) who will be able to afford a $75,000 home t years from now. Suppose that t years from now $A(t)$ will be changing at the rate of

$$0.4t - 4$$

million per year, where $0 \leq t \leq 10$. Presently, 80 million families can afford a $75,000 home. How many families will be able to afford such a home 7 years from now?

48. **Job Training** Let $F(x)$ represent the amount of piping (in feet) an average worker can install per day after receiving x hours of training. Suppose

$$F'(x) = -x^2 + 30x$$

for $5 \leq x \leq 20$. After receiving 9 hours of training, an average worker can install 1025 feet of piping per day. How much piping can an average worker install per day after receiving 15 hours of training?

49. **Memory** Let $F(x)$ represent the number of days Lynette will remember a particular collection of information after studying it x minutes. Suppose

$$F'(x) = \frac{2}{(\sqrt[3]{x})^2}$$

for $5 \leq x \leq 30$. Also, suppose Lynette remembers the information 16 days after studying it 8 minutes. Find a formula for $F(x)$. Then determine how long Lynette will remember the information after studying it 27 minutes.

50. **Propensity to Consume** Let $C(x)$ represent the total monthly amount, in millions of dollars, the residents of a community spend when their aggregate monthly income is x million. Suppose

$$C'(x) = 0.6 + \frac{0.2}{\sqrt{x}}$$

for $x \geq 1$. Also, suppose the residents spend $17 million when their aggregate income is $25 million. Find a formula for $C(x)$. Then determine how much the residents spend when their aggregate income is $36 million. [$C'(x)$ is called the **marginal propensity to consume**.]

51. **Propensity to Save** Let $S(x)$ represent the total monthly amount, in millions of dollars, the residents of a community save when their aggregate monthly income is x million. Suppose

$$S'(x) = \frac{1}{3} - \frac{1}{6\sqrt{x}}$$

for $x \geq 1$. Also, suppose the residents save $2 million when their aggregate income is $9 million. Find a formula for $S(x)$. Then determine how much the residents save when their aggregate income is $16 million. [$S'(x)$ is called the **marginal propensity to save**.]

52. **Vertical Motion** An object is thrown straight downward from the top of a tall building with an initial velocity of 60 feet per second. Suppose the object's velocity increases at the constant rate of 32 feet per second per second. Then, if $v(t)$ represents the object's velocity t seconds after it is thrown,

$$v'(t) = 32$$

for each permissible value of t.

  a. Find a formula for $v(t)$.
     Hint: $v(0) = 60$.

  b. Let $d(t)$ represent how far (in feet) the object is from where it is thrown t seconds after it is thrown. Find a formula for $d(t)$.
     Hint: $d'(t) = v(t)$ and $d(0) = 0$.

c. Suppose the object is thrown from a point 1836 feet above the ground. Determine the object's velocity at the instant it strikes the ground.
   Hint: First, use the formula obtained in part (b) to determine when the object strikes the ground.

53. Suppose $A(x)$ and $F(x)$ are antiderivatives of a function that is continuous on an interval. Explain why

$$F(b) - F(a)$$

represents the amount $A(x)$ changes if x increases from a to b.

54. Suppose $A(x)$ and $F(x)$ are as described in Problem 53. How are the graphs of $A(x)$ and $F(x)$ related?

Graphing Calculator Problems

55. **Depreciation** Let $V(x)$ represent a machine's value (in thousands of dollars) x years after the beginning of 1988. The following table shows the values of $V(x)$ for several values of x:

x	0	1	2	3	4	5	6
$V(x)$	70	62.1	54.8	48	41.8	36	31

a. Use an approach like in Example 2.13 in Section 2.4 to find a function that approximates the derivative of $V(x)$.

b. Use the function you found in part (a) to approximate the rate at which the machine will be depreciating at the beginning of 1998.

c. The table shows that the machine's value was $31 thousand at the beginning of 1994. Use this information and the function you found in part (a) to derive a function that approximates the function $V(x)$.

d. Use the function you derived in part (c) to approximate the machine's value at the beginning of 1998.

56. **Production Cost** The following table shows a firm's total monthly production cost at several production levels:

Production level, tons	3	4	5	8	9	14	15	16	17
Total cost, $ thousands	15.3	15.43	15.68	16.6	16.98	19.48	20.08	20.74	21.47

a. Use an approach like in Example 2.13 in Section 2.4 to find a function that approximates the firm's marginal cost function.

b. Use the function you found in part (a) to approximate the firm's marginal cost when the monthly production level is 12 tons. Then interpret your answer.

c. The table shows that the firm's total production cost is $16.6 thousand when the production level is 8 tons. Use this information and the function you found in part (a) to derive a function that approximates the firm's total production cost function.

d. Use the function you derived in part (c) to approximate the firm's total production cost when the production level is 12 tons.

5.2 The Definite Integral

Suppose $A(x)$ is a function that is continuous on an interval and a and b are numbers in the interval such that $a < b$. If we have a formula for $A(x)$, we can find the *change* $A(x)$ undergoes if x increases from a to b by simply computing

$$A(b) - A(a)$$

In this section, we will see that sometimes we can find such a change in $A(x)$ without using a formula for $A(x)$. We introduce the method through a problem about a country's timberland.

A country currently has 98 million acres of timberland. Forestry officials predict that t years from now the amount of timberland in the country will be changing at the rate of

$$0.1t - 4$$

million acres per year, where $0 \leq t \leq 20$. Note that $0.1t - 4$ is negative for each t in the interval $[0, 20]$. Thus, the amount of timberland will decrease during the next 20 years. We will find how much it will decrease during the 6-year interval of time from $t = 2$ to $t = 8$.

If the amount of timberland decreases at a constant rate, we would compute the decrease in question by merely multiplying the constant rate by 6. But, according to the formula $0.1t - 4$, the amount of timberland will not decrease at a constant rate. So we must use a different approach.

Let $Q(t)$ represent the amount of timberland t years from now. Since $0.1t - 4$ is the rate at which $Q(t)$ will be changing t years from now,

$$Q'(t) = 0.1t - 4$$

Since the country currently has 98 million acres of timberland,

$$Q(0) = 98$$

Therefore, $Q(t)$ is the particular antiderivative of $0.1t - 4$ that satisfies the condition $Q(0) = 98$. The general antiderivative of $0.1t - 4$ is

$$\int (0.1t - 4)\, dt = 0.05t^2 - 4t + C$$

Thus, $$Q(t) = 0.05t^2 - 4t + C$$

for a certain value of C. Since $Q(0) = 98$, the value of C must satisfy the equation

$$0.05(0^2) - 4(0) + C = 98$$

So $$C = 98$$

Consequently, $$Q(t) = 0.05t^2 - 4t + 98$$

for each value of t in the interval $[0, 20]$.

Therefore, 2 years from now the amount of timberland will be

$$Q(2) = 0.05(2^2) - 4(2) + 98$$
$$= 90.2$$

million acres, and 8 years from now it will be

$$Q(8) = 0.05(8^2) - 4(8) + 98$$
$$= 69.2$$

million acres. Since

$$Q(8) - Q(2) = 69.2 - 90.2$$
$$= -21$$

we conclude that the amount of timberland will decrease 21 million acres during the period of time from $t = 2$ to $t = 8$.

Note that
$$H(t) = 0.05t^2 - 4t$$
is another particular antiderivative of $0.1t - 4$. (In this case, $C = 0$.) It turns out that
$$H(8) - H(2) = -21$$
also. Actually, if $H(t)$ is *any* antiderivative of $0.1t - 4$,
$$H(8) - H(2) = -21$$
This fact illustrates Theorem 5.7.

Theorem 5.7

Let $f(x)$ be a function that is continuous on an interval, and let a and b be numbers in the interval such that $a < b$. Suppose $A(x)$ is a function whose derivative is $f(x)$. If $F(x)$ is *any* antiderivative of $f(x)$, the amount $A(x)$ changes if x increases from a to b is
$$F(b) - F(a)$$

Theorem 5.7 provides us with a method for computing the change in $A(x)$ without using a formula for $A(x)$. To use the method, we must have a formula for $f(x)$ [the derivative of $A(x)$], and we must be able to find an antiderivative of $f(x)$. The proof of the theorem uses the fact that any two antiderivatives of a function differ by a constant.

Proof of Theorem 5.7

Since the derivative of $A(x)$ is $f(x)$, it follows that $A(x)$ is an antiderivative of $f(x)$. Suppose $F(x)$ is another antiderivative of $f(x)$. Since any two antiderivatives of a function differ by a constant, there exists a constant C such that
$$A(x) = F(x) + C$$
So
$$A(a) = F(a) + C$$
and
$$A(b) = F(b) + C$$
Thus,
$$A(b) - A(a) = (F(b) + C) - (F(a) + C)$$
$$= F(b) - F(a)$$

We already know that $A(b) - A(a)$ represents the amount $A(x)$ changes if x increases from a to b. Therefore, by showing that

$$A(b) - A(a) = F(b) - F(a)$$

we established the validity of Theorem 5.7.

Theorem 5.7 is a basis for Definition 5.3.

Definition 5.3

Let $f(x)$ be a function that is continuous on an interval, and let a and b be numbers in the interval such that $a < b$. Suppose $F(x)$ is an antiderivative of $f(x)$. The **definite integral** of $f(x)$ from $x = a$ to $x = b$ is the *number* obtained by subtracting $F(a)$ from $F(b)$ and is represented by

$$\int_a^b f(x)\, dx$$

In Definition 5.3, if we let $F(x)\Big|_a^b$ represent $F(b) - F(a)$, we can write

$$\int_a^b f(x)\, dx = F(x)\Big|_a^b$$

The values of a and b are the lower and upper **limits of integration**.

Theorem 5.7 implies that the value of $\int_a^b f(x)\, dx$ is the same, regardless of which antiderivative of $f(x)$ we use to compute it.

We must guard against confusing $\int_a^b f(x)\, dx$ with $\int f(x)\, dx$. According to Definition 5.3, $\int_a^b f(x)\, dx$ represents a number. If you recall, $\int f(x)\, dx$ represents the general antiderivative of $f(x)$.

In Example 5.7, we compute the definite integral of $7/x$ from $x = 3$ to $x = 15$. To find an antiderivative of $7/x$, we use two integral rules from Section 5.1: the constant multiple rule and the logarithm rule.

Example 5.7

Compute

$$\int_3^{15} \frac{7}{x}\, dx$$

Solution

For $x > 0$, the function $7 \ln x$ is an antiderivative of the function $7/x$. So

$$\int_3^{15} \frac{7}{x}\, dx = 7 \ln x \Big|_3^{15}$$
$$= 7 \ln 15 - 7 \ln 3$$
$$= 7(\ln 15 - \ln 3)$$

$$= 7 \ln \frac{15}{3}$$
$$= 7 \ln 5$$
$$\approx 11.27$$

In Section 5.6, we will present a theorem that characterizes definite integrals in terms of limits. Many books use such a characterization of definite integrals as the definition of definite integrals. Books that use this approach present our definition (Definition 5.3) as a theorem, known as the **fundamental theorem of calculus**.

Definite Integrals Interpreted as Change

The following result is a consequence of Theorem 5.7 and Definition 5.3.

> If $f(x)$ is the derivative of $A(x)$, then
> $$\int_a^b f(x)\, dx$$
> is the amount $A(x)$ changes when x increases from a to b.

It stands to reason, then, that

$$-\int_a^b f(x)\, dx$$

is the amount $A(x)$ changes when x *decreases* from b to a.

The timberland decrease we found at the beginning of this section is the definite integral of $0.1t - 4$ from $t = 2$ to $t = 8$. This definite integral is represented by

$$\int_2^8 (0.1t - 4)\, dt$$

In Example 5.8, we further illustrate the use of definite integrals to represent change.

Example 5.8

Suppose $A(x)$ is a function whose derivative is

$$\frac{-8}{x^2}$$

a. Find a definite integral whose value is the amount $A(x)$ changes when x increases from 4 to 10. Then compute this change in $A(x)$.

b. Use a definite integral to express the amount $A(x)$ changes when x decreases from 8 to 2. Then compute this change in $A(x)$.

Solution for (a)

The change in $A(x)$ when x increases from 4 to 10 is the value of

$$\int_4^{10} \frac{-8}{x^2} \, dx$$

Using the power rule for integrals in Section 5.1, we find that $8/x$ is an antiderivative of $-8/x^2$. Thus,

$$\int_4^{10} \frac{-8}{x^2} \, dx = \frac{8}{x} \bigg|_4^{10}$$
$$= 0.8 - 2$$
$$= -1.2$$

Therefore, $A(x)$ decreases 1.2 units when x increases from 4 to 10.

Solution for (b)

The change in $A(x)$ when x decreases from 8 to 2 is expressed by

$$-\int_2^8 \frac{-8}{x^2} \, dx$$

Thus, since

$$-\int_2^8 \frac{-8}{x^2} \, dx = -\left(\frac{8}{x} \bigg|_2^8\right)$$
$$= -(1 - 4)$$
$$= 3$$

we conclude that $A(x)$ increases 3 units when x decreases from 8 to 2.

In Example 5.9, we compute a definite integral whose value represents an increase in a total production cost.

Example 5.9

Let $C(x)$ represent the total weekly cost (in thousands of dollars) of producing x units of a product per week. Suppose the marginal cost is

$$\tfrac{1}{15}x^2 - 2x + 20$$

when the weekly production level is x units. Find a definite integral whose value is the amount $C(x)$ increases if the weekly production level increases from 9 units to 12 units. Then compute this increase in $C(x)$.

Solution

Since $\tfrac{1}{15}x^2 - 2x + 20$ is the marginal cost and $C(x)$ is the total cost, $\tfrac{1}{15}x^2 - 2x + 20$ is the derivative of $C(x)$. Therefore, the increase in $C(x)$ is the value of

$$\int_9^{12} (\tfrac{1}{15}x^2 - 2x + 20)\, dx$$

Note that $\tfrac{1}{45}x^3 - x^2 + 20x$ is an antiderivative of $\tfrac{1}{15}x^2 - 2x + 20$. So

$$\int_9^{12} (\tfrac{1}{15}x^2 - 2x + 20)\, dx = (\tfrac{1}{45}x^3 - x^2 + 20x)\Big|_9^{12}$$

$$= 266.4 - 187.2$$
$$= 79.2$$

Thus, $C(x)$ increases $79.2 thousand if the weekly production level increases from 9 units to 12 units.

In Example 5.9, we did not derive a formula for $C(x)$. Nevertheless, we were able to compute how much $C(x)$ increases if the weekly production level increases as indicated.

In Example 5.10, we find a definite integral whose value is the amount of a mineral mined during an interval of time.

Example 5.10

The remaining supply of a valuable mineral is increasingly difficult to mine. Thus, the rate at which the mineral can be mined is decreasing. Geologists predict that t years from now the mineral will be mined at the rate of

$$-\tfrac{1}{3}t + 9$$

million tons per year, where $0 \le t \le 25$. Find a definite integral whose value is the amount of the mineral that will be mined during the 10-year interval beginning 2 years from now. Then compute the amount that will be mined during that time.

Solution

Let $A(t)$ represent the amount (in million tons) that will be mined during the next t years. Then the amount that will be mined during the 10-year interval beginning 2 years from now is the amount $A(t)$ increases when t increases from 2 to 12.

Since $-\tfrac{1}{3}t + 9$ is the rate at which the mineral will be mined t years from now, $-\tfrac{1}{3}t + 9$ is the derivative of $A(t)$. Therefore, the amount $A(t)$ increases when t increases from 2 to 12 is the value of

$$\int_2^{12} (-\tfrac{1}{3}t + 9)\, dt$$

So the amount of the mineral that will be mined during the 10 years in question is the value of this integral.

Note that $-\frac{1}{6}t^2 + 9t$ is an antiderivative of $-\frac{1}{3}t + 9$. Thus,

$$\int_2^{12} \left(-\frac{1}{3}t + 9\right) dt = \left(-\frac{1}{6}t^2 + 9t\right)\bigg|_2^{12}$$
$$= 84 - 17\frac{1}{3}$$
$$= 66\frac{2}{3}$$

Therefore, about 67 million tons of the mineral will be mined during the 10-year interval beginning 2 years from now.

A Theorem for the Graphing Calculator Problems

Suppose $\ln f(x)$ and $\ln x$ are linearly related. In other words, suppose constants m and b exist such that

$$\ln f(x) = m \ln x + b$$

Then $\quad f(x) = e^{m \ln x + b}$ Definition 4.4 in Section 4.3
$\quad\quad\quad\quad = (e^{\ln x})^m e^b$ Properties E1 and E3 in Section 4.1
$\quad\quad\quad\quad = e^b x^m$ Definition 4.4 in Section 4.3
$\quad\quad\quad\quad = Ax^m$ letting $A = e^b$

Thus, we have proven Theorem 5.8, which we can use to solve the graphing calculator problems at the end of this section. (See Problems 41 and 42.)

Theorem 5.8

Suppose constants m and b exist such that

$$\ln f(x) = m \ln x + b$$

Then

$$f(x) = Ax^m$$

where $A = e^b$

5.2 Problems

In Problems 1 to 14, compute the definite integral.

1. $\int_1^3 10x^4 \, dx$
2. $\int_{-3}^5 x^3 \, dx$
3. $\int_2^{16} (18 - 2x) \, dx$
4. $\int_5^8 (24 - 4x) \, dx$
5. $\int_0^3 (x^3 - 6x^2 + 12x - 8) \, dx$
6. $\int_2^4 (x^3 - 9x^2 + 27x - 27) \, dx$
7. $\int_2^5 \frac{-3}{x^4} \, dx$
8. $\int_1^7 \frac{7}{x^2} \, dx$
9. $\int_7^{21} \frac{1}{x} \, dx$
10. $\int_4^{28} \frac{8}{x} \, dx$

11. $\int_0^3 e^x \, dx$

12. $\int_4^6 5e^x \, dx$

13. $\int_{64}^{125} 4\sqrt[3]{x} \, dx$

14. $\int_9^{16} \dfrac{3}{\sqrt{x}} \, dx$

In Problems 15 to 20, suppose $A(x)$ is a function whose derivative is the given function $f(x)$. Use a definite integral to express the amount $A(x)$ changes if x changes as indicated. Then compute the change in $A(x)$.

15. $f(x) = 3x^2 + 2x$ x increases from 2 to 5

16. $f(x) = 0.2x - 20$ x increases from 3 to 9

17. $f(x) = -\sqrt{x}$ x decreases from 9 to 4

18. $f(x) = \dfrac{2}{x^3}$ x decreases from 5 to 1

19. $f(x) = \dfrac{3}{x}$ x increases from 8 to 16

20. $f(x) = 2e^x$ x increases from 1 to 5

21. **Productivity of Capital**   Let $Q(x)$ represent the number of barrels of Unbug (an insecticide) that Chemtrex International can produce per day with a capital outlay of $\$x$ thousand. Suppose the marginal productivity of capital is

$$-0.1x + 5$$

for $0 \le x \le 50$. Find a definite integral whose value is the amount $Q(x)$ increases if the capital outlay increases from $x = 20$ to $x = 30$. Then compute this increase in $Q(x)$.

    Hint: $-0.1x + 5$ is the derivative of $Q(x)$ for $0 \le x \le 50$.

22. **Appreciation**   Let $V(t)$ represent the value (in thousands of dollars) of an office building t years from now. An appraiser predicts that t years from now $V(t)$ will be increasing at the rate of

$$5 + \dfrac{90}{t^2}$$

thousand dollars per year, where $t \ge 3$. Find a definite integral whose value is the amount $V(t)$ will increase during the 2-year interval from $t = 3$ to $t = 5$. Then compute this increase in $V(t)$.

    Hint: For $t \ge 3$, $5 + 90/t^2$ is the derivative of $V(t)$.

23. **Depreciation**   Let $V(x)$ represent the value (in thousands of dollars) of a machine x months from now. An accountant projects that x months from now $V(x)$ will be changing at the rate of

$$-\dfrac{5}{\sqrt{x}}$$

thousand dollars per month, where $1 \le x \le 36$. Find a definite integral whose value is the amount $V(x)$ will change during the 7-month period from $x = 9$ to $x = 16$. Then compute this change in $V(x)$.

24. **Propensity to Save**   Let $S(x)$ represent the total monthly amount (in millions of dollars) the residents of Bluestone save when their aggregate monthly income is $\$x$ million. Suppose

$$S'(x) = 0.25 - \dfrac{0.125}{\sqrt{x}}$$

for $x \ge 1$. Find a definite integral whose value is the amount $S(x)$ increases if their aggregate income increases from $x = 9$ to $x = 16$. Then compute this increase in $S(x)$.

25. **Agriculture**   Let $Y(x)$ represent the number of sacks of onions a farmer can produce on his land if he uses x barrels of a special fertilizer. Suppose

$$Y'(x) = -3x^2 + 90x$$

for $5 \le x \le 25$. Use a definite integral to express the change in the yield if the fertilization level is decreased from 20 barrels to 10 barrels. Then compute this change in the yield.

26. **Demand**   Let $D(x)$ represent the monthly demand for a product (in thousands of barrels) when its selling price is $\$x$ per liter. Suppose

$$D'(x) = 0.125x - 2$$

for $0 \le x \le 16$. Use a definite integral to express the amount the monthly demand increases if the selling price decreases from $\$10$ per liter to $\$6$ per liter. Then compute this increase in the demand.

27. **Demand for Cigars**   In about a year, the demand for Prince Justin Cigars will begin to decrease. The manufacturer projects that x years from now the cigars will be selling at the rate of

$$\dfrac{10}{x}$$

million boxes per year, where $x \geq 1$. Find a definite integral whose value is the amount that will be sold during the 4-year interval between $x = 2$ and $x = 6$. Then compute the amount that will be sold during that interval of time.

28. **Coal Importation** A country plans to decrease the amount of coal it imports. Suppose that x years from now the country will be importing coal at the rate of

$$0.2x^2 - 4x + 20$$

million tons per year, where $0 \leq x \leq 10$. Find a definite integral whose value is the amount of coal the country will import during the 6-year interval beginning 3 years from now. Then compute the amount it will import during that interval of time.

29. **Demography** Let $S(x)$ represent the number of senior citizens (in millions) residing in Grandville x years after the beginning of 1996. If x years after that date $S(x)$ changes at the rate of

$$-0.01x^2 + 0.1x + 0.5$$

million per year, how much does $S(x)$ change during the interval between the beginning of 1996 and the beginning of 2006?

30. **Demography** Let $T(x)$ represent the number of teenagers (in millions) residing in Grandville x years after the beginning of 1996. If x years after that date $T(x)$ changes at the rate of

$$0.2x - 2$$

million per year, how much does $T(x)$ change during the interval between the beginning of 1996 and the beginning of 2006?

31. **Acid Rain** The pH of the rainwater in a certain region of the world is decreasing. (Thus, the acidity of the rainwater is increasing.) Environmentalists predict that t years from now, the rainwater's pH will be changing at the rate of

$$0.036t - 0.36$$

units per year, where $0 \leq t \leq 10$. Find a definite integral whose value is the amount the rainwater's pH will change during the next 5 years. Then compute this change.

32. **Learning** The work force of a factory will soon begin producing a new product. As the workers become more experienced, the time required to produce successive units decreases. Let $T(x)$ represent the total number of hours it takes to produce the first x units of the product. The foreman believes that

$$T'(x) = -0.05x + 5$$

for $0 \leq x \leq 75$. Find a definite integral whose value is the time it will take to produce the first 20 units of the product. Then compute this amount of time.

33. **Population** Demographers project that x years from now the population of a city will be increasing at the rate of

$$0.2x + 2$$

thousand per year. How many years will it take for the population to become 80,000 more than the current population?

Hint: For each $t > 0$,

$$\int_0^t (0.2x + 2)\, dx$$

represents the amount (in thousands) the population will increase during the next t years. Thus, you want to find the value of t for which

$$\int_0^t (0.2x + 2)\, dx = 80$$

34. Suppose $F(x)$ is a function of x according to the rule

$$F(x) = \int_0^x (4t^3 + 2t)\, dt$$

for $x > 0$. For instance, $F(3) = 90$, since

$$\int_0^3 (4t^3 + 2t)\, dt = (t^4 + t^2)\Big|_0^3 = 90$$

Show that for each $x > 0$,

$$F'(x) = 4x^3 + 2x$$

35. Suppose $F(x)$ and $G(x)$ are antiderivatives of $f(x)$ and $g(x)$, respectively.

a. Prove that for any real number k,

$$\int_a^b k \cdot f(x)\, dx = k \int_a^b f(x)\, dx$$

b. Prove that

$$\int_a^b (f(x) + g(x))\, dx = \int_a^b f(x)\, dx + \int_a^b g(x)\, dx$$

36. Suppose $F(x)$ is an antiderivative of $f(x)$ and $a < b < c$. Prove that
$$\int_a^c f(x)\,dx = \int_a^b f(x)\,dx + \int_b^c f(x)\,dx$$

37. Suppose $f(x)$ is a function that is continuous on an interval, and a and b are numbers in the interval such that $a < b$. The **average** of the $f(x)$ values for $a \le x \le b$ is
$$\frac{\int_a^b f(x)\,dx}{b-a}$$

   a. Find the average of the $f(x)$ values for $3 \le x \le 8$ if
$$f(x) = x^3$$

   b. Prove that if $f(x)$ is a linear function, the average of the $f(x)$ values for $a \le x \le b$ is
$$\frac{f(a) + f(b)}{2}$$
   [Recall that a linear function is a function that can be expressed in the form
$$f(x) = mx + k$$
   where m and k are constants.]

38. **Portfolio Value** A stockbroker projects that x days from now the market value of a stock portfolio will be
$$V(x) = 0.6\sqrt{x} + 10$$
   thousand dollars. Use the definition of "average" presented in Problem 37 to find what the average market value of the portfolio will be during the next 25 days.

39. Suppose $f(x)$ is continuous on an interval whose left endpoint is a, and $F(x)$ is an antiderivative of $f(x)$. Explain why
$$F(t) = F(a) + \int_a^t f(x)\,dx$$
   for each value of t greater than a and in the interval.

40. Suppose $f(x)$ is continuous on an interval, and a and b are numbers in the interval such that $a < b$. Also, suppose $A(x)$ is a function whose derivative is $f(x)$. If
$$\int_a^b f(x)\,dx = 0$$
can we conclude that the value of $A(x)$ remains constant as x increases from a to b? Why?

Graphing Calculator Problems

41. **Waste Production** In a European country, the per capita rate at which waste is generated has been declining. The following table shows what this rate was at six different times:

Date, Jan. 1	1989	1990	1991	1992	1993	1994
Per capita rate, tons per year	1.28	1.06	0.86	0.78	0.72	0.67

Let $W(t)$ represent the per capita amount of waste (in tons) generated during the t years following January 1, 1987.

   a. Set the viewing screen on your graphing calculator to $[0, 3] \times [-1, 1]$ with a scale of 1 on both axes. Then plot the data points of the form
$$(\ln t, \ln W'(t))$$
   for $t = 2, 3, 4, 5, 6,$ and 7.

   b. Draw the regression line for the data points you plotted in part (a).

   c. Find a formula for the least-squares linear function that corresponds to the regression line you drew in part (b).

   d. The formula you found in part (c) approximates the functional relationship between $\ln W'(t)$ and $\ln t$. Use this fact and Theorem 5.8 to find a formula that approximates the functional relationship between $W'(t)$ and t.

   e. Find a definite integral that approximates the per capita amount of waste that will be generated during the last 5 years of the twentieth century.

   f. Use the definite integral you found in part (e) to estimate the per capita amount of waste that will be generated during the last 5 years of the twentieth century.

42. **Water Purification** The yearly rate at which a Third World country spends money purifying its drinking water has been declining. The following table shows what this rate was at seven different times:

Date, Jan. 1	1988	1989	1990	1991	1992	1993	1994
Yearly rate, $ millions	3.22	2.87	2.63	2.44	2.29	2.18	2.08

This desalinization plant in Santa Barbara, California can supply 30% of the water for 190,000 people (Courtesy Ionics, Inc.).

Let $P(t)$ represent the amount the country spends during the t years following January 1, 1985.

a. Set the viewing screen on your graphing calculator to $[0, 3] \times [0, 2]$ with a scale of 1 on both axes. Then plot the data points of the form

$$(\ln t, \ln P'(t))$$

for $t = 3, 4, 5, 6, 7, 8,$ and 9.

b. Draw the regression line for the data points you plotted in part (a).

c. Find a formula for the least-squares linear function that corresponds to the regression line you drew in part (b).

d. The formula you found in part (c) approximates the functional relationship between $\ln P'(t)$ and $\ln t$. Use this fact and Theorem 5.8 to find a formula that approximates the functional relationship between $P'(t)$ and t.

e. Find a definite integral that approximates the amount the country will spend purifying its water during the last decade of the twentieth century.

f. Use the definite integral you found in part (e) to estimate the amount the country will spend purifying its water during the last decade of the twentieth century.

5.3 Area and Definite Integrals

In this section, we characterize definite integrals in terms of areas. We begin with the function

$$g(x) = 2x - 1$$

Let $A(x)$ be the function whose domain is the closed interval $[3, 9]$ and for which the following two conditions are satisfied:

1. For each t in the half-open interval $(3, 9]$, $A(t)$ is the *area* of the region between the graph of $g(x) = 2x - 1$ and the x axis from $x = 3$ to $x = t$. (Such a region is the shaded region in Figure 5.1.)
2. $A(3) = 0$.

Let us derive a formula for the function $A(x)$. The area of a trapezoidal region can be obtained by multiplying the distance between the parallel sides and the average of the lengths of the parallel sides. The shaded region in Figure 5.1 is a trapezoidal region whose parallel sides have

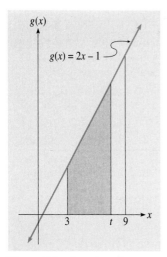

Figure 5.1 A region whose area is $A(t)$.

lengths 5 [since $g(3) = 5$] and $2t - 1$ [since $g(t) = 2t - 1$]. Therefore, since the distance between the parallel sides is $t - 3$, the area of the shaded region is

$$(t - 3) \frac{5 + (2t - 1)}{2} = (t - 3)(t + 2)$$
$$= t^2 - t - 6$$

So, for the function $A(x)$ to satisfy the first of the two conditions listed above, the formula for $A(x)$ must be

$$A(x) = x^2 - x - 6$$

In fact, using this formula for $A(x)$, it follows that $A(x)$ also satisfies the second of the two listed conditions, because

$$A(3) = 3^2 - 3 - 6$$
$$= 0$$

Note that the function

$$A(x) = x^2 - x - 6$$

is an antiderivative of the function

$$g(x) = 2x - 1$$

for values of x in the interval $[3, 9]$. According to Theorem 5.9, this result is not a coincidence.

Theorem 5.9

Let $f(x)$ be a function that is continuous on an interval that contains the closed interval $[a, b]$. Suppose the graph of $f(x)$ is nowhere below the x axis on the interval $[a, b]$. Let $A(x)$ be the function whose domain is the interval $[a, b]$ and for which the following two conditions are satisfied:

1. For each t in the half-open interval $(a, b]$, $A(t)$ is the *area* of the region between the graph of $f(x)$ and the x axis from $x = a$ to $x = t$. (Such a region is the shaded region in Figure 5.2.)
2. $A(a) = 0$.

Then the function $A(x)$ is an antiderivative of $f(x)$ for values of x in the interval $[a, b]$.

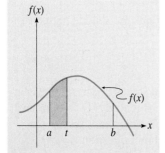

Figure 5.2 A region as described in Theorem 5.9.

Areas of Regions above the Horizontal Axis

Theorem 5.10 is a consequence of Theorem 5.9.

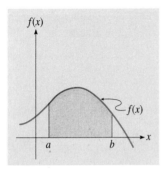

Figure 5.3 A region whose area is $\int_a^b f(x)\,dx$.

Theorem 5.10

Let $f(x)$ be a function that is continuous on an interval that contains the closed interval $[a, b]$. Suppose the graph of $f(x)$ is nowhere below the x axis on the interval $[a, b]$. Then the area of the region between the graph of $f(x)$ and the x axis from $x = a$ to $x = b$ is

$$\int_a^b f(x)\,dx$$

According to Theorem 5.10, the area of the shaded region in Figure 5.3 is

$$\int_a^b f(x)\,dx$$

Proof of Theorem 5.10

Let $A(x)$ be the function described in Theorem 5.9. Then, according to that theorem, $A(x)$ is an antiderivative of $f(x)$ for values of x in the interval $[a, b]$. Therefore,

$$\int_a^b f(x)\,dx = A(b) - A(a) \quad \text{Definition 5.3 in Section 5.2}$$

$$= A(b) \quad \text{since } A(a) = 0$$

Consequently, since $A(b)$ is the area of the region between the graph of $f(x)$ and the x axis from $x = a$ to $x = b$, we have shown that this area is indeed equal to $\int_a^b f(x)\,dx$.

We use Theorem 5.10 in Example 5.11.

Example 5.11

Find the area of the region between the graph of

$$f(x) = 2 - \tfrac{1}{2}x^2$$

and the x axis from $x = -1$ to $x = 2$. This region is the shaded region in Figure 5.4.

Solution

Since $f(x)$ is a polynomial function, it is continuous everywhere. In particular, $f(x)$ is continuous on the interval $[-1, 2]$. Note that $f(x) \geq 0$ for each x in $[-1, 2]$. Thus, the graph of $f(x)$ is nowhere below the x axis on the interval $[-1, 2]$. Therefore, the area of the region in question is

$$\int_{-1}^{2} (2 - \tfrac{1}{2}x^2)\,dx = (2x - \tfrac{1}{6}x^3)\Big|_{-1}^{2}$$

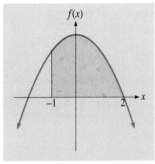

Figure 5.4 A region between the graph of $f(x) = 2 - \tfrac{1}{2}x^2$ and the x axis.

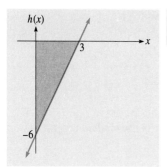

Figure 5.5 A region between the graph of $h(x) = 2x - 6$ and the x axis.

$$= \frac{16}{6} - \left(-\frac{11}{6}\right)$$
$$= 4.5$$

Areas of Regions below the Horizontal Axis

Let us now consider the region between the graph of

$$h(x) = 2x - 6$$

and the x axis from $x = 0$ to $x = 3$. This region is the shaded region in Figure 5.5.

We can think of the shaded region in Figure 5.5 as a triangular region whose base is 3 and whose height is 6. Since the area of a triangular region is $\frac{1}{2}$ times the product of the base and height, the area of the triangular region in Figure 5.5 is

$$\left(\tfrac{1}{2}\right)(3)(6) = 9$$

Note that

$$\int_0^3 (2x - 6)\, dx = (x^2 - 6x)\Big|_0^3$$
$$= -9$$

Therefore,

$$\left| \int_0^3 (2x - 6)\, dx \right| = 9$$

which, as we saw, is the area of the region in Figure 5.5. According to Theorem 5.11, this result is not a coincidence.

Theorem 5.11

> Let $f(x)$ be a function that is continuous on an interval that contains the closed interval $[a, b]$. Suppose the graph of $f(x)$ is nowhere above the x axis on the interval $[a, b]$. Then the magnitude (absolute value) of
>
> $$\int_a^b f(x)\, dx$$
>
> is the area of the region between the graph of $f(x)$ and the x axis from $x = a$ to $x = b$.

According to Theorem 5.11, the area of the shaded region in Figure 5.6 is

$$\left| \int_a^b f(x)\, dx \right|$$

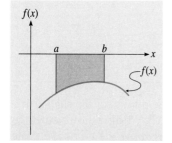

Figure 5.6 A region whose area is $\left| \int_a^b f(x)\, dx \right|$.

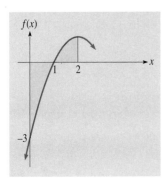

Figure 5.7 A region between the graph of $f(x) = -x^2 + 4x - 3$ and the x axis.

In Example 5.12, we use Theorems 5.10 and 5.11.

Example 5.12

Find the area of the region between the graph of

$$f(x) = -x^2 + 4x - 3$$

and the x axis from $x = 0$ to $x = 2$. This region is the shaded region in Figure 5.7.

Solution

Since

$$-x^2 + 4x - 3 = -(x - 1)(x - 3)$$

we can write

$$f(x) = -(x - 1)(x - 3)$$

Thus, we see that $f(x) \leq 0$ for each x in $[0, 1]$ and $f(x) \geq 0$ for each x in $[1, 2]$. So the graph of $f(x)$ is nowhere above the x axis on the interval $[0, 1]$ and nowhere below the x axis on the interval $[1, 2]$.

According to Theorem 5.11, the area of the region between the graph of $f(x)$ and the x axis from $x = 0$ to $x = 1$ is

$$\left| \int_0^1 (-x^2 + 4x - 3) \, dx \right| = \left| -\frac{4}{3} \right|$$

$$= \frac{4}{3}$$

According to Theorem 5.10, the area of the region between the graph of $f(x)$ and the x axis from $x = 1$ to $x = 2$ is

$$\int_1^2 (-x^2 + 4x - 3) \, dx = \frac{2}{3}$$

Therefore, the area of the region between the graph of $f(x)$ and the x axis from $x = 0$ to $x = 2$ is

$$\frac{4}{3} + \frac{2}{3} = 2$$

Definite Integrals Characterized in Terms of Areas

In Example 5.12, we showed that the area of the portion of the shaded region in Figure 5.7 that lies *below* the x axis is $\frac{4}{3}$. We also showed that the area of the portion of the shaded region that lies *above* the x axis is $\frac{2}{3}$. Note that

$$\frac{2}{3} - \frac{4}{3} = -\frac{2}{3}$$

It turns out that

$$\int_0^2 (-x^2 + 4x - 3)\, dx = -\tfrac{2}{3}$$

also. This result suggests Theorem 5.12.

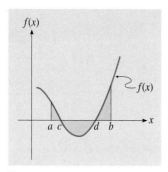

Figure 5.8 A region whose area is the sum of $\int_a^c f(x)\,dx$, $\left|\int_c^d f(x)\,dx\right|$, and $\int_d^b f(x)\,dx$.

Theorem 5.12

Let $f(x)$ be a function that is continuous on an interval that contains the closed interval $[a, b]$. Consider the region between the graph of $f(x)$ and the x axis from $x = a$ to $x = b$. (Such a region is the shaded region in Figure 5.8.) If A represents the total area of the portions of the region *above* the x axis and B represents the total area of the portions of the region *below* the x axis, then

$$\int_a^b f(x)\, dx = A - B$$

In Theorem 5.12, the sign of $\int_a^b f(x)\, dx$ depends on how much of the region lies below the x axis. If most of the region lies above the x axis, then $\int_a^b f(x)\, dx$ is positive. If most of the region lies below the x axis, then $\int_a^b f(x)\, dx$ is negative. If just as much of the region lies below the x axis as above it, then $\int_a^b f(x)\, dx$ is zero. If the region lies entirely above the x axis, then $\int_a^b f(x)\, dx$ is the area of the region. But if some of the region lies below the x axis, then $\int_a^b f(x)\, dx$ is less than the area of the region.

Theorems 5.10 and 5.11 are special cases of Theorem 5.12. Theorem 5.10 is the special case where the region lies entirely *above* the x axis. Theorem 5.11 is the special case where the region lies entirely *below* the x axis.

In Example 5.13, we use the area characterization of definite integrals to find the value of a definite integral.

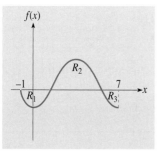

Figure 5.9 The graph of $f(x)$.

Example 5.13

Let $f(x)$ be the function whose graph is the curve in Figure 5.9. Suppose the area of the region labeled R_1 is 4, the area of the region labeled R_2 is 9, and the area of the region labeled R_3 is 2.

a. Find the value of

$$\int_{-1}^7 f(x)\, dx$$

b. Find the area of the region between the graph of $f(x)$ and the x axis from $x = -1$ to $x = 7$.

Solution for (a)

Using Theorem 5.12,

$$\int_{-1}^{7} f(x)\,dx = 9 - (4 + 2)$$
$$= 3$$

Solution for (b)

The area of the region is

$$4 + 9 + 2 = 15$$

Evidently, the derivative concept (via the antiderivative concept) is linked to the problem of finding areas of regions of the type discussed in this section. Since the derivative concept is linked also to the problem of finding slopes of tangents, we see that these two apparently unrelated problems are, in fact, related.

A Theorem for the Graphing Calculator Problems

Sometimes we cannot express an antiderivative with a formula, even if we know its value at a point in its domain. Fortunately, using Theorem 5.13 and a calculator that approximates definite integrals, we can at least approximate the values of such an antiderivative.

Theorem 5.13

Suppose $F(x)$ is an antiderivative of $f(x)$.

1. If $t > a$,

$$F(t) = F(a) + \int_{a}^{t} f(x)\,dx$$

2. If $t < a$,

$$F(t) = F(a) - \int_{t}^{a} f(x)\,dx$$

In Theorem 5.13, the values of a and t, of course, must be in the domain of $F(x)$.

Proof of Part (1) of Theorem 5.13

Suppose $t > a$. The amount $F(x)$ changes if x increases from a to t is

$$F(t) - F(a)$$

Since $f(x)$ is the derivative of $F(x)$, we can express this change also by

$$\int_a^t f(x)\, dx$$

Therefore,

$$F(t) - F(a) = \int_a^t f(x)\, dx$$

which implies that

$$F(t) = F(a) + \int_a^t f(x)\, dx$$

So we have established the validity of part (1) of Theorem 5.13. We can establish the validity of part (2) using a similar approach.

We can use Theorem 5.13 to solve the graphing calculator problems at the end of this section. (See Problems 29 and 30.)

5.3 Problems

In Problems 1 to 12, the graph of the function is nowhere below the x axis between the given pair of x values. Find the area of the region bounded by the graph of the function and the x axis between the given pair of x values.

1. $f(x) = \frac{1}{4}x^2 + 1$ $x = 3$ and $x = 6$
2. $f(x) = -x^2 + 10x$ $x = 1$ and $x = 5$
3. $f(x) = -3x^2 - 12x$ $x = -4$ and $x = 0$
4. $f(x) = 6x^2$ $x = -2$ and $x = 1$
5. $f(x) = x^2 - 6x + 9$ $x = 2$ and $x = 5$
6. $f(x) = 4x + 1$ $x = 1$ and $x = 6$
7. $f(x) = 3\sqrt{x}$ $x = 0$ and $x = 9$
8. $f(x) = \dfrac{21}{x^2}$ $x = 3$ and $x = 7$
9. $f(x) = 4x^3 + 2$ $x = 1$ and $x = 4$
10. $f(x) = 27 - x^3$ $x = 0$ and $x = 3$
11. $f(x) = e^x$ $x = 0$ and $x = 1$
12. $f(x) = \dfrac{8}{x}$ $x = 2$ and $x = 4$

In Problems 13 to 18, the graph of the function is nowhere above the x axis between the given pair of x values. Find the area of the region bounded by the graph of the function and the x axis between the given pair of x values.

13. $f(x) = 3x^2 - 6x$ $x = 0$ and $x = 2$
14. $f(x) = -\frac{1}{4}x^2$ $x = 4$ and $x = 6$
15. $f(x) = \dfrac{-24}{x^2}$ $x = 2$ and $x = 8$
16. $f(x) = -6\sqrt{x} + 12$ $x = 4$ and $x = 16$
17. $f(x) = 2x - 7$ $x = -4$ and $x = -1$
18. $f(x) = -x^2 + 8x - 16$ $x = 1$ and $x = 5$

In Problems 19 to 22, the graph of the function is partially above and partially below the x axis between the given pair of x values. Find the area of the region bounded by the graph of the function and the x axis between the given pair of x values.

Hint: First determine the intervals on which $f(x) \geq 0$ and the intervals on which $f(x) \leq 0$.

19. $f(x) = x^2 - x$ $x = 0$ and $x = 6$
20. $f(x) = -x^2 + 4x$ $x = 3$ and $x = 6$
21. $f(x) = x^3 + x^2 - 2x$ $x = -4$ and $x = -1$
22. $f(x) = x^3 + x^2 - 2x$ $x = -2$ and $x = 1$

23. Let $f(x)$ be the function whose graph is the curve in the figure for this problem. Suppose the area of region R_1 is 1.875, the area of region R_2 is 2.75, and the area of region R_3 is 0.875.

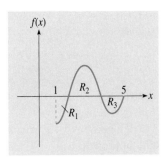

  a. Find the value of $\int_1^5 f(x)\,dx$.

  b. Find the area of the region between the graph of $f(x)$ and the x axis from $x = 1$ to $x = 5$.

24. Let $f(x)$ be the function whose graph is the curve in the figure for this problem. Suppose the area of region R_1 is 2 and the area of region R_2 is 0.625.

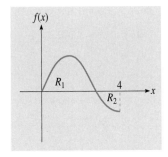

  a. Find the value of $\int_0^4 f(x)\,dx$.

  b. Find the area of the region between the graph of $f(x)$ and the x axis from $x = 0$ to $x = 4$.

25. **Productivity of Labor** Decholest is a shortening that contains no cholesterol. The number of tons of Decholest that can be produced daily depends on the size of the labor force. Suppose the marginal productivity of labor is

$$\frac{1}{225}x^2 - \frac{4}{15}x + 5$$

when x units of labor is used daily, where $0 \leq x \leq 30$. Give the boundaries of a region whose area represents the amount the daily production level increases if the size of the labor force increases from 20 units to 25 units.

26. **Propensity to Save** Let $S(x)$ represent the total monthly amount (in millions of dollars) the residents of Bluestone save when their aggregate monthly income is $\$x$ million. Suppose

$$S'(x) = 0.25 - \frac{0.125}{\sqrt{x}}$$

for $x \geq 1$. Give the boundaries of a region whose area represents the amount $S(x)$ increases if the aggregate monthly income increases from \$9 million to \$16 million.

27. Suppose $f(x)$ is as described in Theorem 5.9 and $F(x)$ is such that

$$F(t) = \begin{cases} 0 & \text{when } t = a \\ \int_a^t f(x)\,dx & \text{when } a < t \leq b \end{cases}$$

Explain how Theorems 5.9 and 5.10 imply that $F(x)$ is an antiderivative of $f(x)$ for values of x in the interval $[a, b]$.

28. Suppose $f(x)$ is continuous on an interval that contains the closed interval $[a, b]$. Explain why

$$\left| \int_a^b f(x)\,dx \right| \leq \int_a^b |f(x)|\,dx$$

 Graphing Calculator Problems

29. Suppose $F(x)$ represents the particular antiderivative of

$$\frac{40}{\sqrt{144 - x^2}}$$

that satisfies the condition $F(4) = 30$. Use Theorem 5.13 and a calculator that approximates definite integrals to approximate the value of $F(x)$ for each of the following values of x. If your calculator displays regions whose areas are definite integrals, set the viewing screen to $[0, 9] \times [0, 6]$ with a scale of 1 on both axes.

  a. $x = 8$ b. $x = 1$

30. **Productivity of Labor** Yogurt Unlimited can produce 25.67 hundred gallons of yogurt daily with 5 units of labor. Suppose the marginal productivity of labor is

$$6e^{-0.02x^2}$$

when the company uses x units of labor per day. Use Theorem 5.13 and a calculator that approximates definite integrals to approximate how much yogurt the company can produce daily with each of the following amounts of labor. If your calculator displays regions whose areas are definite integrals, set the viewing screen to $[0, 12] \times [0, 8]$ with a scale of 1 on both axes.

a. 9 units   b. 3 units

Computer Problems

The Basic program INT listed below uses Theorem 5.12 to approximate the value of a definite integral of a continuous function $f(x)$ that has *at least one* zero. Let a and b represent the lower and upper limits of integration. Also, let MN and MX represent the minimum and maximum values of $f(x)$ on the closed interval from a to b. The approach uses the rectangular region R whose vertices are (a, MN), (a, MX), (b, MX), and (b, MN). This rectangular region contains the region between the graph of $f(x)$ and the x axis from $x = a$ to $x = b$. In general, portions of this region lie above the x axis, and portions of it lie below the x axis. Let UA represent the total area of the portions above the x axis, and let LA represent the total area of the portions below the x axis. The program approximates UA and LA. Then it subtracts LA from UA.

According to Theorem 5.12, the result approximates the value of

$$\int_a^b f(x)\, dx$$

Here is how the program INT approximates UA. (It approximates LA in a similar manner.) The program randomly selects points in the rectangular region R. Let N represent the number of selected points. Also, let UC represent the number of selected points which are, in fact, in the region above the x axis and below the graph of $f(x)$. The program assumes that the ratio of UC to N is approximately equal to the ratio of UA to the area of the region R. This assumption implies that UA is approximately equal to

$$\frac{UC(b-a)(MX-MN)}{N}$$

[Note that $(b-a)(MX-MN)$ is the area of the region R.]

Execution of INT prompts you to enter the lower and upper limits of integration and the number of points to be randomly selected from the rectangular region R. (The more points you select, the better the accuracy of the result.) Then it prompts you to *seed* the set of randomly selected points with an integer between $-32{,}768$ and $32{,}767$. (Different seeds produce different sets of random points.) Finally, it prompts you to enter the minimum and maximum values of $f(x)$ on the closed interval that extends from the lower limit to the upper limit of integration.

As listed, INT applies to the function

$$f(x) = 3x^2 - 12x$$

But you can make it applicable to other functions by editing line 60.

The Program INT and a Sample Run

```
LIST
10   REM                            INT
20   REM
30   REM A PROGRAM THAT USES THE RANDOM NUMBER GENERATOR TO APPROXIMATE A DEFINITE
40   REM INTEGRAL
50   REM
60   DEF FNF(X)=3*X^2-12*X
70   PRINT "ENTER THE LOWER AND UPPER LIMITS OF INTEGRATION."
80   INPUT A,B
90   PRINT
```

```
100  PRINT "ENTER THE NUMBER OF POINTS TO BE RANDOMLY SELECTED. (ABOUT 10,000"
110  PRINT "POINTS ARE USUALLY NEEDED.)"
120  INPUT N
130  PRINT
140  RANDOMIZE
150  PRINT
160  PRINT "ENTER THE MINIMUM AND MAXIMUM VALUES OF THE FUNCTION ON THE CLOSED"
170  PRINT "INTERVAL FROM";A;"TO";B;"."
180  INPUT MN,MX
190  PRINT
200  LET UC=0
210  LET LC=0
220  FOR J=1 TO N
230  LET X=(B-A)*RND+A
240  LET Y=(MX-MN)*RND+MN
250  IF 0>Y OR Y>FNF(X) THEN 270
260  LET UC=UC+1
270  IF FNF(X)>Y OR Y>0 THEN 290
280  LET LC=LC+1
290  NEXT J
300  LET UA=UC*(B-A)*(MX-MN)/N
310  LET LA=LC*(B-A)*(MX-MN)/N
320  LET I=UA-LA
330  PRINT "THE VALUE OF THE DEFINITE INTEGRAL IS APPROXIMATELY";I;"."
340  END
OK
RUN
ENTER THE LOWER AND UPPER LIMITS OF INTEGRATION.
? 1,5

ENTER THE NUMBER OF POINTS TO BE RANDOMLY SELECTED. (ABOUT 10,000
POINTS ARE USUALLY NEEDED.)
?8000

Random number seed (-32768 to 32767)? 7

ENTER THE MINIMUM AND MAXIMUM VALUES OF THE FUNCTION ON THE CLOSED
INTERVAL FROM 1 TO 5.
? -12, 15

THE VALUE OF THE DEFINITE INTEGRAL IS APPROXIMATELY -19.71.
OK
```

In Problems 31 and 32, use the program INT to approximate the definite integral.

31. $\int_{1}^{12} (x^2 - 12x + 20)\, dx$

32. $\int_{0}^{8} (-x^2 + 10x - 21)\, dx$

5.4 Integration by Substitution

Note that
$$(x^2 + 5)^2 \, 6x = 6x^5 + 60x^3 + 150x$$

So we can use two of the integral rules in Section 5.1 to conclude that
$$\int (x^2 + 5)^2 \, 6x \, dx = \int (6x^5 + 60x^3 + 150x) \, dx$$
$$= x^6 + 15x^4 + 75x^2 + C$$

(Which rules did we use?) We could use the same method to find the indefinite integral
$$\int (x^2 + 5)^7 \, 6x \, dx$$

But it takes much more work to expand $(x^2 + 5)^7 \, 6x$ than it took to expand $(x^2 + 5)^2 \, 6x$. Fortunately, we can find the indefinite integral of $(x^2 + 5)^7 \, 6x$ using an alternative method. The alternative method is based on the chain rule. So we begin by deriving the integral version of the chain rule.

Suppose $G(x)$ is an antiderivative of $g(x)$. Then, according to the chain rule (Section 2.5), the derivative of the composite function

$$G[u(x)]$$

is

$$g[u(x)]u'(x)$$

This result implies the following integral rule:

Chain Rule for Integrals

If $G(x)$ is an antiderivative of $g(x)$, then
$$\int g[u(x)]u'(x) \, dx = G[u(x)] + C$$

In this section, we will use u and du as abbreviations for $u(x)$ and $u'(x) \, dx$, respectively. Using this notation, we can restate the above rule as follows:

Chain Rule for Integrals (Abbreviated Form)

If $G(x)$ is an antiderivative of $g(x)$, then
$$\int g(u) \, du = G(u) + C$$
where $u = u(x)$ and $du = u'(x) \, dx$.

The above rule sometimes can help us apply the integral rules in Section 5.1 to integrands having the following forms:

$$[u(x)]^n \, h(x) \qquad \frac{h(x)}{u(x)} \qquad h(x) e^{u(x)}$$

Integrands of the Form $[u(x)]^n \, h(x)$

We can always use the chain rule for integrals to help us find indefinite integrals of functions of the form

$$[u(x)]^n \, h(x)$$

if $h(x)$ is a constant multiple of $u'(x)$ and $n \neq -1$. Example 5.14 illustrates that we can express the indefinite integrals of such functions in the form

$$\int k u^n \, du$$

where k is a constant. Remember that $u = u(x)$ and $du = u'(x) \, dx$.

Example 5.14

Find

$$\int (x^2 + 5)^7 \, 6x \, dx$$

Solution

Note that $6x$ is a constant multiple of the derivative of $x^2 + 5$, which is $2x$. Thus, we can use the chain rule for integrals. We begin by letting

$$u = x^2 + 5$$

Then,

$$du = 2x \, dx$$

So

$$3 \, du = 6x \, dx$$

Substituting, we get

$$\int (x^2 + 5)^7 \, 6x \, dx = \int u^7 \, 3 du$$
$$= \int 3u^7 \, du$$

Thus, we have an indefinite integral of the form

$$\int g(u) \, du$$

where $\quad g(x) = 3x^7$

Using the power rule for integrals (Theorem 5.2 in Section 5.1), an antiderivative of $g(x)$ is

$$G(x) = \tfrac{3}{8} x^8$$

So, using the chain rule for integrals,
$$\int 3u^7 \, du = \frac{3}{8}u^8 + C$$
$$= \frac{3}{8}(x^2 + 5)^8 + C \quad \text{since } u = x^2 + 5$$

Therefore, $\int (x^2 + 5)^7 \, 6x \, dx = \frac{3}{8}(x^2 + 5)^8 + C$

which, of course, means that the antiderivatives of the function
$$(x^2 + 5)^7 \, 6x$$
are the functions of the form
$$\frac{3}{8}(x^2 + 5)^8 + C$$

We can verify the correctness of the indefinite integral we found in Example 5.14 by showing that its derivative is the integrand.

We can think of a function of the form
$$[u(x)]^n$$
as being a function of the form $[u(x)]^n \, h(x)$, where $h(x) = 1$. So, if 1 is a constant multiple of $u'(x)$ and $n \neq -1$, we can find the indefinite integral of such a function using the approach we used in Example 5.14.

In Example 5.15, we use the chain rule for integrals to find the indefinite integral of a function of the form
$$[u(x)]^n \, h(x)$$
even though $h(x)$ is *not* a constant multiple of $u'(x)$.

Example 5.15

Find
$$\int 4x\sqrt{x - 9} \, dx$$

Solution

Note that
$$4x\sqrt{x - 9} = (x - 9)^{1/2} \, 4x$$

But note also that $4x$ is *not* a constant multiple of the derivative of $x - 9$. (The derivative of $x - 9$ is 1.) Nevertheless, in this case, we can use the chain rule for integrals. We begin by letting
$$u = x - 9$$

Then, $du = 1 \, dx = dx$

We need to also express $4x$ in terms of u:

Since
$$u = x - 9$$

it follows that $\quad u + 9 = x$

So $\quad 4u + 36 = 4x$

Substituting, we get
$$\int 4x\sqrt{x-9}\, dx = \int (x-9)^{1/2}\, 4x\, dx$$
$$= \int u^{1/2}\, (4u + 36)\, du$$
$$= \int (4u^{3/2} + 36u^{1/2})\, du$$

Thus, we have an indefinite integral of the form
$$\int g(u)\, du$$
where $\quad g(x) = 4x^{3/2} + 36x^{1/2}$

Using the integral versions of the power rule and the sum and difference rule (Theorems 5.2 and 5.3), an antiderivative of $g(x)$ is
$$G(x) = 1.6x^{5/2} + 24x^{3/2}$$

So, using the chain rule for integrals,
$$\int (4u^{3/2} + 36u^{1/2})\, du = 1.6u^{5/2} + 24u^{3/2} + C$$
$$= 1.6(x-9)^{5/2} + 24(x-9)^{3/2} + C \qquad u = x - 9$$
$$= 1.6(\sqrt{x-9})^5 + 24(\sqrt{x-9})^3 + C$$

Therefore,
$$\int 4x\sqrt{x-9}\, dx = 1.6(\sqrt{x-9})^5 + 24(\sqrt{x-9})^3 + C$$

In this book, we will not be able to find the indefinite integral of every function of the form
$$[u(x)]^n\, h(x)$$
where $n \neq -1$. For example, we will not be able to find the indefinite integral of the function
$$(2x - x^2)^{1/2}\, x$$

We now consider functions of the form
$$\frac{h(x)}{u(x)}$$

Integrands of the Form $h(x)/u(x)$

We can always use the chain rule for integrals to help us find indefinite integrals of functions of the form
$$\frac{h(x)}{u(x)}$$

if $h(x)$ is a constant multiple of $u'(x)$. As Example 5.16 illustrates, we can express the indefinite integrals of such functions in the form

$$\int \frac{k}{u} \, du$$

where k is a constant. Again, remember that $u = u(x)$ and $du = u'(x) \, dx$.

Example 5.16
Find

$$\int \frac{3x^2 + 2}{4x^3 + 8x} \, dx$$

Solution

Note that $3x^2 + 2$ is a constant multiple of the derivative of $4x^3 + 8x$, which is $12x^2 + 8$. So we can use the chain rule for integrals. To use this rule, we let

$$u = 4x^3 + 8x$$

Then,
$$du = (12x^2 + 8) \, dx$$

So
$$\tfrac{1}{4} \, du = (3x^2 + 2) \, dx$$

Rewriting the integrand and substituting, we get

$$\int \frac{3x^2 + 2}{4x^3 + 8x} \, dx = \int \frac{1}{4x^3 + 8x} (3x^2 + 2) \, dx$$

$$= \int \frac{1}{u} \left(\frac{1}{4} \right) du$$

$$= \int \frac{1/4}{u} \, du$$

Thus, we have an indefinite integral of the form

$$\int g(u) \, du$$

where
$$g(x) = \frac{1/4}{x}$$

Using the integral versions of the constant multiple rule and the logarithm rule (Theorems 5.5 and 5.6), an antiderivative of $g(x)$ is

$$G(x) = \tfrac{1}{4} \ln |x|$$

So, using the chain rule for integrals,

$$\int \frac{1/4}{u} \, du = \tfrac{1}{4} \ln |u| + C$$

$$= \tfrac{1}{4} \ln |4x^3 + 8x| + C \qquad \text{since } u = 4x^3 + 8x$$

Therefore,
$$\int \frac{3x^2 + 2}{4x^3 + 8x} \, dx = \tfrac{1}{4} \ln |4x^3 + 8x| + C$$

In Example 5.17, we use the chain rule for integrals to find the indefinite integral of a function of the form
$$\frac{h(x)}{u(x)}$$
even though $h(x)$ is *not* a constant multiple of $u'(x)$.

Example 5.17

Find
$$\int \frac{2x}{\sqrt{4x + 1}} \, dx$$

Solution

Note that $2x$ is *not* a constant multiple of the derivative of $\sqrt{4x + 1}$. (The derivative of $\sqrt{4x + 1}$ is $2/\sqrt{4x + 1}$.) Nevertheless, in this case, we can use the chain rule for integrals. We begin by letting
$$u = \sqrt{4x + 1}$$

Then,
$$du = \frac{2}{\sqrt{4x + 1}} \, dx = \frac{2}{u} \, dx$$

So
$$\tfrac{1}{2} u \, du = dx$$

We need to also express $2x$ in terms of u:

Since
$$u = \sqrt{4x + 1}$$
it follows that
$$u^2 = 4x + 1$$
which implies that
$$\frac{u^2 - 1}{2} = 2x$$

Substituting, we obtain
$$\int \frac{2x}{\sqrt{4x + 1}} \, dx = \int \frac{1}{u} \left(\frac{u^2 - 1}{2} \right) \tfrac{1}{2} u \, du$$
$$= \int (\tfrac{1}{4} u^2 - \tfrac{1}{4}) \, du$$

Thus, we have an indefinite integral of the form
$$\int g(u) \, du$$

where $g(x) = \frac{1}{4}x^2 - \frac{1}{4}$

Using the integral versions of the power rule and the sum and difference rule (Theorems 5.2 and 5.3), an indefinite integral of $g(x)$ is

$$G(x) = \frac{1}{12}x^3 - \frac{1}{4}x$$

So, using the chain rule for integrals,

$$\int (\tfrac{1}{4}u^2 - \tfrac{1}{4})\, du = \tfrac{1}{12}u^3 - \tfrac{1}{4}u + C$$
$$= \tfrac{1}{12}(\sqrt{4x+1})^3 - \tfrac{1}{4}\sqrt{4x+1} + C \qquad u = \sqrt{4x+1}$$

Therefore,

$$\int \frac{2x}{\sqrt{4x+1}}\, dx = \tfrac{1}{12}(\sqrt{4x+1})^3 - \tfrac{1}{4}\sqrt{4x+1} + C$$

We can rewrite the integrand in Example 5.17 as follows:

$$(4x+1)^{-1/2}\, 2x$$

Then the integrand has the form

$$[u(x)]^n\, h(x)$$

where $n \neq -1$, and we can find the indefinite integral using an approach based on letting $u = 4x+1$, rather than on letting $u = \sqrt{4x+1}$. Try it.

Suppose we want to find the value of the definite integral

$$\int_2^6 \frac{2x}{\sqrt{4x+1}}\, dx$$

In Example 5.17, we found that

$$\int \frac{2x}{\sqrt{4x+1}}\, dx = \tfrac{1}{12}(\sqrt{4x+1})^3 - \tfrac{1}{4}\sqrt{4x+1} + C$$

Therefore,

$$\int_2^6 \frac{2x}{\sqrt{4x+1}}\, dx = \left(\tfrac{1}{12}(\sqrt{4x+1})^3 - \tfrac{1}{4}\sqrt{4x+1}\right)\Big|_2^6$$
$$= \frac{110}{12} - \frac{18}{12}$$
$$= \frac{23}{3}$$

In Example 5.18, we use an alternative approach to find the value of this definite integral.

Example 5.18

Find the value of

$$\int_2^6 \frac{2x}{\sqrt{4x+1}}\, dx$$

Solution

In Example 5.17, using an approach based on letting

$$u = \sqrt{4x+1}$$

we showed that

$$\int \frac{2x}{\sqrt{4x+1}}\, dx = \int (\tfrac{1}{4}u^2 - \tfrac{1}{4})\, du$$

$$= \tfrac{1}{12}u^3 - \tfrac{1}{4}u + C$$

When $x = 2$,

$$u = \sqrt{4(2)+1} = 3$$

and when $x = 6$,

$$u = \sqrt{4(6)+1} = 5$$

Therefore,

$$\int_2^6 \frac{2x}{\sqrt{4x+1}}\, dx = \int_3^5 (\tfrac{1}{4}u^2 - \tfrac{1}{4})\, du$$

$$= \left(\tfrac{1}{12}u^3 - \tfrac{1}{4}u\right)\Big|_3^5$$

$$= \frac{110}{12} - \frac{18}{12}$$

$$= \frac{23}{3}$$

which agrees with the value we obtained just prior to this example.

In this book, we will not be able to find the indefinite integral of every function of the form

$$\frac{h(x)}{u(x)}$$

For example, we will not be able to find the indefinite integral of the function

$$\frac{2x}{x^4+9}$$

Next, we consider functions of the form

$$h(x)e^{u(x)}$$

Integrands of the Form $h(x)e^{u(x)}$

We can always use the chain rule for integrals to help us find indefinite integrals of functions of the form

$$h(x)e^{u(x)}$$

if $h(x)$ is a constant multiple of $u'(x)$. As Example 5.19 illustrates, we can express the indefinite integrals of such functions in the form

$$\int ke^u \, du$$

where k is a constant. Once again, remember that $u = u(x)$ and $du = u'(x) \, dx$.

Example 5.19

A respiratory disease began plaguing the inhabitants of an island a few years ago. So far, the disease has caused 700 deaths. Presently, the deaths are occurring at the rate of 160 per year. Health officials believe that from now on the yearly death rate will decrease exponentially at 8 percent per year. How many deaths will be attributable to the disease by the time 3 more years elapse?

Solution

The above information implies that x years from now the number of deaths will be increasing at the rate of

$$1.6e^{-0.08x}$$

hundred per year. Therefore, if we let $D(x)$ represent the number of deaths (in hundreds) by the time x more years elapse, then

$$D'(x) = 1.6e^{-0.08x}$$

So, since 700 deaths have occurred so far, $D(x)$ is the particular antiderivative of $1.6e^{-0.08x}$ that satisfies the condition $D(0) = 7$.

Since 1.6 is a constant multiple of the derivative of $-0.08x$, we can use the chain rule for integrals to find the general antiderivative (indefinite integral) of

$$1.6e^{-0.08x}$$

We begin by letting

$$u = -0.08x$$

Then, $$du = -0.08\,dx$$
So $$-20\,du = 1.6\,dx$$

Rewriting the integrand and substituting, we get
$$\int 1.6e^{-0.08x}\,dx = \int (e^{-0.08x})\,1.6\,dx$$
$$= \int (e^u)(-20)\,du$$
$$= \int -20e^u\,du$$

Thus, we have an indefinite integral of the form
$$\int g(u)\,du$$
where $$g(x) = -20e^x$$

Using the integral versions of the exponential rule and the constant multiple rule (Theorems 5.4 and 5.5), an antiderivative of $g(x)$ is
$$G(x) = -20e^x$$

So, using the chain rule for integrals,
$$\int -20e^u\,du = -20e^u + C$$
$$= -20e^{-0.08x} + C \quad \text{since } u = -0.08x$$

Therefore,
$$\int 1.6e^{-0.08x}\,dx = -20e^{-0.08x} + C$$

which tells us that
$$D(x) = -20e^{-0.08x} + C$$

for a certain value of C. Since $D(0) = 7$, the value of C must satisfy the equation
$$-20e^{-0.08(0)} + C = 7$$

Solving this equation for C, we obtain
$$C = 27$$

Therefore, $$D(x) = -20e^{-0.08x} + 27$$
$$= 27 - 20e^{-0.08x}$$

In particular,
$$D(3) = 27 - 20e^{-0.08(3)}$$
$$\approx 11$$

which tells us that about 1100 deaths will be attributable to the disease by the time 3 more years elapse. [Theorem 4.13 in Section 4.6 can help us visualize the graph of $D(x)$ for $x \geq 0$.]

We cannot find the indefinite integral of every function of the form

$$h(x)e^{u(x)}$$

using only the chain rule for integrals. For example, we cannot use that rule alone to find the indefinite integral of the function

$$10xe^{5x}$$

In Section 5.5, we will present a method that will help us find the indefinite integral of this function.

5.4 Problems

In Problems 1 to 44, find the indefinite integral. Then check your answer by showing that the derivative of the indefinite integral is the integrand.

1. $\int (x^3 + 7)^5 \, 36x^2 \, dx$
2. $\int (x^2 - 8)^3 \, 40x \, dx$
3. $\int \dfrac{4x^3 - 5}{\sqrt{x^4 - 5x}} \, dx$
4. $\int \dfrac{3x^2 + 2}{\sqrt{x^3 + 2x}} \, dx$
5. $\int (8x - 9)^4 \, dx$
6. $\int (5x + 4)^6 \, dx$
7. $\int (2x + 3) \sqrt{6x^2 + 18x} \, dx$
8. $\int (3x^2 + 4)\sqrt{5x^3 + 20x} \, dx$
9. $\int \dfrac{8x}{(6x^2 + 5)^2} \, dx$
10. $\int \dfrac{12x}{(8x^2 + 7)^2} \, dx$
11. $\int (x + 2)^7 \, 3x \, dx$
12. $\int (x - 3)^4 \, 6x \, dx$
13. $\int \dfrac{\ln 8x}{x} \, dx$
14. $\int \dfrac{\ln (x + 8)}{x + 8} \, dx$
15. $\int \dfrac{24x}{3x^2 + 7} \, dx$
16. $\int \dfrac{15x^2}{x^3 - 5} \, dx$
17. $\int \dfrac{8}{8x - 1} \, dx$
18. $\int \dfrac{6}{6x + 11} \, dx$
19. $\int \dfrac{4x + 5}{4x^2 + 10x - 7} \, dx$
20. $\int \dfrac{2x^2 + 3}{6x^3 + 27x + 4} \, dx$
21. $\int \dfrac{4x}{\sqrt{2x + 5}} \, dx$
22. $\int \dfrac{x^2}{x + 7} \, dx$
23. $\int \dfrac{4e^{4x}}{5 + e^{4x}} \, dx$
24. $\int \dfrac{e^x}{1 + e^x} \, dx$
25. $\int \dfrac{5}{x \ln x} \, dx$
26. $\int \dfrac{7}{x \ln 2x} \, dx$
27. $\int \dfrac{6x}{(x + 4)^2} \, dx$
28. $\int \dfrac{3x}{(x + 5)^2} \, dx$
29. $\int \dfrac{0.5}{\sqrt{x} + 3} \, dx$
30. $\int \dfrac{1}{\sqrt{x} - 0.5} \, dx$
31. $\int 3xe^{6x^2} \, dx$
32. $\int 12x^2 \, e^{8x^3} \, dx$
33. $\int 5e^{-0.2x} \, dx$
34. $\int 30e^{0.5x} \, dx$
35. $\int 0.06e^{0.06x} \, dx$
36. $\int -0.04e^{-0.04x} \, dx$
37. $\int e^{0.08x} \, dx$
38. $\int e^{0.05x} \, dx$
39. $\int (12x - 18)e^{x^2 - 3x} \, dx$
40. $\int (15x^2 + 20)e^{x^3 + 4x} \, dx$
41. $\int \dfrac{x}{x + 7} \, dx$
42. $\int \dfrac{6x}{x - 5} \, dx$
43. $\int \dfrac{(\ln x)^4}{x} \, dx$
44. $\int \dfrac{(\ln 3x)^5}{x} \, dx$

In Problems 45 to 50, compute the definite integral using an approach like in Example 5.18.

45. $\int_5^8 \dfrac{6}{\sqrt{3x + 1}} \, dx$
46. $\int_0^2 (x^2 + 3)^3 \, 8x \, dx$
47. $\int_3^7 \dfrac{2x}{x^2 + 1} \, dx$
48. $\int_1^9 \dfrac{1}{5x + 3} \, dx$
49. $\int_4^8 e^{0.25x} \, dx$
50. $\int_0^{25} 0.04e^{0.04x} \, dx$

In Problems 51 to 54, find the area of the region bounded by the graph of the function and the x axis between the given pair of x values.

51. $f(x) = \dfrac{60}{(5x+1)^2}$ $x = 0$ and $x = 1$

52. $f(x) = (x^2 - 12x)^2 (6x - 36)$ $x = 5$ and $x = 6$

53. $f(x) = (x^2 - 16)^3 \, 8x$ $x = 1$ and $x = 5$

54. $f(x) = (x^2 - 4)^5 \, 12x$ $x = 0$ and $x = 3$

In Problems 55 to 58, suppose $A(x)$ is a function whose derivative is the given function $f(x)$. Find the amount $A(x)$ changes if x changes as indicated.

55. $f(x) = (x^2 - 9x)^2 (6x - 27)$ x increases from 1 to 5

56. $f(x) = \sqrt{x^2 + 8x} \, (3x + 12)$ x increases from 0 to 1

57. $f(x) = \dfrac{3x^2}{x^3 + 6}$ x decreases from 4 to 2

58. $f(x) = \dfrac{-4x}{(x^2 - 8)^3}$ x decreases from 5 to 3

In Problems 59 to 62, for the function $f(x)$, find the particular antiderivative $F(x)$ that satisfies the given condition.

59. $f(x) = \dfrac{3}{3x-5}$ $F(2) = 9$

60. $f(x) = \dfrac{x^2}{\sqrt{x^3+3}}$ $F(1) = \dfrac{10}{3}$

61. $f(x) = \dfrac{1}{(4x+3)^2}$ $F(0) = \dfrac{11}{12}$

62. $f(x) = (x^3 + 2)^4 \, 15x^2$ $F(2) = 100{,}000$

63. **Acid Rain** Due to pollutants, the pH of rainwater is decreasing in a portion of southeast Asia. (Thus, the rainwater's acidity is increasing.) Officials predict that t years from now the pH of the rainwater will be changing at the rate of

$$-0.34e^{-0.2t}$$

units per year.

a. How much will the pH of the rainwater decrease during the next 5 years?

b. Presently, the pH of the rainwater is 5.5. Let $F(t)$ represent its pH t years from now. Find a formula for $F(t)$.
    Hint: $F'(t) = -0.34e^{-0.2t}$ and $F(0) = 5.5$. Thus, $F(t)$ is the particular antiderivative of $-0.34e^{-0.2t}$ that satisfies the condition $F(0) = 5.5$.

c. What will be the pH of the rainwater 2 years from now?

64. **Advertising** Let $D(x)$ represent the number of tons of fish a fishmonger sells per month if he spends $\$x$ thousand monthly on advertising. Suppose

$$D'(x) = \dfrac{162}{(2x+9)^2}$$

a. How much does $D(x)$ increase if the fishmonger increases the monthly advertising expenditure from $\$9000$ to $\$13{,}000$?

b. The fishmonger sells 28 tons of fish per month when he spends no money on advertising. Find a formula for $D(x)$.
    Hint: $D(x)$ is the particular antiderivative of $162/(2x+9)^2$ that satisfies the condition $D(0) = 28$.

c. How much fish does the fishmonger sell per month when he spends $\$8000$ monthly on advertising?

65. **Offshore Oil Leakage** An offshore oil well developed a leak. Suppose that t days after the leak developed the oil spills at the rate of

$$\dfrac{27}{3t+1}$$

thousand barrels per day.

a. How much oil spills during the 7-day period beginning 2 days after the leak developed?

b. Let $F(t)$ represent the amount of oil (in thousands of barrels) that spills during the first t days. Find a formula for $F(t)$.
    Hint: $F(0) = 0$.

66. **Production Cost** Suppose the marginal cost for the production of a commodity is

$$\dfrac{40}{\sqrt{2x+25}}$$

when the daily production level is x tons. (The total daily production cost is measured in thousands of dollars.)

a. How much does the total daily production cost increase if the daily production level is increased from 19.5 tons to 28 tons?

b. Let $C(x)$ represent the total daily cost of producing x tons of the commodity per day. Suppose $C(12) = 108$. Find a formula for $C(x)$.
   Hint: $C(x)$ is an antiderivative of $40/\sqrt{2x + 25}$.

67. **Productivity of Robotization** Let $Q(x)$ represent the number of Teddy Bears (in thousands) Santa Toy Company can produce monthly if it spends $\$x$ million to robotize. Suppose the marginal productivity of this capital is

$$8e^{(-1/3)x}$$

if the company spends $\$x$ million to robotize.

a. How much will $Q(x)$ increase if the company increases the expenditure to robotize from $\$3$ million to $\$6$ million?

b. The company can produce 26,000 Teddy Bears per month without robotizing. Find a formula for $Q(x)$. Then find how many Teddy Bears the company can produce monthly if it spends $\$9$ million to robotize.

68. **Housing Cost** In Pardonia, realtors project that t years from now the average monthly housing cost will be increasing at the rate of

$$56e^{-0.16t}$$

dollars per year.

a. How much will the average monthly housing cost increase during the next 4 years?

b. Presently, the average monthly housing cost is $\$370$. Find a formula for the average monthly housing cost t years from now. Then find what the average monthly housing cost will be 6 years from now.

69. **Depreciation** Presently, a machine is depreciating at the rate of $\$5000$ per year. Suppose that from now on the yearly depreciation rate decreases exponentially at 4 percent per year. How much will the machine depreciate during the next 25 years?

70. **Temperature Change** A cold object is placed in a warm environment. Presently, the object's temperature is rising at the rate of 0.5 of a degree per minute. Suppose that from now on its temperature rises at a rate that decreases exponentially at 2 percent per minute. How much warmer will the object be 100 minutes from now?

Graphing Calculator Problems

71. **Intravenous Feeding** A nurse has been feeding glucose intravenously to a patient since noon. The following table shows the rate at which the patient's glucose level was increasing at nine different times:

Time	Noon	1:00 p.m.	2:00 p.m.	3:00 p.m.	4:00 p.m.	5:00 p.m.	6:00 p.m.	7:00 p.m.	8:00 p.m.
Rate, grams per hour	1.2	0.8	0.7	0.5	0.3	0.3	0.2	0.1	0.1

Let $G(t)$ represent the glucose level after t hours of intravenous feeding.

a. Set the viewing screen on your graphing calculator to $[0, 9] \times [-4, 2]$ with a scale of 1 on both axes. Then plot the data points of the form

$$(t, \ln G'(t))$$

for $t = 0, 1, 2, 3, 4, 5, 6, 7,$ and 8.

b. Draw the regression line for the data points you plotted in part (a).

c. Find a formula for the least-squares linear function that corresponds to the regression line you drew in part (b).

d. The formula you found in part (c) approximates the functional relationship between $\ln G'(t)$ and t. Use this fact and Theorem 4.7 in Section 4.3 to find a formula that approximates the functional relationship between $G'(t)$ and t.

e. At noon, the patient's bloodstream contained 3 grams of glucose. Use this fact and the formula you found in part (d) to derive a function that approximates the function $G(t)$.

f. Use the function you derived in part (e) to estimate what the patient's glucose level will be at 11:30 p.m. if the nurse continues the intravenous feeding. How fast will the glucose level be increasing at that time?

72. **Rain Forest Shrinkage** A country's rain forest is shrinking. The following table shows the rate at which it was shrinking at seven different times:

Date, Jan. 1	1988	1989	1990	1991	1992	1993	1994
Shrinkage rate, millions of acres per year	10	8.2	6.7	5.5	4.5	3.7	3.1

Let $F(t)$ represent the size of the country's rain forest (in millions of acres) t years after January 1, 1988.

a. Set the viewing screen on your graphing calculator to $[0, 6] \times [0, 4]$ with a scale of 1 on both axes. Then plot the data points of the form

$$(t, \ln|F'(t)|)$$

for $t = 0, 1, 2, 3, 4, 5,$ and 6.

[Note: Since the rain forest is shrinking, $F'(t)$ is negative.]

b. Draw the regression line for the data points you plotted in part (a).

c. Find a formula for the least-squares linear function that corresponds to the regression line you drew in part (b).

d. The formula you found in part (c) approximates the functional relationship between $\ln|F'(t)|$ and t. Use this fact and Theorem 4.7 in Section 4.3 to find a formula that approximates the functional relationship between $F'(t)$ and t. [Keep in mind that $F'(t) = -|F'(t)|$.]

e. On January 1, 1988, the size of the country's rain forest was 90 million acres. Use this fact and the formula you found in part (d) to derive a function that approximates the function $F(t)$.

f. Use the function you derived in part (e) to estimate what the size of the rain forest will be on January 1, 2000. How rapidly will the rain forest be shrinking at that time?

5.5 Integration by Parts

In Section 5.4, we said that we cannot find the antiderivatives of the function

$$10xe^{5x}$$

using only the chain rule for integrals. The following integral rule will help us find the antiderivatives of this function. Whenever we apply the rule, we think of the integrand as being the product of two functions.

Theorem 5.14 (Integration by Parts Rule)

Suppose $G(x)$ is an antiderivative of $g(x)$. Then

$$\int f(x)g(x)\, dx = f(x)G(x) - \int G(x)f'(x)\, dx$$

The above rule is useful whenever we can find

$$\int G(x)f'(x)\, dx$$

more easily than we can find
$$\int f(x)g(x)\,dx$$

Proof of Theorem 5.14

We only need to show that the derivative of
$$f(x)G(x) - \int G(x)f'(x)\,dx \tag{1}$$
is $f(x)g(x)$. Since the derivative of a difference is the difference of the derivatives, we can express the derivative of formula (1) as follows:
$$D_x[f(x)G(x)] - D_x\left[\int G(x)f'(x)\,dx\right] \tag{2}$$
Now, according to the product rule,
$$\begin{aligned}D_x[f(x)G(x)] &= f(x)\,G'(x) + G(x)f'(x)\\ &= f(x)g(x) + G(x)f'(x) \quad \text{since } G'(x) = g(x)\end{aligned}$$
and according to the definition of indefinite integrals,
$$D_x\left[\int G(x)f'(x)\,dx\right] = G(x)f'(x)$$
Therefore, formula (2) is equal to
$$f(x)g(x) + G(x)f'(x) - G(x)f'(x)$$
which simplifies to
$$f(x)g(x)$$
Therefore, the derivative of formula (1) is indeed $f(x)g(x)$, which completes our proof.

In applying the integration by parts rule, it does not matter which antiderivative of $g(x)$ we use. Therefore, for the sake of simplicity, we will always use the antiderivative of $g(x)$ whose constant of integration is zero.

If we let $u = f(x)$, $dv = g(x)\,dx$, $v = G(x)$, and $du = f'(x)\,dx$, we can express the integration by parts rule as follows:
$$\int u\,dv = uv - \int v\,du$$
Some books use this version of the rule.

In Example 5.20, we use the integration by parts rule to find the antiderivatives of $10xe^{5x}$.

Example 5.20

Find
$$\int 10xe^{5x}\,dx$$

Solution

Let $f(x) = 10x$ and $g(x) = e^{5x}$

Then $f'(x) = 10$ and $G(x) = \frac{1}{5}e^{5x}$ chain rule for integrals

According to the integration by parts rule,

$$\int 10xe^{5x}\,dx = (10x)\tfrac{1}{5}e^{5x} - \int \tfrac{1}{5}e^{5x}(10)\,dx$$

$$= 2xe^{5x} - \int 2e^{5x}\,dx$$

$$= 2xe^{5x} - \tfrac{2}{5}e^{5x} + C \quad \text{chain rule for integrals}$$

Choosing $f(x)$ and $g(x)$

In Example 5.20, we chose $10x$ for $f(x)$. If we had instead chosen e^{5x} for $f(x)$, then $G(x)f'(x)$ would have been equal to $25x^2e^{5x}$, which is more complicated than $10xe^{5x}$. In general, whenever we use the integration by parts rule, we choose formulas for $f(x)$ and $g(x)$ hoping we can find $\int G(x)f'(x)\,dx$ more easily than we can find $\int f(x)g(x)\,dx$. One approach is to choose $f(x)$ and $g(x)$ so that $f'(x)$ is less complicated than $f(x)$. Of course, in making our choice, we also have to keep in mind that $G(x)$, an antiderivative of $g(x)$, must be obtainable.

In Example 5.21, we use the integration by parts rule to find the indefinite integral of a function that has $\ln x$ as a factor.

Example 5.21
Find
$$\int 5x^4 \ln x\,dx$$

Solution

If we let $f(x) = 5x^4$ and $g(x) = \ln x$, we will have difficulty finding $G(x)$. (Can you think of a function whose derivative is $\ln x$?) Therefore, we let

$$f(x) = \ln x \quad \text{and} \quad g(x) = 5x^4$$

Then $\quad f'(x) = \dfrac{1}{x} \quad$ and $\quad G(x) = x^5$

Using the integration by parts rule, we get

$$\int 5x^4 \ln x\,dx = (\ln x)x^5 - \int x^5 \left(\frac{1}{x}\right) dx$$

$$= x^5 \ln x - \int x^4\,dx$$

$$= x^5 \ln x - \tfrac{1}{5}x^5 + C$$

We can find the indefinite integral in Example 5.22 using the substitution method we introduced in Section 5.4. Nevertheless, we will employ the integration by parts rule.

Example 5.22

Find

$$\int (x^2 + 7)^9 \, 2x^3 \, dx$$

Solution

If we let $f(x) = (x^2 + 7)^9$ and $g(x) = 2x^3$, then $G(x)f'(x) = (x^2 + 7)^8 \, 9x^5$, which is more complicated than $(x^2 + 7)^9 \, 2x^3$. If we let $f(x) = 2x^3$ and $g(x) = (x^2 + 7)^9$, finding $G(x)$ will be difficult.

Workable results occur if we think of $(x^2 + 7)^9 \, 2x^3$ as the product of x^2 and $(x^2 + 7)^9 \, 2x$ and let

$$f(x) = x^2 \quad \text{and} \quad g(x) = (x^2 + 7)^9 \, 2x$$

Then $f'(x) = 2x$ and $G(x) = \frac{1}{10}(x^2 + 7)^{10}$ chain rule for integrals

Applying the integration by parts rule, we obtain

$$\int (x^2 + 7)^9 \, 2x^3 \, dx = \int x^2 (x^2 + 7)^9 \, 2x \, dx$$

$$= (x^2) \frac{1}{10}(x^2 + 7)^{10} - \int \frac{1}{10}(x^2 + 7)^{10} \, 2x \, dx$$

$$= \frac{1}{10} x^2 (x^2 + 7)^{10} - \frac{1}{110}(x^2 + 7)^{11} + C \quad \text{chain rule for integrals}$$

As we said, the indefinite integral in Example 5.22 can be found using the substitution method. We can begin this approach by letting

$$u = x^2 + 7$$

Eventually, we would use the chain rule for integrals to conclude that

$$\int (x^2 + 7)^9 \, 2x^3 \, dx = \frac{1}{11}(x^2 + 7)^{11} - \frac{7}{10}(x^2 + 7)^{10} + C$$

It appears that this result is not equivalent to the result in Example 5.22. But the two results are equivalent. (How could we show that they are equivalent?)

In Example 5.23, we find the antiderivatives of the natural logarithmic function.

Example 5.23

Find

$$\int \ln x \, dx$$

Solution

If we think of ln x as 1(ln x), we can let

$$f(x) = \ln x \quad \text{and} \quad g(x) = 1$$

Then

$$f'(x) = \frac{1}{x} \quad \text{and} \quad G(x) = x$$

According to the integration by parts rule,

$$\int \ln x \, dx = (\ln x)x - \int x\left(\frac{1}{x}\right) dx$$
$$= x \ln x - \int 1 \, dx$$
$$= x \ln x - x + C$$

Multiple Applications of the Integration by Parts Rule

To find the indefinite integral in Example 5.24, we apply the integration by parts rule twice.

Example 5.24

Find

$$\int x^2 e^x \, dx$$

Solution

Let

$$f(x) = x^2 \quad \text{and} \quad g(x) = e^x$$

Then

$$f'(x) = 2x \quad \text{and} \quad G(x) = e^x$$

Using the integration by parts rule, we get

$$\int x^2 e^x \, dx = x^2 e^x - \int e^x (2x) \, dx$$
$$= x^2 e^x - \int 2x e^x \, dx \qquad (3)$$

Although $2xe^x$ is a simpler function than $x^2 e^x$, we need to apply the integration by parts rule again. This time we apply the rule to

$$\int 2x e^x \, dx$$

Let

$$f(x) = 2x \quad \text{and} \quad g(x) = e^x$$

Then

$$f'(x) = 2 \quad \text{and} \quad G(x) = e^x$$

Using the integration by parts rule, we obtain

$$\int 2xe^x \, dx = 2xe^x - \int e^x (2) \, dx$$

$$= 2xe^x - \int 2e^x \, dx$$
$$= 2xe^x - 2e^x + C \qquad (4)$$

If we substitute formula (4) for $\int 2xe^x \, dx$ into formula (3), we get

$$\int x^2 e^x \, dx = x^2 e^x - (2xe^x - 2e^x + C)$$
$$= x^2 e^x - 2xe^x + 2e^x + C$$

In Example 5.25, we use the integration by parts rule to help compute a definite integral.

Example 5.25

Let $P(x)$ represent the quantity of bottles (measured in tons) that a container factory can produce daily with a capital outlay of $\$x$ million. Suppose the marginal productivity of capital is

$$\frac{20x}{(x+2)^3}$$

How much will $P(x)$ increase if the capital outlay is increased from $3 million to $6 million?

Solution

Since $20x/(x+2)^3$ is the derivative of $P(x)$, the increase in $P(x)$ is the value of the definite integral

$$\int_3^6 \frac{20x}{(x+2)^3} \, dx$$

To compute this definite integral, we need an antiderivative of $20x/(x+2)^3$. Thus, we want to find the indefinite integral

$$\int \frac{20x}{(x+2)^3} \, dx$$

We view $20x/(x+2)^3$ as the product of $20x$ and $(x+2)^{-3}$, and let

$$f(x) = 20x \quad \text{and} \quad g(x) = (x+2)^{-3}$$

Then $f'(x) = 20$ and $G(x) = -\tfrac{1}{2}(x+2)^{-2}$ chain rule for integrals

Using the integration by parts rule, we get

$$\int \frac{20x}{(x+2)^3} \, dx = 20x(-\tfrac{1}{2})(x+2)^{-2} - \int -\tfrac{1}{2}(x+2)^{-2} \cdot 20 \, dx$$

$$= \frac{-10x}{(x+2)^2} - \int (x+2)^{-2}(-10) \, dx$$

$$= \frac{-10x}{(x+2)^2} - 10(x+2)^{-1} + C \qquad \text{chain rule for integrals}$$

$$= \frac{-(20x+20)}{(x+2)^2} + C$$

Therefore

$$\frac{-(20x+20)}{(x+2)^2}$$

is an antiderivative of $20x/(x+2)^3$, and so

$$\int_3^6 \frac{20x}{(x+2)^3} dx = \left. \frac{-(20x+20)}{(x+2)^2} \right|_3^6$$
$$= -2.1875 - (-3.2)$$
$$= 1.0125$$

Thus, $P(x)$ will increase 1.0125 tons if the capital outlay is increased from $3 million to $6 million.

As in Example 5.22, we can alternatively find the indefinite integral in Example 5.25 using the substitution method we introduced in Section 5.4.

5.5 Problems

In Problems 1 to 16, use the integration by parts rule to find the indefinite integral. Then check your answer by showing that the derivative of the indefinite integral is the integrand.

1. $\int (x+3)^4 x \, dx$
2. $\int \frac{x}{\sqrt{x+5}} dx$
3. $\int 8x^3 \ln x \, dx$
4. $\int 2x \ln 5x \, dx$
5. $\int 3x^5 \sqrt{x^3 - 1} \, dx$
6. $\int (2x^5 + 3)^7 \, 10x^9 \, dx$
7. $\int \frac{4x}{(x+6)^2} dx$
8. $\int \frac{-5x}{(5x+8)^2} dx$
9. $\int \frac{1}{3} x^3 e^x \, dx$
10. $\int \frac{1}{12} x^4 e^x \, dx$
11. $\int 5x e^{0.02x} \, dx$
12. $\int x e^{-0.04x} \, dx$
13. $\int \frac{14x}{(7x+4)^3} dx$
14. $\int \frac{-24x^5}{(x^3+1)^9} dx$
15. $\int \ln(2x+1) \, dx$
16. $\int \ln(5x-20) \, dx$

In Problems 17 to 22, compute the definite integral.

17. $\int_0^2 \frac{12x}{\sqrt{6x+4}} dx$
18. $\int_4^8 18x \sqrt{6x+1} \, dx$
19. $\int_1^8 \ln x \, dx$
20. $\int_0^{100} x e^{-0.01x} \, dx$
21. $\int_1^4 (x-3)^5 \, 42x \, dx$
22. $\int_0^3 (x-2)^3 \, 20x \, dx$

23. Find the area of the region bounded by the graph of
$$h(x) = (x-4)^5 \, 42x$$
and the x axis between each of the following pairs of x values: (a) $x = 2$ and $x = 4$; (b) $x = 5$ and $x = 6$; (c) $x = 3$ and $x = 6$; (d) $x = 0$ and $x = 5$.

24. Find the area of the region bounded by the graph of
$$h(x) = (x^2 - 9)^3 \, 40x^3$$
and the x axis between each of the following pairs of x values: (a) $x = 3$ and $x = 4$; (b) $x = 2$ and $x = 3$; (c) $x = 0$ and $x = 4$; (d) $x = 2$ and $x = 4$.

In Problems 25 and 26, suppose $A(x)$ is a function whose derivative is the given function $f(x)$. Find the amount $A(x)$ changes if x increases as indicated.

25. $f(x) = \dfrac{-2x}{(x-5)^3}$ x increases from 6 to 8

26. $f(x) = \dfrac{135x}{\sqrt{3x+1}}$ x increases from 1 to 8

In Problems 27 to 30, for the function $f(x)$, find the particular antiderivative $F(x)$ that satisfies the given condition.

27. $f(x) = \dfrac{-3x}{(3x-5)^2}$ $F(2) = 7$

28. $f(x) = \dfrac{4x}{(2x+1)^3}$ $F(2) = \dfrac{41}{50}$

29. $f(x) = xe^{0.5x}$ $F(0) = 2$

30. $f(x) = \ln x$ $F(1) = 3$

31. **Productivity of Labor** Insulex is a product used for insulating. Let $Q(x)$ represent the number of tons of Insulex a firm can produce daily using x units of labor. Suppose the marginal productivity of labor is

$$\dfrac{16x}{(x+1)^3}$$

when x units of labor is used.

a. How much will $Q(x)$ increase if the firm increases the labor force from 7 units to 11 units?

b. The firm can produce 6.48 tons of the product daily if it uses 9 units of labor. Find a formula for $Q(x)$.

Hint: $Q(x)$ is the particular antiderivative of $16x/(x+1)^3$ that satisfies the condition $Q(9) = 6.48$.

32. **Population** A demographer projects that the population of a community will increase during the next few years. Suppose that t years from now the population will be increasing at the rate of

$$\dfrac{3t}{\sqrt{t+1}}$$

thousand people per year.

a. How much will the population increase during the next 8 years?

b. Let $P(t)$ represent the population (in thousands) t years from now. Presently, the population is 85,000. Find a formula for $P(t)$.

Hint: $P(t)$ is the particular antiderivative of $3t/\sqrt{t+1}$ that satisfies the condition $P(0) = 85$.

5.6 Riemann Sums

Suppose $f(x)$ is a function that is continuous on an interval, and a and b are numbers in the interval such that $a < b$. According to our definition of definite integrals (Definition 5.3 in Section 5.2),

$$\int_a^b f(x)\, dx = F(b) - F(a)$$

where $F(x)$ is an antiderivative of $f(x)$.

In Section 5.3, we saw that definite integrals can be characterized in terms of areas. Let R represent the region between the graph of $f(x)$ and the x axis from $x = a$ to $x = b$. According to Theorem 5.12 in Section 5.3,

$$\int_a^b f(x)\, dx$$

is the total area of the portions of the region R above the x axis *minus* the total area of the portions of the region R below the x axis.

In this section, we characterize definite integrals in terms of limits. We begin by supposing the graph of $f(x)$ is nowhere below the x axis on the

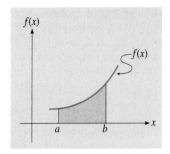

Figure 5.10 The region R.

interval [a, b]. In fact, we begin by supposing f(x), a, and b are as shown in Figure 5.10. Then the region R is the shaded region in the figure. In this case, $\int_a^b f(x)\,dx$ is the area of the region R. (See Theorem 5.10 in Section 5.3.)

The length of the interval [a, b] is b − a. Therefore, if we divide this interval into three subintervals of equal lengths, the length of each subinterval is

$$\Delta x = \frac{b-a}{3}$$

(Δx is read "delta x.") Moreover, the right endpoints of these three subintervals are

$$a + \Delta x \qquad a + 2\Delta x \qquad a + 3\Delta x$$

(Note that $a + 3\Delta x = b$.) Figure 5.11 shows these three points.

Figure 5.11 also shows three rectangles, each with a base of length Δx. The height of the shortest rectangle is $f(a + \Delta x)$, the height of the next to the shortest rectangle is $f(a + 2\Delta x)$, and the height of the tallest rectangle is $f(a + 3\Delta x)$.

Summation Notation

Since the area of a rectangle is the product of its height and its base, the sum of the areas of the three rectangles in Figure 5.11 is

$$f(a + \Delta x)\,\Delta x + f(a + 2\Delta x)\,\Delta x + f(a + 3\Delta x)\,\Delta x$$

If we substitute the positive integers from 1 to 3 for k in the expression

$$f(a + k\,\Delta x)\,\Delta x$$

we generate the areas of the three rectangles. We can express the sum of these three areas by writing

$$\sum_{k=1}^{3} f(a + k\,\Delta x)\,\Delta x$$

[Remember that $\Delta x = (b-a)/3$.]

We can think of

$$f(a + k\,\Delta x)\,\Delta x$$

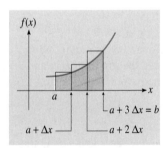

Figure 5.11 Three rectangles superimposed over the region R.

as a function whose independent variable is k and whose domain is the first three positive integers, namely, 1, 2, and 3. With this function, we illustrated Definition 5.4.

Definition 5.4

Suppose $u(k)$ is a function whose domain is the first n positive integers. The sum

$$u(1) + u(2) + u(3) + \cdots + u(n)$$

is expressed by writing

$$\sum_{k=1}^{n} u(k)$$

which is read "the summation of the $u(k)$'s as k varies from 1 to n."

In our area example,

$$u(k) = f(a + k\,\Delta x)\,\Delta x$$

and $n = 3$. To further illustrate the above definition, if

$$u(k) = k^2 + 1$$

and $n = 6$, then

$$\sum_{k=1}^{6} u(k) = \sum_{k=1}^{6} (k^2 + 1)$$
$$= 2 + 5 + 10 + 17 + 26 + 37$$
$$= 97$$

Figure 5.11 suggests that we can think of the sum

$$\sum_{k=1}^{3} f(a + k\,\Delta x)\,\Delta x$$

as an approximation for the area of region R (the shaded region). A better approximation for the area of region R is the sum of the areas of the five rectangles in Figure 5.12. In this case, the interval $[a, b]$ is divided into five subintervals, each of length

$$\Delta x = \frac{b - a}{5}$$

Thus, the sum of the areas of the five rectangles is

$$\sum_{k=1}^{5} f(a + k\,\Delta x)\,\Delta x$$

In more general terms, if n is a positive integer, we can divide the interval $[a, b]$ into n subintervals, each of length

$$\Delta x = \frac{b - a}{n}$$

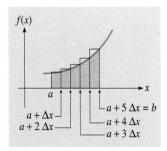

Figure 5.12 Five rectangles superimposed over the region R.

and form the sum

$$\sum_{k=1}^{n} f(a + k\,\Delta x)\,\Delta x \tag{5}$$

By substituting 3 for n in formula (5), we get the sum of the areas of the three rectangles in Figure 5.11. By substituting 5 for n in formula (5), we get the sum of the areas of the five rectangles in Figure 5.12.

Definite Integrals Characterized in Terms of Limits

Figures 5.11 and 5.12 suggest that, as n increases, the values of

$$\sum_{k=1}^{n} f(a + k\,\Delta x)\,\Delta x$$

get closer and closer to the area of region R. Since we are supposing the graph of $f(x)$ is nowhere below the x axis on the interval $[a, b]$, the area of region R is

$$\int_{a}^{b} f(x)\,dx$$

Consequently, Figures 5.11 and 5.12 suggest that, as n increases, the values of

$$\sum_{k=1}^{n} f(a + k\,\Delta x)\,\Delta x$$

get closer and closer to the value of $\int_{a}^{b} f(x)\,dx$. This conclusion is true, even if the graph of $f(x)$ dips below the x axis on the interval $[a, b]$. In fact, as n increases without bound, the values of

$$\sum_{k=1}^{n} f(a + k\,\Delta x)\,\Delta x$$

eventually get and stay arbitrarily close to the value of $\int_{a}^{b} f(x)\,dx$. Thus, $\int_{a}^{b} f(x)\,dx$ is the *limit* of

$$\sum_{k=1}^{n} f(a + k\,\Delta x)\,\Delta x$$

as n approaches positive infinity. We express this result by writing

$$\int_{a}^{b} f(x)\,dx = \lim_{n \to +\infty} \sum_{k=1}^{n} f(a + k\,\Delta x)\,\Delta x$$

Therefore, we have Theorem 5.15.

Theorem 5.15

Suppose $f(x)$ is a function that is continuous on an interval, and a and b are numbers in the interval such that $a < b$. Then

$$\int_a^b f(x)\, dx = \lim_{n \to +\infty} \sum_{k=1}^n f(a + k\,\Delta x)\,\Delta x$$

where $\Delta x = (b - a)/n$.

Riemann Sums

Theorem 5.15 tells us that a definite integral can be characterized as the limit of a sequence of sums. Such sums are called Riemann sums, in honor of the renowned mathematician Georg Friedrich Riemann (1826–1866).

Definition 5.5

Suppose n is a positive integer and $a < b$. The nth **Riemann sum** of $f(x)$ on the interval $[a, b]$ is

$$\sum_{k=1}^n f(a + k\,\Delta x)\,\Delta x$$

where $\Delta x = (b - a)/n$.

Actually, if x_k is *any* number such that

$$a + (k - 1)\,\Delta x \leq x_k \leq a + k\,\Delta x$$

then

$$\sum_{k=1}^n f(x_k)\,\Delta x$$

is also an nth Riemann sum of $f(x)$ on the interval $[a, b]$. Evidently, we are restricting ourselves to those Riemann sums for which

$$x_k = a + k\,\Delta x$$

Therefore, using our approach, there is no ambiguity in referring to "the nth Riemann sum" of a function on an interval.

Example 5.26

Using the symbol $\sum$, write the 15th Riemann sum of

$$f(x) = x^2 - 8x$$

on the interval $[2, 6]$.

Solution

In this case, $a = 2$, $b = 6$, and $n = 15$. So

$$\Delta x = \frac{6 - 2}{15}$$

$$= \frac{4}{15}$$

Thus, the Riemann sum in question is

$$\sum_{k=1}^{15} [(2 + k(\tfrac{4}{15}))^2 - 8(2 + k(\tfrac{4}{15}))]\tfrac{4}{15}$$

Using Definite Integrals to Approximate Riemann Sums

We can use Riemann sums to approximate definite integrals. For example, we can think of the Riemann sum in Example 5.26 as an approximation for

$$\int_2^6 (x^2 - 8x)\, dx$$

However, in this section, we will use definite integrals to approximate Riemann sums.

Example 5.27

Use a definite integral to approximate

$$\sum_{k=1}^{17} \left(6\sqrt{4 + k(\tfrac{5}{17})} + 2\right)\tfrac{5}{17}$$

Solution

The above sum has the form

$$\sum_{k=1}^{n} f(a + k\,\Delta x)\,\Delta x$$

with $f(x) = 6\sqrt{x} + 2$, $a = 4$, $n = 17$, and $\Delta x = \tfrac{5}{17}$. So, according to Definition 5.5, the sum is the 17th Riemann sum of the function $f(x) = 6\sqrt{x} + 2$ on the interval $[4, 9]$. Therefore, we can approximate the sum with the definite integral

$$\int_4^9 (6\sqrt{x} + 2)\, dx$$

Since
$$\int_4^9 (6\sqrt{x} + 2)\, dx = (4(\sqrt{x})^3 + 2x)\Big|_4^9$$
$$= 86$$

the value of the sum is approximately 86.

Example 5.28

Use a definite integral to approximate
$$\sum_{k=1}^{30} 4(k(\tfrac{7}{30}))^3 \tfrac{7}{30}$$

Solution

Note that
$$\sum_{k=1}^{30} 4(k(\tfrac{7}{30}))^3 \tfrac{7}{30} = \sum_{k=1}^{30} 4(0 + k(\tfrac{7}{30}))^3 \tfrac{7}{30}$$

Thus, we recognize the sum as the 30th Riemann sum of the function $4x^3$ on the interval $[0, 7]$. So we can approximate the sum with the definite integral
$$\int_0^7 4x^3\, dx$$

Since
$$\int_0^7 4x^3\, dx = x^4\Big|_0^7 = 2401$$

the value of the sum is approximately 2401.

Riemann sums arise in many applications. Example 5.29 involves such an application.

Example 5.29

A pharmaceutical firm plans to use a rare alga in the production of one of its drugs. When placed in a tank containing a special solution, the alga multiplies exponentially at the rate of 4 percent per week. How many kilograms of the alga must be placed in the tank *now* so that during the next 5 weeks 20 kilograms can be harvested at the end of each day?

Solution

Suppose P kilograms of the alga is placed in the tank now. Then
$$A = Pe^{0.04t}$$

kilograms of the alga will be in the tank t weeks from now. Solving for P, we get

$$P = Ae^{-0.04t} \tag{6}$$

This formula tells us how much of the alga must be placed in the tank now so that t weeks from now there will be A kilograms of the alga in the tank.

Since the kth harvest will be made k days from now and each day is $\frac{1}{7}$ of a week, the kth harvest will be made $k/7$ weeks from now. Therefore, according to formula (6),

$$20e^{-0.04(k/7)}$$

kilograms of the alga must be placed in the tank now so that 20 kilograms will be available for the kth harvest. During the 5-week period there will be 35 harvests. Consequently,

$$\sum_{k=1}^{35} 20e^{-0.04(k/7)}$$

kilograms of the alga must be placed in the tank now so that during the next 5 weeks 20 kilograms can be harvested at the end of each day. Since it is laborious to compute the above sum, we settle for an approximation.

Although it may not be apparent, the above sum is a Riemann sum. In fact, since the alga will be harvested at the end of each day during the next 5 weeks, the above sum is the 35th Riemann sum of a certain function on the interval [0, 5]. Thus, since

$$20 = 140\left(\tfrac{5}{35}\right)$$

and
$$\frac{k}{7} = 0 + k\left(\tfrac{5}{35}\right)$$

we are prompted to rewrite the above sum as follows:

$$\sum_{k=1}^{35} (140e^{-0.04(0 + k5/35)}) \tfrac{5}{35}$$

Now we can see that the sum is indeed a Riemann sum. It is the 35th Riemann sum of the function

$$140e^{-0.04t}$$

on the interval [0, 5]. Therefore, we can approximate the sum with the definite integral

$$\int_0^5 140e^{-0.04t}\, dt$$

Since

$$\int_0^5 140e^{-0.04t}\, dt = -3500e^{-0.04t}\Big|_0^5$$
$$\approx 634$$

approximately 634 kilograms of the alga must be placed in the tank now so that during the next 5 weeks 20 kilograms can be harvested at the end of each day.

The Present Value of an Annuity

An **annuity** is a sequence of payments made regularly during a specified time period. The **present value of an annuity** is the amount that must be invested at the beginning of the specified time period to generate the annuity.

Suppose an annuity consists of payments of A dollars made n times per year during a period of T years. Using an approach similar to the approach we used in Example 5.29, we can show that the present value of the annuity is a Riemann sum. To be more specific, if money draws interest at the annual rate of $100r$ percent compounded continuously, the present value of the annuity is the (Tn)th Riemann sum of the function

$$nAe^{-rt}$$

on the interval $[0, T]$. So we can approximate the present value of the annuity with the definite integral

$$\int_0^T nAe^{-rt}\, dt$$

Our conclusion is partly based on the assumption that the first payment is made $1/n$ year after the present value of the annuity is invested.

In Example 5.30, we use a definite integral to approximate the present value of an annuity.

Example 5.30

An account earns interest at the annual rate of 8 percent compounded continuously. Use a definite integral to approximate the amount that must be deposited into the account now so that during the next 4 years $500 can be withdrawn at the end of each month.

Solution

The amount that must be deposited now is the present value of an annuity consisting of $500 payments made 12 times per year during a 4-year period. Since the account earns interest at the annual rate of 8 percent and since there are 48 months in 4 years, the present value of this annuity is the 48th Riemann sum of the function

$$6000e^{-0.08t}$$

on the interval [0, 4]. [*Note*: 6000 = (12)(500).] We can approximate this Riemann sum with the definite integral

$$\int_0^4 6000e^{-0.08t}\,dt$$

Therefore, since

$$\int_0^4 6000e^{-0.08t}\,dt = -75{,}000e^{-0.08t}\Big|_0^4$$
$$\approx 20{,}538.82$$

approximately \$20,538.82 must be deposited now to generate the sequence of payments in question.

The Future Value of an Annuity

As we said, an annuity is a sequence of payments made regularly during a specified time period. The **future value of an annuity** (sometimes called the **amount of an annuity**) is the sum of the payments in the annuity plus the interest the payments earn by the end of the specified time period.

In Example 5.31, we use a definite integral to approximate the future value of an annuity. The annuity consists of \$250 payments made 12 times per year during a 3-year period.

Example 5.31

Beginning 1 month from now, each month \$250 will be deposited into an account where the interest is compounded continuously at the annual rate of 9 percent. Use a definite integral to approximate the amount of money in the account immediately after the 36th deposit.

Solution

First note that the 36th deposit will be made 3 years from now. Therefore, since the kth deposit will be made $k/12$ years from now, the kth deposit will draw interest for $3 - k/12$ years. So the kth deposit will be worth

$$250e^{0.09(3 - k/12)}$$

dollars 3 years from now. Since 36 deposits will be made, there will be

$$\sum_{k=1}^{36} 250e^{0.09(3 - k/12)}$$

dollars in the account immediately after the 36th deposit is made.

Although it may not be apparent, the above sum is a Riemann sum. In fact, since the deposits will be made at the end of each month during the next 3 years, the above sum is the 36th Riemann sum of a certain function on the interval [0, 3]. Therefore, since

$$250 = 3000(\tfrac{3}{36})$$

and

$$\frac{k}{12} = 0 + k(\tfrac{3}{36})$$

we are prompted to rewrite the above sum as follows:

$$\sum_{k=1}^{36} (3000e^{0.09(3 - [0 + k3/36])}) \tfrac{3}{36}$$

Now we can see that the sum is indeed a Riemann sum. It is the 36th Riemann sum of the function

$$3000e^{0.09(3 - t)}$$

on the interval [0, 3]. So we can approximate the sum with the definite integral

$$\int_0^3 3000e^{0.09(3 - t)}\, dt$$

Since

$$\int_0^3 3000e^{0.09(3 - t)}\, dt = \frac{-100{,}000}{3} e^{0.09(3 - t)} \Big|_0^3$$
$$\approx 10{,}332.15$$

there will be approximately \$10,332.15 in the account immediately after the 36th deposit.

With Example 5.31 we illustrated that the future value of an annuity is a Riemann sum. To be more specific, suppose an annuity consists of payments of A dollars made n times per year during a period of T years. If money earns interest at the annual rate of $100r$ percent compounded continuously, the future value of the annuity is the (Tn)th Riemann sum of the function

$$nAe^{r(T - t)}$$

on the interval $[0, T]$. Thus, we can approximate the future value of the annuity with the definite integral

$$\int_0^T nAe^{r(T - t)}\, dt$$

Our conclusion is partly based on the assumption that the first payment is made $1/n$ year after the T-year period begins.

5.6 Problems

In Problems 1 to 4, find the sum.

1. $\sum_{k=1}^{6} (3k - 2)$
2. $\sum_{k=1}^{4} \frac{1}{k}$
3. $\sum_{k=1}^{5} \frac{12}{k}$
4. $\sum_{k=1}^{7} (2k + 5)$

In Problems 5 to 12, use a definite integral to approximate the Riemann sum.

5. $\sum_{k=1}^{67} [4(3 + k(\frac{2}{67}))^3 + 1] \frac{2}{67}$

6. $\sum_{k=1}^{16} \left(\sqrt{1 + k(\frac{3}{16})}\right) \frac{3}{16}$

7. $\sum_{k=1}^{50} \left(\sqrt{k(\frac{9}{50})}\right) \frac{9}{50}$

8. $\sum_{k=1}^{37} [3(k(\frac{4}{37}))^2 + 2(k(\frac{4}{37}))] \frac{4}{37}$

9. $\sum_{k=1}^{80} \frac{1}{(2 + k(\frac{7}{80}))^2} \cdot \frac{7}{80}$

10. $\sum_{k=1}^{15} \frac{-2}{(5 + k(\frac{1}{15}))^3} \cdot \frac{1}{15}$

11. $\sum_{k=1}^{23} (40e^{0.05(k/23)}) \frac{1}{23}$

12. $\sum_{k=1}^{55} (24e^{0.06(4 - k2/55)}) \frac{2}{55}$

In Problems 13 to 16, express the Riemann sum using the symbol $\sum$. Then use a definite integral to approximate the Riemann sum.

13. The 25th Riemann sum of
$$f(x) = 4x^3$$
on the interval [5, 9]

14. The 17th Riemann sum of
$$f(x) = 3\sqrt{x}$$
on the interval [9, 16]

15. The 10th Riemann sum of
$$f(x) = 3x^2 + 2x$$
on the interval [0, 7]

16. The 36th Riemann sum of
$$f(x) = 6x + 5$$
on the interval [0, 13]

17. **Present and Future Values of an Annuity** Suppose money draws interest at the annual rate of 8.5 percent compounded continuously. Use definite integrals to approximate the present and future values of an annuity consisting of $340 payments made 6 times per year during a 7-year period. (The first payment is made $\frac{1}{6}$ year after the 7-year period begins.)

18. **Present and Future Values of an Annuity** Suppose money draws interest at the annual rate of 9.2 percent compounded continuously. Use definite integrals to approximate the present and future values of an annuity consisting of $680 payments made 8 times per year during a 5-year period. (The first payment is made $\frac{1}{8}$ year after the 5-year period begins.)

19. **An Annuity's Present Value** An account earns interest at the annual rate of 12 percent compounded continuously. Use a definite integral to approximate the amount that must be deposited into the account now so that during the next 15 years $3000 can be withdrawn at the end of every 3 months.

20. **An Annuity's Present Value** An account earns interest at the annual rate of 5 percent compounded continuously. Use a definite integral to approximate the amount that must be deposited into the account now so that during the next 9 years $700 can be withdrawn at the end of each month.

21. **An Annuity's Future Value** Beginning one month from now, each month $425 will be deposited into an account where the interest is compounded continuously at the annual rate of 7 percent. Use a definite integral to approximate the amount of money in the account immediately after the 96th deposit.

22. **An Annuity's Future Value** Beginning 4 months from now, every 4 months $820 will be deposited into an account where the interest is compounded continuously at the annual rate of 6 percent. Use a definite integral to approximate the amount of money in the account immediately after the 30th deposit.

23. **A Single Payment Amount** Beginning 1 month from now, Agapito will pay George $530 per month for a period of 6 years. Suppose money earns inter-

est at the annual rate of 9 percent compounded continuously. Use a definite integral to approximate the single payment which if made now is equivalent to the sequence of payments.

24. **A Single Payment Amount** Beginning 1 month from now, Sara will pay Rosa $80 per month for a period of 2 years. Suppose money draws interest at the annual rate of 8 percent compounded continuously. Use a definite integral to approximate the single payment which if made 2 years from now is equivalent to the sequence of payments.

25. **A Drug's Concentration in a Bloodstream** Beginning a half hour from now, every half hour 150 milligrams of a drug will be injected into the bloodstream of a woman. The amount of the drug in the woman's bloodstream decreases exponentially at the rate of 20 percent per hour. Use a definite integral to approximate how much of the drug will be in the woman's bloodstream immediately after she receives the 8th injection.
    Hint: See Example 5.31.

26. **Comparing Two Deals** Yvette wants to sell her car. Her uncle is offering her $4000 in cash. Her cousin is offering her a total of $4,320 in installments of $360 due at the end of every 4 months for a period of 4 years. If money earns interest at the annual rate of 8 percent compounded continuously, who is offering Yvette the better deal?

In Problems 27 to 32, find the limit.
Hint: Each limit is a definite integral.

27. $\lim\limits_{n \to +\infty} \sum\limits_{k=1}^{n} 6\left(3 + k\dfrac{4}{n}\right)^2 \dfrac{4}{n}$

28. $\lim\limits_{n \to +\infty} \sum\limits_{k=1}^{n} \dfrac{3}{2}\left(\sqrt{4 + k\dfrac{5}{n}}\right) \dfrac{5}{n}$

29. $\lim\limits_{n \to +\infty} \sum\limits_{k=1}^{n} \left[3\left(k\dfrac{3}{n}\right)^2 + 2\left(k\dfrac{3}{n}\right)\right] \dfrac{3}{n}$

30. $\lim\limits_{n \to +\infty} \sum\limits_{k=1}^{n} \left[4\left(k\dfrac{1}{n}\right)^3 + 7\right] \dfrac{1}{n}$

31. $\lim\limits_{n \to +\infty} \sum\limits_{k=1}^{n} (60e^{-0.03(k/n)}) \dfrac{1}{n}$

32. $\lim\limits_{n \to +\infty} \sum\limits_{k=1}^{n} (48e^{0.08(k4/n)}) \dfrac{4}{n}$

33. In Theorem 5.15, as n approaches positive infinity, the Δx values get closer and closer to zero. So the addends in the Riemann sums also get closer and closer to zero. Explain why this fact does not cause the limit of the Riemann sums to necessarily be zero.

34. In Theorem 5.15, as n approaches positive infinity, the number of addends in the Riemann sums increases without bound. Explain why this fact does not cause the values of the Riemann sums to necessarily increase without bound.

Chapter 5 Important Terms

Section 5.1

antiderivative (279)
indefinite integral (280)
general antiderivative (280)
integrand (280)
constant of integration (280)
particular antiderivative (280)
constant rule for integrals (281)
power rule for integrals (282)
sum or difference rule for integrals (283)
exponential rule for integrals (285)
constant multiple rule for integrals (285)

logarithm rule for integrals (287)

Section 5.2

definite integral (294)
limits of integration (294)
fundamental theorem of calculus (295)
definite integrals as change (295)

Section 5.3

definite integrals as areas (304, 305, and 307)

Section 5.4

chain rule for integrals (313)

Section 5.5

integration by parts rule (326)

Section 5.6

definite integrals as limits (337)
Riemann sums (337)
present value of an annuity (341)
future value of an annuity (342)

Chapter 5 Review Problems

Section 5.1

1. Find the indefinite integral.

   a. $\int 15x^4 \, dx$
   b. $\int 6x \, dx$
   c. $\int (x^3 + 7x - 5) \, dx$
   d. $\int \left(\frac{9x^3 - 3}{x^4}\right) dx$
   e. $\int (5e^x + 6) \, dx$
   f. $\int \left(e^x + \frac{9}{\sqrt[4]{x}}\right) dx$

2. For the function $f(x)$, find the particular antiderivative $F(x)$ that satisfies the given condition.

   a. $f(x) = 5x^4 + 3$ $F(2) = 30$
   b. $f(x) = 6\sqrt{x}$ $F(1) = 17$
   c. $f(x) = \frac{1}{x} + \frac{3}{x^2}$ $F(1) = -3$
   d. $f(x) = 8e^x - x^4 + 1$ $F(0) = 2$

3. **Demography** Presently, there are 3 million octogenarians in an Asian country. (An octogenarian is a person whose age is in the eighties.) Demographers project that x years from now the number of octogenarians will be increasing at the rate of

   $$0.25 - 0.002x$$

   million per year, where $0 \leq x \leq 50$. For such values of x, let $P(x)$ represent the number of octogenarians x years from now. Find a formula for $P(x)$. Then determine what the number of octogenarians will be 4 years from now.

4. **Production Cost** The management of a manufacturing company finds that the total monthly cost of producing 6 units of its best-selling product per month is $3020. The management also finds that the marginal cost for the production of the product is

   $$x^2 - 24x + 210$$

   when the monthly production level is x units. Let $C(x)$ represent the total monthly cost of producing x units per month. Find a formula for $C(x)$. Then determine the total monthly cost of producing 18 units per month.

Section 5.2

5. Compute the definite integral.

   a. $\int_3^5 2x^3 \, dx$
   b. $\int_1^9 (3x^2 - 39x + 90) \, dx$
   c. $\int_2^6 \frac{12}{x^2} \, dx$
   d. $\int_0^4 3\sqrt{x} \, dx$
   e. $\int_2^{14} \frac{1}{x} \, dx$
   f. $\int_{-1}^2 3e^x \, dx$

6. Suppose $A(x)$ is a function whose derivative is the given function $f(x)$. Use a definite integral to express the amount $A(x)$ changes if x changes as indicated. Then compute the change in $A(x)$.

   a. $f(x) = \frac{3}{\sqrt{x}}$ x increases from 4 to 9
   b. $f(x) = 0.4x - 25$ x decreases from 20 to 14
   c. $f(x) = -0.06x^2 + 1.2x - 6$ x increases from 0 to 7

7. **Rain Forest Shrinkage** Rain forests are disappearing at an alarming rate. Let $F(x)$ represent the total area of rain forests (in millions of acres) x years after the beginning of 1995. Suppose that x years after that date $F(x)$ changes at the rate of

   $$0.28x - 52$$

   million acres per year. Find a definite integral whose value is the amount $F(x)$ changes during the interval between the beginning of 1995 and the beginning of 2015. Then compute this change in $F(x)$.

8. **Blood Cholesterol Level** Let $C(x)$ represent a man's blood cholesterol level if his daily dosage of a special drug is x units. The man's physician believes that

   $$C'(x) = 0.03x^2 - 25$$

   for $15 \leq x \leq 25$. Use a definite integral to express the increase in the man's blood cholesterol level if his daily drug dosage is decreased from 22 units to 18 units. Then compute this increase in his blood cholesterol level.

Section 5.3

9. Find the area of the region bounded by the graph of the function $f(x)$ and the x axis between the given pair of x values.

   Hint: First determine whether the graph is nowhere below the x axis, nowhere above the x axis, or partially above and partially below the x axis between the given pair of x values.

   a. $f(x) = \dfrac{18}{x^2}$ $x = 3$ and $x = 9$

   b. $f(x) = 3x^2 - 48x + 180$ $x = 6$ and $x = 9$

   c. $f(x) = 6 - 3\sqrt{x}$ $x = 1$ and $x = 16$

10. Let $f(x)$ be the function whose graph is the curve in the figure for this problem. Suppose the area of region R_1 is 4.1, the area of region R_2 is 2.4, and the area of region R_3 is 3.2.

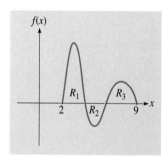

   a. Find the value of $\displaystyle\int_2^9 f(x)\,dx$.

   b. Find the area of the region between the graph of $f(x)$ and the x axis from $x = 2$ to $x = 9$.

Section 5.4

11. Find the indefinite integral. Then check your answer by showing that the derivative of the indefinite integral is the integrand.

   a. $\displaystyle\int 7x^2 \sqrt{x^3 + 4}\,dx$ b. $\displaystyle\int \dfrac{4x^3 - 5}{(x^4 - 5x)^2}\,dx$

   c. $\displaystyle\int \dfrac{3x^2 + 2}{2x^3 + 4x - 1}\,dx$ d. $\displaystyle\int \dfrac{4}{4x + 1}\,dx$

   e. $\displaystyle\int 24e^{0.16x}\,dx$ f. $\displaystyle\int 2xe^{x^2 - 6}\,dx$

g. $\displaystyle\int \dfrac{8x}{(2x - 3)^2}\,dx$ h. $\displaystyle\int \dfrac{5x}{x + 7}\,dx$

i. $\displaystyle\int (x + 8)^5 x\,dx$ j. $\displaystyle\int \dfrac{3}{\sqrt{x} - 8}\,dx$

Section 5.5

12. Use the integration by parts rule to find the indefinite integral. Then check your answer by showing that the derivative of the indefinite integral is the integrand.

   a. $\displaystyle\int 8xe^{-0.05x}\,dx$ b. $\displaystyle\int (4x + 5)\ln x\,dx$

   c. $\displaystyle\int 3x^2 e^{4x}\,dx$ d. $\displaystyle\int \dfrac{8x}{(2x + 5)^3}\,dx$

   e. $\displaystyle\int (x^5 + 1)^7 40x^9\,dx$ f. $\displaystyle\int \ln(4x + 8)\,dx$

Section 5.6

13. Use a definite integral to approximate the Riemann sum.

   a. $\displaystyle\sum_{k=1}^{27} 6(3 + k(\tfrac{4}{27}))^2 \tfrac{4}{27}$ b. $\displaystyle\sum_{k=1}^{59}\left(\sqrt{k(\tfrac{16}{59})}\right)\tfrac{16}{59}$

   c. $\displaystyle\sum_{k=1}^{34} [2(1 + k(\tfrac{5}{34})) + 9]\tfrac{5}{34}$

   d. $\displaystyle\sum_{k=1}^{25} \dfrac{-4}{(3 + k(\tfrac{1}{25}))^2} \cdot \dfrac{1}{25}$

   e. $\displaystyle\sum_{k=1}^{40} (24e^{0.02(k3/40)})\tfrac{3}{40}$

   f. $\displaystyle\sum_{k=1}^{52} (60e^{0.03(1 - k/52)})\tfrac{1}{52}$

14. Express the Riemann sum using the symbol $\sum$. Then use a definite integral to approximate the Riemann sum.

   a. The 120th Riemann sum of
   $$f(x) = 480e^{-0.08x}$$
   on the interval $[0, 10]$

   b. The 50th Riemann sum of
   $$f(x) = \dfrac{1}{x}$$
   on the interval $[2, 9]$

15. **Present and Future Values of an Annuity** Suppose money earns interest at the annual rate of 6 percent compounded continuously. Use definite integrals to approximate the present and future values of an annuity consisting of $160 payments made each month during a 4-year period. (The first payment is made 1 month after the 4-year period begins.)

16. **An Annuity's Present Value** An account draws interest at the annual rate of 8 percent compounded continuously. Use a definite integral to approximate the amount that must be deposited into the account now so that during the next 3 years $4000 can be withdrawn at the end of every 2 months.

17. **An Annuity's Future Value** Beginning 1 month from now, each month $180 will be deposited into an account where the interest is compounded continuously at the annual rate of 9 percent. Use a definite integral to approximate the amount of money in the account immediately after the 60th deposit.

CHAPTER 6

More on Integration

6.1 Simpson's Rule

6.2 Consumers' and Producers' Surpluses

6.3 Probability

6.4 Expected Value

6.5 Differential Equations for Exponential Change

6.6 Differential Equations for Bounded Change

Important Terms

Review Problems

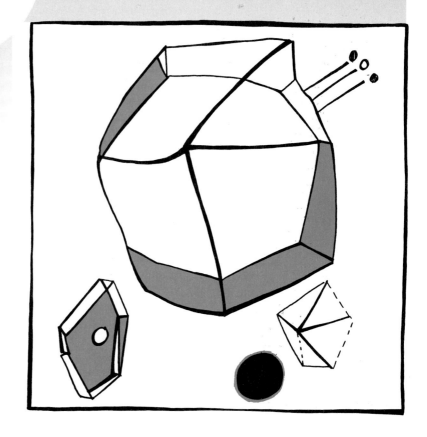

We begin this chapter with an introduction to a numerical method for approximating definite integrals. Then we use definite integrals to find the amount consumers save and the extra amount producers receive when a product is sold at a fixed price. We also use definite integrals to determine the likelihood a certain result will occur and what result to expect when an element is randomly selected from a set. In Sections 6.5 and 6.6, we solve equations whose solutions are functions rather than numbers.

6.1 Simpson's Rule

Suppose $f(x)$ is a function that is continuous on an interval, and a and b are numbers in the interval such that $a < b$. According to our definition of definite integrals (Definition 5.3 in Section 5.2),

$$\int_a^b f(x)\, dx = F(b) - F(a)$$

where $F(x)$ is an antiderivative of $f(x)$. But it may happen that we cannot find an antiderivative of $f(x)$. In such a case, how do we compute $\int_a^b f(x)\, dx$? In theory, we could use Theorem 5.15 in Section 5.6, which says that the value of a definite integral is the limit of a sequence of Riemann sums. To be more specific,

$$\int_a^b f(x)\, dx = \lim_{n \to +\infty} \sum_{k=1}^n f(a + k\,\Delta x)\, \Delta x$$

where $\Delta x = (b - a)/n$. In practice, however, it is often very difficult to compute such a limit.

In this section, we present a method for approximating definite integrals known as **Simpson's rule**. The method, named after the English mathematician Thomas Simpson (1710–1761), is based on the idea of approximating a function with several quadratic functions. Since the graph of a quadratic function is a parabola, Simpson's rule is sometimes called the **parabolic rule**.

Two Preliminary Results for Simpson's Rule

Before stating Simpson's rule, we present two results necessary for a discussion of the rule. Theorem 6.1 provides the first of these results.

Theorem 6.1

If three points do not lie on the same line, there is a unique quadratic function whose graph includes the three points.

Instead of proving Theorem 6.1, we outline the procedure for finding the quadratic function for the special case where the three points are (2, 4), (5, 7), and (6, 12). A general proof of the theorem involves similar steps.

First note that the slope of the line through the points (2, 4) and (5, 7) is

$$\frac{7-4}{5-2} = 1$$

whereas the slope of the line through the points (5, 7) and (6, 12) is

$$\frac{12-7}{6-5} = 5$$

Therefore, the three points do not lie on the same line.

We want to find values for a, b, and c such that the graph of

$$g(x) = ax^2 + bx + c$$

includes the points (2, 4), (5, 7), and (6, 12). Thus, we want to find the values of a, b, and c that satisfy the following system of equations:

$$g(2) = a(2^2) + b(2) + c = 4$$
$$g(5) = a(5^2) + b(5) + c = 7$$
$$g(6) = a(6^2) + b(6) + c = 12$$

This system of equations is satisfied by

$$a = 1$$
$$b = -6$$
and $$c = 12$$

So the graph of the quadratic function

$$g(x) = x^2 - 6x + 12$$

includes the points (2, 4), (5, 7), and (6, 12). Figure 6.1 shows the three points and the graph of the function, which is a parabola.

Theorem 6.2 provides the second result necessary for a discussion of Simpson's rule.

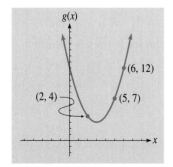

Figure 6.1
$g(x) = x^2 - 6x + 12$

Theorem 6.2

Suppose a, b, and c are real numbers such that $a < b$ and c is halfway between a and b. If $g(x)$ is a quadratic function, then

$$\int_a^b g(x)\, dx = \frac{b-a}{6}(g(a) + 4g(c) + g(b))$$

The proof of Theorem 6.2 is conceptually simple. It does, however, contain numerous algebraic steps. Instead of proving the theorem, we verify it for the special case where

$$g(x) = 6x^2 + 2x + 9$$

$a = 3$, $b = 7$, and $c = 5$.

According to our definition of definite integrals,

$$\int_3^7 (6x^2 + 2x + 9)\, dx = (2x^3 + x^2 + 9x)\Big|_3^7$$
$$= 798 - 90$$
$$= 708$$

We now show that we get the same result if we instead compute $\int_3^7 (6x^2 + 2x + 9)\, dx$ using Theorem 6.2. First note that

$$g(3) = 69$$
$$g(5) = 169$$
and
$$g(7) = 317$$

Thus, using Theorem 6.2,

$$\int_3^7 (6x^2 + 2x + 9)\, dx = \frac{7-3}{6}(69 + 4(169) + 317)$$
$$= 708$$

which agrees with the result we obtained when we used the definition of definite integrals.

Now we are ready to discuss Simpson's rule. For now, we assume that $f(x)$ is a function whose graph is nowhere below the x axis on the interval $[a, b]$. In fact, we assume that $f(x)$, a, and b are as Figure 6.2 shows.

The length of the interval $[a, b]$ is $b - a$. Therefore, if we divide this interval into six subintervals of equal lengths, the length of each subinterval is

$$\frac{b-a}{6}$$

Figure 6.2 shows that x_k represents the right endpoint of the kth subinterval. Thus,

$$x_k = a + \frac{k(b-a)}{6}$$

Note that $x_0 = a$ and $x_6 = b$.

Figure 6.2 also shows that R_1 represents the region between the graph of $f(x)$ and the x axis from $x = x_0$ to $x = x_2$, R_2 represents the region between the graph of $f(x)$ and the x axis from $x = x_2$ to $x = x_4$, and R_3 represents the region between the graph of $f(x)$ and the x axis from $x = x_4$ to $x = x_6$.

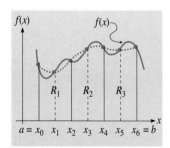

Figure 6.2 The regions R_1, R_2, and R_3.

Theorem 6.1 tells us that there is a quadratic function whose graph includes the points $(x_0, f(x_0))$, $(x_1, f(x_1))$, and $(x_2, f(x_2))$. Let $p(x)$ represent this quadratic function. In Figure 6.2, the dotted parabolic arc above the interval $[x_0, x_2]$ is a portion of the graph of $p(x)$.

Note that x_1 is halfway between x_0 and x_2. Therefore, according to Theorem 6.2,

$$\int_{x_0}^{x_2} p(x)\, dx = \frac{x_2 - x_0}{6} (p(x_0) + 4p(x_1) + p(x_2)) \tag{1}$$

Since the graph of $p(x)$ includes the points $(x_0, f(x_0))$, $(x_1, f(x_1))$, and $(x_2, f(x_2))$, it follows that

$$p(x_0) = f(x_0)$$
$$p(x_1) = f(x_1)$$

and
$$p(x_2) = f(x_2)$$

Since the length of each of the six subintervals of the interval $[a, b]$ is $(b - a)/6$,

$$x_2 - x_0 = 2(b - a)/6$$
$$= (b - a)/3$$

Thus, we can rewrite formula (1) as follows:

$$\int_{x_0}^{x_2} p(x)\, dx = \frac{(b-a)/3}{6} (f(x_0) + 4f(x_1) + f(x_2))$$

$$= \frac{b - a}{3(6)} (f(x_0) + 4f(x_1) + f(x_2)) \tag{2}$$

The value of $\int_{x_0}^{x_2} p(x)\, dx$ is the area of the region between the graph of $p(x)$ and the x axis from $x = x_0$ to $x = x_2$. Figure 6.2 suggests that this area is a reasonably accurate approximation for the area of region R_1. Consequently, expression (2) is a reasonably accurate approximation for the area of region R_1.

Similarly, we could show that

$$\frac{b - a}{3(6)} (f(x_2) + 4f(x_3) + f(x_4)) \tag{3}$$

is a reasonably accurate approximation for the area of region R_2, and

$$\frac{b - a}{3(6)} (f(x_4) + 4f(x_5) + f(x_6)) \tag{4}$$

is a reasonably accurate approximation for the area of region R_3.

By adding expressions (2), (3), and (4), we get

$$\frac{b - a}{3(6)} [f(x_0) + 4f(x_1) + 2f(x_2) + 4f(x_3)$$
$$+ 2f(x_4) + 4f(x_5) + f(x_6)] \tag{5}$$

which is a reasonably accurate approximation for the area of the region between the graph of $f(x)$ and the x axis from $x = a$ to $x = b$. But this area is

$$\int_a^b f(x)\, dx$$

Therefore, we can think of expression (5) as an approximation for $\int_a^b f(x)\, dx$. According to Definition 6.1, this approximation is the 6th Simpson's rule approximation of $\int_a^b f(x)\, dx$.

Definition 6.1

Suppose $f(x)$ is continuous on an interval and a and b are numbers in the interval such that $a < b$. Divide the interval $[a, b]$ into n subintervals of equal lengths, where n is *even*. Let $x_0 = a$ and let x_k be the right endpoint of the kth subinterval. ($x_n = b$.) The nth **Simpson's rule** approximation of

$$\int_a^b f(x)\, dx$$

is

$$\frac{b-a}{3n}[f(x_0) + 4f(x_1) + 2f(x_2) + 4f(x_3) + 2f(x_4) + \cdots + 2f(x_{n-2}) + 4f(x_{n-1}) + f(x_n)]$$

Except for $f(x_0)$ and $f(x_n)$, note that $f(x_k)$ is multiplied by 4 if k is odd, whereas $f(x_k)$ is multiplied by 2 if k is even.

The approximations for $\int_a^b f(x)\, dx$ using Simpson's rule become more accurate as n increases. In fact,

$$\int_a^b f(x)\, dx = \lim_{n \to +\infty} \frac{b-a}{3n}[f(x_0) + 4f(x_1) + 2f(x_2) + 4f(x_3) + 2f(x_4) + \cdots + 2f(x_{n-2}) + 4f(x_{n-1}) + f(x_n)]$$

In the example we used to introduce Simpson's rule, we assumed that the graph of $f(x)$ is nowhere below the x axis on the interval $[a, b]$. Simpson's rule, however, is valid even if the graph of $f(x)$ dips below the x axis on the interval $[a, b]$.

The procedure in using Simpson's rule is readily programmable. Consequently, the rule is usually applied with the help of electronic computing.

In Example 6.1, we use Simpson's rule to approximate a population increase.

Example 6.1

The rabbit population of an island is expected to increase. A biologist predicts that x months from now the population will be increasing at the rate of

$$\frac{12}{x^2 + 1}$$

thousand per month. How much will the population increase during the next 2 months?

Solution

The increase in the rabbit population during the next 2 months is the value of

$$\int_0^2 \frac{12}{x^2 + 1} \, dx$$

We could easily find the value of this definite integral by computing

$$F(2) - F(0)$$

if $F(x)$ is an antiderivative of $12/(x^2 + 1)$. But unfortunately the antiderivatives of $12/(x^2 + 1)$ involve a function which we do not discuss in this book. So we will approximate the value of the definite integral using Simpson's rule, say with $n = 8$.

Since in this problem $a = 0$, $b = 2$, and $n = 8$, the length of each of the eight subintervals is

$$\frac{b - a}{n} = \frac{2 - 0}{8} = 0.25$$

Thus, $x_0 = 0$, $x_1 = 0.25$, $x_2 = 0.5$, $x_3 = 0.75$, $x_4 = 1$, $x_5 = 1.25$, $x_6 = 1.5$, $x_7 = 1.75$, and $x_8 = 2$. So, if we let $f(x) = 12/(x^2 + 1)$,

$$f(x_0) = f(0) = 12$$
$$4f(x_1) = 4f(0.25) \approx 45.18$$
$$2f(x_2) = 2f(0.5) = 19.2$$
$$4f(x_3) = 4f(0.75) = 30.72$$
$$2f(x_4) = 2f(1) = 12$$
$$4f(x_5) = 4f(1.25) \approx 18.73$$
$$2f(x_6) = 2f(1.5) \approx 7.385$$
$$4f(x_7) = 4f(1.75) \approx 11.82$$
$$f(x_8) = f(2) = 2.4$$

Therefore, the 8th Simpson's rule approximation of

$$\int_0^2 \frac{12}{x^2 + 1} \, dx$$

is

$$\tfrac{1}{12}(12 + 45.18 + 19.2 + 30.72 + 12 + 18.73 + 7.385 + 11.82 + 2.4)$$

which is about 13.29. So the rabbit population will increase by about 13,290 rabbits during the next 2 months.

The actual value of

$$\int_0^2 \frac{12}{x^2 + 1}\, dx$$

correct through five decimal places is 13.28578. Thus, the approximation for this definite integral that we obtained in the above example using Simpson's rule with $n = 8$ is quite accurate.

A Simpson's rule approximation of a definite integral involves approximating the graph of a function with several parabolic arcs. A Riemann sum approximation of a definite integral involves approximating the graph of a function with several horizontal line segments. Usually, parabolic arcs approximate the graph of a function better than horizontal line segments. Therefore, the nth Simpson's rule approximation of a definite integral is usually more accurate than the nth Riemann sum approximation of the definite integral.

In Example 6.2, we use Simpson's rule to approximate a definite integral of a function even though we are not given a formula for the function.

Example 6.2

Suppose $f(x)$ is a function that is continuous on an interval containing 5 and 8. Table 6.1 shows the values of $f(x)$ for several values of x.

Table 6.1

x	5	5.5	6	6.5	7	7.5	8
$f(x)$	9.5	9	8.8	8.4	8.6	9.4	10

Use Simpson's rule to approximate

$$\int_5^8 f(x)\, dx$$

Solution

The numbers 5, 5.5, 6, 6.5, 7, 7.5, and 8 divide the interval [5, 8] into six subintervals, each of length 0.5. Therefore, we can approximate

$$\int_5^8 f(x)\, dx$$

using Simpson's rule with $n = 6$. The approximation is

$$\frac{8-5}{3(6)}[9.5 + 4(9) + 2(8.8) + 4(8.4) + 2(8.6) + 4(9.4) + 10]$$

which simplifies to 26.9 (rounded to one decimal place).

6.1 Problems

In Problems 1 to 8, use Simpson's rule with the indicated value of n to approximate the definite integral. Round your answers to three decimal places.

1. $\int_0^2 x^5 \, dx$ $n = 4$

2. $\int_0^4 x^6 \, dx$ $n = 4$

3. $\int_3^7 \sqrt{x} \, dx$ $n = 4$

4. $\int_1^3 \frac{1}{x} \, dx$ $n = 4$

5. $\int_0^3 \sqrt{81 - x^2} \, dx$ $n = 6$

6. $\int_0^3 \sqrt{36 - x^2} \, dx$ $n = 6$

7. $\int_4^6 \frac{24}{x^2 + 5} \, dx$ $n = 8$

8. $\int_2^4 \frac{16}{x^2 + 3} \, dx$ $n = 8$

9. Suppose $f(x)$ is a function that is continuous on an interval containing 4 and 6. The following table shows the values of $f(x)$ for several values of x:

x	4	$4\frac{1}{3}$	$4\frac{2}{3}$	5	$5\frac{1}{3}$	$5\frac{2}{3}$	6
$f(x)$	5.2	5	4.7	4.3	4.2	3.8	3.6

a. Use Simpson's rule to approximate $\int_4^6 f(x) \, dx$.

b. Suppose $F(x)$ is the particular antiderivative of $f(x)$ for which $F(4) = 13$. Approximate the value of $F(6)$.

10. Suppose $f(x)$ is a function that is continuous on an interval containing 0 and 1. The following table shows the values of $f(x)$ for several values of x:

x	0	0.25	0.5	0.75	1
$f(x)$	7	7.3	7.4	7.6	7.7

a. Use Simpson's rule to approximate $\int_0^1 f(x) \, dx$.

b. Suppose $F(x)$ is the particular antiderivative of $f(x)$ for which $F(0) = 20$. Approximate the value of $F(1)$.

11. Suppose $f(x)$ is a function that is continuous and nonnegative on an interval containing 5 and 7. The following table shows the values of $f(x)$ for several values of x:

x	5	5.25	5.5	5.75	6	6.25	6.5	6.75	7
$f(x)$	8.1	8.3	8.35	8.5	8.6	8.75	9	9.1	9.3

Use Simpson's rule to approximate the area of the region between the graph of $f(x)$ and the x axis from $x = 5$ to $x = 7$.

12. Suppose $f(x)$ is a function that is continuous and nonnegative on an interval containing 6 and 9. The following table shows the values of $f(x)$ for several values of x:

x	6	6.5	7	7.5	8	8.5	9
$f(x)$	1.5	2	2.8	3	2.9	2.3	1.8

Use Simpson's rule to approximate the area of the region between the graph of $f(x)$ and the x axis from $x = 6$ to $x = 9$.

13. **Production Cost** The following table shows the marginal cost for the production of a product at several weekly production levels:

Production level, tons	0	0.5	1	1.5	2	2.5	3	3.5	4
Marginal cost, $	3010	3000	2080	2065	2070	2078	2085	3000	3020

The fixed weekly cost is $25,000. Use Simpson's rule to find an approximation for the total weekly cost of producing 4 tons per week.

14. **Population** The following table shows what a demographer projects about the growth rate of a population at several future times:

Time, years from now	0	0.25	0.5	0.75	1	1.25	1.5	1.75	2	2.25	2.5	2.75	3
Growth rate, millions per year	7.2	7	6.7	6.3	5.7	5.2	4.8	4	3.5	3.2	3	2.5	2.2

Presently, the population is 38 million. Use Simpson's rule to estimate what the population will be 3 years from now.

15. The maximum possible error in the nth Simpson's rule approximation of

$$\int_a^b f(x)\,dx$$

is

$$\frac{M(b-a)^5}{180n^4}$$

where M is the maximum value attained by the absolute value of the *fourth derivative* of $f(x)$ on the interval $[a, b]$. Suppose $g(x)$ is a function for which $M = 45$ on the interval $[8, 12]$. Using Simpson's rule, how large must n be to approximate

$$\int_8^{12} g(x)\,dx$$

(a) with an error less than 0.0001; (b) with an error less than 0.00001?

16. Suppose $g(x)$ is a function for which $M = 60$ on the interval $[8, 11]$, where M is defined as in Problem 15. Use the formula presented in Problem 15 to determine how large n must be if Simpson's rule is used to approximate

$$\int_8^{11} g(x)\,dx$$

(a) with an error less than 0.001; (b) with an error less than 0.0001.

17. Suppose $f(x)$ is a *quadratic function* and $a < b$. Then for *any* even positive integer n, the nth Simpson's rule approximation of

$$\int_a^b f(x)\,dx$$

is, in fact, the *exact* value of this definite integral. Explain why.

18. Explain why in Definition 6.1 we require n to be even.

Electronic Computing

The Basic program SIMP listed below uses Simpson's rule to approximate the value of a definite integral of a continuous function $f(x)$. Execution of the program prompts you to enter the lower and upper limits of integration. After you enter these two values, the program prints the approximation obtained with $n = 4$. You can obtain more accurate approximations by striking the space bar. Specifically, each stroke of the space bar prints the approximation obtained with n increased by 4. You can exit the program by striking E.

As listed, SIMP applies to the function

$$f(x) = \frac{12}{x^2 + 1}$$

But you can make the program applicable to other functions by editing line 50.

The Program SIMP and a Sample Run

```
10   REM                         SIMP
20   REM
30   REM A PROGRAM THAT USES SIMPSON'S RULE TO APPROXIMATE A DEFINITE INTEGRAL
40   REM
50   DEF FNF(X)=12/(X^2+1)
60   PRINT "ENTER THE LOWER AND UPPER LIMITS OF INTEGRATION. THE PROGRAM WILL"
70   PRINT "PRINT THE APPROXIMATION OBTAINED USING SIMPSON'S RULE WITH N=4."
80   PRINT "AFTERWARDS, EACH STROKE OF THE SPACE BAR PRINTS THE APPROXIMATION"
90   PRINT "OBTAINED WITH N INCREASED BY 4. WHEN TWO APPROXIMATIONS ARE EQUAL,"
100  PRINT "EXIT THE PROGRAM BY STRIKING E."
110  PRINT
120  INPUT A,B
130  PRINT
140  LET N=4
150  LET I=(B-A)/N
160  LET S=FNF(A)+FNF(B)
170  FOR J=1 TO (N-1) STEP 2
180  LET S=S+4*FNF(A+J*I)
190  NEXT J
200  FOR J=2 TO (N-2) STEP 2
210  LET S=S+2*FNF(A+J*I)
220  NEXT J
230  LET S=I*S/3
240  PRINT S
250  LET N=N+4
260  LET B$=INKEY$
270  IF B$="E" THEN 300
280  IF B$=" " THEN 150
290  GOTO 260
300  END
Ok
RUN
ENTER THE LOWER AND UPPER LIMITS OF INTEGRATION. THE PROGRAM WILL
PRINT THE APPROXIMATION OBTAINED USING SIMPSON'S RULE WITH N=4.
AFTERWARDS, EACH STROKE OF THE SPACE BAR PRINTS THE APPROXIMATION
OBTAINED WITH N INCREASED BY 4. WHEN TWO APPROXIMATIONS ARE EQUAL,
EXIT THE PROGRAM BY STRIKING E.

? 1,5

 7.095023
 7.057449
 7.056177
 7.056059
 7.056038
 7.056034
 7.056032
 7.056031
 7.056031
Ok
```

Computer Problems

Use the program SIMP to approximate the definite integrals in Problems 19 to 24.

19. $\int_{1}^{7} (12x^5 - 4x + 7) \, dx$

20. $\int_{2}^{9} (21x^6 - x^2 - 1) \, dx$

21. $\int_{3}^{5} \sqrt{49 - x^2} \, dx$

22. $\int_{1}^{4} \sqrt{25 - x^2} \, dx$

23. $\int_{2}^{4} \frac{48}{x^2 + 4} \, dx$

24. $\int_{3}^{6} \frac{37}{x^2 + 1} \, dx$

6.2 Consumers' and Producers' Surpluses

For most products, demand increases as price decreases, and supply increases as price increases. When such a product sells at a fixed price, consumers collectively pay less than what they are willing to pay for the amount they consume, and producers collectively receive more revenue than what they are willing to receive for the amount they supply. In this section, we use definite integrals to find how much consumers save and how much extra producers receive. We begin with an example involving a product used as a substitute for salt.

Tasty is a low-sodium salt substitute. Let $D(x)$ represent the price of Tasty (in $ hundreds per ton) at which consumers buy x tons per week. The management of the company that produces the product finds that

$$D(x) = -\tfrac{1}{36}x^2 + 9$$

for $0 \leq x \leq 15$.

The owner of the company wants to sell 12 tons of Tasty next week. Therefore, since

$$D(12) = -\tfrac{1}{36}(12^2) + 9 = 5$$

management will set the price at $500 per ton. Since

$$12 \cdot 5 = 60$$

consumers will pay a total of $6000 for the 12 tons of the product. The area of the shaded rectangular region in Figure 6.3 represents this amount of money.

Consumers would buy portions of the 12 tons of Tasty at prices higher than $500 per ton. For instance, since

$$D(6) = -\tfrac{1}{36}(6^2) + 9 = 8$$

consumers would buy 6 tons at $800 per ton. So consumers save by buying 12 tons at $500 per ton. To determine how much they save, we first have to determine how much they are willing to pay for the 12 tons.

To determine how much consumers are willing to pay for 12 tons of Tasty, we consider an idealistic approach to selling the product. Suppose

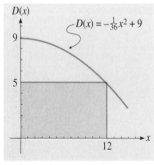

Figure 6.3 Total amount consumers pay for 12 tons of Tasty at $500 per ton.

that next week, instead of setting the price at $500 per ton, management gradually lowers the price from $900 per ton (the lowest price at which consumers buy none of the product) until consumers buy a total of 12 tons. Also, suppose management does not lower a price until it sells all that can be sold at that price. Then the company accumulates the revenue at a varying rate with respect to the demand. To be more specific, when the company has sold x tons, it accumulates the revenue at the rate of

$$D(x) = -\tfrac{1}{36}x^2 + 9$$

hundred dollars per ton. So, if we let $R(x)$ represent the total revenue the company accumulates from selling the first x tons, $D(x) = -\tfrac{1}{36}x^2 + 9$ is the derivative of $R(x)$. Thus, the value of

$$\int_0^{12} (-\tfrac{1}{36}x^2 + 9)\,dx$$

is the increase in $R(x)$ that results when x increases from zero to 12. In fact, since $R(0) = 0$, the value of this integral is equal to $R(12)$, which is the total revenue the company collects from selling 12 tons using the idealistic approach.

The idealistic approach takes advantage of the fact consumers would buy portions of the 12 tons of Tasty at prices higher than $500 per ton. So we can think of the value of

$$\int_0^{12} (-\tfrac{1}{36}x^2 + 9)\,dx$$

as the total amount consumers are willing to pay for the 12 tons. Therefore, since

$$\int_0^{12} (-\tfrac{1}{36}x^2 + 9)\,dx = -\tfrac{1}{108}x^3 + 9x \Big|_0^{12}$$
$$= 92$$

the total amount consumers are willing to pay for the 12 tons is $9200. The area of the shaded region in Figure 6.4 represents this amount of money.

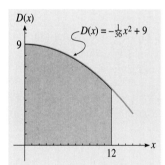

Figure 6.4 Total amount consumers are willing to pay for 12 tons of Tasty.

Consumers' Surplus

Since management will set the price of Tasty at $500 per ton, consumers will actually pay a total of only $6000, rather than $9200, for the 12 tons of Tasty. So, since

$$9200 - 6000 = 3200$$

consumers will pay $3200 *less* than what they are willing to pay for the 12 tons. The area of the shaded region in Figure 6.5 represents this *saving*, which we call the consumers' surplus.

In general, suppose $D(x)$ is the unit price of a product at which consumers buy x units of the product during a designated time period (for

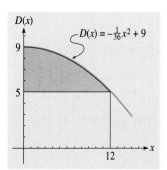

Figure 6.5 The consumers' surplus.

example, a week or a month). Let us assume $D(x)$ decreases as x increases, which usually is the case. The salt substitute example illustrates that when the price is $D(n)$, consumers pay *less* than what they are willing to pay for the n units they consume during the designated time period. The amount consumers save during the time period is the consumers' surplus that results when the price is $D(n)$.

The salt substitute example also illustrates that the total amount consumers are willing to pay for consuming n units of the product during the designated time period is

$$\int_0^n D(x)\, dx$$

Therefore, since $n \cdot D(n)$ is the total amount consumers actually pay for the n units when the price is $D(n)$, the consumers' surplus is

$$\int_0^n D(x)\, dx - n \cdot D(n)$$

when the price is $D(n)$. This result is restated in Theorem 6.3.

Theorem 6.3

Let $D(x)$ be the unit price at which consumers buy x units of a product during a designated time period. Suppose $D(x)$ decreases as x increases. When the price is $D(n)$, the consumers' surplus for the time period is

$$\int_0^n D(x)\, dx - n \cdot D(n)$$

We use Theorem 6.3 in Example 6.3.

Example 6.3

Let $D(x)$ represent the price of Divino Wine (in hundreds of dollars per barrel) at which consumers buy x barrels of the wine per day. Suppose

$$D(x) = \sqrt{81 - x}$$

for $0 \le x \le 50$. Find the daily consumers' surplus when the price is $700 per barrel.

Solution

First we find how much of the wine consumers buy per day when the price is $700 per barrel:

$$\sqrt{81 - x} = 7$$
$$81 - x = 49$$
$$x = 32$$

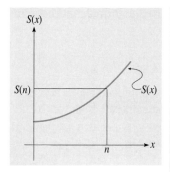

Figure 6.6 The graph of $S(x)$.

Thus, consumers buy 32 barrels of the wine per day when the price is $700 per barrel.
Since

$$\int_0^{32} \sqrt{81-x}\, dx = -\tfrac{2}{3}(\sqrt{81-x})^3 \Big|_0^{32}$$
$$= 257\tfrac{1}{3}$$

consumers are willing to pay $25,733.33 for the 32 barrels. Since

$$32 \cdot 7 = 224$$

consumers actually pay only $22,400 for the 32 barrels when the price is $700 per barrel. Therefore, since

$$25{,}733.33 - 22{,}400 = 3333.33$$

the daily consumers' surplus is $3333.33 when the price of the wine is $700 per barrel.

Producers' Surplus

Suppose $S(x)$ is the unit price of a product at which producers supply x units of the product during a designated time period. We assume that producers sell whatever quantity they supply and that $S(x)$ increases as x increases. (See Figure 6.6.)

When the unit price is $S(n)$, producers receive

$$n \cdot S(n)$$

units of money for the n units of the product they supply during the designated time period. In Figure 6.6, the area of the rectangle with base n and height $S(n)$ represents this amount of money.

In the salt substitute example, we showed that the value of

$$\int_0^{12} (-\tfrac{1}{36}x^2 + 9)\, dx$$

is the total amount consumers are willing to pay for 12 tons of Tasty. Similarly, we can show that the value of

$$\int_0^n S(x)\, dx$$

is the total revenue producers are willing to receive for supplying n units of the product during the designated time period. In Figure 6.6, the area of the region between the graph of $S(x)$ and the x axis from $x = 0$ to $x = n$ represents this revenue. Note that this area is less than the area of the rectangle in the figure.

Thus, when the unit price is $S(n)$, producers receive *more* revenue than what they are willing to receive for the n units they supply during the designated time period. The *extra* revenue producers receive during the time period is the producers' surplus that results when the price is $S(n)$. The area of the shaded region in Figure 6.7 represents this extra revenue.

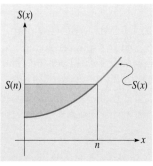

Figure 6.7 The producers' surplus.

The above discussion implies that we can use the formula in Theorem 6.4 to find the producers' surplus.

Theorem 6.4

Let $S(x)$ be the unit price at which producers supply x units of a product during a designated time period. Suppose producers sell whatever quantity they supply and $S(x)$ increases as x increases. When the price is $S(n)$, the producers' surplus for the time period is

$$n \cdot S(n) - \int_0^n S(x)\, dx$$

We use Theorem 6.4 in Example 6.4.

Example 6.4

Tobax is a low nicotine tobacco substitute. Table 6.2 shows the quantity of this product producers supply monthly at several price levels.

Table 6.2

Quantity, tons	0	0.25	0.5	0.75	1	1.25	1.5	1.75	2
Price per ton, $ thousands	1	1.14	1.24	1.38	1.54	1.74	1.96	2.22	2.5

Producers sell whatever quantity of Tobax they supply.

a. Use Simpson's rule to approximate how much revenue producers are willing to receive for the quantity of Tobax they supply monthly when the price is $2500 per ton ($2.5 thousand per ton).

b. Use the result in part (a) to find an approximation for the monthly producers' surplus when the price of Tobax is $2500 per ton.

Solution for (a)

Let $S(x)$ represent the price per ton (in thousands of dollars) at which producers supply x tons of Tobax monthly. Table 6.2 shows that producers supply 2 tons of the product monthly when the price is $2500 per ton. The revenue (in thousands of dollars) producers are willing to receive for supplying the 2 tons of Tobax is the value of

$$\int_0^2 S(x)\, dx$$

Table 6.2 also shows the values of $S(x)$ for $x = 0, 0.25, 0.5, 0.75, 1, 1.25, 1.5, 1.75,$ and 2. These values of x divide the interval $[0, 2]$ into eight subintervals, each of length 0.25. So we can use the information in the table to find the 8th Simpson's rule approximation of

$$\int_0^2 S(x)\, dx$$

This approximation is

$$\frac{2-0}{3(8)}[1 + 4(1.14) + 2(1.24) + 4(1.38)$$
$$+ 2(1.54) + 4(1.74) + 2(1.96) + 4(2.22) + 2.5]$$

which simplifies to 3.24 (rounded to two decimal places). Thus, producers are willing to receive about $3240 for supplying 2 tons of Tobax monthly.

Solution for (b)

As we saw, producers supply 2 tons of Tobax monthly when the price is $2.5 thousand per ton. Therefore, since

$$2(2.5) = 5$$

producers actually receive $5000 for the 2 tons. But in part (a) we showed that producers are willing to supply this quantity of Tobax for about $3240. So, since

$$5000 - 3240 = 1760$$

the monthly producers' surplus is approximately $1760 when the price of Tobax is $2500 per ton.

Surpluses When the Market Is in Equilibrium

The market for a product is in equilibrium when the demand is equal to the supply. In other words, the market is in equilibrium when the quantity purchased is equal to the quantity supplied.

Example 6.5

Suppose the unit price (in thousands of dollars) at which consumers purchase (demand) x units of a product monthly is

$$D(x) = -\tfrac{5}{144}x^2 + 10$$

while the unit price at which producers supply x units of the product monthly is

$$S(x) = \tfrac{1}{48}x^2 + 2$$

Find the monthly consumers' surplus and the monthly producers' surplus when the market is in equilibrium.

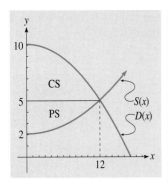

Figure 6.8 A consumers' surplus (CS) and a producers' surplus (PS) at market equilibrium.

Solution

First we find how much of the product consumers purchase and producers supply when the market is in equilibrium:

$$-\tfrac{5}{144}x^2 + 10 = \tfrac{1}{48}x^2 + 2$$
$$8 = \tfrac{1}{18}x^2$$
$$144 = x^2$$
$$x = \pm 12$$

So, when the market is in equilibrium, consumers purchase and producers supply 12 units of the product monthly. Since

$$D(12) = S(12) = 5$$

the price is $5000 per unit when the market is in equilibrium.
Thus, since

$$\int_0^{12} (-\tfrac{5}{144}x^2 + 10)\, dx - 12 \cdot 5 = 40$$

the monthly consumers' surplus is $40,000 when the market is in equilibrium. The area of the region labeled CS in Figure 6.8 represents this consumers' surplus.
Also, since

$$12 \cdot 5 - \int_0^{12} (\tfrac{1}{48}x^2 + 2)\, dx = 24$$

the monthly producers' surplus is $24,000 when the market is in equilibrium. The area of the region labeled PS in Figure 6.8 represents this producers' surplus.

6.2 Problems

1. **Consumers' Surplus** Suppose the unit price (in $ thousands) at which consumers buy x units of a product weekly is

$$D(x) = 8 - 0.04x^2$$

for $0 \le x \le 12$.

a. Find how much of the product consumers buy weekly when the unit price is $4000.

b. Find the total amount consumers are willing to pay for the quantity they buy weekly when the unit price is $4000.

c. Find the total amount consumers actually pay for the quantity they buy weekly when the unit price is $4000.

d. Find the weekly consumers' surplus when the unit price is $4000.

2. **Consumers' Surplus** Suppose the unit price (in $ hundreds) at which consumers purchase x units of a commodity daily is

$$D(x) = \frac{75}{x + 2}$$

a. Find how much of the commodity consumers purchase daily when the unit price is $300.

b. Find the total amount consumers are willing to pay for the quantity they purchase daily when the unit price is $300.

c. Find the total amount consumers actually pay for the quantity they purchase daily when the unit price is $300.

d. Find the daily consumers' surplus when the unit price is $300.

3. **Producers' Surplus** Suppose the unit price (in $ hundreds) at which producers supply x units of a commodity monthly is

$$S(x) = \sqrt{x+9}$$

(Producers sell whatever quantity they supply.)

a. Find how much of the commodity producers supply monthly when the unit price is $500.

b. Find the total revenue producers are willing to receive for the quantity they supply monthly when the unit price is $500.

c. Find the total revenue producers actually receive for the quantity they supply monthly when the unit price is $500.

d. Find the monthly producers' surplus when the unit price is $500.

4. **Producers' Surplus** Suppose the unit price (in $ thousands) at which producers supply x units of a product weekly is

$$S(x) = 0.03x^2 + 1$$

(Producers sell whatever quantity they supply.)

a. Find how much of the product producers supply weekly when the unit price is $13,000.

b. Find the total revenue producers are willing to receive for the quantity they supply weekly when the unit price is $13,000.

c. Find the total revenue producers actually receive for the quantity they supply weekly when the unit price is $13,000.

d. Find the weekly producers' surplus when the unit price is $13,000.

5. **Surpluses at Market Equilibrium** Suppose the unit price (in $ thousands) at which consumers purchase x units of a product daily is

$$D(x) = -\tfrac{1}{32}x^2 + 19$$

and the unit price at which producers supply x units of the product daily is

$$S(x) = \tfrac{15}{64}x^2 + 2$$

a. Find how much of the product consumers purchase daily and producers supply daily when the market is in equilibrium.

b. Find the unit price when the market is in equilibrium.

c. Find the daily consumers' surplus and the daily producers' surplus when the market is in equilibrium.

6. **Surpluses at Market Equilibrium** Suppose the unit price (in $ hundreds) at which consumers buy x units of a commodity monthly is

$$D(x) = -\tfrac{1}{15}x^2 + 13$$

while the unit price at which producers supply x units of the commodity monthly is

$$S(x) = \tfrac{4}{75}x^2 + 1$$

a. Find how much of the commodity consumers purchase monthly and producers supply monthly when the market is in equilibrium.

b. Find the unit price when the market is in equilibrium.

c. Find the monthly consumers' surplus and the monthly producers' surplus when the market is in equilibrium.

7. **Consumers' Surplus** If the unit price (in $ hundreds) at which consumers buy x units of a product weekly is

$$D(x) = \frac{144}{(x+2)^2}$$

what is the weekly consumers' surplus when the unit price is $400?

8. **Consumers' Surplus** If the unit price (in $ thousands) at which consumers purchase x units of a product daily is

$$D(x) = \frac{12}{\sqrt{3x+1}}$$

what is the daily consumers' surplus when the unit price is $3000?

9. **Producers' Surplus** Suppose the unit price (in $ thousands) at which producers supply x units of a product monthly is

$$S(x) = \tfrac{1}{3}x\sqrt{x+4}$$

(Producers sell whatever quantity they supply.) Find the monthly producers' surplus when the unit price is $16,000.

Note: $S(x) = 16$ when $x = 12$.

10. **Producers' Surplus** Suppose the unit price (in $ hundreds) at which producers supply x units of a commodity weekly is

$$S(x) = \frac{3x}{\sqrt{x+1}}$$

(Producers sell whatever quantity they supply.) Find the weekly producers' surplus when the unit price is $1125.

11. **Consumers' Surplus** Cinco Equis is a nonalcoholic beer. The following table shows the quantity of this beer the patrons of Jake's Tavern consume weekly at several price levels:

Quantity, kegs	0	0.5	1	1.5	2	2.5	3	3.5	4
Price per keg, $ hundreds	5	3.14	2.4	2	1.75	1.58	1.45	1.36	1.29

a. Use Simpson's rule to approximate how much the patrons are willing to pay for the quantity of Cinco Equis they consume weekly when the price is $129 per keg ($1.29 hundred per keg).

b. Use the result in part (a) to find an approximation for the weekly consumers' surplus when the price of Cinco Equis is $129 per keg.

12. **Producers' Surplus** Fibrex is a hybrid cereal high in dietary fiber. The following table shows the quantity of this cereal producers supply yearly at several price levels:

Quantity, thousands of tons	0	0.25	0.5	0.75	1	1.25	1.5	1.75	2
Price per thousand tons, $ millions	2	2.5	2.71	2.87	3	3.12	3.22	3.32	3.41

Producers sell whatever quantity of Fibrex they supply.

a. Use Simpson's rule to approximate how much revenue producers are willing to receive for the quantity of Fibrex they supply yearly when the price is $3,410,000 per thousand tons ($3.41 million per thousand tons).

b. Use the result in part (a) to find an approximation for the yearly producers' surplus when the price of Fibrex is $3,410,000 per thousand tons.

13. Use the area representation of a consumers' surplus to explain why in Theorem 6.3 we could alternatively conclude that the consumers' surplus is

$$\int_0^n (D(x) - D(n))\,dx$$

14. Use the area representation of a producers' surplus to explain why in Theorem 6.4 we could alternatively conclude that the producers' surplus is

$$\int_0^n (S(n) - S(x))\,dx$$

Graphing Calculator Problems

15. **Consumers' Surplus** Suppose the unit price (in $ thousands) at which consumers buy x units of a product monthly is

$$D(x) = \frac{31x + 40}{x^2 + 6x + 5}$$

for $0 \le x \le 15$.

a. Use a calculator that approximates definite integrals to approximate how much consumers are willing to pay for the quantity they buy monthly when the unit price is $3250. If your calculator displays regions whose areas are definite integrals, set the viewing screen to $[0, 15] \times [-2, 8]$ with a scale of 1 on both axes.

b. Use the result in part (a) to find an approximation for the monthly consumers' surplus when the unit price is $3250.

16. **Producers' Surplus** Suppose the unit price (in $ thousands) at which producers supply x units of a commodity weekly is

$$S(x) = \frac{x^2 + 52}{36 - x^2}$$

for $0 \leq x \leq 3$. (Producers sell whatever quantity they supply.)

a. Use a calculator that approximates definite integrals to approximate how much revenue producers are willing to receive for the quantity they supply weekly when the unit price is $1750. If your calculator displays regions whose areas are definite integrals, set the viewing screen to $[0, 3] \times [-0.5, 2.5]$ with a scale of 1 on both axes.

b. Use the result in part (a) to find an approximation for the weekly producers' surplus when the unit price is $1750.

6.3 Probability

In this section, we will see that the value of a definite integral sometimes can be interpreted as a probability.

Definition 6.2

The **probability** of an event is a number in the interval [0, 1] that measures the event's likelihood.

The closer the probability of an event is to 1, the greater the event's likelihood. If the probability is 1, the event is certain. If the probability is zero, there is no chance whatsoever for the event. If the probability is $\frac{7}{8}$, the event is likely, but not certain.

Consider, for example, the event of obtaining a number less than 6 when we toss a die. If the die is fair (i.e., not loaded), the probability of this event is $\frac{5}{6}$, because there are five numbers less than 6 among the six numbers on the die. Suppose, on the other hand, the die is loaded. In this case, to determine the probability of the event, we would toss the die a large number of times. If we find, for example, that in $\frac{4}{9}$ of the tosses we obtain a number less than 6, it would be reasonable for us to conclude that the probability of obtaining such a number is $\frac{4}{9}$.

By tossing a die, we, in effect, randomly select a toss from the set of all possible tosses of the die. Each toss is associated with a number, namely, the number of spots on the top side of the die. Thus, each toss is associated with a number from a discrete set of numbers, namely, 1, 2, 3, 4, 5, and 6. In this section, however, we are primarily interested in problems where the outcome of a random selection is associated with a real number that can lie *anywhere* in an interval. We illustrate this type of problem with a problem involving the idea of a "Solar Index."

The Solar Index of a region on a given day is the percentage of heat the sun could provide to produce 80 gallons of hot water on that day. For instance, if the Solar Index of a region is 55 on a certain day, then 55 percent of the heat necessary to produce 80 gallons of hot water could be obtained from the sun on that day.

Solar One, a solar energy plant in the Mojave Desert. (Courtesy Southern California Edison Company)

A company specializing in solar water-heating systems is contemplating opening a branch in a southwestern city. The management will open the branch if the probability is high that on a randomly selected day the city's Solar Index is at least 60.

Let x represent the city's Solar Index on a randomly selected day. Then, since a Solar Index is a percentage, it is possible for the value of x to be any real number in the closed interval [0, 100]. The management of the company wants to know the probability that the value of x lies in the subinterval [60, 100]. So the management hired a consulting firm.

The consulting firm claims that the function

$$g(x) = 0.0006x - 0.000006x^2$$

has the following property: If J is *any* subinterval of the interval [0, 100], the probability that x lies in the interval J is given by

$$\int_c^d (0.0006x - 0.000006x^2)\, dx$$

where c and d are the endpoints of the interval J. Therefore, since

$$\int_{60}^{100} (0.0006x - 0.000006x^2)\, dx = (0.0003x^2 - 0.000002x^3)\Big|_{60}^{100}$$
$$= 1 - 0.648$$
$$= 0.352$$

the probability that x lies in the interval [60, 100] is 0.352.

Since 0.352 is closer to zero than to 1, the likelihood is somewhat small that on a randomly selected day the Solar Index of the southwestern city is at least 60. So the management of the solar water-heating company will probably decide against opening a branch in that city.

Probability Density Functions

The function

$$g(x) = 0.0006x - 0.000006x^2$$

is a polynomial function. Thus, $g(x)$ is continuous everywhere. In particular, $g(x)$ is continuous on the interval $[0, 100]$. Note that

$$0.0006x - 0.000006x^2 = 0.0006x(1 - 0.01x)$$

Thus, we see that $g(x) \geq 0$ for each x in $[0, 100]$. Note also that

$$\int_0^{100} (0.0006x - 0.000006x^2)\, dx = (0.0003x^2 - 0.000002x^3) \Big|_0^{100}$$
$$= 1$$

Thus, the area of the region between the graph of $g(x)$ and the interval $[0, 100]$ is 1. According to Definition 6.3, $g(x)$ is therefore a probability density function on the interval $[0, 100]$.

Definition 6.3

Suppose I is an interval and $f(x)$ is a function that is continuous on I. The function $f(x)$ is a **probability density function** on I if $f(x)$ has the following two properties:

1. $f(x) \geq 0$ for each x in I.
2. The area of the region between the graph of $f(x)$ and the interval I is 1.

Example 6.6

Verify that

$$h(x) = \frac{3}{(x + 3)^2}$$

is a probability density function on the interval $[0, +\infty)$.

Solution

The denominator of $3/(x + 3)^2$ is different from zero for each value of x in $[0, +\infty)$. Thus, since $h(x)$ is a rational function, $h(x)$ is continuous on the interval $[0, +\infty)$.

The value of $3/(x + 3)^2$ is nonnegative for each value of x in the interval $[0, +\infty)$. So $h(x)$ satisfies property 1 in Definition 6.3.

Let R represent the region to the right of the y axis that lies between the graph of $h(x)$ and the x axis. This region is not bounded on the right. Nevertheless, in what follows, we will show that its area is 1, and, thus, conclude that $h(x)$ satisfies property 2 in Definition 6.3.

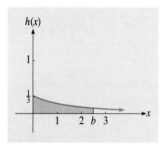

Figure 6.9 A region between the graph of $h(x) = 3/(x + 3)^2$ and the x axis.

For each positive real number b, the value of

$$\int_0^b \frac{3}{(x + 3)^2} \, dx$$

is the area of the portion of the region R that extends from $x = 0$ to $x = b$. Such a portion of the region R is the shaded region in Figure 6.9. The figure suggests that, as b increases, the values of

$$\int_0^b \frac{3}{(x + 3)^2} \, dx$$

get closer and closer to the area of the *entire* region R. In fact, the area of the region R is

$$\lim_{b \to +\infty} \int_0^b \frac{3}{(x + 3)^2} \, dx$$

To find the value of the above limit, we first note that, for each positive real number b,

$$\int_0^b \frac{3}{(x + 3)^2} \, dx = \left. \frac{-3}{x + 3} \right|_0^b$$

$$= \frac{-3}{b + 3} - (-1)$$

$$= 1 - \frac{3}{b + 3}$$

Therefore,

$$\lim_{b \to +\infty} \int_0^b \frac{3}{(x + 3)^2} \, dx = \lim_{b \to +\infty} \left(1 - \frac{3}{b + 3} \right)$$

$$= 1 - 0$$

$$= 1$$

So the area of the region R is indeed 1, which means that $h(x)$ satisfies property 2 in Definition 6.3.

Thus, according to Definition 6.3, we have verified that the function

$$h(x) = \frac{3}{(x + 3)^2}$$

is a probability density function on the interval $[0, +\infty)$.

Improper Integrals

According to Definition 6.4, the limit we found in Example 6.6 is an example of an improper integral.

Definition 6.4

Suppose $f(x)$ is continuous on the interval $[a, +\infty)$. The **improper integral** of $f(x)$ on the interval $[a, +\infty)$ is

$$\lim_{b \to +\infty} \int_a^b f(x)\, dx$$

if the limit exists, and is represented by the symbol

$$\int_a^\infty f(x)\, dx$$

In the above definition, if the limit exists, we say that the improper integral **converges**. If the limit does not exist, we say that the improper integral **diverges**. In this book, we deal only with improper integrals that converge.

Other types of improper integrals exist. However, in this book, we deal only with improper integrals of the type described in Definition 6.4.

The relationship between improper integrals and area is similar to the relationship between definite integrals and area. Suppose $f(x)$ is a function that is continuous and nowhere below the x axis on the interval $[a, +\infty)$. Let R represent the region to the right of the vertical line $x = a$ and that lies between the graph of $f(x)$ and the x axis. Then, as we illustrated in Example 6.6 with the function $h(x)$, the value of the improper integral

$$\int_a^\infty f(x)\, dx$$

if it converges, is the area of the region R.

In Example 6.7, we verify that an exponential function is a probability density function on the interval $[0, +\infty)$.

Example 6.7

Verify that

$$f(x) = 0.09e^{-0.09x}$$

is a probability density function on the interval $[0, +\infty)$.

Solution

Since $f(x)$ is an exponential function, we already know that $f(x)$ is continuous on $[0, +\infty)$ and that it satisfies property 1 of Definition 6.3. We will show that $f(x)$ satisfies property 2 of that definition by showing that

$$\int_0^\infty 0.09e^{-0.09x} = 1$$

For each positive real number b,

$$\int_0^b 0.09e^{-0.09x}\,dx = -e^{-0.09x}\Big|_0^b$$
$$= -e^{-0.09(b)} - (-e^{-0.09(0)})$$
$$= 1 - e^{-0.09b}$$

Therefore,

$$\int_0^\infty 0.09e^{-0.09x}\,dx = \lim_{b\to+\infty}\int_0^b 0.09e^{-0.09x}\,dx$$
$$= \lim_{b\to+\infty}(1 - e^{-0.09b})$$
$$= 1 - 0$$
$$= 1$$

Thus, we have shown that $f(x)$ satisfies property 2 of Definition 6.3, because the value of the above improper integral is the area of the region between the graph of $f(x)$ and the interval $[0, +\infty)$.

So, according to Definition 6.3, we have verified that the function

$$f(x) = 0.09e^{-0.09x}$$

is a probability density function on the interval $[0, +\infty)$.

Not all exponential functions are probability density functions on the interval $[0, +\infty)$.

Random Variables

In the solar heating example, we let x represent the Solar Index of a city on a randomly selected day. Thus, according to Definition 6.5, x is a random variable distributed on the interval $[0, 100]$.

Definition 6.5

Suppose S is a set for which each element is uniquely associated with a real number. Let x represent the real number associated with an element randomly selected from the set S. If it is possible for the value of x to be any real number in an interval I, then x is a **random variable** distributed on the interval I.

In the solar heating example, the elements of the set S are days. Each day is uniquely associated with a city's Solar Index on that day.

As another example, let x represent the weight of a student randomly selected from a large group of students. If no one in the group weighs less than 80 pounds or more than 240 pounds, then x is a random variable distributed on the interval $[80, 240]$. Here the elements of the set S are students. Each student is uniquely associated with his/her weight.

Probability Density Functions and Random Variables

The solar heating example involves a probability density function that is related to a random variable in a special way. The following discussion provides us with a label for the relationship.

Suppose x is a random variable distributed on an interval I, and $f(x)$ is a function that is continuous on I. Also, suppose $f(x)$ is related to x in the following way:

For any subinterval J of the interval I, the probability that x lies in J is given by
$$\int_c^d f(x)\,dx$$
where c and d are the endpoints of the interval J.

Then $f(x)$ is a probability density function on I. (In Problem 23 at the end of this section, you will be asked to explain why we can arrive at this conclusion.) Furthermore, we can say that the random variable x is distributed on the interval I according to the probability density function $f(x)$. Alternatively, we can say that $f(x)$ is the probability density function of the random variable x.

Using the above terminology in the solar heating example, we can say that the random variable x is distributed on the interval [0, 100] according to the probability density function

$$g(x) = 0.0006x - 0.000006x^2$$

We can express the above idea in terms of area. Suppose x is a random variable distributed on an interval $[a, b]$. Also, suppose the curve in Figure 6.10 is the graph of the probability density function of the random variable x. Then the area of the shaded region in the figure represents the probability that x lies in the subinterval $[c, d]$, because the area of this region is

$$\int_c^d f(x)\,dx$$

where $f(x)$ is the probability density function.

Example 6.8 involves a random variable distributed on the interval $[0, +\infty)$.

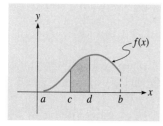

Figure 6.10 A region whose area represents the probability x lies in $[c, d]$.

Example 6.8

Suppose x is a random variable distributed on the interval $[0, +\infty)$ according to the probability density function

$$h(x) = \frac{3}{(x+3)^2}$$

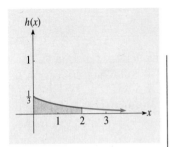

Figure 6.11 A region whose area represents the probability in Example 6.8.

What is the probability that x is less than 2? That is, what is the probability that x lies in the interval $[0, 2)$?

Solution

Since

$$\int_0^2 \frac{3}{(x+3)^2}\,dx = \frac{-3}{x+3}\Big|_0^2$$
$$= 0.4$$

the probability that x lies in the interval $[0, 2)$ is 0.4. The area of the shaded region in Figure 6.11 represents this probability. The curve in the figure is the graph of $h(x) = 3/(x+3)^2$ for $x \geq 0$.

If the probability that a random variable lies in an interval is p, then the probability that it does not lie in the interval is $1 - p$. We use this fact in Example 6.9.

Example 6.9

Traffic engineers want to determine if a traffic light needs to be installed at the intersection of a bicycle path and a busy boulevard. Let x represent the number of minutes a randomly selected bicyclist has to wait before crossing the boulevard. Then x is a random variable distributed on the interval $[0, +\infty)$. Suppose the distribution is according to the probability density function

$$f(x) = 0.09e^{-0.09x}$$

Find the probability that a randomly selected bicyclist has to wait more than 5 minutes before crossing the boulevard.

Solution

Since

$$\int_0^5 0.09e^{-0.09x}\,dx = -e^{-0.09x}\Big|_0^5$$
$$\approx 0.36$$

the probability that x lies in the interval $[0, 5]$ is approximately 0.36. Therefore, since

$$1 - 0.36 = 0.64$$

the probability that x lies in the interval $(5, +\infty)$ is approximately 0.64. So the probability that a randomly selected bicyclist has to wait more than 5 minutes is about 0.64.

The result in the above example implies that about 64 percent of the bicyclists have to wait more than 5 minutes before crossing the boulevard. Therefore, should the traffic light be installed?

Finding a probability density function that models the distribution of a random variable in an actual application is a problem beyond the scope of this book.

6.3 Problems

In Problems 1 and 2, verify that the function is a probability density function on the interval [1, 6].

1. $f(x) = \dfrac{1.2}{x^2}$
2. $f(x) = \dfrac{3}{215}x^2$

In Problems 3 and 4, verify that the function is a probability density function on the interval $[0, +\infty)$.

3. $f(x) = \dfrac{4}{(x+4)^2}$
4. $f(x) = \dfrac{6x}{(x^2+3)^2}$

5. Suppose x is a random variable distributed on the interval [0, 6] according to the probability density function
$$f(x) = \tfrac{1}{72}x^2$$
   a. What is the probability that x lies in the interval [2, 4]?
   b. What is the probability that x does *not* lie in the interval [1, 5]?

6. Suppose x is a random variable distributed on the interval [0, 9] according to the probability density function
$$f(x) = \tfrac{1}{243}x^2$$
   a. What is the probability that x lies in the interval [1, 6]?
   b. What is the probability that x does *not* lie in the interval [3, 7]?

7. Suppose x is a random variable distributed on the interval [1, 5] according to the probability density function
$$f(x) = \dfrac{1.25}{x^2}$$
Find the probability that x is between 1 and 4.

8. Suppose x is a random variable distributed on the interval [2, 10] according to the probability density function
$$f(x) = \dfrac{2.5}{x^2}$$
Find the probability that x is between 5 and 10.

9. If x is a random variable distributed on the interval $[0, +\infty)$ according to the probability density function
$$f(x) = \dfrac{4}{(x+4)^2}$$
what is the probability that x is greater than 8?

10. If x is a random variable distributed on the interval $[0, +\infty)$ according to the probability density function
$$f(x) = \dfrac{5}{(x+5)^2}$$
what is the probability that x is greater than 15?

11. Suppose x is a random variable distributed on the interval $[0, +\infty)$ according to the probability density function
$$f(x) = \dfrac{10x}{(x^2+5)^2}$$
Find the probability that x is less than 2.

12. Suppose x is a random variable distributed on the interval $[0, +\infty)$ according to the probability density function
$$f(x) = \dfrac{6x}{(x^2+3)^2}$$
Find the probability that x is less than 4.

13. Suppose x is a random variable distributed on the interval $[0, +\infty)$ according to the probability density function
$$f(x) = 4xe^{-2x}$$
   a. What is the probability that x lies in the interval (0, 2]?
   b. What is the probability that x is less than 1 or greater than 3?

14. Suppose x is a random variable distributed on the interval $[0, +\infty)$ according to the probability density function
$$f(x) = xe^{-x}$$

a. What is the probability that x lies in the interval $[0, 1)$?

b. What is the probability that x is less than 2 or greater than 5?

15. **Testing Time** Let x represent the number of minutes that a randomly selected student spends taking a mathematics placement test. The professor who proctors the test allows students at most 75 minutes to complete the test. So x is a random variable distributed on the interval $[0, 75]$. Suppose the distribution is according to the probability density function

$$f(x) = \frac{1}{140{,}625} x^2$$

a. What is the probability that a randomly selected student spends at least 50 minutes taking the test?

b. What is the probability that a randomly selected student completes the test in less than 45 minutes?

16. **Lifespan of a Product** Ratno is a product that wards off rodents by emitting a high-frequency sound. The maximum lifespan of this product is 8 months. Therefore, if x represents the lifespan (in months) of a randomly selected Ratno, then x is a random variable distributed on the interval $[0, 8]$. Suppose the distribution is according to the probability density function

$$f(x) = \frac{3}{32} x - \frac{3}{256} x^2$$

a. Find the probability that a randomly selected Ratno has a lifespan of more than 5 months.

b. Find the probability that a randomly selected Ratno has a lifespan of at most 3 months.

17. **Annual Incomes** Let x represent the annual income (in $ thousands) of a household randomly selected from those in a midwestern city. None of the households has an annual income that exceeds $60,000. Therefore, x is a random variable distributed on the interval $[0, 60]$. Suppose the distribution is according to the probability density function

$$f(x) = \frac{1}{900} x - \frac{1}{3{,}240{,}000} x^3$$

What percent of the households have an annual income less than $30,000?

18. **Weights of Chocolate Bars** Let x represent the weight (in ounces) of a randomly selected chocolate bar. The machine that produces the bars always forms bars that weigh at least 6 ounces but not more than 8 ounces. Thus, x is a random variable distributed on the interval $[6, 8]$. Suppose the probability density function of x is

$$f(x) = \frac{24}{x^2}$$

Approximately what percent of the chocolate bars in a large batch weigh at most 7 ounces?

19. **Efficiency of a Truck** Let x represent the average number of miles a delivery truck travels per gallon of gasoline during a randomly selected day. Under the worst possible conditions, x is 7. Under the best possible conditions, x is 12. So x is a random variable distributed on the interval $[7, 12]$. The truck's owner will replace the truck if during a randomly selected day the probability of averaging less than 9 miles per gallon exceeds 0.75. If the probability density function of x is

$$f(x) = \frac{60}{(x + 8)^2}$$

will the truck's owner replace the truck?

20. **Efficiency of a Refrigerator** On Monday mornings, the chef at Jim's Restaurant prepares 400 pounds of a Greek delicacy that requires no refrigeration until closing time. On Tuesday nights, whatever amount of the delicacy remains is thrown away. Thus, if x represents the amount (in hundreds of pounds) that must be refrigerated at closing time on a randomly selected Monday, then x is a random variable distributed on the interval $[0, 4]$. Jim would like to purchase a special refrigerator that stores up to 250 pounds of the delicacy. The refrigerator is not cost-efficient if less than 100 pounds of the delicacy is stored in it. So Jim has decided to purchase the refrigerator only if the probability is at least 0.9 that between 100 and 250 pounds of the delicacy will require refrigeration on a randomly selected Monday. If the probability density function of x is

$$f(x) = \frac{3}{64} x^2 - \frac{3}{8} x + \frac{3}{4}$$

will Jim purchase the refrigerator?

In Problems 21 and 22, find the value of k that makes $f(x)$ a probability density function on the interval $[2, 5]$.

21. $f(x) = kx^2$ **22.** $f(x) = \dfrac{k}{x^2}$

23. Suppose x is a random variable distributed on an interval I, and $f(x)$ is a function that is continuous on I. Also, suppose that for any subinterval J of the interval I, the probability that x lies in J is given by

$$\int_c^d f(x)\, dx$$

where c and d are the endpoints of the interval J. Then $f(x)$ *must* be a probability density function on I. Explain why.

24. Suppose $f(x)$ is a function that is continuous on a closed interval $[a, b]$. Also, suppose $f(x) \geq 0$ for each x in the interval $[a, b]$. Explain how we can use the function $f(x)$ to obtain a probability density function on the interval $[a, b]$.

6.4 Expected Value

In this section, we will see that the value of a definite integral sometimes can be used as a measure of the *central tendency* of a random variable.

Consider the following game. A fair die is tossed. If the number obtained is less than 3, the player wins \$15. If the number obtained is greater than 2 but less than 6, the player pays \$7 (wins \$−7). If the number obtained is 6, the player pays \$11 (wins \$−11). Table 6.3 summarizes this information.

Table 6.3

Spots on top side	1	2	3	4	5	6
Win	\$15	\$15	\$−7	\$−7	\$−7	\$−11

There are two ways of winning \$15, three ways of winning \$−7, and one way of winning \$−11. Thus, the *weighted average* of the possible winnings is

$$\frac{2(15) + 3(-7) + 1(-11)}{6} = -\frac{1}{3} \tag{6}$$

This result tells us that a player who plays the game many times should *expect* to lose an amount that *averages out* to about \$0.33 per game. So, for example, if a player plays the game 1000 times, the player should expect to lose about \$333.

We can rewrite Equation (6) as follows:

$$(15)\tfrac{2}{6} + (-7)\tfrac{3}{6} + (-11)\tfrac{1}{6} = -\tfrac{1}{3} \tag{7}$$

Since the die is fair and there are two ways of winning \$15, the probability of winning \$15 (from a single toss) is $\tfrac{2}{6}$. Similarly, the probability of winning \$−7 (losing \$7) is $\tfrac{3}{6}$, and the probability of winning \$−11 (losing \$11) is $\tfrac{1}{6}$. Therefore, if we let $f(x)$ represent the probability of winning \$$x$ from a single toss of the die, then

$$f(15) = \tfrac{2}{6}$$
$$f(-7) = \tfrac{3}{6}$$
and $$f(-11) = \tfrac{1}{6}$$

Substituting in Equation (7), we get

$$(15)f(15) + (-7)f(-7) + (-11)f(-11) = -\tfrac{1}{3} \qquad (8)$$

If we let
$$x_1 = 15$$
$$x_2 = -7$$
and $$x_3 = -11$$

we can use the summation notation introduced in Section 5.6 to rewrite Equation (8) as follows:

$$\sum_{k=1}^{3} x_k f(x_k) = -\tfrac{1}{3}$$

This result suggests that Definition 6.6 is reasonable.

Definition 6.6

Suppose x is a random variable and $f(x)$ is the probability density function of x. If x is distributed on the closed interval $[a, b]$, the **expected value** of x is

$$\int_a^b x f(x)\,dx$$

If x is distributed on the interval $[a, +\infty)$, the **expected value** of x is

$$\int_a^\infty x f(x)\,dx$$

Either way, the expected value is denoted by the Greek letter μ (mu) and is alternatively called the **average value** or **mean** of x.

In Example 6.10, we refer to the expected value of a random variable as the average of the random variable.

Example 6.10

The Taco Factory produces tacos that weigh at least 9 ounces but not more than 12 ounces. So, if x represents the weight, in ounces, of a randomly selected taco, then x is a random variable distributed on the interval $[9, 12]$. Suppose the probability density function of x is

$$f(x) = -\tfrac{1}{9}x^2 + \tfrac{20}{9}x - \tfrac{32}{3}$$

Find the average weight of the tacos.

Solution

According to Definition 6.6, the average weight of the tacos is

$$\mu = \int_9^{12} x(-\tfrac{1}{9}x^2 + \tfrac{20}{9}x - \tfrac{32}{3})\, dx$$

$$= \int_9^{12} (-\tfrac{1}{9}x^3 + \tfrac{20}{9}x^2 - \tfrac{32}{3}x)\, dx$$

$$= (-\tfrac{1}{36}x^4 + \tfrac{20}{27}x^3 - \tfrac{16}{3}x^2)\Big|_9^{12}$$

$$= 10.25$$

ounces. Figure 6.12 shows this average and the probability density function.

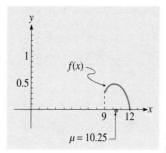

Figure 6.12 A probability density function and the corresponding expected value.

Exponentially Distributed Random Variables

In Example 6.7 of Section 6.3, we verified that

$$f(x) = 0.09e^{-0.09x}$$

is a probability density function on the interval $[0, +\infty)$. Therefore, we verified the special case of Theorem 6.5 in which $k = 0.09$.

Theorem 6.5

Suppose k is a positive constant. If

$$f(x) = ke^{-kx}$$

then $f(x)$ is a probability density function on $[0, +\infty)$.

Theorem 6.5 is a basis for Definition 6.7.

Definition 6.7

A random variable x is **exponentially distributed** if it is distributed on $[0, +\infty)$ according to a function of the form

$$f(x) = ke^{-kx}$$

where k is a positive constant.

Thus, the random variable in Example 6.9 of Section 6.3 is exponentially distributed.

Expected Values of Exponentially Distributed Random Variables

Suppose x is an exponentially distributed random variable. Then we can express the probability density function of x in the form

$$f(x) = ke^{-kx}$$

where k is a positive constant. According to Definition 6.6, the expected value of x is

$$\mu = \int_0^\infty xke^{-kx}\,dx \tag{9}$$

Using the integration by parts rule, we can show that, for each positive real number b,

$$\int_0^b xke^{-kx}\,dx = \frac{1}{k} - \frac{b}{e^{kb}} - \frac{1}{ke^{kb}}$$

So

$$\int_0^\infty xke^{-kx}\,dx = \lim_{b\to+\infty} \int_0^b xke^{-kx}\,dx$$

$$= \lim_{b\to+\infty} \left(\frac{1}{k} - \frac{b}{e^{kb}} - \frac{1}{ke^{kb}}\right)$$

Thus, we can rewrite Equation (9) as follows:

$$\mu = \lim_{b\to+\infty} \left(\frac{1}{k} - \frac{b}{e^{kb}} - \frac{1}{ke^{kb}}\right)$$

Using two properties of limits from Section 1.5 and L'Hôpital's rule, we can show that

$$\lim_{b\to+\infty} \left(\frac{1}{k} - \frac{b}{e^{kb}} - \frac{1}{ke^{kb}}\right) = \frac{1}{k}$$

(As we said in Section 4.5, L'Hôpital's rule is a theorem usually found in more advanced calculus books.) Therefore,

$$\mu = \frac{1}{k}$$

which establishes the validity of Theorem 6.6.

Theorem 6.6

Suppose x is a random variable distributed (exponentially) on $[0, +\infty)$ according to a function of the form

$$f(x) = ke^{-kx}$$

where k is a positive constant. Then the expected value of x is

$$\mu = \frac{1}{k}$$

So, according to Theorem 6.6, the expected value of the random variable in Example 6.9 of Section 6.3 is

$$\mu = \frac{1}{0.09}$$
$$\approx 11$$

This result means that a bicyclist should expect to wait about 11 minutes before crossing the busy boulevard.

In Example 6.11, we also use Theorem 6.6.

Example 6.11

The time (in hours) it takes a randomly selected person to memorize a certain poem is exponentially distributed. If the expected time is 2 hours, what is the probability that it takes at least 1 hour but not more than 4 hours for a randomly selected person to memorize the poem?

Solution

Let x represent the time it takes a randomly selected person to memorize the poem. Since x is exponentially distributed, we can express the probability density function of x in the form

$$f(x) = ke^{-kx} \qquad (10)$$

where k is a positive constant. According to Theorem 6.6, the expected value of x is

$$\frac{1}{k}$$

Thus, since in this example the expected value of x is 2,

$$\frac{1}{k} = 2$$

which implies that

$$k = 0.5$$

Substituting 0.5 for k in formula (10), we get

$$f(x) = 0.5e^{-0.5x}$$

Since

$$\int_1^4 0.5e^{-0.5x}\, dx = -e^{-0.5x}\Big|_1^4$$
$$\approx 0.47$$

the probability that x lies in $[1, 4]$ is approximately 0.47. So the probability that it takes at least 1 hour but not more than 4 hours for a randomly selected person to memorize the poem is about 0.47. The area of the shaded region in Figure 6.13 represents this probability. The curve in the figure is the graph of $f(x) = 0.5e^{-0.5x}$ for $x \geq 0$.

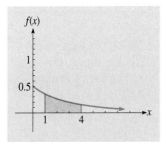

Figure 6.13 A region whose area represents the probability in Example 6.11.

The result in Example 6.11 implies that about 47 percent of the people in a large group require at least 1 hour but not more than 4 hours to memorize the poem. In such a group, the average time it takes to memorize the poem should be close to 2 hours.

The next subsection is a basis for the graphing calculator problems at the end of this section.

Normally Distributed Random Variables

Suppose x is a random variable distributed on an interval. The **standard deviation** of x is a number, denoted by the Greek letter σ (sigma), that measures the *dispersion* of x. If the standard deviation is small, the probability is high that x is close to its expected value. If the standard deviation is large, the probability is high that x is far from its expected value. We could use the idea of a definite integral to define standard deviation more precisely. But our approach does not require such a definition.

Functions that can be expressed in the form

$$f(x) = \frac{1}{s\sqrt{2\pi}} e^{-0.5(x-m)^2/s^2}$$

where m is any constant and s is a positive constant, are probability density functions on the set of real numbers. This fact, whose proof is beyond the scope of this book, is a basis for Definition 6.8.

Definition 6.8

A random variable x is **normally distributed** if it is distributed on the set of real numbers according to a function of the form

$$f(x) = \frac{1}{s\sqrt{2\pi}} e^{-0.5(x-m)^2/s^2}$$

where m is any constant and s is a positive constant.

In Definition 6.8, although we will not prove it, m is the expected value of x and s is the standard deviation of x. So, for example, if the expected value is 7 and the standard deviation is 3, then the probability density function of x is

$$f(x) = \frac{1}{3\sqrt{2\pi}} e^{-0.5(x-7)^2/9}$$

Now suppose x is a random variable distributed on an interval which is not necessarily the set of real numbers. If we can closely approximate the probability density function of x with a function of the form described in Definition 6.8, we take the liberty of saying that x is normally distributed. We take such liberty in the graphing calculator problems at the end of this section.

6.4 Problems

1. Suppose x is a random variable distributed on the interval $[1, 5]$ according to the probability density function
$$f(x) = \frac{25}{12x^3}$$
What is the expected value of x?

2. Suppose x is a random variable distributed on the interval $[2, 8]$ according to the probability density function
$$f(x) = \frac{128}{15x^3}$$
What is the expected value of x?

3. Suppose x is a random variable distributed on the interval $[0, +\infty)$ according to the probability density function
$$f(x) = 0.25e^{-0.25x}$$
Find the average value of x.

4. Suppose x is a random variable distributed on the interval $[0, +\infty)$ according to the probability density function
$$f(x) = 8e^{-8x}$$
Find the average value of x.

5. If x is a random variable distributed on the interval $[0, 3]$ according to the probability density function
$$f(x) = \frac{2}{3} - \frac{2}{9}x$$
what is the mean of x?

6. If x is a random variable distributed on the interval $[0, 6]$ according to the probability density function
$$f(x) = \frac{1}{3} - \frac{1}{18}x$$
what is the mean of x?

7. Suppose x is a random variable distributed on the interval $[0, 5]$ according to the probability density function
$$f(x) = \frac{7.2}{(x + 4)^2}$$
Find the expected value of x.

8. Suppose x is a random variable distributed on the interval $[0, 4]$ according to the probability density function
$$f(x) = \frac{1.25}{(x + 1)^2}$$
Find the expected value of x.

9. Suppose x is an exponentially distributed random variable. If the expected value of x is 5, what is the probability that x is not greater than 7?

10. Suppose x is an exponentially distributed random variable. If the expected value of x is 0.8, what is the probability that x is not greater than 2?

11. **Average Testing Time** Problem 15 in Section 6.3 deals with the time students spend taking a mathematics placement test. Find the average time students spend taking the test.

12. **Expected Lifespan of a Product** Problem 16 in Section 6.3 deals with the lifespan of Ratno, a product that wards off rodents. Find the expected lifespan of a randomly selected Ratno.

13. **Average Annual Income** Problem 17 in Section 6.3 deals with the annual income of the households in a city. Find the average annual income of the households.

14. **Expected Weight of a Chocolate Bar** Problem 18 in Section 6.3 deals with the weights of chocolate bars. Find the expected weight of a randomly selected chocolate bar.

15. **Overdue Time of Library Books** The overdue time (in days) of a randomly selected overdue library book is exponentially distributed. If the average overdue time is 8 days, approximately what percentage of the overdue books is more than 4 days overdue?

16. **Waiting Time at a Restaurant** The number of minutes a randomly selected patron waits before being seated at a popular San Francisco restaurant is exponentially distributed. If the average waiting time is 16 minutes, approximately what percentage of the restaurant's patrons waits more than 10 minutes?

17. Suppose x is a random variable distributed on an interval I, and $f(x)$ is the probability density function of x. Furthermore, suppose there is only one number in I at which $f(x)$ achieves its maximum value. Then such a number is the **mode** of x.

    a. Give an example of a random variable whose mode is equal to its expected value.

    b. Give an example of a random variable whose mode is not equal to its expected value.

18. Suppose x is a random variable distributed on an interval I. The **median** of x is the number m in I for which the probability is $\frac{1}{2}$ that x is less than m.

    a. Give an example of a random variable whose median is equal to its expected value.

    b. Give an example of a random variable whose median is not equal to its expected value.

Graphing Calculator Problems

19. Suppose x is a normally distributed random variable for which μ (the expected value) is 12.

    a. Set the viewing screen on your graphing calculator to $[0, 24] \times [0, 0.15]$ with a scale of 1 on both axes. Then draw the graph of the probability density function of x for the case in which σ (the standard deviation) is 5. Use the trace function to locate the point on the graph whose x coordinate is the expected value. Finally, superimpose the graph of the probability density function of x for the case in which $\sigma = 3$.

    b. Explain how the graphs in part (*a*) support the fact that a random variable with a small standard deviation has a higher probability of being close to its expected value than a random variable with a large standard deviation.

20. **Cholesterol Levels** The cholesterol level of a randomly selected citizen of an African country is normally distributed with an expected value of 170 and a standard deviation of 18. Use a calculator that approximates definite integrals to answer the following questions. If your calculator displays regions whose areas are definite integrals, set the viewing screen to $[100, 240] \times [0.005, 0.025]$ with a scale of 10 on both axes.

    a. Approximately what percentage of the citizens has a cholesterol level between 160 and 190?

    b. Approximately what percentage of the citizens has a cholesterol level greater than 180?

6.5 Differential Equations for Exponential Change

Differential equations are equations that involve derivatives. The solutions of differential equations are *functions* rather than numbers.

An example of a simple differential equation is

$$f'(x) = 2x + 5$$

Using the techniques introduced in Chapter 5, the solutions of this differential equation are the functions of the form

$$f(x) = x^2 + 5x + C$$

Sec. 6.5 Differential Equations for Exponential Change

In this section, we deal with differential equations whose solutions are functions of the form

$$f(x) = Ce^{rx}$$

We begin by assuming that $f(x)$ is a *positive* function for which

$$f'(x) = rf(x)$$

where r is a nonzero constant. [To say that $f(x)$ is a positive function means that $f(x) > 0$ for each x in the domain.] If we divide both sides by $f(x)$, we get

$$\frac{f'(x)}{f(x)} = r$$

Using the integral versions of the logarithm rule and the chain rule, we can show that $\ln f(x)$ is an antiderivative of $f'(x)/f(x)$. According to the constant rule for integrals, rx is an antiderivative of r. Therefore, since $f'(x)/f(x) = r$, the functions $\ln f(x)$ and rx are antiderivatives of the same function, and so they must differ by a constant. In other words, there exists a constant d such that

$$\ln f(x) = rx + d$$
$$f(x) = e^{rx+d} \qquad \text{definition of ln } x$$
$$= e^{rx}e^{d} \qquad \text{a property of exponents}$$
$$= e^{d}e^{rx} \qquad \text{multiplication is commutative}$$

Since d is a constant, e^d is also a constant. If we let $C = e^d$, we can write

$$f(x) = Ce^{rx}$$

Therefore, we have shown that any positive function $f(x)$ that satisfies a differential equation of the form

$$f'(x) = rf(x)$$

where r is a nonzero constant, must have the form

$$f(x) = Ce^{rx}$$

Actually, this result is valid even if $f(x)$ is not a positive function.

The converse of the above result is also valid. That is to say, any function of the form

$$f(x) = Ce^{rx}$$

where r is a nonzero constant, satisfies the differential equation

$$f'(x) = rf(x)$$

To establish the validity of this claim, suppose

$$f(x) = Ce^{rx}$$

where r is a nonzero constant. Using the generalized exponential rule and the constant multiple rule,

$$\begin{aligned} f'(x) &= Cre^{rx} \\ &= rCe^{rx} \\ &= rf(x) \qquad \text{since } f(x) = Ce^{rx} \end{aligned}$$

Thus, any function of the form

$$f(x) = Ce^{rx}$$

where r is a nonzero constant, indeed satisfies the differential equation

$$f'(x) = rf(x)$$

Theorem 6.7 summarizes our conclusions regarding differential equations of the form $f'(x) = rf(x)$.

Theorem 6.7

Suppose r is a nonzero constant. The solutions of the differential equation

$$f'(x) = rf(x)$$

are the functions of the form

$$f(x) = Ce^{rx}$$

General and Particular Solutions

We can paraphrase the above theorem by saying that the **general solution** of

$$f'(x) = rf(x)$$

is

$$f(x) = Ce^{rx}$$

where C is an arbitrary constant. If we assign particular values to C, we obtain **particular solutions**.

Using the above terminology, the general solution of

$$f'(x) = 0.14 f(x)$$

is

$$f(x) = Ce^{0.14x}$$

But the particular solution that satisfies the condition

$$f(0) = 2$$

is

$$f(x) = 2e^{0.14x}$$

while the particular solution that satisfies the condition
$$f(0) = -3$$
is
$$f(x) = -3e^{0.14x}$$

Figure 6.14 shows the graphs of these two solutions.

Figure 6.15 shows the graphs of two particular solutions of the differential equation
$$f'(x) = -0.09f(x)$$
namely,
$$f(x) = 3e^{-0.09x}$$
and
$$f(x) = -4e^{-0.09x}$$

Differential equations of the type $f'(x) = rf(x)$ arise in many situations. Example 6.12 deals with such a situation.

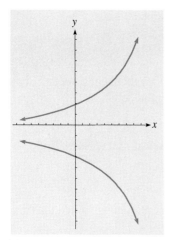

Figure 6.14 Two solutions of $f'(x) = 0.14f(x)$.

Example 6.12

A construction accident caused 900 pounds of a contaminant to spill into a swimming pool containing 35,000 gallons of water. To alleviate the problem, the contaminated water is pumped out of the pool at the constant rate of 7000 gallons per hour. Uncontaminated water is pumped simultaneously into the pool at the same rate. Thus, the pool always contains 35,000 gallons of water.

The contaminant remains evenly dispersed. So the pumping procedure gradually reduces the amount of contaminant in the pool. Health officials warn that the pool is unsafe until less than 50 pounds of the contaminant remains. According to this requirement, is the pool safe after 10 hours of pumping?

Solution

Let $A(t)$ represent the amount of the contaminant in the pool after t hours of pumping. We will derive a formula for $A(t)$. Then we will substitute 10 for t in the formula. The result will tell us if the pool is safe after 10 hours of pumping.

Since the contaminant remains evenly dispersed and the pool always contains 35,000 gallons of water, each gallon of water contains
$$\frac{A(t)}{35,000}$$
pounds of the contaminant after t hours of pumping. Therefore, since the water is pumped out at the constant rate of 7000 gallons per hour, the contaminant leaves the pool at the rate of
$$7000 \cdot \frac{A(t)}{35,000}$$

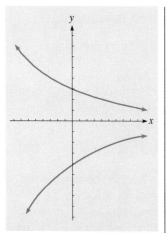

Figure 6.15 Two solutions of $f'(x) = -0.09f(x)$.

pounds per hour after t hours of pumping. So, since

$$7000 \cdot \frac{A(t)}{35{,}000} = 0.2A(t)$$

and $A'(t)$ is the rate at which the contamination level changes, we can conclude that the function $A(t)$ satisfies the differential equation

$$A'(t) = -0.2A(t)$$

[We use the negative sign because $A(t)$ decreases as t increases.] Thus, according to Theorem 6.7,

$$A(t) = Ce^{-0.2t}$$

for some constant C. Since 900 pounds of the contaminant spilled into the pool,

$$A(0) = 900$$

which implies that $C = 900$. Therefore,

$$A(t) = 900e^{-0.2t}$$

Substituting 10 for t, we get

$$A(10) = 900e^{-0.2(10)}$$
$$\approx 122$$

This result tells us that about 122 pounds of the contaminant remains in the pool after 10 hours of pumping. So, since $122 > 50$, the pool is still unsafe after 10 hours of pumping.

The Language of Proportionality

A variable y is **proportional** to a variable x if there exists a nonzero constant c such that

$$y = cx$$

Thus, if $f(x)$ is a function that satisfies a differential equation of the type

$$f'(x) = rf(x)$$

where r is a nonzero constant, we can say that $f(x)$ changes at a rate proportional to $f(x)$. We use this terminology in Example 6.13.

Example 6.13

A woman started a special diet 5 weeks ago. As long as she remains on the diet, she loses her excess weight at a rate proportional to her excess weight. Two weeks ago she was 24 pounds overweight. At that time she was losing her excess weight at the rate of 3 pounds per week. If she continues the diet, how many pounds overweight will she be 4 weeks from now?

Solution

Let $W(t)$ represent the woman's excess weight after t weeks of dieting. Since she loses her excess weight at a rate proportional to her excess weight, there exists a nonzero constant r such that

$$W'(t) = rW(t) \qquad (11)$$

Since she was losing her excess weight at the rate of 3 pounds per week when she was 24 pounds overweight, $W'(t) = -3$ when $W(t) = 24$. Substituting these values in formula (11), we get

$$-3 = r(24)$$

So
$$r = -0.125$$

Substituting -0.125 for r in formula (11), we obtain

$$W'(t) = -0.125W(t)$$

Thus, according to Theorem 6.7,

$$W(t) = Ce^{-0.125t} \qquad (12)$$

for some constant C.

Two weeks ago the woman had been on the diet for 3 weeks. Therefore, since 2 weeks ago she was 24 pounds overweight, $W(3) = 24$, which tells us that

$$24 = Ce^{-0.125(3)}$$

So
$$C = 24e^{0.375}$$
$$\approx 35$$

Substituting 35 for C in formula (12), we get

$$W(t) = 35e^{-0.125t}$$

We can use this formula to approximate the woman's excess weight at any time while she diets. Since she started the diet 5 weeks ago, 4 weeks from now she will be on the diet 9 weeks. Consequently, since

$$W(9) = 35e^{-0.125(9)}$$
$$\approx 11$$

the woman will be approximately 11 pounds overweight 4 weeks from now.

6.5 Problems

In Problems 1 to 6, find the general solution of the differential equation. Then find the particular solution that satisfies the given condition.

1. $f'(x) = -0.15f(x)$ $f(0) = 45$

2. $f'(x) = 0.24f(x)$ $f(0) = 130$

3. $f'(x) = 0.7f(x)$ $f(0) = 60$

4. $f'(x) = -0.3f(x)$ $f(0) = 48$

5. $f'(x) = -0.04f(x)$ $f(8) = 75$
6. $f'(x) = 0.08f(x)$ $f(6) = 90$

In Problems 7 to 10, find the function that satisfies the given set of conditions. Then sketch the graph of the function for $x \geq 0$.

7. $f'(x) = rf(x)$
   $f(0) = 25$
   $f(10) = 40$

8. $f'(x) = rf(x)$
   $f(0) = 75$
   $f(9) = 25$

9. $f'(x) = rf(x)$
   $f'(x) = -2$ when $f(x) = 10$
   $f(0) = 15$

10. $f'(x) = rf(x)$
    $f'(x) = 4$ when $f(x) = 16$
    $f(0) = 7$

11. **World Population** According to a Washington research group known as The Population Reference Bureau, the world population was about 4.8 billion at the beginning of May 1984. Suppose the population increases at a rate proportional to its size. Also, suppose it increases at the rate of 0.085 billion per year when it is 5 billion. What will the population be at the beginning of May 2034?

12. **Concentration of a Drug in a Bloodstream** A man's bloodstream contained 450 milligrams of a drug 8 hours ago. His body absorbs the drug at a rate proportional to the amount of the drug in his bloodstream. When his bloodstream contained 390 milligrams of the drug, his body was absorbing it at the rate of 35.1 milligrams per hour. How much of the drug will remain in his bloodstream 2 hours from now?

13. **Depreciation** An accountant believes that a machine depreciates at a rate proportional to its value. Four years ago, the machine was worth $48,000. Presently, it is worth $32,000. How much will it be worth 2 years from now?

14. **Physics** A physicist has been observing a substance that decays at a rate proportional to its amount. She had 12 grams of the substance 5 hours ago. If she had 4 grams of the substance 2 hours ago, how much will she have 1 hour from now?

15. **Concentration of a Pollutant** Six months ago, a chemical plant installed an antipollution device. At that time, the concentration of a pollutant in a nearby lake was 80 milligrams per liter. Since then, the concentration has been decreasing at a rate proportional to the concentration. Suppose the concentration continues decreasing in this manner. If the concentration is presently 20 milligrams per liter, what will it be 3 months from now?

16. **Finance** Three years ago, $25,000 was placed in an account where the balance increases at a rate proportional to the balance. If the balance is presently $32,000, what will it be 2 years from now?

17. **Illiteracy** Ten years ago, a country implemented a new system of education. At that time, the country had 70 million illiterates. Since the implementation, illiteracy has been decreasing at a rate proportional to its level. Suppose illiteracy continues decreasing in this manner. Also, suppose that the number of illiterates was decreasing at the rate of 3 million per year when the country had 60 million illiterates. How many illiterates will the country have 5 years from now?

18. **Production Discontinuance of a Car** An automobile manufacturer recently discontinued the production of Star Trak, one of its sports cars. Presently, there are 40,000 Star Traks in use. A car magazine predicts that the number of Star Traks in use will decrease at a rate proportional to the number in use. The magazine also predicts that the number in use will be decreasing at the rate of 1000 Star Traks per year when there are 25,000 Star Traks in use. How many Star Traks will still be in use 15 years from now?

19. **Dilution of a Solution** Presently, a tank contains 12,000 gallons of a solution of water and 500 pounds of a chemical. The solution is diluted by pumping pure water into the tank at the rate of 3000 gallons per hour and simultaneously draining the tank at the same rate. If the chemical is always evenly dispersed, when will there be only 100 pounds of the chemical in the tank?

20. **Dilution of a Solution** Suppose the tank in Problem 19 presently contains 20,000 gallons of the solution, and pure water is pumped into the tank at the rate of 2000 gallons per hour while the tank is drained at the same rate. How long will it take the 500 pounds of the chemical to be reduced to 100 pounds?

21. We can perceive that several points lie on a line more easily than we can perceive that several points lie on the graph of an exponential function. Explain how we can use this fact to discover that a quantity changes exponentially.

22. Explain how Theorem 6.7 implies the converse of Theorem 4.11 in Section 4.5. In other words, explain how Theorem 6.7 implies that if a quantity, say $f(x)$, changes in a manner whereby its percentage rate of change remains constant as x increases, then the quantity changes exponentially.

6.6 Differential Equations for Bounded Change

Recall that *differential equations* are equations that involve derivatives and whose solutions are functions rather than numbers. In this section, we deal with differential equations whose solutions can be used to model quantities that approach a fixed level at a decreasing rate.

We begin by assuming that $f(x)$ is a function for which

$$f'(x) = 0.3(5 - f(x))$$

For now, we also assume that $f(x) < 5$ for each value of x in the domain of $f(x)$. If we divide both sides of the above equation by $5 - f(x)$, we get

$$\frac{f'(x)}{5 - f(x)} = 0.3$$

Using the integral versions of the logarithm rule and the chain rule, we can show that $-\ln(5 - f(x))$ is an antiderivative of

$$\frac{f'(x)}{5 - f(x)}$$

According to the constant rule for integrals, $0.3x$ is an antiderivative of 0.3. Therefore, since $f'(x)/(5 - f(x)) = 0.3$, the functions $-\ln(5 - f(x))$ and $0.3x$ are antiderivatives of the same function, and so they must differ by a constant. In other words, there exists a constant d such that

$$-\ln(5 - f(x)) = 0.3x + d$$
$$\ln(5 - f(x)) = -0.3x - d \qquad \text{multiplying by } -1$$
$$e^{-0.3x - d} = 5 - f(x) \qquad \text{definition of } \ln x$$
$$f(x) = 5 - e^{-0.3x - d} \qquad \text{solving for } f(x)$$
$$ = 5 - e^{-d}e^{-0.3x} \qquad \text{a property of exponents}$$

Since d is a constant, e^{-d} is also a constant. If we let $c = e^{-d}$, we can write

$$f(x) = 5 - ce^{-0.3x}$$

Therefore, we have proven that any function $f(x)$ that satisfies the differential equation

$$f'(x) = 0.3(5 - f(x))$$

must have the form
$$f(x) = 5 - ce^{-0.3x}$$
In proving this result, we used the assumption that $f(x) < 5$ for each value of x in the domain of $f(x)$. Actually, the result is valid even if $f(x)$ does not satisfy this requirement.

The converse of the above result is also valid. That is to say, any function of the form
$$f(x) = 5 - ce^{-0.3x}$$
satisfies the differential equation
$$f'(x) = 0.3(5 - f(x))$$
To establish the validity of this claim, suppose
$$f(x) = 5 - ce^{-0.3x}$$
Then $\qquad\qquad\qquad f'(x) = 0.3ce^{-0.3x}$

But $\qquad\qquad\qquad ce^{-0.3x} = 5 - f(x) \qquad$ because $f(x) = 5 - ce^{-0.3x}$

Therefore, $\qquad\qquad f'(x) = 0.3(5 - f(x))$

Thus, any function of the form
$$f(x) = 5 - ce^{-0.3x}$$
indeed satisfies the differential equation
$$f'(x) = 0.3(5 - f(x))$$
With the above differential equation we illustrated Theorem 6.8.

Theorem 6.8

Suppose r is a positive constant and b is any constant. The solutions of the differential equation
$$f'(x) = r(b - f(x))$$
are the functions of the form
$$f(x) = b - ce^{-rx}$$

General and Particular Solutions

We can paraphrase the above theorem by saying that the general solution of
$$f'(x) = r(b - f(x))$$
is
$$f(x) = b - ce^{-rx}$$

where c is an arbitrary constant. By assigning particular values to c, we obtain particular solutions. The particular solution that satisfies the condition

$$f(0) = a$$

is
$$f(x) = b - (b - a)e^{-rx}$$

To realize the validity of this claim, note that if $f(0) = a$, then

$$a = b - ce^{-r(0)}$$
$$a = b - c \quad \text{since } e^0 = 1$$
$$c = b - a \quad \text{solving for } c$$

According to the above terminology, the general solution of the differential equation

$$f'(x) = 0.3(5 - f(x))$$

is
$$f(x) = 5 - ce^{-0.3x}$$

But the particular solution that satisfies the condition

$$f(0) = 1$$

is
$$f(x) = 5 - (5 - 1)e^{-0.3x}$$
$$= 5 - 4e^{-0.3x}$$

Similarly, the particular solution that satisfies the condition

$$f(0) = 8$$

is
$$f(x) = 5 - (5 - 8)e^{-0.3x}$$
$$= 5 - (-3)e^{-0.3x}$$
$$= 5 + 3e^{-0.3x}$$

Figure 6.16 shows the graphs of these two solutions.

Suppose $f(x)$ is a solution of a differential equation of the type

$$f'(x) = r(b - f(x))$$

where r is a positive constant and b is any constant. Figure 6.16 illustrates that, if $f(0) > b$, then the graph of $f(x)$ is above the horizontal line $y = b$. The figure also illustrates that, if $f(0) < b$, then the graph of $f(x)$ is below the horizontal line $y = b$. In either case, the horizontal line is a horizontal asymptote of the graph of $f(x)$.

Example 6.14 deals with a situation in which a differential equation of the type $f'(x) = r(b - f(x))$ arises.

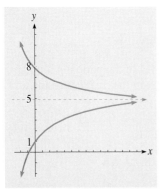

Figure 6.16 Two solutions of $f'(x) = 0.3(5 - f(x))$.

Example 6.14

A nurse began feeding glucose intravenously to a patient when the patient's bloodstream contained 5 grams of glucose. Let $G(t)$ represent the number of grams of glucose in the patient's bloodstream after t minutes of intravenous feeding. The patient's body absorbs glucose *from the bloodstream* at the rate of

$$0.01G(t)$$

grams per minute after t minutes of intravenous feeding. If the glucose is fed *into the bloodstream* at the constant rate of 0.075 gram per minute, how long does it take for the glucose level in the patient's bloodstream to reach 7 grams?

Solution

The glucose *enters* the bloodstream at the constant rate of 0.075 gram per minute. It *leaves* the bloodstream at the rate of $0.01G(t)$ grams per minute after t minutes of intravenous feeding. Therefore, the glucose level in the bloodstream changes at the rate of

$$0.075 - 0.01G(t)$$

grams per minute after t minutes of intravenous feeding. This result tells us that the function $G(t)$ satisfies the differential equation

$$G'(t) = 0.075 - 0.01G(t)$$
$$= 0.01(7.5 - G(t)) \quad \text{factoring}$$

So, according to Theorem 6.8,

$$G(t) = 7.5 - ce^{-0.01t}$$

for some constant c. Since the nurse began the intravenous feeding when the patient's bloodstream contained 5 grams of glucose,

$$G(0) = 5$$

which implies that

$$c = 7.5 - 5$$
$$= 2.5$$

Therefore,

$$G(t) = 7.5 - 2.5e^{-0.01t}$$

To find how long it takes for the glucose level in the patient's bloodstream to reach 7 grams, we substitute 7 for $G(t)$ in the above formula and solve for t:

$$7 = 7.5 - 2.5e^{-0.01t}$$
$$0.2 = e^{-0.01t}$$
$$-0.01t = \ln 0.2 \quad \text{definition of ln } x$$
$$t = \frac{\ln 0.2}{-0.01}$$
$$\approx 161$$

Therefore, the glucose level in the patient's bloodstream reaches 7 grams after approximately 161 minutes of intravenous feeding.

Figure 6.17 shows the graph of

$$G(t) = 7.5 - 2.5e^{-0.01t}$$

for $t \geq 0$. The graph shows that the glucose level approaches 7.5 grams while increasing at a decreasing rate.

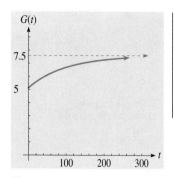

Figure 6.17
$G(t) = 7.5 - 2.5e^{-0.01t}$ for $t \geq 0$.

The Language of Proportionality

If $f(x)$ is a function that satisfies a differential equation of the type

$$f'(x) = r(b - f(x))$$

we can say that $f(x)$ changes at a rate *proportional* to $b - f(x)$. We use this terminology in Example 6.15.

Example 6.15

A pathologist is trying to establish when a man died. The body is located in the basement of an office building where the temperature is always 70 degrees Fahrenheit. The pathologist believes that the body has been at that location since the man died. Therefore, according to a law known as Newton's law of cooling, the body's temperature has been decreasing at a rate proportional to the difference between 70 degrees Fahrenheit and the body's temperature. The pathologist observed that, when the body's temperature was 82 degrees Fahrenheit, it was dropping at the rate of 3 degrees per hour. Suppose the man had a body temperature of 98.6 degrees Fahrenheit when he died. If the temperature of the body is now 74 degrees Fahrenheit, when did the man die?

Solution

Let $T(x)$ represent the temperature of the body x hours after the man died. Since the body's temperature decreases at a rate proportional to the difference between 70 degrees Fahrenheit and the body's temperature, there exists a positive constant r such that

$$T'(x) = r(70 - T(x)) \tag{13}$$

Since the body's temperature was dropping at the rate of 3 degrees per hour when it was 82 degrees Fahrenheit, we have that $T'(x) = -3$ when $T(x) = 82$. Substituting these values in formula (13), we get

$$-3 = r(70 - 82)$$

which implies that

$$r = 0.25$$

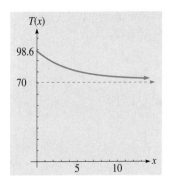

Figure 6.18
$T(x) = 70 + 28.6e^{-0.25x}$
for $x \geq 0$.

Substituting 0.25 for r in formula (13), we obtain

$$T'(x) = 0.25(70 - T(x))$$

Therefore, according to Theorem 6.8,

$$T(x) = 70 - ce^{-0.25x}$$

for some constant c. Since the man had a body temperature of 98.6 degrees Fahrenheit when he died,

$$T(0) = 98.6$$

which implies that

$$c = 70 - 98.6$$
$$= -28.6$$

So
$$T(x) = 70 - (-28.6)e^{-0.25x}$$
$$= 70 + 28.6e^{-0.25x}$$

We want to find the value of x for which $T(x) = 74$. So we want to solve the following equation:

$$74 = 70 + 28.6e^{-0.25x}$$

$$\frac{4}{28.6} = e^{-0.25x}$$

$$-0.25x = \ln\left(\frac{4}{28.6}\right) \quad \text{definition of } \ln x$$

$$x = \frac{\ln(4/28.6)}{-0.25}$$

$$\approx 7.8$$

Therefore, it took nearly 8 hours for the temperature of the body to drop to 74 degrees Fahrenheit. This result means that the man died almost 8 hours ago.

Figure 6.18 shows the graph of

$$T(x) = 70 + 28.6e^{-0.25x}$$

for $x \geq 0$. The graph shows that the body's temperature approaches 70 degrees Fahrenheit while decreasing at a decreasing rate.

6.6 Problems

In Problems 1 to 6, find the general solution of the differential equation. Then find the particular solution that satisfies the given condition.

1. $f'(x) = 0.24(6 - f(x))$ $f(0) = 2$
2. $f'(x) = 0.3(1 - f(x))$ $f(0) = 5$
3. $f'(x) = 0.18(2 - f(x))$ $f(0) = 8$
4. $f'(x) = 0.4(7 - f(x))$ $f(0) = 6$
5. $f'(x) = 0.26(3 - f(x))$ $f(5) = 4$
6. $f'(x) = 0.32(8 - f(x))$ $f(4) = 6$

In Problems 7 to 12, find the function that satisfies the given set of conditions. Then sketch the graph of the function for $x \geq 0$.

7. $f'(x) = r(9 - f(x))$
   $f'(x) = 1$ when $f(x) = 4$
   $f(0) = 3$

8. $f'(x) = r(12 - f(x))$
   $f'(x) = 1$ when $f(x) = 10$
   $f(0) = 0$

9. $f'(x) = r(4 - f(x))$
   $f'(x) = -0.8$ when $f(x) = 6$
   $f(0) = 9$

10. $f'(x) = r(2 - f(x))$
    $f'(x) = -0.6$ when $f(x) = 4$
    $f(0) = 10$

11. $f'(x) = r(11 - f(x))$
    $f(0) = 2$
    $f(9) = 10$

12. $f'(x) = r(3 - f(x))$
    $f(0) = 7$
    $f(5) = 4$

13. **Depreciation** A machine used in manufacturing shoe soles is worth $54,000 when new. An accountant assumes the machine depreciates at a rate proportional to the difference between its scrap value and its value. If the machine is worth $40,000 when it is 3 years old and has a scrap value of $10,000, how much will it be worth when it is 8 years old?

14. **Consumer Debt** Four years ago, the consumer debt in a country was $6 billion. Since then, this debt has been increasing. Although economists believe that the debt will continue to increase, they predict that it will never surpass $14 billion. To be more specific, economists believe that the consumer debt increases at a rate proportional to the difference between $14 billion and the debt. If presently the debt is $8 billion, when will it be $12 billion?

15. **Dieting** A man started a diet to gain weight 8 weeks ago when his weight was 90 pounds. The physician who recommended the diet believes that, as long as the man remains on the diet, his weight will gradually increase. However, the physician predicts that the man's weight will never surpass 140 pounds. In fact, the physician believes that as long as the man remains on the diet, his weight will increase at a rate proportional to the difference between 140 pounds and his weight. If the man's weight was increasing at the rate of 4 pounds per week when he weighed 100 pounds, when will he weigh 130 pounds?

16. **Law of Cooling** Five hours ago, a woman removed a vase from a kiln and placed it in a room where the temperature is always 74 degrees Fahrenheit. At that time, the temperature of the vase was 180 degrees Fahrenheit. According to Newton's law of cooling, the temperature of the vase decreases at a rate proportional to the difference between the room temperature and the temperature of the vase. If the temperature of the vase was decreasing at the rate of 12 degrees per hour when it was 134 degrees Fahrenheit, what will be its temperature 2 hours from now?

    Caution: Two hours from now is 7 hours after the vase was placed in the room.

17. **Aerodynamics** An aerodynamicist plans to drop an object from a point high above the ground. If the air in the object's path were somehow eliminated, the velocity of the object would increase at the constant rate of 32 feet per second per second. In reality, however, due to the resistance of air, the velocity will increase at a decreasing rate while it approaches what is called the "terminal velocity" of the object. Therefore, it is reasonable to assume that the actual velocity will increase at a rate proportional to the difference between the terminal velocity and the velocity. The aerodynamicist determined that the terminal velocity of the object is 800 feet per second. She predicts that, when the velocity is 200 feet per second, the velocity will be increasing at the rate of 24 feet per second per second. Find a formula for the velocity of the object t seconds after it is dropped.

18. **Aerodynamics** Do Problem 17 again. But this time assume that the terminal velocity of the object is 1600 feet per second and that the velocity will be increasing at the rate of 20 feet per second per second when it is 600 feet per second.

19. **Enrollment in an Insurance Program** Two years ago, 1 million people were enrolled in a health maintenance insurance program. Experts predict that the enrollment (in millions) will increase at a rate proportional to the difference between 6 million and the enrollment. If presently the enrollment is 1.75 million, when will it reach 4 million?

20. **Depreciation** A machine used in manufacturing steering wheels is worth $80,000 when new. An

accountant assumes the machine depreciates at a rate proportional to the difference between its scrap value and its value. If the machine is worth $32,000 when it is 5 years old and has a scrap value of $16,000, how old will it be when it is worth $24,000?

21. **Public Transportation** For convenience, transportation officials in a large metropolitan area refer to people who use the public transportation system at least once a week as "users" and people who use the system less often as "nonusers." Ten years ago, of the 12 million people in the area, only 2 million were users. Since then, the number of users has been increasing at a rate proportional to the number of nonusers. Suppose this trend continues. When will there be 7 million users, if the number of users was increasing at the rate of 0.2 million per year when there were 4 million users? (Assume the population of the metropolitan area never changes.)

22. **Cable Television** Five months ago, of the 74,000 households in a city, 10,000 subscribed to cable television. Since then, the number of subscribers has been increasing at a rate proportional to the number of nonsubscribers. Suppose this trend continues. When will there be 60,000 subscribers, if the number of subscribers was increasing at the rate of 3000 per month when there were 24,000 subscribers? (Assume the number of households never changes.)

23. **Concentration of a Solution** A chemist will pump a solution containing 600 grams of potassium bromide per liter into a tank that contains 500 liters of pure water. He will pump the solution into the tank at the rate of 20 liters per hour while simultaneously draining the tank at the same rate. Thus, the tank will always contain 500 liters of liquid. However, the amount of potassium bromide in the tank will gradually increase. Show that it will increase at a rate proportional to the difference between 300,000 and the amount in the tank. Then find a formula for the amount of the bromide in the tank after t hours. How long will it take for the tank to contain 120,000 grams of the bromide? (The bromide is kept evenly dispersed.)

24. **Concentration of a Solution** Do Problem 23 again. This time, however, assume the chemist will pump the solution into the tank at the rate of 25 liters per hour. (He will drain the tank also at this rate.)

25. Suppose $f(x)$ is a solution of a differential equation of the form

$$f'(x) = r(b - f(x))$$

where r is a positive constant and b is any constant.

a. If $f(0) < b$, how does $f(x)$ behave as x increases?

b. If $f(0) > b$, how does $f(x)$ behave as x increases?

c. If $f(0) = b$, how does $f(x)$ behave as x increases?

26. We can perceive that several points lie on a line more easily than we can perceive that several points lie on the graph of a nonlinear function. Explain how we can use this fact to discover that a variable y is a function of another variable x according to a formula of the form

$$y = b - ce^{-rx}$$

Chapter 6 Important Terms

Section 6.1
Simpson's rule (354)

Section 6.2
consumers' surplus (361, 362)
producers' surplus (363, 364)
market equilibrium (365)

Section 6.3
probability (369)
probability density function (371)
improper integral (373)
random variable (374)

Section 6.4
expected value (380)
average value (380)
mean (380)
exponentially distributed random variable (381)
normally distributed random variable (384)

Section 6.5
differential equation (386)
general and particular solutions of a differential equation (388)

Chapter 6 Review Problems

Section 6.1

1. Use Simpson's rule with the indicated value of n to approximate the definite integral. Round your answer to three decimal places.

   a. $\int_3^5 \dfrac{2}{x}\, dx$ $n = 8$

   b. $\int_0^3 \dfrac{8}{x^2 + 9}\, dx$ $n = 6$

   c. $\int_{-3}^3 \sqrt{16 - x^2}\, dx$ $n = 6$

   d. $\int_2^3 \dfrac{1}{\sqrt{12 - x^2}}\, dx$ $n = 8$

2. Suppose $f(x)$ is a function that is continuous on an interval containing 1 and 3. The following table shows the values of $f(x)$ for several values of x:

x	1	$1\tfrac{1}{3}$	$1\tfrac{2}{3}$	2	$2\tfrac{1}{3}$	$2\tfrac{2}{3}$	3
$f(x)$	3	0.4	-1.4	-2	-1.3	1	5

   a. Use Simpson's rule to approximate $\int_1^3 f(x)\, dx$.

   b. Suppose $F(x)$ is the particular antiderivative of $f(x)$ for which $F(1) = 9$. Approximate the value of $F(3)$.

Section 6.2

3. **Consumers' and Producers' Surpluses** Suppose the unit price (in $ thousands) at which consumers buy x units of a product weekly is

   $$D(x) = -\tfrac{1}{12}x^2 + 6$$

   and the unit price at which producers supply x units of the product weekly is

   $$S(x) = 7 - 0.5\sqrt{100 - x^2}$$

   (These formulas are valid for $0 \le x \le 8$.)

   a. Find the weekly consumers' surplus when the market is in equilibrium.

   b. Use Simpson's rule with $n = 12$ to approximate the weekly producers' surplus when the market is in equilibrium.

Section 6.3

4. Verify that the function

   $$f(x) = \dfrac{30}{(x + 7)^2}$$

   is a probability density function on the interval $[5, 13]$.

5. Verify that the function

   $$f(x) = \dfrac{8x}{(x^2 + 4)^2}$$

   is a probability density function on the interval $[0, +\infty)$.

6. **Lifespan of a Car** A major automobile manufacturer recently began producing an economy sports car called Ecomo. Let x represent the lifespan (in years) of a randomly selected Ecomo. Then x is a random variable distributed on the interval $[0, +\infty)$. Suppose the probability density function of x is

   $$f(x) = 0.36xe^{-0.6x}$$

   a. Find the probability that a randomly selected Ecomo has a lifespan of at most 4 years.

   b. Approximately what percentage of the Ecomos in a large batch lasts more than 6 years?

Section 6.4

7. If x is a random variable distributed on the interval $[0, 6]$ according to the probability density function

   $$f(x) = \tfrac{1}{3}x - \tfrac{1}{9}x^2 + \tfrac{1}{108}x^3$$

   what is the expected value of x?

8. Suppose x is a random variable distributed on the interval $[0, +\infty)$. If the probability density function of x is

   $$f(x) = 0.32e^{-0.32x}$$

   what is the average value of x?

9. **Average Outstanding Balance** Plasto is a charge card that has a credit limit of $8000. Therefore, if x represents the outstanding balance (in $ thousands) on a randomly selected Plasto, then x is a random variable distributed on the interval $[0, 8]$. Suppose the distribution is according to the probability density function

$$f(x) = \frac{10.5}{(x+6)^2}$$

Find the average outstanding balance on the charge cards.

Section 6.5

10. Find the general solution of the differential equation. Then find the particular solution that satisfies the given condition.

    a. $f'(x) = -0.25f(x)$ $f(0) = 38$
    b. $f'(x) = 0.16f(x)$ $f(4) = 57$

11. Find the function that satisfies the following set of conditions:

    $$f'(x) = rf(x)$$
    $$f(0) = 32$$
    $$f(5) = 40$$

12. **Eradication of a Disease** Two years ago, a research group discovered a cure for Ro's disease, a communicable disease that causes itchy abdominal rashes. At that time, 5.25 million people had the disease. Since the discovery, the disease has been disappearing at a rate proportional to the number of remaining cases. The number of cases was decreasing at the rate of 0.86 million per year when 4 million people had the disease. How many people will have the disease 3 years from now?

Section 6.6

13. Find the general solution of the differential equation. Then find the particular solution that satisfies the given condition.

    a. $f'(x) = 0.14(5 - f(x))$ $f(0) = 2$
    b. $f'(x) = 0.19(3 - f(x))$ $f(0) = 7$
    c. $f'(x) = 0.22(4 - f(x))$ $f(3) = 9$
    d. $f'(x) = 0.34(6 - f(x))$ $f(0) = 0$

14. Find the function that satisfies the given set of conditions. Then sketch the graph of the function for $x \geq 0$.

    a. $f'(x) = r(1 - f(x))$
       $f'(x) = -0.9$ when $f(x) = 5$
       $f(0) = 6$

    b. $f'(x) = r(7 - f(x))$
       $f'(x) = 2$ when $f(x) = 3$
       $f(0) = 1$

    c. $f'(x) = r(2 - f(x))$
       $f(0) = 5$
       $f(4) = 3$

15. **Magazine Subscriptions** A marketing expert believes that in any city the number of subscribers to *The Vita Journal* increases at a rate proportional to the number of nonsubscribers. Three years ago, of the 7 million people in a northeastern city, only 1.25 million subscribed to the magazine. In that city, the number of subscribers was increasing at the rate of 0.6 million per year when there were 2 million subscribers. When will the city have 4.5 million subscribers? (Assume that the population of the city is always 7 million.)

CHAPTER 7

Differentiation of Functions of More than One Variable

- **7.1** Functions of More than One Variable
- **7.2** Partial Derivatives
- **7.3** Optimization
- **7.4** The Second Derivative Test
- **7.5** Curve Fitting
- **7.6** Constrained Optimization
- **7.7** The Method of Lagrange Multipliers

Important Terms

Review Problems

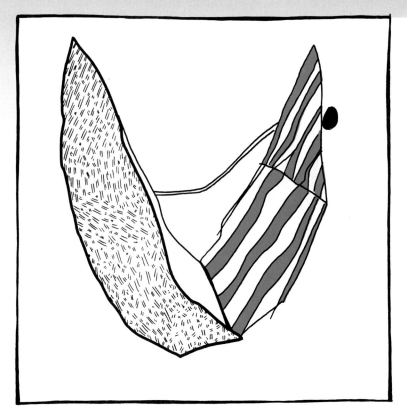

In this chapter, we adapt the derivative concept to functions that have more than one independent variable. We then use the adaptation to solve optimization problems and to approximate the change in the dependent variable if one of the independent variables increases 1 unit.

7.1 Functions of More than One Variable

In previous chapters, we restricted ourselves to situations that involve a variable whose values depend on the values of *only one* other variable. Many situations, however, involve a variable whose values depend on the values of *more than one* other variable. The following example illustrates such a situation.

Since many people substitute broccoli for cauliflower in their meals, agriculture officials believe that any increase in the price of cauliflower causes an increase in the demand for broccoli. To be more specific, they believe that consumers buy

$$z = -\tfrac{1}{18}x^2 + \tfrac{1}{2}y + 8 \tag{1}$$

million crates of broccoli monthly if the price of broccoli is $x per crate and the price of cauliflower is $y per crate. Thus, for example, if the price of broccoli is $6 per crate and the price of cauliflower is $10 per crate, consumers buy 11 million crates of broccoli monthly. We obtain this result by substituting 6 for x and 10 for y in formula (1):

$$z = -\tfrac{1}{18}(6^2) + \tfrac{1}{2}(10) + 8$$
$$= 11$$

Let us agree that the first number in an ordered pair of nonnegative real numbers represents the price of broccoli and the second number represents the price of cauliflower. For example, in the ordered pair (6, 10), the number 6 is the price of broccoli, while the number 10 is the price of cauliflower. According to this agreement, formula (1) is a rule that associates exactly one real number with each ordered pair of nonnegative real numbers. For instance, the formula associates the number 11 with the ordered pair (6, 10).

Definition 7.1

A **function of two independent variables** is a rule that associates exactly one real number z with each ordered pair (x, y) in a set S of ordered pairs of real numbers.

In the above definition, the set S is the **domain** of the function, while the set of real numbers associated with the ordered pairs in the set S is the **range** of the function.

According to Definition 7.1, formula (1) is a function of two independent variables. The **independent variables** are x and y. The **dependent variable** is z. The domain is the set of ordered pairs of nonnegative real numbers. We impose this restriction on the domain because x and y represent prices. Actually, since z represents the demand for a product, we restrict the domain to those ordered pairs of nonnegative real numbers for which the corresponding values of z are also nonnegative real numbers. For example, (15, 4) is *not* in the domain because the corresponding z value is a negative number, namely, -2.5. We can verify this claim by substituting 15 for x and 4 for y in formula (1).

Notation

We can use a symbol such as $f(x, y)$ (read "f of x comma y") to represent our demand function. Then we can write formula (1) as follows:

$$f(x, y) = -\tfrac{1}{18}x^2 + \tfrac{1}{2}y + 8 \tag{2}$$

This notation expresses our agreement that the first number in an ordered pair of numbers represents the price of broccoli and the second number represents the price of cauliflower. So $f(5, 8)$, for example, represents the amount of broccoli consumers buy monthly if the price of broccoli is $5 per crate and the price of cauliflower is $8 per crate.

Example 7.1

Suppose $f(x, y)$ is the demand function given by formula (2). Compute $f(9, 12)$ and interpret the result.

Solution

Substituting 9 for x and 12 for y in formula (2), we get

$$f(9, 12) = -\tfrac{1}{18}(9^2) + \tfrac{1}{2}(12) + 8$$
$$= 18.5$$

Thus, consumers buy 18.5 million crates of broccoli monthly if the price of broccoli is $9 per crate and the price of cauliflower is $12 per crate.

Change

We often want to know how much the value of the dependent variable changes if the value of one of the independent variables increases 1 unit.

Example 7.2

How much does the monthly demand for broccoli increase if the price of cauliflower increases from $13 to $14 per crate and the price of broccoli remains at $11 per crate?

Solution

When the price of broccoli is $11 per crate and the price of cauliflower is $13 per crate, the monthly demand for broccoli is

$$f(11, 13) = -\frac{1}{18}(11^2) + \frac{1}{2}(13) + 8$$
$$\approx 7.8$$

million crates. If the price of cauliflower increases $1 per crate and the price of broccoli remains at $11 per crate, the monthly demand for broccoli is

$$f(11, 14) = -\frac{1}{18}(11^2) + \frac{1}{2}(14) + 8$$
$$\approx 8.3$$

million crates. Therefore, if the price of cauliflower increases from $13 to $14 per crate and the price of broccoli remains fixed at $11 per crate, the monthly demand for broccoli increases

$$f(11, 14) - f(11, 13) \approx 8.3 - 7.8$$
$$= 0.5$$

million crates.

Example 7.2 illustrates that

$$f(a, b+1) - f(a, b)$$

represents the amount $f(x, y)$ changes if y increases from b to $b+1$ and x remains fixed at a. Similarly,

$$f(a+1, b) - f(a, b)$$

represents the amount $f(x, y)$ changes if x increases from a to $a+1$ and y remains fixed at b.

Example 7.3

How much does the monthly demand for broccoli decrease if its price increases from $11 to $12 per crate and the price of cauliflower remains fixed at $13 per crate?

Solution

We already know that $f(11, 13) \approx 7.8$. Therefore, since

$$f(12, 13) = -\frac{1}{18}(12^2) + \frac{1}{2}(13) + 8$$
$$= 6.5$$

we conclude that

$$f(12, 13) - f(11, 13) \approx 6.5 - 7.8$$
$$= -1.3$$

So the monthly demand for broccoli decreases about 1.3 million crates if its price increases from $11 to $12 per crate and the price of cauliflower remains at $13 per crate.

Three-Dimensional Coordinate Systems

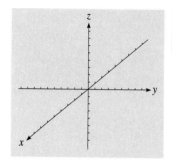

Figure 7.1 A three-dimensional coordinate system.

We can visualize functions of two independent variables geometrically in the three-dimensional expanse we live in, namely, *space*. Such visualizations require three mutually perpendicular number lines that meet at their respective origins. (See Figure 7.1.) If we call these lines the x axis, the y axis, and the z axis, we orient the xy plane horizontally. (The xy plane is the plane determined by the x axis and the y axis.) Thus, we orient the z axis vertically. The points on the x axis that have positive coordinates are in front of the yz plane. (Those that have negative coordinates are behind the yz plane.) On the y axis, the points that have positive coordinates are to the right of the xz plane, while on the z axis such points are above the xy plane.

We visualize the ordered pairs of numbers in the domain of a function of two independent variables as points in the xy plane. For instance, in the context of our demand example, we visualize the domain of

$$f(x, y) = -\tfrac{1}{18}x^2 + \tfrac{1}{2}y + 8$$

as a certain collection of points in the xy plane. From now on, however, we consider this function from a strictly mathematical point of view. So now its domain is the set of *all* ordered pairs of real numbers. Geometrically, now the domain is the set of all points in the xy plane.

Functions of two independent variables generated *ordered triples* of real numbers. For example, the function

$$f(x, y) = -\tfrac{1}{18}x^2 + \tfrac{1}{2}y + 8$$

generates ordered triples of the form (x, y, z), where $z = f(x, y)$. Since $f(9, 8) = 7.5$, one of the infinitely many ordered triples generated by this function is $(9, 8, 7.5)$.

The three axes in Figure 7.1 enable us to visualize ordered triples of real numbers as points in space. For example, we can visualize the ordered triple $(9, 8, 7.5)$ as the point located 7.5 units above the point $(9, 8)$ on the xy plane. Figure 7.2 shows the result of plotting this point. Figure 7.2 also shows the result of plotting the point that corresponds to the ordered triple $(6, -16, -2)$, which is another ordered triple of numbers generated by the function $f(x, y)$. This point is located 2 units below the point $(6, -16)$ on the xy plane.

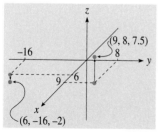

Figure 7.2 The points $(9, 8, 7.5)$ and $(6, -16, -2)$.

Definition 7.2

The **graph** of a function of two independent variables is the set of points in space that corresponds to the set of ordered triples of numbers generated by the function.

The graph of the function

$$f(x, y) = -\tfrac{1}{18}x^2 + \tfrac{1}{2}y + 8$$

is a **surface**. Figure 7.3 shows a portion of this surface.

The first, second, and third numbers in an ordered triple of numbers are, respectively, the x, y, and z coordinates of the point in space that represents the ordered triple of numbers. For example, in Figure 7.2, the y coordinate of point $(9, 8, 7.5)$ is 8.

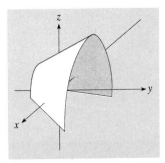

Figure 7.3 A portion of the graph of $f(x, y) = -\tfrac{1}{18}x^2 + \tfrac{1}{2}y + 8$.

Example 7.4

Find the z coordinate of the point on the graph of

$$f(x, y) = -\tfrac{1}{18}x^2 + \tfrac{1}{2}y + 8$$

whose x and y coordinates are 12 and 2, respectively.

Solution

Substituting 12 for x and 2 for y in the formula for $f(x, y)$, we get

$$f(12, 2) = -\tfrac{1}{18}(12^2) + \tfrac{1}{2}(2) + 8$$
$$= 1$$

So the z coordinate of the point is 1.

Linear Functions

Linear functions are functions that can be expressed in the form

$$z = ax + by + c$$

The graphs of such functions are nonvertical *planes*, not lines. Example 7.5 illustrates how to sketch graphs of linear functions.

Example 7.5

Sketch the graph of the linear function

$$z = -2x - 4y + 6$$

Solution

We begin by determining where the graph crosses the x, y, and z axes. Substituting 0 for y and z in the formula, we get

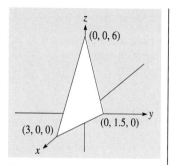

Figure 7.4 A triangular portion of the graph of $z = -2x - 4y + 6$.

$$0 = -2x - 4(0) + 6$$
$$x = 3$$

Thus, the graph crosses the x axis at the point $(3, 0, 0)$.

Similarly, substituting 0 for x and z in the formula, we learn that the graph crosses the y axis at the point $(0, 1.5, 0)$.

Finally, substituting 0 for x and y in the formula, we see that the graph crosses the z axis at $(0, 0, 6)$.

The graph is the plane that passes through the three points we found above. Connecting the three points with line segments, we obtain a triangular portion of the plane. (See Figure 7.4.)

Functions of More than Two Variables

The definition of a function of *more* than two independent variables is similar to the definition of a function of two independent variables. For instance, a **function of three independent variables** is a *rule* that associates exactly one real number with each ordered triple in a set of ordered triples of real numbers called the **domain**. In this case, if we use x, y, and z to represent the three independent variables, we can use a symbol such as $f(x, y, z)$ to represent the function. Note that here z represents an independent variable, rather than a dependent variable.

As we saw, functions of two independent variables generate ordered triples of numbers. The graphs of these functions are three dimensional and can be visualized as surfaces in space. Functions of three independent variables, however, generate *ordered quadruples* of numbers. It is impossible to visualize such functions graphically. In fact, it is impossible to visualize graphically any function of more than two independent variables. Nevertheless, we can use such functions to model real world situations, as Example 7.6 illustrates.

Example 7.6

If P dollars is invested at an annual interest rate of $100r$ percent compounded n times per year, the balance at the end of t years is a function of *four* independent variables, namely, P, r, t, and n. Let B represent the balance. Find a formula for this function. Then find the value of B that corresponds to the following:

$$P = 1000$$
$$r = 0.08$$
$$t = 6$$
$$n = 4$$

Solution

According to Theorem 4.2 in Section 4.1, a formula that expresses B as a function of P, r, t, and n is

$$B = P\left(1 + \frac{r}{n}\right)^{nt}$$

Consequently, when $P = 1000$, $r = 0.08$, $t = 6$, and $n = 4$, we get

$$B = 1000\left(1 + \frac{0.08}{4}\right)^{4(6)}$$

$$= 1608.44$$

So, after 6 years, the balance is $1608.44 if $1000 is invested at an annual interest rate of 8 percent compounded quarterly.

7.1 Problems

1. Find $f(4, 2)$ if $f(x, y) = 5x^2 - 3y + 1$.

2. Find $f(1, 6)$ if $f(x, y) = 7x + y - 10$.

3. If $f(x, y) = \dfrac{y^2}{x + 3}$, compute $f(5, 4)$.

4. If $f(x, y) = \dfrac{20y}{x + 1}$, compute $f(14, 3)$.

5. Given $f(x, y) = 25x^{1/3}y^{2/3}$, find $f(8, 27)$.

6. Given $f(x, y) = 10x^{1/4}y^{3/4}$, find $f(81, 16)$.

7. Compute $f(1, 7, 5)$ if $f(x, y, z) = x^2z + 3xy$.

8. Compute $f(5, 2, 9)$ if $f(x, y, z) = 5y - xz^2$.

9. Suppose

$$f(x, y) = \frac{4y^2}{x + 3}$$

How much does $f(x, y)$ change if x increases from 2 to 3 and y remains fixed at 6?

10. Suppose

$$f(x, y) = 2y^2 - 3x$$

How much does $f(x, y)$ change if y increases from 3 to 4 and x remains fixed at 5?

11. If

$$f(x, y) = 3xy + y$$

what is the change in $f(x, y)$ when y increases from 7 to 8 and x remains fixed at 13?

12. If

$$f(x, y) = \frac{3y + 1}{x^2 + 5}$$

what is the change in $f(x, y)$ if x increases from 6 to 7 and y remains fixed at 4?

13. Let

$$f(x, y) = 17x - y^2$$

Find the change in $f(x, y)$ if x increases from 5 to 6 and y remains fixed at 9.

14. Let

$$f(x, y) = \frac{30}{x^2 + 3y}$$

Find the change $f(x, y)$ if y increases from 1 to 2 and x remains fixed at 5.

For each function in Problems 15 to 22, find the z coordinate of the point on the graph with the given x and y coordinates. Then plot the point.

15. $f(x, y) = 5x + 3y - 11$ $\quad x = 4$ and $y = -1$

16. $f(x, y) = \dfrac{y^2 - 5}{x + 3}$ $\quad x = -2$ and $y = 3$

17. $f(x, y) = -2y \ln x$ $\quad x = 3$ and $y = 4$

18. $f(x, y) = e^{x+y}$ $\quad x = 1$ and $y = 3$

19. $f(x, y) = 3xy + y^2$ $\quad x = 0$ and $y = 3$

20. $f(x, y) = x^2y - 2x$ $\quad x = 3$ and $y = 0$

21. $f(x, y) = 3x - 5y + 24$ $x = 2$ and $y = 6$
22. $f(x, y) = \sqrt{2y^2 - 4x}$ $x = -1$ and $y = -4$

For each linear function in Problems 23 to 26, sketch a triangular portion of the graph.

23. $z = 0.75x + 0.5y - 3$
24. $z = 0.5x - y + 3$
25. $z = -2x + 4y + 8$
26. $z = 3x - 1.5y - 6$

27. **Demand** Let $D(x, y)$ represent the number of million barrels of Chuck-a-lot (a chocolate drink) consumers buy weekly if its price is x cents per ounce and the price of Smiley (another chocolate drink) is y cents per ounce. Suppose $D(x, y)$ is a function of x and y according to

$$D(x, y) = \frac{8y + 50}{x + 2}$$

  a. How much Chuck-a-lot do consumers buy weekly if its price is 3 cents per ounce and the price of Smiley is 5 cents per ounce?

  b. How much does the weekly demand for Chuck-a-lot change if its price increases from 3 cents to 4 cents per ounce and the price of Smiley remains fixed at 5 cents per ounce?

  c. How much does the weekly demand for Chuck-a-lot change if the price of Smiley increases from 5 cents to 6 cents per ounce and the price of Chuck-a-lot remains fixed at 3 cents per ounce?

28. **Production Cost** Suppose the joint weekly cost of producing x barrels of commodity A and y barrels of commodity B per week is

$$C(x, y) = \tfrac{1}{40}x^2 + \tfrac{1}{15}xy + 20$$

  a. Find the joint weekly production cost if the weekly production level of commodity A is 80 barrels and the weekly production level of commodity B is 75 barrels.

  b. How much does the joint weekly production cost increase if the weekly production level of commodity A is increased from 80 barrels to 81 barrels and the weekly production level of commodity B is kept fixed at 75 barrels?

  c. How much does the joint weekly production cost increase if the weekly production level of commodity B is increased from 75 barrels to 76 barrels and the weekly production level of commodity A is kept fixed at 80 barrels?

29. **Productivity of Labor and Capital** Let $P(x, y)$ represent the number of gallons of ice cream Gulf Coast Dairy Products produces daily with x workers and a capital outlay of $\$y$ thousand. Suppose $P(x, y)$ is a function of x and y according to

$$P(x, y) = 100x^{2/3}y^{1/3}$$

  a. Find the amount of ice cream the company produces daily with 343 workers and a capital outlay of $729,000.

  b. If the management doubles the work force and the capital outlay, the daily production level doubles. That is,

$$P(2x, 2y) = 2 \cdot P(x, y)$$

  Verify this result.
    Comment: Actually, $P(kx, ky) = k \cdot P(x, y)$ for any positive constant k. Economists express this result by saying that the returns to scale are constant.

30. **Productivity of Labor and Capital** Parisian Mist is a perfume produced by The Fragrance Company. Let $P(x, y)$ represent the amount of Parisian Mist (in kiloliters) the company produces weekly with x workers and a capital outlay of $\$y$ million. Suppose $P(x, y)$ is a function of x and y according to

$$P(x, y) = \sqrt{3x^2 + 2.92y^2}$$

  a. How much of the perfume does the company produce weekly with six workers and a capital outlay of $10 million?

  b. If the management doubles the work force and the capital outlay, the weekly production level doubles. That is,

$$P(2x, 2y) = 2 \cdot P(x, y)$$

  Verify this result.
    Comment: Actually, $P(kx, ky) = k \cdot P(x, y)$ for any positive constant k.

31. **Budget Constraints** A buyer for a cigar manufacturing firm budgeted $8000 to purchase tobacco. Only three types of tobacco are available: type A at $2 per pound, type B at $3 per pound, and type C at $4 per pound. Let x, y, and z, respectively, represent the number of pounds of each type the buyer can purchase with the entire amount budgeted.

a. Show that z is a linear function of x and y. That is, find constants a, b, and c such that
$$z = ax + by + c$$

b. If the buyer purchases 1000 pounds of type A and 800 pounds of type B, how many pounds of type C must she purchase to spend the entire $8000?

c. If the buyer purchases 700 pounds of type A and 1110 pounds of type C, how many pounds of type B must she purchase to spend the entire $8000?

32. **Box Problem** A carpenter plans to build a box whose sides, bottom, and top are rectangular in shape. (See the figure for this problem.) The material for the side panels costs $4 per square foot, while the material for the bottom and top panels costs $7 per square foot.

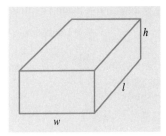

a. Suppose the carpenter decides that the volume of the box must be 200 cubic feet. Find a formula that expresses the cost of the material as a function of w and l.
   Suggestion: First express h in terms of w and l.

b. Suppose the carpenter decides to spend exactly $300 on the material. Find a formula that expresses the volume of the box as a function of w and l.

7.2 Partial Derivatives

In this section, we adapt the derivative concept to functions that have more than one independent variable. We begin with a problem where the demand for a product is expressed as a function of two types of advertising expenditures.

Let $f(x, y)$ represent the number of barrels of orange juice The Juice Company sells daily if it spends x thousand on newspaper advertising and y thousand on television advertising. The management believes that $f(x, y)$ is a function of x and y according to the formula

$$f(x, y) = \tfrac{1}{60}x^2y^3 + 4y + 200$$

Currently, the company spends $8000 daily on newspaper advertising and $5000 daily on television advertising. The management wants to know how the daily demand for the orange juice would be affected if it increases the daily television advertising expenditure $1000. That is, the management wants to know the amount $f(x, y)$ would change if y is increased from 5 to 6, but x is kept fixed at 8.

If we treat x as a constant, we can think of the function $f(x, y)$ as having only one independent variable, namely, y. Furthermore, we can think of the formula for $f(x, y)$ as having the form

$$ay^3 + by + c$$

where a, b, and c are constants whose values are as follows:

$$a = \tfrac{1}{60}x^2$$
$$b = 4$$
$$c = 200$$

Then we can compute the derivative of $f(x, y)$ using the rules in Chapter 2 and obtain

$$3ay^2 + b$$

Substituting $\tfrac{1}{60}x^2$ for a and 4 for b, we get

$$3(\tfrac{1}{60}x^2)y^2 + 4$$

which simplifies to

$$\tfrac{1}{20}x^2y^2 + 4$$

This expression is a formula for the partial derivative of $f(x, y)$ with respect to y. Using the symbol $f_y(x, y)$ to represent this partial derivative, we can write

$$f_y(x, y) = \tfrac{1}{20}x^2y^2 + 4$$

We can think of $f_y(8, 5)$ as an approximation of the amount $f(x, y)$ changes if y increases from 5 to 6 and x remains at 8. Therefore, since

$$f_y(8, 5) = \tfrac{1}{20}(8^2)(5^2) + 4$$
$$= 84$$

$f(x, y)$ increases approximately 84 units if y increases from 5 to 6 and x remains at 8. So the daily demand for the orange juice would increase by approximately 84 barrels if the management increases the daily television advertising expenditure $1000. We can obtain the actual demand increase by computing

$$f(8, 6) - f(8, 5)$$

Would the increase in the daily demand be greater if the management spends the extra $1000 per day on newspaper advertising, rather than on television advertising? To answer this question, we treat y as a constant and think of $f(x, y)$ as a function of x alone. Also, we think of the formula for $f(x, y)$ as having the form

$$ax^2 + b$$

where a and b are constants whose values are as follows:

$$a = \tfrac{1}{60}y^3$$
$$b = 4y + 200$$

Then we compute the derivative of $f(x, y)$ using the rules in Chapter 2 and obtain

$$2ax$$

Substituting $\frac{1}{60}y^3$ for a, we get

$$2(\tfrac{1}{60}y^3)x$$

which simplifies to

$$\tfrac{1}{30}xy^3$$

This expression is a formula for the partial derivative of $f(x, y)$ with respect to x. Using the symbol $f_x(x, y)$ to represent this partial derivative, we can write

$$f_x(x, y) = \tfrac{1}{30}xy^3$$

We can think of $f_x(8, 5)$ as an approximation of the amount $f(x, y)$ changes if x increases from 8 to 9 and y remains at 5. Consequently, since

$$f_x(8, 5) = \tfrac{1}{30}(8)(5^3)$$
$$= 33\tfrac{1}{3}$$

$f(x, y)$ increases approximately $33\tfrac{1}{3}$ units if x increases from 8 to 9 and y remains at 5. So a $1000 increase in the daily newspaper advertising expenditure would have less impact on the daily demand than a $1000 increase in the daily television advertising expenditure. We can obtain the actual demand increase due to a $1000 increase in the daily newspaper advertising expenditure by computing

$$f(9, 5) - f(8, 5)$$

The advertising example illustrates Definition 7.3.

Definition 7.3

The **partial derivative of $f(x, y)$ with respect to x** is the function obtained by computing the derivative of $f(x, y)$ while treating y as a constant and $f(x, y)$ as a function of x alone. Such a derivative can be represented by

$$f_x(x, y)$$

Similarly, the **partial derivative of $f(x, y)$ with respect to y** is the function obtained by computing the derivative of $f(x, y)$ while treating x as a constant and $f(x, y)$ as a function of y alone. Such a derivative can be represented by

$$f_y(x, y)$$

A function of two independent variables is **partial differentiable** if both partial derivatives exist at every point in the domain. Thus,

$$f(x, y) = \tfrac{1}{60}x^2y^3 + 4y + 200$$

is a partial differentiable function of two independent variables. The function in Example 7.7 is also partial differentiable.

Example 7.7

For the function

$$f(x, y) = x^2(xy + 7)^5$$

find formulas for $f_x(x, y)$ and $f_y(x, y)$.

Solution

To find a formula for $f_x(x, y)$, we treat y as a constant and $f(x, y)$ as a function of x alone. Then we apply the product rule and the generalized power rule and obtain

$$\begin{aligned}f_x(x, y) &= x^2 \cdot 5(xy + 7)^4 y + (xy + 7)^5 \cdot 2x \\ &= 5x^2y(xy + 7)^4 + 2x(xy + 7)^5\end{aligned}$$

To find a formula for $f_y(x, y)$, we treat x as a constant and $f(x, y)$ as a function of y alone. Since y appears only in the factor $(xy + 7)^5$, we do not need to use the product rule. Applying the generalized power rule, we obtain

$$\begin{aligned}f_y(x, y) &= 5x^2(xy + 7)^4 x \\ &= 5x^3(xy + 7)^4\end{aligned}$$

In Example 7.8, we use the generalized exponential rule.

Example 7.8

Suppose

$$f(x, y) = e^{3x^2y}$$

Compute both partial derivatives of $f(x, y)$.

Solution

To find a formula for $f_x(x, y)$, we compute the derivative of $f(x, y)$ while treating y as a constant and $f(x, y)$ as a function of x alone:

$$f_x(x, y) = 6xye^{3x^2y}$$

To find a formula for $f_y(x, y)$, we compute the derivative of $f(x, y)$ while treating x as a constant and $f(x, y)$ as a function of y alone:

$$f_y(x, y) = 3x^2e^{3x^2y}$$

Alternative Notation

If the dependent variable is z, we can represent the partial derivative with respect to x using the symbol

$$\frac{\partial z}{\partial x}$$

and the partial derivative with respect to y using the symbol

$$\frac{\partial z}{\partial y}$$

We use this notation in Example 7.9.

Example 7.9

For the function

$$z = \frac{x^5 y^2}{2y^3 + 8}$$

find formulas for $\partial z/\partial x$ and $\partial z/\partial y$.

Solution

Since x appears only in the numerator, we can find a formula for $\partial z/\partial x$ without using the quotient rule. Note that

$$\frac{x^5 y^2}{2y^3 + 8} = \frac{y^2}{2y^3 + 8}(x^5)$$

So

$$z = \frac{y^2}{2y^3 + 8}(x^5)$$

Thus, we can use the power rule while treating y as a constant:

$$\frac{\partial z}{\partial x} = \frac{5y^2}{2y^3 + 8}(x^4)$$

$$= \frac{5x^4 y^2}{2y^3 + 8}$$

Note that y appears in both the numerator and denominator. So, to find a formula for $\partial z/\partial y$, we use the quotient rule while treating x as a constant:

$$\frac{\partial z}{\partial y} = \frac{(2y^3 + 8)2x^5 y - (x^5 y^2)6y^2}{(2y^3 + 8)^2}$$

$$= \frac{16x^5 y - 2x^5 y^4}{(2y^3 + 8)^2}$$

Approximating Change

In the advertising example, we used partial derivatives to approximate change.

> In many cases, $f_x(a, b)$ is a close approximation for the amount $f(x, y)$ changes if x increases from a to $a + 1$ and y remains fixed at b. Similarly, $f_y(a, b)$ is frequently a close approximation for the amount $f(x, y)$ changes if y increases from b to $b + 1$ and x remains fixed at a.

Example 7.10

Suppose

$$f(x, y) = \ln(x^3 - 7y)$$

Use a partial derivative to approximate the amount $f(x, y)$ changes if y increases from 9 to 10 and x remains fixed at 4.

Solution

Since y is the variable that increases, we compute the partial derivative with respect to y. Using the generalized logarithm rule, we get

$$f_y(x, y) = \frac{-7}{x^3 - 7y}$$

Substituting 4 for x and 9 for y, we obtain

$$f_y(4, 9) = \frac{-7}{4^3 - 7(9)}$$

$$= -7$$

Thus, $f(x, y)$ *decreases* approximately 7 units if y increases from 9 to 10 and x remains fixed at 4.

Marginal Productivity

Suppose $P(x, y)$ is the amount of a product (or service) that a company can produce (or render) during a specified interval of time using x units of labor and y units of capital. Then

$$P_x(x, y)$$

is the **marginal productivity of labor** and

$$P_y(x, y)$$

is the **marginal productivity of capital**.

If the formula for $P(x, y)$ can be expressed in the form

$$Ax^m y^n$$

where $m + n = 1$, $P(x, y)$ is a **Cobb-Douglas production function**. Example 7.11 involves such a function.

Example 7.11

Suppose Spartan Courier Services delivers

$$P(x, y) = 1000 x^{2/3} y^{1/3}$$

pounds of packages weekly with x workers and a capital outlay of $\$y$ thousand. Find the marginal productivity of labor when the company uses 343 workers and a capital outlay of \$729,000.

Solution

The partial derivative of $P(x, y)$ with respect to x is

$$P_x(x, y) = \frac{2000}{3} x^{-1/3} y^{1/3}$$

$$= \frac{2000 \sqrt[3]{y}}{3 \sqrt[3]{x}}$$

So, when the company uses 343 workers and a capital outlay of \$729,000, the marginal productivity of labor is

$$P_x(343, 729) = \frac{2000 \sqrt[3]{729}}{3 \sqrt[3]{343}}$$

$$\approx 857$$

This result tells us that the company increases its weekly ability to deliver packages by approximately 857 pounds if it increases its labor force from 343 workers to 344 workers and keeps the capital outlay fixed at \$729,000.

Geometric Interpretation

We can interpret partial derivatives of functions of two independent variables geometrically. Such an interpretation is similar to the interpretation we used for derivatives of functions of only one independent variable. Suppose $f(x, y)$ is the partial differentiable function whose graph is the surface in Figure 7.5. The figure shows the intersection of the surface with the plane that passes through the point $(a, b, f(a, b))$ and is parallel to the xz plane. As you can see, this intersection is a curve that passes through the point $(a, b, f(a, b))$. The value of $f_x(a, b)$ is the *slope* of the tangent to the curve at the point $(a, b, f(a, b))$. This tangent is labeled T in Figure 7.5.

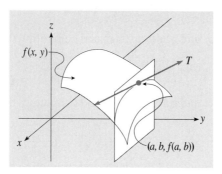

Figure 7.5 $f_x(a, b)$ geometrically interpreted.

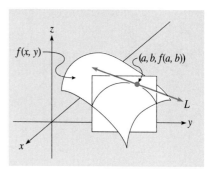

Figure 7.6 $f_y(a, b)$ geometrically interpreted.

Figure 7.6 shows the intersection of the graph of $f(x, y)$ with the plane that passes through the point $(a, b, f(a, b))$ and is parallel to the yz plane, rather than the xz plane. This intersection also is a curve that passes through the point $(a, b, f(a, b))$. The value of $f_y(a, b)$ is the *slope* of the tangent to the curve at the point $(a, b, f(a, b))$. This tangent is labeled L in Figure 7.6.

Functions of More than Two Independent Variables

The definition of a partial derivative of a function of more than two independent variables is similar to the definition of a partial derivative of a function of two independent variables. For example, **the partial derivative of $f(x, y, z)$ with respect to x** is the function obtained by computing the derivative of $f(x, y, z)$ while treating y and z as constants and $f(x, y, z)$ as a function of x alone. As expected, such a derivative can be represented by

$$f_x(x, y, z)$$

or

$$\frac{\partial w}{\partial x}$$

if the dependent variable is represented by w.

Example 7.12

For the function

$$f(x, y, z) = 3xy^3z^2 - x^4y^2 + 7z^5$$

find formulas for $f_x(x, y, z), f_y(x, y, z),$ and $f_z(x, y, z)$.

Solution

Treating y and z as constants and computing the derivative of $f(x, y)$, we obtain

$$f_x(x, y, z) = 3y^3z^2 - 4x^3y^2$$

Treating x and z as constants and computing the derivative of $f(x, y, z)$, we get

$$f_y(x, y, z) = 9xy^2z^2 - 2x^4y$$

Treating x and y as constants and computing the derivative of $f(x, y, z)$, we obtain

$$f_z(x, y, z) = 6xy^3z + 35z^4$$

Substitute and Complementary Products

If the relationship between two products is such that an increase in the price of either product causes the demand for the other product to *increase*, the products are **substitutes**. Meat and fish, for example, are substitutes. If, on the other hand, the relationship between the products is such that an increase in the price of either product causes the demand for the other product to *decrease*, the products are **complementary**. Houses and mortgage loans are complementary. (The price of a mortgage loan is its interest rate.)

Let $f(x, y)$ be the demand for product A, and let $g(x, y)$ be the demand for product B when the price of product A is x and the price of product B is y. The products are substitutes if

$$f_y(x, y) > 0 \quad \text{and} \quad g_x(x, y) > 0$$

for all allowable values of x and y. On the other hand, the products are complementary if

$$f_y(x, y) < 0 \quad \text{and} \quad g_x(x, y) < 0$$

for all allowable values of x and y.

Example 7.13

When the price of product M is x cents per pound and the price of product N is y cents per pound, the daily demand for product M is

$$f(x, y) = \frac{3y + 16}{x + 2}$$

tons and the daily demand for product N is

$$g(x, y) = \frac{x + 15}{4y + 3}$$

tons. Determine whether the products are substitutes or complementary.

Solution

The partial derivative of $f(x, y)$ with respect to y is
$$f_y(x, y) = \frac{3}{x + 2}$$
The partial derivative of $g(x, y)$ with respect to x is
$$g_x(x, y) = \frac{1}{4y + 3}$$
Note that for all permissible values of x and y
$$f_y(x, y) > 0 \quad \text{and} \quad g_x(x, y) > 0$$
Consequently, the two products are substitutes.

7.2 Problems

In Problems 1 to 32, find formulas for $f_x(x, y)$ and $f_y(x, y)$.

1. $f(x, y) = 2x^3y + 7y^2$
2. $f(x, y) = xy^2 - 5x^3 + 7$
3. $f(x, y) = x^4y^3 - 3x + 9$
4. $f(x, y) = 3xy^5 + x^2 + y^3$
5. $f(x, y) = 100x^{1/4}y^{3/4}$
6. $f(x, y) = 40x^{3/5}y^{2/5}$
7. $f(x, y) = 3x - 7y + 1$
8. $f(x, y) = x + 5y - 8$
9. $f(x, y) = \frac{3y^2}{x^3} - \frac{7x}{y}$
10. $f(x, y) = \frac{2}{y} + \frac{1}{x} + x^3$
11. $f(x, y) = (3x^2 + y^3)^7$
12. $f(x, y) = (x^3y^2 + 6)^4$
13. $f(x, y) = \sqrt{2x^3y + x}$
14. $f(x, y) = \sqrt{x^2 - y^2}$
15. $f(x, y) = \frac{1}{xy + 5y}$
16. $f(x, y) = \frac{-5}{x^3y + 4}$
17. $f(x, y) = y^2(2xy^3 - 4)^3$
18. $f(x, y) = 4x(x^2y + y^2)^5$
19. $f(x, y) = \frac{xy^2}{x^3 + 5}$
20. $f(x, y) = \frac{3y + 7x}{y^2 - 1}$
21. $f(x, y) = \ln(x^3 + 5y)$
22. $f(x, y) = \ln(8x + y^2 + 4)$
23. $f(x, y) = y^2 \ln(3x + y)$
24. $f(x, y) = 5x \ln(x + 7y)$
25. $f(x, y) = xy^3 \ln(2x + 1)$
26. $f(x, y) = x^2y \ln(4y + 6)$
27. $f(x, y) = e^{xy+5y}$
28. $f(x, y) = e^{xy^2-7x}$
29. $f(x, y) = x^3 e^{8x-y^2}$
30. $f(x, y) = (3y^2 + 4)e^{x^2+5y}$
31. $f(x, y) = (e^{xy} + 2y)^4$
32. $f(x, y) = \sqrt{e^{6xy} + x}$

In Problems 33 and 34, find formulas for $\partial w/\partial x$, $\partial w/\partial y$, and $\partial w/\partial z$.

33. $w = x^2z^3 - 7xy + yz^2$
34. $w = 3y^2 + xy + 5x^2z$

35. Suppose
$$f(x, y) = x^2y - 23x$$
Use a partial derivative to approximate the amount $f(x, y)$ changes if x increases from 5 to 6 and y remains fixed at 2.

36. Suppose
$$f(x, y) = x^3y^2 - 430y$$
Use a partial derivative to approximate the amount $f(x, y)$ changes if y increases from 8 to 9 and x remains fixed at 3.

37. Suppose
$$f(x, y) = \frac{y^3}{6x}$$
Use a partial derivative to approximate the amount $f(x, y)$ changes if y increases from 4 to 5 and x remains fixed at 2.

38. Suppose
$$f(x, y) = \frac{3y}{x^2}$$
Use a partial derivative to approximate the amount $f(x, y)$ changes if x increases from 1 to 2 and y remains fixed at $\frac{1}{2}$.

39. Suppose
$$f(x, y) = \frac{3}{x^2 + y}$$
Use a partial derivative to approximate the amount $f(x, y)$ changes if x increases from 3 to 4 and y remains fixed at 1.

40. Suppose
$$f(x, y) = \sqrt{xy + 2x}$$
Use a partial derivative to approximate the amount $f(x, y)$ changes if y increases from 7 to 8 and x remains fixed at 9.

41. For
$$f(x, y) = \frac{x^2}{y}$$
compute $f_x(5, 10)$ and $f_y(5, 10)$. Then interpret your answers in terms of change.

42. For
$$f(x, y) = -\frac{1}{20}x^2 + \frac{1}{72}y^3$$
compute $f_x(20, 12)$ and $f_y(20, 12)$. Then interpret your answers in terms of change.

43. **Pleasure Derived from Activities** Let $f(x, y)$ be a measure of the pleasure Wilma gets from jogging x hours per week and playing racket ball y hours per week. Wilma has been jogging 4 hours per week and playing racket ball 7 hours per week. Suppose $f(x, y)$ is a function of x and y according to the formula

$$f(x, y) = \frac{28xy + 57x + 40y}{2xy + 5y + 8x + 20}$$

Use partial derivatives to decide which change would increase Wilma's pleasure more: a 1-hour increase in the weekly time spent jogging or a 1-hour increase in the weekly time spent playing racket ball?

44. **Demand** Soda Unlimited produces two kinds of soda pop: Valley Mist and Irish Green. Let $D(x, y)$ represent the number of million barrels of Valley Mist the company sells daily if its price is x cents per ounce and the price of Irish Green is y cents per ounce. Suppose $D(x, y)$ is a function of x and y according to the formula

$$D(x, y) = \frac{72\sqrt{y + 1}}{x + 1}$$

a. Use a partial derivative to approximate how much the daily demand for Valley Mist changes if its price increases from 5 cents to 6 cents per ounce and the price of Irish Green remains fixed at 8 cents per ounce.

b. Use a partial derivative to approximate how much the daily demand for Valley Mist changes if the price of Irish Green increases from 8 cents to 9 cents per ounce and the price of Valley Mist remains fixed at 5 cents per ounce.

45. **Production Cost** The management of Tecno, a company that manufactures car radios, believes that the joint weekly cost of producing x hundred AM radios and y hundred AM-FM radios per week is

$$C(x, y) = 20 \ln(2x + 3y + 6)$$

thousand dollars. The company has been producing 700 AM radios and 500 AM-FM radios weekly. Use a partial derivative to approximate the amount the joint weekly production cost increases if the company increases the weekly production level of the AM-FM radios by 100.

46. **Advertising** Tropico is an ice cream company in Graceville that makes ice cream flavored with exotic fruits such as zapote, anon, tamarindo, and mamey. Tropico's owner believes she can sell

$$D(x, y) = 24 \ln(4x + y + 2)$$

hundred gallons of ice cream monthly if she spends $x thousand per month advertising in the *Graceville Times* and $y thousand per month advertising in *La Prensa*, Graceville's Spanish newspaper. She now spends $5000 per month advertising in the *Graceville Times* and $3000 per month advertising in *La Prensa*. Use a partial derivative to approximate how much more ice cream she would sell if she increased the monthly amount she spends advertising in the *Graceville Times* by $1000.

47. **Advertising** A distributor for Featherweight, a new light beer, believes that x months after its introduction his district will consume

$$D(x, y) = 80 - 78e^{-0.01x - 0.02y}$$

thousand barrels of the beer per month if the monthly advertising expenditure is $y million. Naturally, 6 months after the beer's introduction, the monthly demand would be greater if the monthly advertising expenditure were $45 million rather than $44 million. Use a partial derivative to approximate how much greater it would be.

48. **Advertising** The makers of No-Sweat, an oven cleaner, believe that if they spend $x million per week on newspaper advertising and $y million per week on radio advertising they will sell

$$D(x, y) = 250 - 200e^{-0.04x - 0.03y}$$

thousand cans of No-Sweat weekly. They now spend $35 million per week on newspaper advertising and $27 million per week on radio advertising. Use a partial derivative to approximate how much the weekly demand for No-Sweat changes if they increase the weekly expenditure for newspaper advertising by $1 million.

Marginal Productivities of Labor and Capital In Problems 49 to 54, $P(x, y)$ represents the amount of a product that can be produced daily with x units of labor and y units of capital. In each problem, compute the marginal productivity of labor and the marginal productivity of capital when the labor force and the capital outlay are as given. Then interpret your answers.

49. $P(x, y) = \sqrt{0.97x^2 + 2y^2}$.
Labor force is 10 and capital outlay is 8.

50. $P(x, y) = 650x^{3/5}y^{2/5}$.
Labor force is 32 and capital outlay is 243.

51. $P(x, y) = \frac{1}{5}x^2y + \frac{1}{7}xy^2$.
Labor force is 5 and capital outlay is 7.

52. $P(x, y) = y\sqrt{x} + x\sqrt{y}$.
Labor force is 36 and capital outlay is 25.

53. $P(x, y) = 10x^{1/3}y^{2/3}$.
Labor force is 64 and capital outlay is 125.

54. $P(x, y) = \sqrt{x^2 + 2.56y^2}$.
Labor force is 12 and capital outlay is 10.

Substitute and Complementary Products In Problems 55 to 60, $f(x, y)$ is the weekly demand for product A and $g(x, y)$ is the weekly demand for product B when the price of product A is x and the price of product B is y. In each problem, use partial derivatives to determine whether the products are substitutes or complementary.

55. $f(x, y) = \dfrac{y + 10}{2x + 1}$ $g(x, y) = \dfrac{5x + 36}{y + 3}$

56. $f(x, y) = \dfrac{45}{x + 36y + 1}$ $g(x, y) = \dfrac{100}{2x + 4y + 5}$

57. $f(x, y) = \dfrac{62}{3x + 5y + 2}$ $g(x, y) = \dfrac{50}{4x + y + 1}$

58. $f(x, y) = \dfrac{3y + 20}{x + 3}$ $g(x, y) = \dfrac{x + 7}{5y + 1}$

59. $f(x, y) = 150 - \sqrt{x} + 0.25y$
$g(x, y) = 135 + 0.2x - \sqrt{y}$

60. $f(x, y) = 200 - 0.5x - 2\sqrt{y + 1}$
$g(x, y) = 250 - \sqrt{x + 2} - \frac{1}{3}y$

In Problems 61 and 62, $f(x, y)$ is the demand for product A and $g(x, y)$ is the demand for product B when the price of product A is x and the price of product B is y.

61. **Substitute Products** In this section, we said that the products are substitutes if

$$f_y(x, y) > 0 \quad \text{and} \quad g_x(x, y) > 0$$

for all allowable values of x and y. Explain why this statement is valid.

62. **Complementary Products** In this section, we said that the products are complementary if

$$f_y(x, y) < 0 \quad \text{and} \quad g_x(x, y) < 0$$

for all allowable values of x and y. Explain why this statement is valid.

7.3 Optimization

In this section, we will see that the basic idea behind optimization of functions of two independent variables is similar to the basic idea behind optimization of functions of one independent variable.

High and Low Points

Suppose $f(x)$ is a differentiable function of one independent variable. As we saw in Section 2.7, excluding endpoints, the high and low points on the graph are among the points where $f'(x) = 0$. Theorem 7.1 tells us that a similar result is valid for partial differentiable functions of two independent variables.

Theorem 7.1

> Suppose $f(x, y)$ is a partial differentiable function. Excluding points on the edge of the graph, the high and low points on the graph are among the points where *both* $f_x(x, y)$ and $f_y(x, y)$ are equal to zero.

As with functions of one independent variable, the graph of a function of two independent variables may have a high point that is not the highest point. In fact, the graph of a function of two independent variables may have several high points and no highest point. We can make a similar comment about the low points on the graph of such a function.

Critical Points

In Section 2.7 we also saw that for a differentiable function of one independent variable, the critical points are the points where the derivative is zero. For a partial differentiable function of two independent variables, the definition of a critical point is similar.

Definition 7.4

> Suppose $f(x, y)$ is a partial differentiable function. The point $(a, b, f(a, b))$ is a **critical point** on the graph if *both* $f_x(a, b)$ and $f_y(a, b)$ are equal to zero.

Using Definition 7.4, we can rephrase Theorem 7.1 as follows:

Theorem 7.1 (Alternative Version)

> Suppose $f(x, y)$ is a partial differentiable function. Excluding points on the edge of the graph, the high and low points on the graph are among the critical points.

Maximum and Minimum Values

The point (c, d, e) in Figure 7.7 is a critical point that is a high point on the graph of a partial differentiable function $g(x, y)$. In fact, (c, d, e) is the *highest* point on the graph. Thus, $g(x, y)$ is **maximum** when $x = c$ and $y = d$, and the maximum value of $g(x, y)$ is e.

The point (r, s, t) in Figure 7.8 is a critical point that is a low point on the graph of a partial differentiable function $h(x, y)$. Actually, (r, s, t) is the *lowest* point on the graph. So $h(x, y)$ is **minimum** when $x = r$ and $y = s$, and the minimum value of $h(x, y)$ is t.

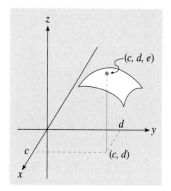

Figure 7.7 A critical point that is a high point.

Saddle Points

It is possible for a critical point to be neither a high point nor a low point on the graph of a function of two independent variables. Definition 7.5 labels such a point.

Definition 7.5

A **saddle point** is a critical point that is neither a high point nor a low point on the graph of a function of two independent variables.

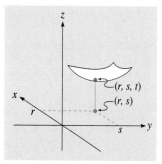

Figure 7.8 A critical point that is a low point.

Figure 7.9 shows a saddle point on the graph of a function of two independent variables.

Theorem 7.1 and Definition 7.5 imply that every critical point is a high point (which may or may not be the highest point), a low point (which may or may not be the lowest point), or a saddle point. In Section 7.4, we will introduce a test to determine whether a critical point is a high point, a low point, or a saddle point.

Example 7.14 involves a function whose graph has one critical point, and the critical point is the highest point on the graph.

Example 7.14

The management of a winery finds that if the winery's yearly production consists of x thousand bottles of red wine and y thousand bottles of white wine, the winery's yearly profit is

$$P(x, y) = -2x^2 - 3y^2 + 2xy + 100x + 90y - 3075$$

thousand dollars. The highest point on the graph of this function is the critical point. Find how much of each wine the winery should produce yearly to maximize the profit.

Solution

The partial derivatives of $P(x, y)$ are

$$P_x(x, y) = -4x + 2y + 100$$

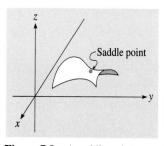

Figure 7.9 A saddle point.

and $P_y(x, y) = -6y + 2x + 90$

Therefore, to find the critical point, we solve the following system of two equations in two unknowns:

$$-4x + 2y + 100 = 0$$
$$-6y + 2x + 90 = 0$$

If we solve the first equation for y, we get

$$y = 2x - 50$$

Then, if we substitute $2x - 50$ for y in the second equation and solve for x, we obtain

$$x = 39$$

Finally, if we substitute 39 for x in $y = 2x - 50$, we get

$$y = 28$$

So the critical point is the point where $x = 39$ and $y = 28$. Since the critical point is the highest point on the graph, to maximize the profit, the winery's yearly production should consist of 39,000 bottles of red wine and 28,000 bottles of white wine.

In Example 7.14

$$P(39, 28) = 135$$

Thus, the winery's maximum yearly profit is $135,000.

Example 7.15 involves a function whose graph has one critical point, and the critical point is the lowest point on the graph.

Example 7.15

The lowest point on the graph of

$$f(x, y) = 70 - e^{-x^2 - y^2 + 10x + 4y - 25}$$

is the critical point. Find the value of x and the value of y where $f(x, y)$ is minimum. Then find the minimum value of $f(x, y)$.

Solution

Using the generalized exponential rule, the partial derivatives of $f(x, y)$ are

$$f_x(x, y) = 0 - (-2x + 10)e^{-x^2 - y^2 + 10x + 4y - 25}$$
$$= (2x - 10)e^{-x^2 - y^2 + 10x + 4y - 25}$$

and
$$f_y(x, y) = 0 - (-2y + 4)e^{-x^2 - y^2 + 10x + 4y - 25}$$
$$= (2y - 4)e^{-x^2 - y^2 + 10x + 4y - 25}$$

Since e raised to any power is positive, we see that both partial derivatives are zero only when $x = 5$ and $y = 2$.

So the critical point is the point where $x = 5$ and $y = 2$. Since the critical point is the lowest point on the graph, $f(x, y)$ is minimum when $x = 5$ and $y = 2$. Since

$$f(5, 2) \approx 15.4$$

the minimum value of $f(x, y)$ is approximately 15.4.

Examples 7.16 and 7.17 involve functions whose graphs have more than one critical point.

Example 7.16

Find the critical points of

$$f(x, y) = 2x^3 + \tfrac{1}{3}y^3 - y^2 - 24x - 3y$$

Solution

The partial derivatives of $f(x, y)$ are

$$f_x(x, y) = 6x^2 - 24$$
$$= 6(x - 2)(x + 2)$$

and

$$f_y(x, y) = y^2 - 2y - 3$$
$$= (y - 3)(y + 1)$$

Therefore, to find the critical points, we solve the following system of two equations:

$$6(x - 2)(x + 2) = 0$$
$$(y - 3)(y + 1) = 0$$

Both equations are satisfied if

$x = 2$	and	$y = 3$
$x = -2$	and	$y = 3$
$x = 2$	and	$y = -1$
$x = -2$	and	$y = -1$

Substituting these values in the formula for $f(x, y)$, we get

$$f(2, 3) = -41 \qquad f(2, -1) = -\tfrac{91}{3}$$
$$f(-2, 3) = 23 \qquad f(-2, -1) = \tfrac{101}{3}$$

Consequently, the critical points are $(2, 3, -41)$, $(2, -1, -\tfrac{91}{3})$, $(-2, 3, 23)$, and $(-2, -1, \tfrac{101}{3})$.

Example 7.17

Find the critical points of

$$f(x, y) = 8x^3 + y^3 - 6xy + 5$$

Solution

To find the critical points, we solve the following system of two equations:
$$f_x(x, y) = 24x^2 - 6y = 0$$
$$f_y(x, y) = 3y^2 - 6x = 0$$

Solving the second equation for x in terms of y, we get
$$x = 0.5y^2$$

Then, substituting $0.5y^2$ for x in the first equation and factoring, we obtain
$$6y(y^3 - 1) = 0$$

This equation is satisfied by $y = 1$ and $y = 0$. Substituting 1 for y in $x = 0.5y^2$, we get $x = 0.5$. Substituting zero for y in $x = 0.5y^2$, we get $x = 0$. So both partial derivatives are equal to zero when

$x = 0.5$ and $y = 1$

and when $x = 0$ and $y = 0$

Thus, since $f(0.5, 1) = 4$

and $f(0, 0) = 5$

the critical points are $(0.5, 1, 4)$ and $(0, 0, 5)$.

7.3 Problems

For each function in Problems 1 to 12, find the critical points.

1. $f(x, y) = 4x^2 + y^2 - 6xy + 6x + 8y$
2. $f(x, y) = 3x^2 + 2y^2 + xy - 26x - 12y + 70$
3. $f(x, y) = \ln(-3x^2 - 0.5y^2 + 30x + 7y + 5)$
4. $f(x, y) = \ln(2.5x^2 + y^2 - 10x - 18y + 100)$
5. $f(x, y) = x^2 + \frac{1}{3}y^3 - 14x - \frac{7}{2}y^2 + 10y$
6. $f(x, y) = \frac{1}{3}x^3 - 3y^2 - 9x + 24y$
7. $f(x, y) = 2x^2 + y^3 - 2xy + 3$
8. $f(x, y) = 9xy - x^3 - 8y^3$
9. $f(x, y) = \frac{1}{3}x^3 + \frac{1}{3}y^3 - 4x^2 - \frac{5}{2}y^2 + 15x + 4y$
10. $f(x, y) = x^3 + y^3 - 15x^2 - 15y^2 + 72x + 48y$
11. $f(x, y) = \sqrt{2x + 6y - x^2 - y^2 - 1}$
12. $f(x, y) = \sqrt{4x + 8y - x^2 - y^2 - 4}$

13. The lowest point on the graph of
$$f(x, y) = x^2 + 5y^2 + 4xy - 50x - 120y + 800$$
is the critical point. Find the value of x and the value of y where $f(x, y)$ is minimum. Then find the minimum value of $f(x, y)$.

14. The highest point on the graph of
$$f(x, y) = -6x^2 - 4y^2 + 36x + 32y$$
is the critical point. Find the value of x and the value of y where $f(x, y)$ is maximum. Then find the maximum value of $f(x, y)$.

15. The highest point on the graph of
$$f(x, y) = -2x^2 - 9y^2 + 56x + 108y$$
is the critical point. Find the value of x and the value of y where $f(x, y)$ is maximum.

16. The lowest point on the graph of
$$f(x, y) = 4x^2 + y^2 + 2xy - 66x - 18y + 210$$
is the critical point. Find the value of x and the value of y where $f(x, y)$ is minimum.

17. The lowest point on the graph of
$$f(x, y) = \ln(x^2 + y^2 - 16x - 20y + 200)$$
is the critical point. Find the value of x and the value of y where $f(x, y)$ is minimum.

18. The highest point on the graph of
$$f(x, y) = e^{-x^2 - y^2 + 4x + 12y - 35}$$
is the critical point. Find the value of x and the value of y where $f(x, y)$ is maximum.

19. **Cost of Producing Pencils** A firm received an order to produce 15,000 of its inexpensive mechanical pencils. The management believes that if the firm uses x workers and y machines to produce the 15,000 pencils, the production cost will be
$$C(x, y) = \tfrac{1}{15}x^2 + \tfrac{1}{10}y^2 - 4x - 4y + 108$$
thousand dollars. The lowest point on the graph of this cost function is the critical point. How many workers and how many machines should the firm use to minimize the production cost?

20. **Cost of Producing Onions** A farmer finds that when he uses x barrels of an insecticide and y sacks of fertilizer, the average cost of producing each ton of onions is
$$A(x, y) = \tfrac{1}{18}x^2 + \tfrac{1}{45}y^2 - 4x - 4y + 254$$
hundred dollars. The lowest point on the graph of this average cost function is the critical point. How much insecticide and how much fertilizer should the farmer use to minimize the average cost of producing each ton of onions?

21. **Optimal Production Levels** A cattleman finds that the average cost of producing each ton of beef depends not only on its production level, but also on the production level of pork. To be more specific, the cattleman believes that the average cost of producing each ton of beef is
$$A(x, y) = 8 - e^{-x^2 - y^2 + 180x + 60y - 9000}$$
hundred dollars when he produces x tons of beef and y tons of pork per year. The lowest point on the graph of this function is the critical point. How much of each type of meat should the cattleman produce per year to minimize the average cost of producing each ton of beef?

22. **Cost of Producing Blue Jeans** The management of Nice-Fit, a company that manufactures blue jeans, believes that with a capital outlay of $\$x$ million and a daily work force of y hundred hours the average cost of producing each pair of blue jeans is
$$A(x, y) = \ln(x^2 + y^2 - 12x - 40y + 10{,}000)$$
dollars. The lowest point on the graph of this function is the critical point. Determine the amount of capital and the amount of labor that minimizes the average cost of producing each pair of blue jeans.

23. **Optimal Prices** A company produces product G and product H. The fixed costs associated with the production of these two products amount to $7000 per week. It costs $35 to produce each ton of product G and $28 to produce each ton of product H. The management believes that if it sells product G for $\$x$ per ton and product H for $\$y$ per ton, the weekly demand for product G will be
$$166 - 2x + y$$
tons and the weekly demand for product H will be
$$191 - 3y + 2x$$
tons.

  a. Express the weekly profit from selling the two products as a function of their prices.

  b. The highest point on the graph of the profit function in part (a) is the critical point. At what prices should the company sell the products to maximize the profit?

24. **Optimal Production Levels** The joint monthly cost of producing x units of product A and y units of product B is
$$75x + 80y + xy + 10{,}000$$
dollars. The producer sells x units of product A monthly at
$$875 - x$$
dollars per unit and y units of product B monthly at
$$1530 - 2y$$
dollars per unit. Each month the producer sets the prices to sell whatever amounts of the two products it produces.

  a. Express the monthly profit from selling the two products as a function of their monthly production levels.

  b. The highest point on the graph of the profit function in part (a) is the critical point. How many units of each product should the producer produce monthly to maximize the profit?

In Problems 25 and 26, $f(x, y)$ is a partial differentiable function and $(a, b, f(a, b))$ is a point on the graph of $f(x, y)$. Also, L_1 and L_2 are the lines defined as follows: Consider the intersection of the graph of $f(x, y)$ with the plane that passes through the point $(a, b, f(a, b))$ and is parallel to the xz plane. This intersection is a curve. Line L_1 is the tangent to this curve at the point $(a, b, f(a, b))$. Similarly, consider the intersection of the graph of $f(x, y)$ with the plane that passes through the point $(a, b, f(a, b))$ and is parallel to the yz plane, rather than the xz plane. This intersection also is a curve. Line L_2 is the tangent to this curve at the point $(a, b, f(a, b))$.

25. Explain why lines L_1 and L_2 must be horizontal if $(a, b, f(a, b))$ is a high point or a low point on the graph of $f(x, y)$.

26. The plane tangent to the graph of $f(x, y)$ at the point $(a, b, f(a, b))$ is the unique plane that contains lines L_1 and L_2. Explain why this plane must be horizontal if $(a, b, f(a, b))$ is a high point or a low point on the graph of $f(x, y)$.

7.4 The Second Derivative Test

Suppose $f(x, y)$ is a partial differentiable function. As we saw in Section 7.3, excluding points on the edge of the graph, the high and low points on the graph are among the critical points. In this section, we present a test to determine whether a critical point is a low point, a high point, or a saddle point. The test involves partial derivatives of partial derivatives.

Second Partial Derivatives

The partial derivative $f_x(x, y)$ is a function of two independent variables. Therefore, we can compute its partial derivatives. If we compute its partial derivative with respect to x, we obtain a function which we represent using the symbol

$$f_{xx}(x, y)$$

If we compute its partial derivative with respect to y, we obtain a function which we represent using the symbol

$$f_{xy}(x, y)$$

The partial derivative $f_y(x, y)$ is also a function of two independent variables. If we compute its partial derivative with respect to x, we obtain a function which we represent using the symbol

$$f_{yx}(x, y)$$

If we compute its partial derivative with respect to y, we obtain a function which we represent using the symbol

$$f_{yy}(x, y)$$

The functions $f_{xx}(x, y), f_{xy}(x, y), f_{yx}(x, y),$ and $f_{yy}(x, y)$ are the **second partial derivatives** of $f(x, y)$.

Example 7.18

Find formulas for the second partial derivatives of
$$f(x, y) = 4x^3y^5 - 7x + y^2 + 9$$

Solution

The partial derivative of $f(x, y)$ with respect to x is
$$f_x(x, y) = 12x^2y^5 - 7$$
Thus,
$$f_{xx}(x, y) = 24xy^5$$
and
$$f_{xy}(x, y) = 60x^2y^4$$

The partial derivative of $f(x, y)$ with respect to y is
$$f_y(x, y) = 20x^3y^4 + 2y$$
So
$$f_{yx}(x, y) = 60x^2y^4$$
and
$$f_{yy}(x, y) = 80x^3y^3 + 2$$

In Example 7.18,
$$f_{xy}(x, y) = f_{yx}(x, y)$$
This type of equality occurs for all the functions of two independent variables in this chapter.

A Test for Classifying Critical Points

Theorem 7.2 provides us with a test to determine whether a critical point is a low point, a high point, or a saddle point. Advanced calculus books usually include a proof of the theorem.

Theorem 7.2 (The Second Derivative Test)

Suppose $(a, b, f(a, b))$ is a critical point on the graph of $f(x, y)$ and
$$D(x, y) = f_{xx}(x, y) \cdot f_{yy}(x, y) - (f_{xy}(x, y))^2$$

1. If $D(a, b)$ and $f_{xx}(a, b)$ are both positive, then $(a, b, f(a, b))$ is a *low point* on the graph of $f(x, y)$.
2. If $D(a, b)$ is positive and $f_{xx}(a, b)$ is negative, then $(a, b, f(a, b))$ is a *high point* on the graph of $f(x, y)$.
3. If $D(a, b)$ is negative, then $(a, b, f(a, b))$ is a *saddle point* on the graph of $f(x, y)$.

We cannot use the above test if $D(a, b) = 0$.

Example 7.19

Find the critical points of

$$f(x, y) = 4x^3 + 4y^3 - 12xy + 7$$

Then use the second derivative test to determine which of the critical points are high points, which are low points, and which are saddle points.

Solution

The (first) partial derivatives of $f(x, y)$ are

$$f_x(x, y) = 12x^2 - 12y$$

and

$$f_y(x, y) = 12y^2 - 12x$$

Therefore, to find the critical points, we solve the following system of two equations:

$$12x^2 - 12y = 0$$
$$12y^2 - 12x = 0$$

If we solve the first equation for y, we get

$$y = x^2$$

Then, substituting x^2 for y in the second equation, we obtain

$$12x^4 - 12x = 0$$

which is equivalent to

$$12x(x^3 - 1) = 0$$

The solutions of this equation are 1 and 0. Substituting 1 for x in $y = x^2$, we get $y = 1$. Substituting 0 for x in $y = x^2$, we get $y = 0$. So both partial derivatives are equal to zero when

$$x = 1 \quad \text{and} \quad y = 1$$

and when

$$x = 0 \quad \text{and} \quad y = 0$$

Thus, since

$$f(1, 1) = 3$$

and

$$f(0, 0) = 7$$

the critical points are $(1, 1, 3)$ and $(0, 0, 7)$.

To apply the second derivative test, we need formulas for $f_{xx}(x, y)$, $f_{yy}(x, y)$, and $f_{xy}(x, y)$. Note that

$$f_{xx}(x, y) = 24x$$
$$f_{yy}(x, y) = 24y$$

and

$$f_{xy}(x, y) = -12$$

So

$$D(x, y) = (24x)(24y) - (-12)^2$$
$$= 576xy - 144$$

First we test the point (1, 1, 3). Since
$$D(1, 1) = 576(1)(1) - 144$$
$$= 432$$

and
$$f_{xx}(1, 1) = 24(1)$$
$$= 24$$

part 1 of the test tells us that (1, 1, 3) is a *low point* on the graph of $f(x, y)$. Next we test the point (0, 0, 7). Since
$$D(0, 0) = 576(0)(0) - 144$$
$$= -144$$

part 3 of the test tells us that (0, 0, 7) is a *saddle point* on the graph of $f(x, y)$.

In Example 7.19, we used the second derivative test to conclude that (1, 1, 3) is a low point on the graph of
$$f(x, y) = 4x^3 + 4y^3 - 12xy + 7$$
This result does not necessarily mean that (1, 1, 3) is the lowest point on the graph. In fact, $(-10, 2, -3721)$ is a point on the graph that is lower than (1, 1, 3). Therefore, the minimum value of $f(x, y)$ does *not* occur when $x = 1$ and $y = 1$.

We can use the second derivative test to conclude that a critical point is a low point on a graph. But, as we illustrated above, we cannot use the test to conclude that a critical point is the lowest point on the graph. Similarly, we can use the second derivative test to conclude that a critical point is a high point on a graph. But we cannot use the test to conclude that a critical point is the highest point on the graph.

Example 7.20 involves a function whose graph has one high point, and the high point is the highest point on the graph.

Example 7.20

Suppose a company earns a daily profit of
$$P(x, y) = -3x^2 - 2y^2 + 4xy + 50x + 64y - 600$$
dollars if it uses x workers and a capital outlay of y thousand. The high point on the graph of $P(x, y)$ is the highest point. How many workers and how much capital should the company use to maximize the daily profit?

Solution

The partial derivatives of $P(x, y)$ are
$$P_x(x, y) = -6x + 4y + 50$$

and
$$P_y(x, y) = -4y + 4x + 64$$

Using an approach like in Example 7.14 of Section 7.3, we can show that both partial derivatives are zero only when $x = 57$ and $y = 73$. So the only critical point is $(57, 73, P(57, 73))$.

Note that $P_{xx}(x, y) = -6$, $P_{yy}(x, y) = -4$, and $P_{xy}(x, y) = 4$. Therefore,

$$D(x, y) = (-6)(-4) - 4^2$$
$$= 8$$

So $D(57, 73)$ is positive. Since $P_{xx}(57, 73)$ is negative, part 2 of the second derivative test tells us that $(57, 73, P(57, 73))$ is a high point on the graph of $P(x, y)$. Since the high point is the highest point, to maximize the daily profit, the company should use 57 workers and a capital outlay of $73,000.

7.4 Problems

In Problems 1 to 14, find the critical points on the graph of the given function. Then use the second derivative test to determine which of the critical points are high points, which are low points, and which are saddle points.

1. $f(x, y) = 3xy - 3x^3 - \frac{1}{9}y^3$
2. $f(x, y) = x^3 + 125y^3 - 450xy + 28,000$
3. $f(x, y) = x^2 + 2y^3 - 15y^2 - 10x + 36y$
4. $f(x, y) = -2x^3 + 15x^2 - 6y^2 - 24x + 36y$
5. $f(x, y) = 18xy - 3x^2 - 2y^3 + 12x - 156y + 200$
6. $f(x, y) = 2x^3 + 3y^2 - 12xy - 54x + 36y + 110$
7. $f(x, y) = 2x^3 + y^3 - 24x^2 - 12y^2 + 42x + 36y + 200$
8. $f(x, y) = x^3 + y^3 - 3x^2 - 12y^2 - 24x + 45y$
9. $f(x, y) = x^2 - 4y^2 + 12xy - 30x + 60y$
10. $f(x, y) = 3x^2 + 4y^2 - 2xy + 4x - 60y + 250$
11. $f(x, y) = -3x^2 - 2y^2 + 30x + 12y$
12. $f(x, y) = 2x^2 + y^2 - 24x - 18y + 160$
13. $f(x, y) = 27xy + \frac{1}{x} + \frac{1}{y}$
14. $f(x, y) = 125xy - \frac{1}{x} + \frac{1}{y}$

15. **Horticulture** A horticulturist developed an enclosure with controllable climatic conditions. The enclosure allows farmers to produce

$$Y(x, y) = -2x^2 - 0.5y^2 - xy + 360x + 150y$$

crates of a special lettuce monthly if they set the temperature at x degrees Fahrenheit and the humidity at y percent. The high point on the graph of $Y(x, y)$ is the highest point. Determine the temperature and humidity that result in the greatest monthly yield.

16. **Book Publishing** A college textbook publishing firm finds that its yearly profit is

$$P(x, y) = -0.5x^2 - 3y^2 - 2xy + 206x + 612y$$

hundred dollars if each year it sends x thousand complimentary copies of its books to professors and employs y sales representatives. The high point on the graph of $P(x, y)$ is the highest point. How many complimentary copies should the firm send to professors and how many sales representatives should it employ to maximize the yearly profit?

17. **Optimal Production Levels** The average cost of producing each unit of product A depends not only on its production level, but also on the production level of product B. Suppose the average cost of producing each unit of product A is

$$C(x, y) = 0.01x^2 + 0.01y^2 - 6x - 4y + 1400$$

dollars when the producer produces x units of product A and y units of product B weekly. How many units of each product should the producer produce weekly to minimize the average cost of producing each unit of product A? [The lowest point on the graph of $C(x, y)$ is the low point.]

18. **Optimal Inputs** A manufacturing firm earns a weekly profit of

$$P(x, y) = -0.04x^2 - 0.02y^2 + 6x + 2y - 255$$

thousand dollars if it uses x workers and a capital outlay of $\$y$ thousand. How many workers and how much capital should the firm use to earn the greatest possible profit? [The highest point on the graph of $P(x, y)$ is the high point.]

19. **Optimal Dimensions** A carpenter plans to build a rectangular box without a top. The volume of the box will be 4 cubic meters. So, if the length is x meters and the width is y meters, the height has to be $4/(xy)$ meters. (See the figure for this problem.) The price of the material is $\$7$ per square meter. Find a formula that expresses the cost of the material as a function of x and y. Then find the values of x and y that minimize the cost of the material. (The low point on the graph of the cost function is the lowest point.)

20. **Optimal Dimensions** Redo Problem 19 assuming the material for the front, back, and bottom costs $\$4$ per square meter and the material for the remaining two sides costs $\$32$ per square meter.

21. The highest point on the graph of

$$f(x, y) = 25 - (x - 7)^2 - (y - 4)^2$$

is the point where $x = 7$ and $y = 4$. Explain how we can arrive at this conclusion *without* using calculus.

22. The lowest point on the graph of

$$f(x, y) = 17 + (x - 5)^2 + (y - 9)^2$$

is the point where $x = 5$ and $y = 9$. Explain how we can arrive at this conclusion *without* using calculus.

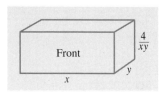

7.5 Curve Fitting

In Sections 6.5 and 6.6, we derived functions from assumptions about their derivatives. In this section, we use the idea of minimizing functions of two independent variables to derive functions from empirical data. We begin by deriving a function that approximates the relationship between a skin lotion's demand and its price.

The price of a liter of Revit, a lotion for revitalizing skin, has changed several times. Table 7.1 shows the monthly demand for each price.

Table 7.1

Price per liter, $	6	10	15	18	20
Monthly demand, millions of liters	9	8	5	4	5

Figure 7.10 depicts the information in Table 7.1 as **data points** in a cartesian plane. The prices are plotted on the horizontal axis (the x axis) and the corresponding demands are plotted on the vertical axis (the y axis).

There is no line that passes through all five data points in Figure 7.10. Therefore, the demand is not a linear function of the price. Nevertheless,

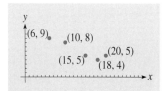

Figure 7.10 The information in Table 7.1 depicted as data points.

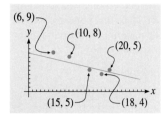

Figure 7.11 A linear function that approximates the relationship between Revit's demand and its price.

Figure 7.11 suggests that it would be reasonable to *approximate* the relationship between the demand and the price with a linear function.

Let $D(x)$ represent the function whose graph is the line (segment) in Figure 7.11. Then we can express the formula for $D(x)$ in the form

$$D(x) = mx + b$$

Proceeding from left to right, let d_1, d_2, d_3, d_4, and d_5 represent the vertical distances between the data points and the line. We want to position the line so that the sum of the squares of these vertical distances is as small as possible. So we will find the values of m and b that minimize

$$d_1^2 + d_2^2 + d_3^2 + d_4^2 + d_5^2$$

This procedure will make the function $D(x)$ a good approximation for the actual relationship between the demand and the price.

Note that

$$d_1^2 = (D(6) - 9)^2$$

But

$$D(6) = m(6) + b$$
$$= 6m + b$$

So we can substitute $6m + b$ for $D(6)$ and get

$$d_1^2 = (6m + b - 9)^2$$

Similarly,

$$d_2^2 = (10m + b - 8)^2$$
$$d_3^2 = (15m + b - 5)^2$$
$$d_4^2 = (18m + b - 4)^2$$

and

$$d_5^2 = (20m + b - 5)^2$$

Therefore, using $S(m, b)$ to represent the sum of these squares, we can write

$$S(m, b) = d_1^2 + d_2^2 + d_3^2 + d_4^2 + d_5^2$$
$$= (6m + b - 9)^2 + (10m + b - 8)^2 + (15m + b - 5)^2$$
$$+ (18m + b - 4)^2 + (20m + b - 5)^2$$

Using the generalized power rule and the sum rule, the partial derivative of $S(m, b)$ with respect to m is

$$S_m(m, b) = 2(6m + b - 9)6 + 2(10m + b - 8)10$$
$$+ 2(15m + b - 5)15 + 2(18m + b - 4)18$$
$$+ 2(20m + b - 5)20$$

Simplifying, we get

$$S_m(m, b) = 2170m + 138b - 762$$

Similarly, $\quad S_b(m, b) = 138m + 10b - 62$

Using an approach like in Example 7.14 of Section 7.3, we can show that the only critical point on the graph of $S(m, b)$ is the point where

$$m = -\frac{117}{332}$$

and
$$b = \frac{3673}{332}$$

because both partial derivatives are zero only for these values of m and b. Using the second derivative test, we can show that this critical point is a low point on the graph of $S(m, b)$. In fact, it is the lowest point. Therefore, the minimum value of $S(m, b)$ occurs when $m = -\frac{117}{332}$ and $b = \frac{3673}{332}$.

So the linear function

$$D(x) = -\frac{117}{332}x + \frac{3673}{332}$$

is a good approximation for the actual relationship between Revit's monthly demand and its price. But this formula is clumsy. Since

$$-\frac{117}{332} \approx -0.35$$

and
$$\frac{3673}{332} \approx 11$$

we find it more practical to use the linear function

$$D(x) = -0.35x + 11$$

to approximate the relationship between the demand and the price.

We can use the above function to estimate Revit's monthly demand for prices other than those listed in Table 7.1. For example, to estimate the monthly demand if the price is $12 per liter, we substitute 12 for x:

$$D(12) = -0.35(12) + 11$$
$$= 6.8$$

So we estimate that 6.8 million liters of Revit will be sold monthly if its price is $12 per liter.

In selecting a function to approximate the relationship between Revit's demand and its price, we followed two steps. First, we decided what type of function to use to approximate the relationship. After examining the data points, we decided to use a function of the type

$$D(x) = mx + b$$

Then we found values for m and b that make the function a good approximation for the relationship.

The functions of the type

$$D(x) = mx + b$$

have two **parameters**, namely, m and b. There exist other functions that have two parameters. For example, the functions of the type

$$f(x) = \frac{1}{cx + d}$$

also have two parameters. In this case, the parameters are c and d. If we assign specific values to these two parameters, we select a specific function among the functions of the type $f(x) = 1/(cx + d)$. Of course, there exist functions that have more than two parameters. For instance, functions of the type

$$f(x) = ax^2 + bx + c$$

have three parameters. Here the parameters are a, b, and c. In this section, we restrict ourselves to functions that have only two parameters.

The skin lotion example illustrates the following method for finding a function that approximates the relationship between two variables.

Method of Least Squares

Table 7.2 shows the values of y that correspond to n values of x.

Table 7.2

x	x_1	x_2	x_3	...	...	...	x_n
y	y_1	y_2	y_3	...	...	...	y_n

Thinking of y as a function of x, suppose we decide to approximate the relationship between x and y with a two-parameter function $f(x)$ of a particular type. Let c and d represent the parameters. Then the sum

$$(f(x_1) - y_1)^2 + (f(x_2) - y_2)^2 + \cdots + (f(x_n) - y_n)^2$$

is a function of c and d, which we represent with the symbol $S(c, d)$. If we assemble the formula for $f(x)$ using the values of c and d that minimize $S(c, d)$, we say that we used the **method of least squares** to select the function $f(x)$. In this case, we refer to $f(x)$ as a **least-squares function** and call the graph of $f(x)$ a **regression curve**.

The graph of the function $S(c, d)$ has exactly one critical point, and the critical point is the lowest point. Therefore, finding the values of c and d that minimize $S(c, d)$ is equivalent to finding the values of c and d where both partial derivatives of $S(c, d)$ are zero.

In Example 7.21, we use the method of least squares to find a quadratic function that approximates the relationship between two variables.

Example 7.21

Last year a farmer planted a new type of legume on 5 acres that contained different amounts of fertilizer. Table 7.3 shows each acre's yield.

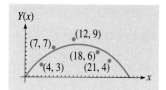

Figure 7.12 A quadratic function that approximates the relationship between the yield and the fertilization level.

Table 7.3

Sacks of fertilizer	4	7	12	18	21
Yield, hundreds of pounds	3	7	9	6	4

Use the method of least squares to find a function that approximates the relationship between the yield and the amount of fertilizer.

Solution

Figure 7.12 shows the data points that correspond to the information in Table 7.3. The figure suggests that it would be reasonable to approximate the relationship between the yield and the fertilization level with a function whose graph is part of a parabola that passes through the origin. Such a function has the form

$$Y(x) = cx^2 + dx$$

For a function of the above type,

$$\begin{aligned} S(c, d) &= (Y(4) - 3)^2 + (Y(7) - 7)^2 + (Y(12) - 9)^2 \\ &\quad + (Y(18) - 6)^2 + (Y(21) - 4)^2 \\ &= (16c + 4d - 3)^2 + (49c + 7d - 7)^2 + (144c + 12d - 9)^2 \\ &\quad + (324c + 18d - 6)^2 + (441c + 21d - 4)^2 \end{aligned}$$

So the partial derivatives of $S(c, d)$ are

$$S_c(c, d) = 645{,}700c + 34{,}456d - 10{,}790$$

and

$$S_d(c, d) = 34{,}456c + 1948d - 722$$

Both partial derivatives of $S(c, d)$ are zero when

$$c = -\frac{692{,}406{,}241}{12{,}671{,}133{,}702} \approx -0.05$$

and

$$d = \frac{11{,}801{,}895}{8{,}825{,}958} \approx 1.3$$

Therefore, the least-squares function of the type $Y(x) = cx^2 + dx$ that approximates the relationship between the yield and the fertilization level is

$$Y(x) = -0.05x^2 + 1.3x$$

A Rule for Least-Squares Linear Functions

We could prove Theorem 7.3 using an approach similar to the approach we used in the skin lotion example.

Theorem 7.3

Suppose the values of y that correspond to several values of x are known. If y is regarded as a function of x, the least-squares *linear function* that approximates the relationship between y and x is

$$f(x) = \frac{nD - AB}{nC - A^2}x + \frac{BC - AD}{nC - A^2}$$

where n, A, B, C, and D are as follows:

- n = the number of data points
- A = the sum of the x coordinates of the data points
- B = the sum of the y coordinates of the data points
- C = the sum of the squares of the x coordinates
- D = the sum of the products of the x coordinates and the corresponding y coordinates

In the skin lotion problem at the beginning of this section, we decided to use the linear function

$$D(x) = -0.35x + 11$$

to approximate the relationship between the lotion's demand and its price. Example 7.22 shows that we arrive at the same decision if we apply Theorem 7.3 to the problem.

Example 7.22

Use Theorem 7.3 to find the least-squares linear function that approximates the relationship between the monthly demand for Revit and its price.

Solution

From Table 7.1, we see that the data points are (6, 9), (10, 8), (15, 5), (18, 4), and (20, 5). If we let x represent the price and y represent the demand, we can organize the information we need to apply Theorem 7.3 as follows.

x	y	x^2	xy
6	9	36	54
10	8	100	80
15	5	225	75
18	4	324	72
20	5	400	100
A = 69	B = 31	C = 1085	D = 381

Using the above values for A, B, C, and D, we get

$$\frac{nD - AB}{nC - A^2} = \frac{5(381) - 69(31)}{5(1085) - 69^2}$$

$$= -\frac{117}{332} \approx -0.35$$

and

$$\frac{BC - AD}{nC - A^2} = \frac{31(1085) - 69(381)}{5(1085) - 69^2}$$

$$= \frac{3673}{332} \approx 11$$

Therefore, according to Theorem 7.3, the least-squares linear function that approximates the relationship between the monthly demand and the price is

$$f(x) = -0.35x + 11$$

7.5 Problems

In Problems 1 to 6, each table shows the values of y that correspond to several values of x. Thinking of y as a function of x, find the least-squares linear function that approximates the relationship between x and y. (Round the parameters to three decimal places.) Then use the function to estimate the value of y that corresponds to the given value of x. Finally, place the data points and the graph of the function on the same cartesian plane.

1.
x	5	9	15
y	3	5	6

Estimate y when $x = 10$.

2.
x	3	7	12
y	7	5	2

Estimate y when $x = 9$.

3.
x	4	6	10	13
y	8	6	5	4

Estimate y when $x = 15$.

4.
x	2	5	9	11
y	1	1	3	4

Estimate y when $x = 7$.

5.
x	5	8	12	16	20
y	10	8	7	4	3

Estimate y when $x = 10$.

6.
x	4	6	9	12	16
y	3	4	6	8	9

Estimate y when $x = 20$.

In Problems 7 to 10, each table shows the values of y that correspond to several values of x. Thinking of y as a function of x, use the method of least squares to find a function of the type

$$f(x) = cx^2 + d$$

that approximates the relationship between x and y. Then sketch the function, and, on the same cartesian plane, plot the data points.

7.
x	3	8	12
y	2	5	7

8.
x	4	10	17
y	10	9	5

9.
x	2	6	9	14
y	12	11	9	6

10.

x	5	8	12	15
y	2	3	7	12

In Problems 11 to 14, each table shows the values of y that correspond to several values of x. Thinking of y as a function of x, find the least-squares function of the type

$$f(x) = cx^2 + dx$$

that approximates the relationship between x and y. (Round the parameters to three decimal places.) Then sketch the function, and, on the same cartesian plane, plot the data points.

11.

x	4	12	18
y	6	10	3

12.

x	5	10	25
y	5	7	3

13.

x	2	7	16	20
y	4	12	12	8

14.

x	3	8	15	20
y	3	6	5	2

15. Demand and Unemployment The management of a firm that makes belts wants a formula that expresses the monthly demand for the belts as a function of the unemployment rate. The management compiled the following data.

Unemployment rate, %	5	7	10	13
Monthly demand, millions of belts	6	5	5	4

a. Use the method of least squares to find a linear function that approximates the relationship between the monthly demand and the unemployment rate.

b. Use the function you found in part (a) to estimate the monthly demand when the unemployment rate is 11 percent.

c. Sketch the graph of the function you found in part (a). Then, on the same cartesian plane, plot the four data points.

16. Demand and Advertising The Soda Company has changed the weekly advertising expenditure for its grape drink four times. The following table shows the drink's weekly demand for each of the four weekly expenditures:

Weekly advertising expenditure, $ millions	4	8	13	16
Weekly demand, millions of barrels	5	7	11	12

a. Thinking of the demand as a function of the advertising expenditure, use the method of least squares to find a linear function that approximates the relationship between the demand and the advertising expenditure.

b. Use the function you found in part (a) to estimate the weekly demand if the company spends $10 million weekly on advertising.

c. Sketch the graph of the function you found in part (a). Then, on the same cartesian plane, plot the four data points.

17. Population The following table shows the population of a country on four different dates:

Date, Jan. 1	1982	1984	1987	1991
Population, millions	5	6	6	8

a. Find the least-squares function of the type

$$P(x) = cx^2 + d$$

that approximates the population x years after the beginning of 1980.

b. Use the derivative of the function you found in part (a) to estimate the rate at which the population will be changing at the beginning of 1998.

18. Supply and Price The price of a product has changed several times. The following table shows how much of the product was supplied daily at each price:

Price per pound, $	4	7	12	14
Daily supply, thousands of tons	1	2	7	10

a. Find the least-squares function of the type
$$S(x) = cx^2 + d$$
that approximates the daily supply when the price is x per pound.

b. Use the derivative of the function you found in part (a) to approximate the amount the daily supply increases if the price increases from $8 to $9 per pound.

19. **Epidemic** A community is in the midst of an epidemic. The following table shows the percent of the population afflicted at four different times:

Number of weeks after the epidemic started	2	6	13	15
Percent of the population afflicted	10	21	19	14

a. Use the method of least squares to find a function of the type
$$P(x) = cx^2 + dx$$
that approximates the percentage of the population afflicted x weeks after the epidemic started.

b. Use the function you found in part (a) to determine when the epidemic will complete its course.

20. **Reproduction** A biologist is studying the relationship between the ability of a particular kind of rodent to reproduce and the temperature of its environment. The following table shows the average number of offspring each female produces for four different temperatures:

Temperature, degrees Celsius	9	16	25	33
Offspring	11	14	14	9

a. Use the method of least squares to find a function of the type
$$R(x) = cx^2 + dx$$
that approximates the number of offspring each female produces when the temperature is x degrees.

b. Use the function you found in part (a) to determine the temperature at which a female is most prolific.

21. The following table shows the values of y that correspond to several values of x:

x	6	11	14
y	7	9	13

Consider y as a function of x.

a. Find the least-squares function of the type
$$f(x) = mx + b$$
that approximates the relationship between x and y.

b. Find the least-squares function of the type
$$f(x) = cx^2 + d$$
that approximates the relationship between x and y.

c. Which of the functions you found in parts (a) and (b) is the better approximation for the relationship between x and y?

22. Suppose y is a function of x. The following table shows the values of y that correspond to several values of x:

x	4	10	13
y	6	7	8

Let $f(x)$ represent the least-squares linear function that approximates the relationship between x and y. Also, let $\bar{x}$ represent the average of the three given values of x. Show that $f(\bar{x})$ is the average of the three given values of y. (This result occurs whenever the least-squares function is linear.)

23. Suppose we know the values of y that correspond to several given values of x. Let s represent the smallest given x value and let g represent the greatest given x value. Also, let $f(x)$ be a least-squares function that approximates the relationship between x and y. Explain why it would be risky to use the function $f(x)$ to estimate the values of y that correspond to x values which are either much smaller than s or much greater than g.

24. Suppose x and y are related according to a formula of the type

$$\ln y = rx + b$$

where b and r are constants ($r \neq 0$). Then

$$y = Ae^{rx}$$

where $A = e^b$. Explain why this fact can be useful in helping us decide to use a least-squares exponential function to approximate the relationship between two variables.

Electronic Computing

The Basic program LESQ listed below uses the method of least squares to find a linear function that approximates the relationship between two variables. Execution of the program prompts you to enter the number of data points. (Line 80 limits this number to at most 1000.) Then it prompts you to enter the x and y coordinates of the data points. After you enter these values, the program prints the least-squares linear function that approximates the relationship between the two variables. (The coefficients are rounded to two decimal places.)

The Program LESQ and a Sample Run

```
LIST
10   REM                           LESQ
20   REM
30   REM A PROGRAM THAT FINDS LEAST SQUARES LINEAR FUNCTIONS
40   REM
50   PRINT "ENTER THE NUMBER OF DATA POINTS."
60   INPUT N
70   PRINT
80   DIM X(1000),Y(1000)
90   PRINT "ENTER THE X AND Y COORDINATES OF A DATA POINT."
100  INPUT A,B
110  PRINT
120  LET C=A^2
130  LET D=A*B
140  FOR J=2 TO N STEP 1
150  PRINT "ENTER THE X AND Y COORDINATES OF YOUR NEXT DATA POINT."
160  INPUT X(J),Y(J)
170  PRINT
180  LET A=A+X(J)
190  LET B=B+Y(J)
200  LET C=C+X(J)^2
210  LET D=D+X(J)*Y(J)
220  NEXT J
230  LET E=(N*D-A*B)/(N*C-A^2)
240  LET F=(B*C-A*D)/(N*C-A^2)
250  PRINT "THE LEAST SQUARES LINEAR FUNCTION IS:"
260  PRINT USING "Y = ##.##";E;
270  PRINT USING "X + ##.##";F
280  END
```

```
Ok
RUN
ENTER THE NUMBER OF DATA POINTS.
? 5

ENTER THE X AND Y COORDINATES OF A DATA POINT.
? 1,6.8

ENTER THE X AND Y COORDINATES OF YOUR NEXT DATA POINT.
? 2,5.3

ENTER THE X AND Y COORDINATES OF YOUR NEXT DATA POINT.
? 3,3.1

ENTER THE X AND Y COORDINATES OF YOUR NEXT DATA POINT.
? 4,.6

ENTER THE X AND Y COORDINATES OF YOUR NEXT DATA POINT.
? 5,-1

THE LEAST SQUARES LINEAR FUNCTION IS:
Y = -2.03X + 9.05
Ok
```

Computer Problems

In Problems 25 to 28, each table shows the values of y that correspond to several values of x. Thinking of y as a function of x, use the program LESQ to find the least-squares linear function that approximates the relationship between x and y.

25.
x	2.1	3.7	5	5.8	6	6.9	9.22	12
y	11.5	16.3	19.6	22	23	26	31	43

26.
x	1	3.8	4.35	6	6.2	7
y	25.4	15	13	5.6	5.2	3

27.
x	3	6	7	12	14	15	20	24	30	32
y	2	2.8	4.1	5.2	6.3	6.7	10	13.6	13.5	17.4

28.
x	5	9	10	12	15	16	20
y	30	50	58	64	91	94	123

7.6 Constrained Optimization

In Sections 7.3 and 7.4, our main objective was to find the point in the domain of a function where the dependent variable achieves its maximum or minimum value. In this section, our main objective is the same, except that the point must be among the points that satisfy a given condition. We begin with an example from the publishing industry.

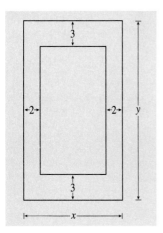

Figure 7.13 A layout for the pages of a calculus book.

A publishing firm is planning the production of a new calculus book. Each page will have an area of 384 square centimeters. The top and bottom margins will be 3 centimeters wide, and the side margins will be 2 centimeters wide. In Figure 7.13, x and y represent the dimensions of the pages. The book designer wants to find the values of x and y that *maximize* the area of the print region.

On each page, the area of the print region will be

$$A(x, y) = (x - 4)(y - 6)$$
$$= xy - 6x - 4y + 24 \tag{3}$$

square centimeters. Since the total area of each page will be 384 square centimeters, x and y must satisfy the condition

$$xy = 384$$

So the book designer wants to find the values of x and y that maximize $A(x, y)$ subject to the constraint $xy = 384$.

Solving the *constraint equation* for y in terms of x, we obtain

$$y = \frac{384}{x} \tag{4}$$

Substituting $384/x$ for y in formula (3), we obtain a formula for the area of the print region that conforms to the constraint and is in terms of x alone. The formula is

$$A(x) = x\left(\frac{384}{x}\right) - 6x - 4\left(\frac{384}{x}\right) + 24$$
$$= 408 - \frac{1536}{x} - 6x$$

The derivative of $A(x)$ is

$$A'(x) = \frac{1536}{x^2} - 6$$

The only positive value of x for which $A'(x) = 0$ is $x = 16$. Moreover, for positive values of x, $A'(x)$ is positive when $x < 16$ and negative when $x > 16$. Therefore, $A(x)$ is maximum when $x = 16$. Substituting 16 for x in formula (4), we obtain

$$y = \frac{384}{16}$$
$$= 24$$

So 16 and 24 are, respectively, the values of x and y that maximize $A(x, y)$ subject to the constraint $xy = 384$. Consequently, the book designer should make each page 16 centimeters wide and 24 centimeters long.

Sec. 7.6 Constrained Optimization

The book production problem is an example of a **constrained optimization problem**. In situations involving two independent variables, constrained optimization problems have the following form:

> Suppose $f(x, y)$ is a function of two independent variables, and $g(x, y) = 0$ is an equation that involves the two variables. Find the values of x and y that maximize (or minimize) $f(x, y)$ subject to the constraint $g(x, y) = 0$.

In the book production problem, the constraint equation is $xy = 384$. Subtracting 384 from both sides, we obtain

$$xy - 384 = 0$$

So, in the book production problem, $g(x, y)$ is $xy - 384$, not xy.

We can adapt the idea of constrained optimization to functions of more than two independent variables.

Graphical Interpretation

For functions of two independent variables, we can interpret the idea of constrained optimization graphically. The points on the graph of $f(x, y)$ whose x and y coordinates satisfy $g(x, y) = 0$ are **constrained points**. Using this terminology, if $(a, b, f(a, b))$ is the *lowest constrained point* on the graph of $f(x, y)$, then $f(x, y)$ attains its **constrained minimum value** when $x = a$ and $y = b$. The constrained minimum value is $f(a, b)$. Similarly, if $(a, b, f(a, b))$ is the *highest constrained point* on the graph of $f(x, y)$, then $f(x, y)$ attains its **constrained maximum value** when $x = a$ and $y = b$. The constrained maximum value is $f(a, b)$.

Figure 7.14 shows the graph of a function $f(x, y)$ that has both a constrained minimum and an unconstrained minimum. Since the constraint equation $g(x, y) = 0$ involves only x and y, its graph is a curve in the xy plane. For the sake of simplicity, we picture the curve as a line. The point $(a, b, f(a, b))$ is the lowest point among the constrained points on the graph of $f(x, y)$. As expected, the lowest constrained point is *not* lower than the lowest unconstrained point, which is $(c, d, f(c, d))$.

Figure 7.14 illustrates that, if a function has both an unconstrained minimum and a constrained minimum, the constrained minimum is not less than the unconstrained minimum. Similarly, if a function has both an unconstrained maximum and a constrained maximum, the constrained maximum is not greater than the unconstrained maximum.

In Example 7.23, we apply the idea of constrained optimization to a function of three independent variables.

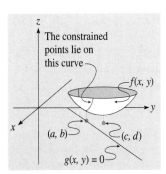

Figure 7.14 A function that has both a constrained and an unconstrained minimum.

Example 7.23

A manufacturing firm produces three versions of a product: versions A, B, and C. The firm earns a weekly profit of

$$P(x, y, z) = 20x + 40y + 30z - x^2 - y^2 - z^2 + 1000$$

dollars if its weekly output consists of x units of version A, y units of version B, and z units of version C. If the firm must produce a total of 30 units of the product per week, how many units of each version should it produce to maximize the profit?

Solution

We want to find the values of x, y, and z that maximize $P(x, y, z)$ subject to the constraint

$$x + y + z = 30$$

Solving this equation for z in terms of x and y, we obtain

$$z = 30 - x - y$$

Substituting $30 - x - y$ for z in the formula for $P(x, y, z)$, we obtain a formula for the total weekly profit that conforms to the constraint and is in terms of only x and y. The formula is

$$P(x, y) = 20x + 40y + 30(30 - x - y) - x^2 - y^2 - (30 - x - y)^2 + 1000$$
$$= 50x + 70y - 2x^2 - 2y^2 - 2xy + 1000$$

We can now use the techniques of Sections 7.3 and 7.4 to find the values of x and y that maximize $P(x, y)$. The partial derivatives of $P(x, y)$ are

$$P_x(x, y) = 50 - 4x - 2y$$

and

$$P_y(x, y) = 70 - 4y - 2x$$

Both of these partial derivatives are zero only when $x = 5$ and $y = 15$. Therefore, the only critical point on the graph of $P(x, y)$ is $(5, 15, P(5, 15))$. Using the second derivative test, we find that $(5, 15, P(5, 15))$ is a high point on the graph of $P(x, y)$. In fact, it is the highest point. Consequently, $P(x, y)$ is maximum when $x = 5$ and $y = 15$. Since $z = 30 - x - y$, the corresponding z value is

$$z = 30 - 5 - 15 = 10$$

Therefore, 5, 15, and 10 are, respectively, the values of x, y, and z that maximize $P(x, y, z)$ subject to the constraint $x + y + z = 30$. So, if the firm must produce 30 units of the product per week, it should produce 5 units of version A, 15 units of version B, and 10 units of version C. These production levels maximize the firm's profit.

In Example 7.23, if we substitute 5 for x, 15 for y, and 10 for z in the formula for $P(x, y, z)$, we obtain

$$P(5, 15, 10) = 1650$$

Thus, subject to the given constraint on the total production level, the maximum weekly profit the firm can earn is $1650. If we eliminate the constraint, $P(x, y, z)$ would be maximum when $x = 10$, $y = 20$, and $z = 15$. In this case, the maximum weekly profit would be

$$P(10, 20, 15) = 1725$$

dollars, rather than $1650.

Substitution Method

In the book production example, we solved the constraint equation for y in terms of x and obtained $y = 384/x$. Then, by substituting $384/x$ for y, we eliminated one of the two independent variables in the area function. Similarly, in Example 7.23, we solved the constraint equation for z in terms of x and y and obtained $z = 30 - x - y$. Then, by substituting $30 - x - y$ for z, we eliminated one of the three independent variables in the profit function. The technique used in both examples is called the **substitution method**.

In Example 7.24, we use the substitution method in a situation that involves a utility function. Suppose A and B represent two products. A function $U(x, y)$ is a **utility function** if $U(x, y)$ ranks the pleasure, benefit, or satisfaction an individual obtains from consuming (buying) x units of A and y units of B during a designated time period. Utility functions can have more than two independent variables.

Example 7.24

Suppose the function

$$U(x, y) = 6x^{1/2}y$$

ranks the satisfaction Connie obtains during a month in which she buys x units of product A, which sells for $8 per unit, and y units of product B, which sells for $4 per unit. Next month Connie plans to spend a total of $216 on the two products. How much of each product, therefore, should Connie buy next month to maximize the satisfaction she obtains from buying the products?

Solution

Since product A sells for $8 per unit, product B sells for $4 per unit, and Connie plans to spend a total of $216 on the products next month, we want to find the values of x and y that maximize $U(x, y)$ subject to the constraint

$$8x + 4y = 216$$

Solving this equation for y, we get

$$y = 54 - 2x$$

Substituting $54 - 2x$ for y in the formula for $U(x, y)$, we obtain a formula for the utility of both products that conforms to the constraint and is in terms of only x. The formula is

$$U(x) = 6x^{1/2}(54 - 2x)$$
$$= 324x^{1/2} - 12x^{3/2}$$

The derivative of $U(x)$ is

$$U'(x) = 162x^{-1/2} - 18x^{1/2}$$
$$= \frac{162}{\sqrt{x}} - 18\sqrt{x}$$

The only value of x for which $U'(x) = 0$ is $x = 9$. Moreover, for positive values of x, $U'(x)$ is positive when $x < 9$ and negative when $x > 9$. Therefore, $U(x)$ is maximum when $x = 9$. Since $y = 54 - 2x$, the corresponding y value is

$$y = 54 - 2(9)$$
$$= 36$$

So 9 and 36 are, respectively, the values of x and y that together maximize $U(x, y)$ subject to the constraint $8x + 4y = 216$. Thus, Connie should use the $216 to purchase 9 units of product A and 36 units of product B.

In Section 7.7, we present a method for solving constrained optimization problems that does not involve isolating one of the variables in the constraint equation.

7.6 Problems

1. Find the values of x and y that

    minimize $f(x, y) = x^2 + 2y^2$
    subject to $x + y = 9$

2. Find the values of x and y that

    maximize $f(x, y) = 200 - 3x^2 - y^2$
    subject to $x + y = 16$

3. Find the values of x and y that

    maximize $f(x, y) = xy$
    subject to $4x + 2y = 24$

4. Find the values of x and y that

    minimize $f(x, y) = 30 - 2xy$
    subject to $3x + y = 12$

5. Suppose the domain of

    $$f(x, y) = 25 - x^2 - 4y^2$$

    consists of the pairs (x, y), where x and y are positive real numbers. Find the pair (x, y) that maximizes $f(x, y)$ subject to

    $$xy = 2$$

6. Suppose the domain of

    $$f(x, y) = 9x^2 - y^2$$

    consists of the pairs (x, y), where x and y are positive real numbers. Find the pair (x, y) that minimizes $f(x, y)$ subject to

    $$xy = 27$$

7. Find the x and y values that
$$\text{minimize } f(x, y) = 3x^2 - y^2 + xy + 2000$$
subject to $y - x = 42$

8. Find the x and y values that
$$\text{maximize } f(x, y) = 2y^2 - x^2 - 5xy$$
subject to $y - 2x = 10$

9. Find the x and y values that
$$\text{maximize } f(x, y) = 50 - x^2 - 3y^2$$
subject to $2x + y = 13$

10. Find the x and y values that
$$\text{minimize } f(x, y) = x^2 + y^2 - 2x - 4y + 6$$
subject to $x + y = 7$

11. Find the values of x and y that
$$\text{minimize } f(x, y) = x^2 + 2y^2 - xy$$
subject to $3x + y = 44$

12. Find the values of x and y that
$$\text{maximize } f(x, y) = 16x + 6y - 4x^2 - y^2 + 2$$
subject to $x + y = 2$

13. Find the values of x, y, and z that
$$\text{minimize } f(x, y, z) = x^2 + y^2 + z^2$$
subject to $x + y + z = 3$

14. Find the x, y, and z values that
$$\text{maximize } f(x, y, z) = 400 - x^2 - y^2 - z^2$$
subject to $2x + 3y + z = 70$

15. Find the x, y, and z values that
$$\text{minimize } f(x, y, z) = x + 8y + z$$
subject to $xyz = 8$

16. Suppose the domain of
$$f(x, y, z) = xyz$$
consists of the triples (x, y, z), where x, y, and z are positive real numbers. Find the triple (x, y, z) that maximizes $f(x, y, z)$ subject to
$$x + y + z = 21$$

17. **Average Production Cost** A manufacturing firm produces three brands of a product: brands A, B, and C. The average cost of producing each unit of brand A depends not only on its monthly production level, but also on the monthly production levels of brands B and C. The management believes that the average cost of producing each unit of brand A is
$$C(x, y, z) = 0.1x^2 + 0.1y^2 + 0.1z^2 - 6x - 4y - 10z + 385$$
dollars when the firm's monthly output consists of x units of brand A, y units of brand B, and z units of brand C. The management has decided to produce a total of 85 units of the product per month. How many units of each brand should the firm produce to minimize the average cost of producing each unit of brand A?

18. **Optimal Advertising Expenditures** A company advertises its product on billboards, on radio, and in magazines. The management believes that the company earns a monthly profit of
$$P(x, y, z) = -0.05x^2 - 0.05y^2 - 0.05z^2 + 4x + 6y + 5z$$
thousand dollars when its monthly advertising expenditure consists of x thousand for billboard advertising, y thousand for radio advertising, and z thousand for magazine advertising. If the company spends a total of $120,000 monthly on advertising, how much should it spend on each of the three forms of advertising to maximize the monthly profit?

19. **Optimal Labor Expenditures** A manufacturing firm produces
$$P(x, y) = 9x^2y$$
units of a product per week if its weekly labor expenditure consists of x thousand for unskilled labor and y thousand for skilled labor.

   a. Suppose the management decides to spend a total of $36,000 weekly on labor. How much should it spend weekly on each of the two types of labor to maximize the output?

   b. Suppose the management decides to produce 121,500 units of the product weekly. How much should it spend weekly on each of the two types of labor to minimize the labor expenditure?
   Suggestion: First express the weekly labor expenditure as a function of x and y. Then find the values of x and y that minimize this expenditure subject to an appropriate constraint.

20. **Optimal Mix of Machines** A manufacturing firm can produce a product using either type A machines or type B machines. Each type A machine costs $30 daily to operate, and each type B machine costs $40 daily to operate. The firm produces

$$P(x, y) = 7xy$$

units of the product per day using x type A machines and y type B machines.

a. Suppose the management decides to spend a total of $1440 daily on the operation of the machines. How many machines of each type should it use daily to maximize the output?

b. Suppose the management decides to produce 2100 units of the product daily. How many machines of each type should it use daily to minimize the daily cost of operating the machines?
Suggestion: First express the daily cost of operating the machines as a function of x and y. Then find the values of x and y that minimize this cost subject to an appropriate constraint.

21. **Architecture** An architect is designing a one-story rectangular building that must have a floor area of 24,000 square feet. The architect wants to situate the building on a rectangular lot. The figure for this problem shows the architect's border requirements. Find the area of the smallest lot the architect can use.
Suggestion: Let x and y represent the dimensions of the building.

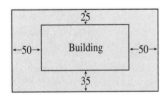

22. **Architecture** An architect is designing a one-story rectangular building that must have a floor area of 268,960 square feet. The front and back walls cost $16 per horizontal foot, and the two side walls cost $10 per horizontal foot. Find the least amount the walls can cost.

23. **Optimal Dimensions** A carpenter plans to build a box with a square base. (See the figure for this problem.) The material for the top and bottom costs $3 per square foot, and the material for sides costs $2 per square foot.

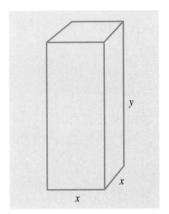

a. Suppose the volume of the box has to be 96 cubic feet. Find the dimensions that minimize the cost of the material.

b. Find the dimensions of the largest box the carpenter can build using $450 worth of material.

24. **Optimal Dimensions** A cabinet maker plans to build a drinking trough having ends that are isosceles right triangles. (See the figure for this problem.) The material for the two rectangular panels costs $0.25 per square foot, and the material for the triangular ends costs $0.50 per square foot.

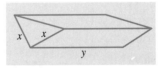

a. Suppose the volume of the trough must be 8 cubic feet. Find the dimensions that minimize the cost of the material.
Comments: The area of each end is $\frac{1}{2}x^2$. The volume is the area of one end multiplied by y.

b. If the cabinet maker spends exactly $13.50 on the material, what dimensions of the trough maximize the volume?

25. **Optimal Dimensions** Peter plans to build a closed rectangular box that has a volume of 64 cubic feet. (See the figure for this problem.) Find the dimensions that minimize the amount of material.

26. **Optimal Dimensions** A cabinet maker plans to build a closed rectangular box that has a volume of 45 cubic feet. (See the figure for Problem 25.) The material for the top and bottom costs $5 per square foot, and the material for the sides costs $3 per square foot. Find the dimensions that minimize the cost of the material.

7.7 The Method of Lagrange Multipliers

In Section 7.6, we solved constrained optimization problems using a method that involves solving the constraint equation for one variable in terms of the other variable (or variables). In this section, we will solve such problems using a more versatile method developed by Joseph L. Lagrange (1736–1813). This method can be applied to functions that have *many* independent variables subject to *more* than one constraint. Nevertheless, for the sake of simplicity, we will apply the method to functions that have only two independent variables subject to only one constraint. We begin with a definition.

Definition 7.6

The **Lagrange function** associated with the function $f(x, y)$ and the equation $g(x, y) = 0$ is

$$L(x, y, \lambda) = f(x, y) + \lambda \cdot g(x, y)$$

So Lagrange functions are functions of *three* independent variables, namely, x, y, and λ. The variable λ (pronounced "lambda") is the **Lagrange multiplier**.

In Section 7.3, we defined the concept of a "critical point" for functions of two *unconstrained* independent variables. Now we define a similar concept for functions of two independent variables *subject to a constraint*.

Definition 7.7

Suppose $L(x, y, \lambda)$ is the Lagrange function associated with $f(x, y)$ and a constraint equation. Also, suppose (a, b, c) is such that

$$L_x(a, b, c) = 0$$
$$L_y(a, b, c) = 0$$
and
$$L_\lambda(a, b, c) = 0$$

Then $(a, b, f(a, b))$ is a **constrained critical point** on the graph of $f(x, y)$.

A constrained critical point is also a constrained point. This result is a consequence of the fact that

$$L_\lambda(x, y, \lambda) = g(x, y)$$

where $g(x, y) = 0$ is the constraint equation.

Suppose $f(x, y)$ is a partial differentiable function. In Section 7.3, we saw that if we exclude the points on the edge of the graph, the *unconstrained* high and low points on the graph are among the *unconstrained* critical points. Theorem 7.4 tells us that a similar result is true if the independent variables are subject to a constraint. You can find a proof of the theorem in more advanced calculus books.

Theorem 7.4

Suppose $f(x, y)$ is a partial differentiable function whose independent variables are subject to a constraint. Excluding points on the edge of the graph, the high and low constrained points on the graph are among the constrained critical points.

In this section, we restrict ourselves to situations where there exists exactly one constrained critical point that is either the highest or the lowest constrained point.

Whenever we solve a constrained optimization problem by finding the constrained critical point, we can say we solved the problem using the **method of Lagrange multipliers**.

Example 7.25

Use the method of Lagrange multipliers to find the values of x and y that

$$\text{minimize } f(x, y) = x^2 + y^2 - 16x - 12y + 105$$
$$\text{subject to the constraint } 2x + y = 7$$

(The constrained critical point is the lowest constrained point on the graph.)

Solution

We can rewrite the constraint equation as

$$2x + y - 7 = 0$$

Thus,
$$g(x, y) = 2x + y - 7$$

So the Lagrange function is

$$L(x, y, \lambda) = f(x, y) + \lambda \cdot g(x, y)$$
$$= x^2 + y^2 - 16x - 12y + 105 + \lambda(2x + y - 7)$$
$$= x^2 + y^2 - 16x - 12y + 2x\lambda + y\lambda - 7\lambda + 105$$

To find the constrained critical point, we solve the following system of three equations:

$$L_x(x, y, \lambda) = 2x - 16 + 2\lambda = 0$$
$$L_y(x, y, \lambda) = 2y - 12 + \lambda = 0$$
$$L_\lambda(x, y, \lambda) = 2x + y - 7 = 0$$

Solving each of the first two equations for λ in terms of x and y, we get

$$\lambda = 8 - x$$
$$\lambda = 12 - 2y$$

Thus, $\qquad 8 - x = 12 - 2y$

So $\qquad x = 2y - 4$

Substituting $2y - 4$ for x in the third equation of the system of three equations, we obtain

$$2(2y - 4) + y - 7 = 0$$

Hence, $\qquad y = 3$

Substituting 3 for y in $x = 2y - 4$, we get

$$x = 2$$

Then, substituting 2 for x in $\lambda = 8 - x$, we get

$$\lambda = 6$$

(This result could also be obtained by substituting 3 for y in $\lambda = 12 - 2y$.) Therefore, $(2, 3, 6)$ is the triple of numbers that satisfies the system of three equations. In other words, $(2, 3, 6)$ is such that

$$L_x(2, 3, 6) = 0$$
$$L_y(2, 3, 6) = 0$$
and $\qquad L_\lambda(2, 3, 6) = 0$

Therefore, according to Definition 7.7, $(2, 3, f(2, 3))$ is the constrained critical point on the graph of $f(x, y)$. Since the constrained critical point is the lowest constrained point on the graph of $f(x, y)$, we conclude that, subject to the constraint $2x + y = 7$, $f(x, y)$ attains its minimum value when $x = 2$ and $y = 3$.

In Example 7.25,
$$f(2, 3) = 2^2 + 3^2 - 16(2) - 12(3) + 105$$
$$= 50$$

Thus, the constrained minimum value of $f(x, y)$ is 50. We could show that the unconstrained minimum occurs when $x = 8$ and $y = 6$. Since
$$f(8, 6) = 8^2 + 6^2 - 16(8) - 12(6) + 105$$
$$= 5$$

the unconstrained minimum is 5.

A Lagrange function will have more than three independent variables if we apply the method of Lagrange multipliers in a situation with more than two independent variables and/or more than one constraint. For example, suppose we use the method to find the values of x, y, and z that maximize a function $f(x, y, z)$ subject to two constraints, say $g(x, y, z) = 0$ and $h(x, y, z) = 0$. Then the Lagrange function would have the form

$$L(x, y, z, \lambda_1, \lambda_2) = f(x, y, z) + \lambda_1 \cdot g(x, y, z) + \lambda_2 \cdot h(x, y, z)$$

Such a Lagrange function has five independent variables. Two of these variables are Lagrange multipliers, namely, λ_1 and λ_2. So, to find the values of x, y, and z that maximize $f(x, y, z)$ subject to the two constraints, we would have to solve a system of five equations, each with five variables. Such a system would have the following form:

$$L_x(x, y, z, \lambda_1, \lambda_2) = 0$$
$$L_y(x, y, z, \lambda_1, \lambda_2) = 0$$
$$L_z(x, y, z, \lambda_1, \lambda_2) = 0$$
$$L_{\lambda_1}(x, y, z, \lambda_1, \lambda_2) = 0$$
$$L_{\lambda_2}(x, y, z, \lambda_1, \lambda_2) = 0$$

7.7 Problems

Use the method of Lagrange multipliers to solve Problems 1 to 12 in Section 7.6. In each problem, the constrained critical point is either the highest or lowest constrained point on the graph, depending on whether the problem refers to a maximum or a minimum.

13. Optimal Buying Levels The function

$$U(x, y) = 4xy^{1/2}$$

ranks the benefit Paul receives during a week in which he buys x units of product A, which sells for $3 per unit, and y units of product B, which sells for $5 per unit. Paul spends a total of $90 per week on the two products. Thus, x and y are subject to a constraint. Use the method of Lagrange multipliers to determine how much of each product Paul should buy each week to maximize the benefit he receives from buying the products. [The constrained critical point is the highest constrained point on the graph of $U(x, y)$.]

14. Optimal Buying Levels Product A sells for $6 per unit and product B sells for $2 per unit. The pleasure Tina derives during a month in which she buys x units of product A and y units of product B is ranked by

$$U(x, y) = 5xy$$

Tina spends a total of $180 per month on the two products. Thus, x and y are subject to a constraint. Use the method of Lagrange multipliers to determine how much of each product Tina should buy monthly to maximize the pleasure she derives from buying the products. [The constrained critical point is the highest constrained point on the graph of $U(x, y)$.]

15. **Optimal Mix of Workers** Able Electric Company is planning to produce 25,000 hair dryers. The management believes that if it uses x experienced workers and y inexperienced workers to produce the 25,000 dryers, the production cost will be

$$C(x, y) = \frac{1}{15}x^2 + \frac{1}{10}y^2 - 4x - 4y + 200$$

thousand dollars. The company will use a total of 100 workers. Thus, x and y are subject to a constraint. Use the method of Lagrange multipliers to determine how many of each kind of worker the company should use to minimize the production cost. [The constrained critical point is the lowest constrained point on the graph of $C(x, y)$.]

16. **Average Production Cost** A manufacturing firm produces two brands of a product: brand A and brand B. The average cost of producing each unit of brand A depends not only on its weekly production level, but also on the weekly production level of brand B. More specifically, the average cost of producing each unit of brand A is

$$C(x, y) = 0.01x^2 + 0.01y^2 - 6x - 4y + 1500$$

dollars when the firm's weekly output consists of x units of brand A and y units of brand B. The firm produces a total of 200 units of the product per week. Thus, x and y are subject to a constraint. Use the method of Lagrange multipliers to determine how many units of each brand the firm should produce weekly to minimize the average cost of producing each unit of brand A. [The constrained critical point is the lowest constrained point on the graph of $C(x, y)$.]

Chapter 7 Important Terms

Section 7.1
function of more than one independent variable (404 and 409)
three-dimensional coordinate system (407)
graph of a function of two independent variables (408)
linear function of two independent variables (408)

Section 7.2
partial derivatives (414 and 419)
partial differentiable function (415)
marginal productivity of labor (417)
marginal productivity of capital (417)
Cobb-Douglas production function (418)
substitute and complementary products (420)

Section 7.3
high and low points (424)
critical point (424)
maximum and minimum values (425)
saddle point (425)

Section 7.4
second partial derivatives (430)
the second derivative test (431)

Section 7.5
data point (435)
two-parameter function (437)
method of least squares (438)
least-squares function (438)
regression curve (438)

Section 7.6
constrained optimization problem (447)
constraint equation (447)
constrained minimum or maximum value (447)
constrained point (447)
substitution method (449)
utility function (449)

Section 7.7
Lagrange function (453 and 456)
Lagrange multipliers (453 and 456)
constrained critical point (453)
method of Lagrange multipliers (454)

Chapter 7 Review Problems

Section 7.1

1. A point has x coordinate 2 and y coordinate -5. Find the z coordinate if the point is on the graph of
$$f(x, y) = \frac{x^3 + 46}{4 - y}$$

2. Sketch a triangular portion of the graph of
$$z = 0.5x + 0.8y - 4$$

3. **Profit** The joint monthly cost of producing x tons of product A and y tons of product B is
$$700x + 600y + xy$$
dollars. The producer sells x tons of product A monthly at
$$4000 - 2x$$
dollars per ton and y tons of product B monthly at
$$3000 - y$$
dollars per ton. Each month the producer sets the prices to sell whatever amounts of the two products it produces.

   a. Express the monthly profit from selling the two products as a function of their monthly production levels.

   b. The producer produces 450 tons of product A and 775 tons of product B each month. The producer, however, wants to increase by 1 ton either the monthly production level of product A or that of product B. Determine which increase would be more profitable.

Section 7.2

4. Suppose
$$f(x, y) = 8x\sqrt{y} + 3x$$
Use a partial derivative to approximate the amount $f(x, y)$ changes if y increases from 9 to 10 and x remains fixed at 6.

5. Suppose
$$f(x, y) = \frac{2y}{\sqrt{x^2 + 3y}}$$
Use a partial derivative to approximate the amount $f(x, y)$ changes if x increases from 7 to 8 and y remains fixed at 5.

6. Suppose
$$w = 2yze^{0.01xz}$$
Use a partial derivative to approximate the amount w changes if z increases from 5 to 6 while x and y remain fixed at 3 and 7, respectively.

7. **Marginal Productivity of Capital** A company can produce
$$P(x, y) = 18 \ln(4x^2y + 1)$$
tons of a product daily using x hundred hours of labor and a capital outlay of y thousand. Find the marginal productivity of capital (rounded to the nearest ton) when the company uses 300 hours of labor and a capital outlay of $8000. Then interpret your answer.

8. **Price Related Products** When the price of product A is x per liter and the price of product B is y per liter, the weekly demand for product A is
$$A(x, y) = \frac{5x + 14}{y + 2}$$
million liters and the weekly demand for product B is
$$B(x, y) = \frac{y + 18}{2x + 3}$$
million liters. Use partial derivatives to determine whether the products are substitutes or complementary.

Section 7.3

9. The lowest point on the graph of
$$f(x, y) = \sqrt{x^2 + y^2 - 8x - 12y + 116}$$

is the critical point. Find the value of x and the value of y where $f(x, y)$ is minimum. Then find the minimum value of $f(x, y)$.

10. **Optimal Production Levels** In part (a) of Review Problem 3, you expressed the monthly profit from selling two products as a function of their monthly production levels. The highest point on the graph of the function is the critical point. How many tons of each product should the producer produce monthly to maximize the profit? What is the maximum profit?

Section 7.4

11. Find the critical points on the graph of
$$f(x, y) = 32xy - 8x^4 - \tfrac{1}{3}y^3$$
Then use the second derivative test to determine which of the critical points are high points, which are low points, and which are saddle points.

12. **Optimal Dimensions** A carpenter plans to build a rectangular box. The volume of the box will be 12 cubic meters. So, if the length is x meters and the width is y meters, the height has to be $12/(xy)$ meters. (See the figure for this problem.) The price of the material for the sides is \$2 per square meter, and the price of the material for the top and bottom is \$3 per square meter. Find a formula that expresses the cost of the material as a function of x and y. Then find the values of x and y that minimize the cost of the material. (The low point on the graph of the cost function is the lowest point.)

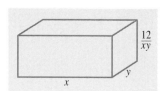

Section 7.5

13. The table shows the values of y that correspond to several values of x. Thinking of y as a function of x, use the method of least squares to find a linear function that approximates the relationship between x and y. Then use the function to estimate the values of y when $x = 5$ and when $x = 17$. Finally, place the data points and the graph of the function on the same cartesian plane.

x	1	3	7	9	12	13
y	9.5	9	7	6.5	5.5	4.5

14. **Medicine** Since the beginning of 1988, doctors have been prescribing a special drug that fights a disease of the nervous system. The following table shows the number of people afflicted with the disease on six different dates:

Date, Jan. 1	1989	1990	1991	1992	1993	1994
Number afflicted, thousands	763	757	740	722	692	660

a. Use the method of least squares to find a function of the type
$$D(x) = cx^2 + d$$
that approximates the number of people afflicted (in thousands) x years after the beginning of 1988.

b. Use the function you found in part (a) to estimate when the disease will be eradicated.

Section 7.6

15. **Architecture** An architect is designing a one-story rectangular building. The front and back walls cost \$20 per horizontal foot, and the two side walls cost \$16 per horizontal foot. The architect wants the total cost of the walls to be \$18,000. Thus, the length and width of the building are subject to a constraint. Use the substitution method to find the maximum area the floor can have.

Section 7.7

16. **Architecture** Use the method of Lagrange multipliers to find the maximum area the floor of the building in Problem 15 can have. (The constrained critical point is the highest constrained point on the graph of the area function.)

APPENDIX A

Algebra Review

A.1 Real Numbers

A.2 Exponents and Radicals

A.3 Polynomial Expressions

A.4 Rational Expressions

A.1 Real Numbers

The set of **integers** consists of the counting numbers, the negatives of the counting numbers, and zero:

$$\ldots, -3, -2, -1, 0, 1, 2, 3, \ldots$$

Definition A.1

The **rational numbers** are the numbers that can be expressed in the form

$$\frac{m}{n}$$

where m and n are integers and $n \neq 0$.

Examples of rational numbers are

$$\frac{2}{3}, \frac{-7}{5}, -13, \text{ and } 0$$

Note that -13 can be expressed as $-13/1$ and 0 can be expressed as $0/1$.

Theorem A.1 gives us an alternative way of conceptualizing the rational numbers.

Theorem A.1

The rational numbers are the numbers whose decimal representations either *terminate* or have a cycle of digits that *repeats* indefinitely.

Thus, the terminating decimal 18.504 and the repeating decimal $0.4\overline{93}$ represent rational numbers. (The bar over 93 indicates that the cycle consisting of 9 followed by 3 repeats indefinitely.)

Definition A.2

The **irrational numbers** are the numbers whose decimal representations neither terminate nor have a cycle of digits that repeats indefinitely.

Examples of irrational numbers are $\sqrt{2}$ and π. ($\sqrt{2}$ is approximately 1.414213, and π is approximately 3.141592.)

Definition A.3

The set of **real numbers** is the collection of all rational and irrational numbers.

So every real number is either a rational number or an irrational number. For brevity, we sometimes refer to real numbers as simply numbers.

Number Lines

We can use the points on a line to represent the real numbers. To establish such a representation on a horizontal line, we select a point on the line and call it the **origin**. We also select a unit for measuring distances. Then we assign a real number to each point on the line according to the following procedure:

1. To each point on the right side of the origin we assign the real number that expresses the distance between the point and the origin.
2. To each point on the left side of the origin we assign the *negative* of the real number that expresses the distance between the point and the origin.
3. To the origin, we assign zero.

The above procedure establishes a **coordinate system** on the line. The numbers we assign to the points are the **coordinates** of the points. A line with a coordinate system is called a **number line**. Figure A.1 shows the coordinates of several points on a number line.

Every point on a number line corresponds to exactly one real number, and every real number corresponds to exactly one point on the number line. So we often refer to numbers as points. For example, we might say that the point -3 is 3 units to the left of the origin.

Inequalities

Definition A.4

Suppose c and d are real numbers. If on a number line the point representing c is to the *left* of the point representing d, then c is **less than** d and we write

$$c < d$$

If the point representing c is to the *right* of the point representing d, then c is **greater than** d and we write

$$c > d$$

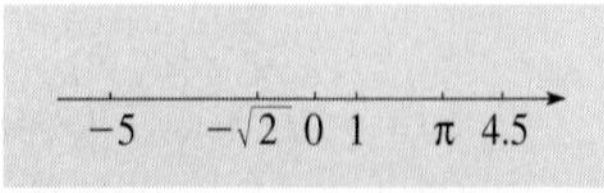

We refer to the symbols $<$ and $>$ as **inequality symbols**.

Figure A.1 A number line.

Example A.1

Place the appropriate inequality symbol between each pair of numbers.

a. 2 and -5
b. -14.2 and -3
c. 0 and -0.6
d. 0 and π

Solutions

a. $2 > -5$
b. $-14.2 < -3$
c. $0 > -0.6$
d. $0 < \pi$

We can express the statement "c is less than or equal to d" by writing

$$c \leq d$$

Similarly, we can express the statement "c is greater than or equal to d" by writing

$$c \geq d$$

Thus, the following statements are true:

$2.75 \leq 8$ In fact, $2.75 < 8$.
$\sqrt{3} \leq \sqrt{3}$ In fact, $\sqrt{3} = \sqrt{3}$.
$-4 \geq -4$ In fact, $-4 = -4$.
$2 \geq -17.6$ In fact, $2 > -17.6$.

We can express the compound statement

$$c < e \text{ and } e < d$$

by writing

$$c < e < d$$

or by saying that e is **between** c and d. We also can say that e is greater than c and less than d. For example, since

$$-5 < -1 \text{ and } -1 < 3$$

we can write

$$-5 < -1 < 3$$

and say that -1 is between -5 and 3. Alternatively, we can say that -1 is greater than -5 and less than 3.

Example A.2

Express each statement using the symbols $<$, $>$, $\leq$, or $\geq$.

a. x is less than or equal to 8.3.
b. z is greater than 5.

c. y is between -4 and 2.9
d. x is not less than zero
e. x is not greater than 0.307
f. x is to the right of 4 on a number line
g. x is greater than or equal to 5 and less than 9

Solutions

a. $x \leq 8.3$
b. $z > 5$
c. $-4 < y < 2.9$
d. Since x is not less than zero, x is greater than or equal to zero. So we can write $x \geq 0$.
e. Since x is not greater than 0.307, x is less than or equal to 0.307. Thus, we can write $x \leq 0.307$.
f. $x > 4$
g. $5 \leq x < 9$

Intervals

Suppose a and b are real numbers such that $a < b$. The set of all real numbers x such that

$$a \leq x \leq b$$

is an **interval**. The **endpoints** of the interval are a and b. Since this interval contains its endpoints, it is a **closed interval** and we can represent it using the symbol $[a, b]$.

Figure A.2 shows the graph of the closed interval $[0, 5]$ on a number line.

If we exclude the endpoints a and b from the closed interval $[a, b]$, we are left with the set of all real numbers x such that

$$a < x < b$$

Figure A.2 The closed interval $[0, 5]$.

This set also is an interval. To be more specific, this set is an **open interval** and we can represent it using the symbol (a, b).

Figure A.3 shows the graph of the open interval $(0, 5)$. Note that we replaced the brackets in Figure A.2 with parentheses in Figure A.3. Even though the interval $(0, 5)$ does not contain zero and 5, these two numbers are still the endpoints of the interval.

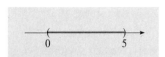

Figure A.3 The open interval $(0, 5)$.

If we exclude only one of the endpoints from the closed interval $[a, b]$, the resulting set is again an interval. Such an interval is a **half-open interval**. If a is the endpoint we exclude, we can represent the resulting interval using the symbol $(a, b]$. If b is the endpoint we exclude, we can represent it using the symbol $[a, b)$.

Figure A.4 shows the half-open interval $[0, 5)$. This interval is the set of all real numbers x such that $0 \leq x < 5$.

Figure A.4 The half-open interval $[0, 5)$.

The following summarizes what we have so far said about intervals:

Figure A.5 The open interval $(0, +\infty)$.

Types of Intervals

Intervals that contain two endpoints are *closed*.
Intervals that contain no endpoints are *open*.
Intervals that contain one endpoint are *half-open*.

The set of all real numbers x such that $x \geq a$ is also an interval. Since this interval contains only one endpoint (namely, a), it is a half-open interval and we can represent it using the symbol $[a, +\infty)$, where ∞ represents *infinity*. If we exclude the endpoint a from this half-open interval, the resulting set is an interval that contains no endpoints. Therefore, it is an open interval and we can represent it using the symbol $(a, +\infty)$.

Figure A.5 shows the open interval $(0, +\infty)$.

The set of all real numbers is another example of an interval. Is this interval closed, open, or half-open?

Absolute Value

Definition A.5

The **absolute value** of a real number c is represented by

$$|c|$$

and is the distance between the point representing c and the origin on a number line.

Sometimes we call the absolute value of a number the **magnitude** of the number.

Since a distance cannot be negative,

$$|c| \geq 0$$

for *any* real number c.

Example A.3

Find the absolute values of the following numbers:

a. 47.3
b. -8
c. 0
d. $-\sqrt{2}$

Solutions

a. $|47.3| = 47.3$
b. $|-8| = 8$
c. $|0| = 0$
d. $|-\sqrt{2}| = \sqrt{2}$

Order of Operations

If an expression involves more than one of the operations of addition, subtraction, multiplication, and division, we find the value of the expression using the following rule:

Order of Operations Rule

1. First, evaluate multiplications and divisions as they appear from left to right.
2. Then, evaluate additions and subtractions as they appear from left to right.
3. If the expression involves parentheses (or brackets, or braces), first evaluate the expressions within parentheses. Expressions above or below fraction bars are considered as being within parentheses.

In Section A.2, we extend the above rule to include powers and radicals.

Example A.4

Find the values of the following expressions:

a. $-7 + 5(-8)$
b. $(-7 + 5)(-8)$
c. $2(9) - 20 + 4(-2)$
d. $2(9) - [20 + 4(-2)]$
e. $\dfrac{14 - 7(4) + 2}{2}$
f. $14 - \dfrac{7(4) + 2}{2}$
g. $\dfrac{126}{12 - 5(6)}$
h. $\dfrac{126}{(12 - 5)6}$

Solutions

a. $-7 + 5(-8) = -7 + (-40)$
$ = -47$

b. $(-7 + 5)(-8) = (-2)(-8)$
$ = 16$

c. $2(9) - 20 + 4(-2) = 18 - 20 + (-8)$
$ = -2 + (-8)$
$ = -10$

d. $2(9) - [20 + 4(-2)] = 2(9) - [20 + (-8)]$
$ = 2(9) - 12$
$ = 18 - 12$
$ = 6$

e. $\dfrac{14 - 7(4) + 2}{2} = \dfrac{14 - 28 + 2}{2}$

$= \dfrac{-14 + 2}{2}$

$= \dfrac{-12}{2}$

$= -6$

f. $14 - \dfrac{7(4) + 2}{2} = 14 - \dfrac{28 + 2}{2}$

$= 14 - \dfrac{30}{2}$

$= 14 - 15$

$= -1$

g. $\dfrac{126}{12 - 5(6)} = \dfrac{126}{12 - 30}$

$= \dfrac{126}{-18}$

$= -7$

h. $\dfrac{126}{(12 - 5)6} = \dfrac{126}{(7)6}$

$= \dfrac{126}{42}$

$= 3$

We conclude this section with a list of the basic properties of the real numbers.

Basic Properties of the Real Numbers

1. $a + b = b + a$
2. $ab = ba$ } commutative properties
3. $(a + b) + c = a + (b + c)$
4. $(ab)c = a(bc)$ } associative properties
5. $a(b + c) = ab + ac$ distributive property
6. $c + 0 = c$
7. $1c = c$ } identity properties
8. $c + (-c) = 0$
9. If $c \neq 0$, $(1/c)c = 1$ } inverse properties

A.1 Problems

In Problems 1 to 9, decide whether the statement is true or false.

1. Every rational number is a real number.
2. Some real numbers are neither rational nor irrational numbers.
3. Every integer is a rational number.
4. Every irrational number is a real number.
5. Negative numbers are not real numbers.
6. If $|x| < 7$, then $-7 < x < 7$.
7. $|c + d| = |c| + |d|$ for any pair of real numbers c and d.
8. $|c| = |-c|$ for any real number c.
9. If $c < 0$, $|c| = -c$.
10. Using a centimeter as a unit of length, label the following points on a number line:
    $$3, -5, 4.5, -0.8, -\pi, \text{ and } \sqrt{5}$$

In Problems 11 to 16, place the appropriate inequality symbol between the pair of numbers.

11. -8 and 3
12. 1.82 and -2
13. -4.7 and -4
14. -1 and $-\sqrt{2}$
15. 7.46 and 7.439
16. -0.4 and 0

In Problems 17 to 30, express the statement using the symbols $<, >, \leq,$ or $\geq$.

17. -12.3 is greater than -20.
18. -8 is less than 1.
19. -4.1 is between -5 and -2.
20. 2.5 is between 0 and 7.34.
21. x is less than or equal to 5.
22. x is greater than or equal to 3.
23. x is negative.
24. x is positive.
25. x is nonnegative.
26. x is nonpositive.
27. x is not greater than 7.8.
28. x is not less than -2.
29. c is to the left of d on a number line.
30. c is to the right of d on a number line.

In Problems 31 to 46, determine whether the given interval is closed, open, or half-open. Then represent it using brackets and/or parentheses. Finally, draw its graph on a number line, labeling the endpoints.

31. The real numbers x such that $2 < x < 7$.
32. The real numbers x such that $5 < x < 9$.
33. The real numbers x such that $0 \leq x \leq 4$.
34. The real numbers x such that $-1 \leq x \leq 3$.
35. The real numbers x such that $x \leq 0$.
36. The real numbers x such that $x \geq 1$.
37. The real numbers x such that $x > -4$.
38. The real numbers x such that $x < 20$.
39. The real numbers x such that $0 < x \leq 12.4$.
40. The real numbers x such that $1.75 \leq x < 5$.
41. The real numbers between 4.2 and 8.
42. The real numbers between -2 and 1.5.
43. The real numbers that are less than 3.
44. The real numbers that are greater than 0.5.
45. The real numbers x such that $|x| < 4$.
46. The real numbers x such that $|x| \leq 2$.

In Problems 47 to 62, find the value of the expression.

47. $8 - 6(4)$
48. $-1 - 3(-5)$
49. $(8 - 6)4$
50. $(-1 - 3)(-5)$
51. $(-3)5 - 7 + 4(-1)$
52. $-13 - 2(-8) - 3(4)$
53. $(-3)5 - [7 + 4(-1)]$
54. $-13 - [2(-8) - 3(4)]$

55. $\dfrac{15 - 2(3)}{3} + 18$ 56. $\dfrac{5(-4) - 32 + 8}{4}$ 59. $\dfrac{80}{1 + 3(5)}$ 60. $\dfrac{20}{2(8) - 6}$

57. $\dfrac{15 - 2(3) + 18}{3}$ 58. $5(-4) - \dfrac{32 + 8}{4}$ 61. $\dfrac{80}{(1 + 3)5}$ 62. $\dfrac{20}{2(8 - 6)}$

A.2 Exponents and Radicals

Positive Integer Exponents

Suppose b is a real number and n is a *positive integer* (counting number). The symbol

$$b^n$$

is read "b **raised to the nth power**" (or the "nth power of b") and is used to compactly represent the product

$$b \cdot b \cdot b \cdots b$$

where the factor b appears n times. Thus, 23 raised to the 4th power is

$$23^4 = (23)(23)(23)(23)$$
$$= 279{,}841$$

In the context of the notation b^n, the number b is the **base** and the number n is the **exponent**. Usually, b^2 is read b **squared**, and b^3 is read b **cubed**.

Sometimes the result of raising a number to a power has more digits than a calculator can display. Example A.5 illustrates how some calculators treat such a result.

Example A.5

Find the value of

$$7004^3$$

Solution

Using a calculator, we obtain

$$3.435883361\text{E}11$$

which means that

$$7004^3 = (3.435883361)10^{11}$$
$$= 343{,}588{,}336{,}100$$

Actually, the result we obtained in Example A.5 is the value of 7004^3 rounded to the nearest hundred. The exact value of 7004^3 is 343,588,336,064.

Theorem A.2 is a consequence of the rule for multiplying numbers that are expressed in decimal notation.

Theorem A.2

If b has m decimal places to the right of the decimal point and n is a positive integer, then

$$b^n$$

has nm decimal places to the right of the decimal point.

We use Theorem A.2 in Example A.6.

Example A.6

Use a calculator to find the values of the following powers:

a. 0.083^6
b. 0.013^4
c. 4.28^5
d. 3.7^4

Solutions

a. We press the appropriate keys and obtain

$$3.269403734\text{E}-7$$

which means that

$$0.083^6 = \frac{3.269403734}{10^7}$$
$$= 0.0000003269403734$$

Actually, this result is the value of 0.083^6 rounded to 16 decimal places. According to Theorem A.2, the exact value of 0.083^6 has 18 decimal places.

b. By pressing the appropriate keys, we get

$$2.8561\text{E}-8$$

which means that

$$0.013^4 = \frac{2.8561}{10^8}$$
$$= 0.000000028561$$

Since this result has 12 decimal places, it is the exact value of 0.013^4. According to Theorem A.2, the value of 0.013^4 should have 12 decimal places.

c. We press the appropriate keys and the screen shows

$$1436.212972$$

This result is the value of 4.28^5 rounded to six decimal places. According to Theorem A.2, the exact value of 4.28^5 has 10 decimal places to the right of the decimal point.

d. By pressing the appropriate keys, we obtain

$$187.4161$$

Since this result has four decimal places to the right of the decimal point, it is the exact value of 3.7^4. According to Theorem A.2, the value of 3.7^4 should have four decimal places to the right of the decimal point.

Radicals

Definition A.6

Suppose b is a real number and n is a positive integer greater than 1. A number c is an **nth root of** b if

$$c^n = b$$

A number can have more than one nth root. For example, both -3 and 3 are 4th roots of 81, because $(-3)^4$ and 3^4 are both equal to 81. In general, if n is *even* and b is *positive*, b has two nth roots among the real numbers, one negative and the other positive. In such a case, we call the positive nth root the **principal nth root** and represent it using the symbol

$$\sqrt[n]{b}$$

So
$$\sqrt[4]{81} = 3$$

rather than -3.

The result of raising a real number to an even power is always a nonnegative number. Therefore, if n is *even* and b is *negative*, b has no nth roots among the real numbers. For example, -81 has no 4th roots among the real numbers. So $\sqrt[4]{-81}$ does not represent a real number.

If n is *odd*, b has exactly one nth root among the real numbers, and, like principal nth roots, we represent it using the symbol $\sqrt[n]{b}$. Such an nth root is positive if b is positive and negative if b is negative. For example,

$$\sqrt[5]{-32} = -2 \quad \text{because } (-2)^5 = -32$$

while
$$\sqrt[5]{32} = 2 \quad \text{because } 2^5 = 32$$

If n is any positive integer greater than 1, the nth root of zero is zero itself. We express this result by writing

$$\sqrt[n]{0} = 0$$

The following statements summarize what we have so far said about the symbol $\sqrt[n]{b}$:

1. If n is even and b is positive, $\sqrt[n]{b}$ represents the positive number whose nth power is b.
2. If n is odd, $\sqrt[n]{b}$ represents the unique number whose nth power is b.
3. $\sqrt[n]{0} = 0$.

We usually call second roots **square roots** and third roots **cube roots**. Also, we usually represent the positive square root of a number b using the symbol $\sqrt{b}$ rather than the symbol $\sqrt[2]{b}$.

We call $\sqrt[n]{b}$ a **radical**. In this context, n is the **index** of the radical, and b is the **radicand**.

Positive Rational Exponents

Definition A.7 gives meaning to the idea of raising a number to a *rational* power.

Definition A.7

Suppose m and n are positive integers with no common factors, $n > 1$, and b is a real number. Then
$$b^{m/n} = (\sqrt[n]{b})^m = \sqrt[n]{b^m}$$
provided $b \geq 0$ if n is even.

With rational exponents we use terminology similar to the terminology we use with integer exponents. For example, we read $8^{2/3}$ as "8 raised to the $\frac{2}{3}$ power."

The following statement is the special case of Definition A.7 where $m = 1$:

If n is a positive integer greater than 1, then
$$b^{1/n} = \sqrt[n]{b}$$
provided $b \geq 0$ if n is even.

Example A.7

Find the values of the following powers:

a. $8^{2/3}$ \qquad\qquad\qquad\qquad b. $25^{1/2}$
c. $(-32)^{1/5}$ \qquad\qquad\qquad d. $(-27)^{4/3}$

Solutions

a. $8^{2/3} = (\sqrt[3]{8})^2$
$= 2^2$
$= 4$

b. $25^{1/2} = \sqrt{25}$
$= 5$

c. $(-32)^{1/5} = \sqrt[5]{-32}$
$= -2$

d. $(-27)^{4/3} = (\sqrt[3]{-27})^4$
$= (-3)^4$
$= 81$

Negative Exponents

Definition A.8 attaches meaning to the idea of raising a number to a *negative* power.

Definition A.8

Suppose r is a positive rational number, and b is a real number different from zero. Then

$$b^{-r} = \frac{1}{b^r}$$

provided b^r is a real number.

Example A.8

Find the values of the following powers:

a. 4^{-2}
b. $25^{-1/2}$
c. $8^{-5/3}$
d. $(-32)^{-3/5}$

Solutions

a. $4^{-2} = \dfrac{1}{4^2}$
$= \dfrac{1}{16}$
$= 0.0625$

b. $25^{-1/2} = \dfrac{1}{25^{1/2}}$
$= \dfrac{1}{\sqrt{25}}$
$= \dfrac{1}{5}$
$= 0.2$

c. $8^{-5/3} = \dfrac{1}{8^{5/3}}$
$= \dfrac{1}{(\sqrt[3]{8})^5}$

d. $(-32)^{-3/5} = \dfrac{1}{(-32)^{3/5}}$
$= \dfrac{1}{(\sqrt[5]{-32})^3}$

$$= \frac{1}{2^5} \qquad\qquad = \frac{1}{(-2)^3}$$
$$= \frac{1}{32} \qquad\qquad = \frac{1}{-8}$$
$$= 0.03125 \qquad\qquad = -0.125$$

Definition A.9 assigns meaning to the idea of raising a number to the *zeroth* power.

Definition A.9

If $b \neq 0$, $b^0 = 1$.

Thus, $(-13)^0 = 1$.

Properties of Exponents

The following five statements are valid provided the indicated powers are real numbers:

Properties of Exponents

Suppose b and c are real numbers, and r and s are rational numbers. Then:

EP1 $b^r \cdot b^s = b^{r+s}$

EP2 $\dfrac{b^r}{b^s} = b^{r-s}$ provided $b \neq 0$

EP3 $(b^r)^s = b^{rs}$

EP4 $(bc)^r = b^r c^r$

EP5 $\left(\dfrac{b}{c}\right)^r = \dfrac{b^r}{c^r}$ provided $c \neq 0$

We can use the above properties to simplify expressions involving powers and radicals. A radical is in **simplest form** if it satisfies the following three conditions:

1. The radicand has no factor whose power is greater than or equal to the index. (So $\sqrt[5]{3x^{17}}$ is not in simplest form.)
2. The radicand is not a fraction. (So $\sqrt[3]{5/x}$ is not in simplest form.)
3. There is no factor common to the index and all of the exponents in the radicand. (So $\sqrt[12]{81x^8}$ is not in simplest form. *Note*: $81 = 3^4$.)

Sec. A.2 Exponents and Radicals 475

If the denominator of an expression is a radical, we can express the denominator in the form $b^{m/n}$. Then we can multiply the numerator and denominator by $b^{p/n}$, where $p/n + m/n = 1$, and apply property EP1 to the denominator. The result is an expression equivalent to the original expression but whose denominator is simply b. This procedure is called **rationalizing the denominator**.

Example A.9

Simplify the following expressions. (Rationalize all denominators.)

a. $\sqrt[5]{3x^{17}}$

b. $\dfrac{4}{\sqrt[7]{x^2}}$

c. $\sqrt[3]{\dfrac{5}{x}}$

d. $\sqrt[12]{81x^8}$

e. $\dfrac{6x^3}{x^7}$

f. $5\sqrt{32x} - \sqrt{18x}$

Solutions

a. $\sqrt[5]{3x^{17}} = (3x^{17})^{1/5}$ Definition A.7
$= (x^{15}\, 3x^2)^{1/5}$ EP1
$= x^3(3x^2)^{1/5}$ EP3 and EP4
$= x^3(\sqrt[5]{3x^2})$ Definition A.7

b. $\dfrac{4}{\sqrt[7]{x^2}} = \dfrac{4}{x^{2/7}}$ Definition A.7
$= \dfrac{4x^{5/7}}{x^{2/7}\, x^{5/7}}$ to rationalize the denominator
$= \dfrac{4\sqrt[7]{x^5}}{x}$ Definition A.7 and EP1

c. $\sqrt[3]{\dfrac{5}{x}} = \left(\dfrac{5}{x}\right)^{1/3}$ Definition A.7
$= \dfrac{5^{1/3}}{x^{1/3}}$ EP5
$= \dfrac{5^{1/3}\, x^{2/3}}{x^{1/3}\, x^{2/3}}$ to rationalize the denominator
$= \dfrac{(5x^2)^{1/3}}{x}$ EP1, EP3, and EP4
$= \dfrac{\sqrt[3]{5x^2}}{x}$ Definition A.7

d. $\sqrt[12]{81x^8} = \sqrt[12]{3^4 \cdot x^8}$ since $81 = 3^4$
$= \sqrt[12]{(3x^2)^4}$ EP3 and EP4
$= (3x^2)^{4/12}$ Definition A.7
$= (3x^2)^{1/3}$ since $\frac{4}{12} = \frac{1}{3}$
$= \sqrt[3]{3x^2}$ Definition A.7

e. $\frac{6x^3}{x^7} = 6x^{-4}$ EP2

$= \frac{6}{x^4}$ Definition A.8

f. $5\sqrt{32x} - \sqrt{18x} = 5\sqrt{16 \cdot 2x} - \sqrt{9 \cdot 2x}$
$= 5(16 \cdot 2x)^{1/2} - (9 \cdot 2x)^{1/2}$ Definition A.7
$= 5 \cdot 16^{1/2}(2x)^{1/2} - 9^{1/2}(2x)^{1/2}$ EP4
$= 5\sqrt{16}\sqrt{2x} - \sqrt{9}\sqrt{2x}$ Definition A.7
$= 20\sqrt{2x} - 3\sqrt{2x}$
$= 17\sqrt{2x}$

Order of Operations Revisited

To evaluate an expression involving powers and/or radicals and at least one of the operations of addition, subtraction, multiplication, and division, first evaluate the powers and radicals as they appear from left to right. Then proceed according to the order of operations rule we presented in Section A.1. Consider expressions within the symbol $\sqrt{}$ as being within parentheses.

Example A.10

Find the values of the following expressions:

a. $100 - 3 \cdot 4^2$ b. $2(-9) + \sqrt[5]{7(5) - 3}$

Solutions

a. $100 - 3 \cdot 4^2 = 100 - 3(16)$
$= 100 - 48$
$= 52$

b. $2(-9) + \sqrt[5]{7(5) - 3} = 2(-9) + \sqrt[5]{35 - 3}$
$= 2(-9) + \sqrt[5]{32}$
$= 2(-9) + 2$
$= -18 + 2$
$= -16$

We conclude this section with a *warning*:
$$-x^2 \neq (-x)^2$$
For example, $-7^2 = -49$, whereas $(-7)^2 = 49$. So -7^2 means $-(7^2)$, not $(-7)^2$.

A.2 Problems

In Problems 1 to 26, find the value of the expression without using a calculator.

1. 8^3
2. 40^2
3. $(-2)^6$
4. $(-3)^4$
5. -2^6
6. -3^4
7. $(-2)^5$
8. $(-3)^3$
9. $\sqrt[4]{16}$
10. $\sqrt{16}$
11. $\sqrt[3]{-27}$
12. $\sqrt[3]{-8}$
13. $49^{1/2}$
14. $16^{1/4}$
15. $(-64)^{2/3}$
16. $36^{3/2}$
17. 3^{-2}
18. 2^{-3}
19. $16^{-1/2}$
20. $(-32)^{-1/5}$
21. 153^0
22. 19^0
23. $125^{-2/3}$
24. $100^{-3/2}$
25. $6 + 2 \cdot 5^3$
26. $23 - 5\sqrt[3]{84 - 4(5)}$

In Problems 27 to 50, use a calculator to find the value of the expression.

27. 13^7
28. 21^6
29. 42^8
30. 215^4
31. 0.028^5
32. 0.009^5
33. $(-6)^7$
34. $(-3)^9$
35. $(-0.5)^6$
36. $(-0.4)^8$
37. $\sqrt[5]{371293}$
38. $\sqrt[3]{-24389}$
39. $\sqrt{46}$
40. $\sqrt[4]{24}$
41. $\sqrt[3]{-5.17}$
42. $\sqrt{22.9}$
43. $20^{2/5}$
44. $10^{5/3}$
45. $18.2^{-1/3}$
46. $12.7^{-1/2}$
47. $34^{-3/2}$
48. $23^{-4/3}$
49. $7\sqrt{8} + 12$
50. $2 \cdot 8^7$

In Problems 51 to 68, simplify the expression. (Rationalize all denominators.)

51. $\sqrt[4]{48x^7}$
52. $\sqrt{9x^5}$
53. $\dfrac{8}{\sqrt[4]{x^3}}$
54. $\dfrac{5}{\sqrt[3]{x}}$
55. $\sqrt[5]{\dfrac{32}{x^3}}$
56. $\sqrt[3]{\dfrac{2}{x^2}}$
57. $\sqrt[8]{x^6}$
58. $\sqrt[6]{25x^4}$
59. $\dfrac{x^9}{4x^3}$
60. $\dfrac{3x^2}{10x^5}$
61. $\dfrac{\sqrt[3]{x}}{\sqrt{x}}$
62. $\dfrac{\sqrt[4]{x}}{\sqrt[3]{x}}$
63. $\sqrt[4]{x}\sqrt{x}$
64. $\sqrt{x}\sqrt[3]{x^2}$
65. $\sqrt[3]{24x^2} + \sqrt[3]{81x^2}$
66. $\sqrt{125x} - \sqrt{45x}$
67. $\sqrt{\dfrac{9}{\sqrt[3]{x}}}$
68. $\sqrt[4]{16\sqrt{x}}$

A.3 Polynomial Expressions

A **variable** is a symbol (usually a letter) that represents any number in a set of real numbers called the **replacement set**. An **algebraic expression** is the result of combining variables and real numbers using addition, subtraction, multiplication, division, and/or exponentiation (raising to a power or taking a root). The following expressions are examples of algebraic expressions:

$$7x^3 \qquad x + 3y - 4 \qquad \frac{x^2 - x}{2x + 1} \qquad \frac{8\sqrt{z - 9}}{y^2}$$

Polynomials

Suppose x is a variable. A **monomial** in x of degree n is an algebraic expression of the form

$$ax^n$$

where a is a nonzero real number, called the **coefficient** of x^n, and n is a positive integer. Thus, x^4 is a monomial of degree 4 with $a = 1$, while $-0.8x$ is a monomial of degree 1 with $a = -0.8$. A nonzero real number c is a monomial of degree 0 with coefficient c.

Definition A.10

A **polynomial** in x is an algebraic expression formed by adding and/or subtracting several monomials in x.

We frequently call monomials polynomials.

The **degree of a polynomial** is the degree of the monomial of highest degree among the monomials used to form the polynomial. The coefficient of the monomial of highest degree is the **leading coefficient** of the polynomial.

A polynomial can have more than one variable. In this section, however, we consider only polynomials with one variable.

The following algebraic expressions are examples of polynomials in x:

$2x^4 + 7x^2 - x$ degree 4 and leading coefficient 2
$8x - x^3$ degree 3 and leading coefficient -1
$4x^6$ degree 6 and leading coefficient 4
$-5x + 9$ degree 1 and leading coefficient -5

Of course, not all algebraic expressions are polynomials. For example, the following algebraic expressions are not polynomials:

$$\frac{3x + 1}{x^2 + 5x} \qquad 4x - \sqrt{x} \qquad x^3 + \frac{2}{x}$$

Arithmetic Operations on Polynomials

Since variables represent real numbers, polynomials represent real numbers. Thus, we can add, subtract, multiply, and divide polynomials using the properties of real numbers. In Example A.11, we subtract and multiply polynomials.

Example A.11

Perform the following computations:

a. $(3x^4 + 9x^2 - 8) - (5x^4 - x^3 + 4x^2)$
b. $(x + 3)(4x^2 - 5x + 6)$

Solutions

a. We remove the parentheses, change the signs of the terms in the subtrahend, and combine like terms:

$$(3x^4 + 9x^2 - 8) - (5x^4 - x^3 + 4x^2)$$
$$= 3x^4 + 9x^2 - 8 - 5x^4 + x^3 - 4x^2$$
$$= (3x^4 - 5x^4) + x^3 + (9x^2 - 4x^2) - 8$$
$$= -2x^4 + x^3 + 5x^2 - 8$$

b. We use the distributive property in the first two steps and then combine like terms:

$$(x + 3)(4x^2 - 5x + 6)$$
$$= (x + 3)4x^2 - (x + 3)5x + (x + 3)6$$
$$= 4x^3 + 12x^2 - (5x^2 + 15x) + 6x + 18$$
$$= 4x^3 + 12x^2 - 5x^2 - 15x + 6x + 18$$
$$= 4x^3 + 7x^2 - 9x + 18$$

Example A.12 illustrates a technique for dividing polynomials.

Example A.12

Perform the following division:

$$\frac{x^3 + 8x^2 - 7}{x^2 - 5x + 2}$$

Solution

Step 1.

$$\begin{array}{r} x \\ x^2 - 5x + 2 \overline{) x^3 + 8x^2 + 0x - 7} \\ \underline{x^3 - 5x^2 + 2x} \\ 13x^2 - 2x - 7 \end{array}$$

dividing x^3 by x^2

multiplying the divisor by x
subtracting

Step 2.

$$\begin{array}{r} x + 13 \\ x^2 - 5x + 2 \overline{) x^3 + 8x^2 + 0x - 7} \\ \underline{x^3 - 5x^2 + 2x} \\ 13x^2 - 2x - 7 \\ \underline{13x^2 - 65x + 26} \\ 63x - 33 \end{array}$$

dividing $13x^2$ by x^2

multiplying the divisor by 13
subtracting

Since the degree of $63x - 33$ is less than the degree of the divisor, we have finished. Moreover, we can write

$$\frac{x^3 + 8x^2 - 7}{x^2 - 5x + 2} = x + 13 + \frac{63x - 33}{x^2 - 5x + 2}$$

and say $x + 13$ is the **quotient** and $63x - 33$ is the **remainder** that results from the division.

Factoring

We often can express a polynomial as a product of other polynomials. The polynomials in such a product are **factors** of the original polynomial. Thus, $x + 3$ and $4x^2 - 5x + 6$ are factors of $4x^3 + 7x^2 - 9x + 18$, because

$$4x^3 + 7x^2 - 9x + 18 = (x + 3)(4x^2 - 5x + 6)$$

[See part (b) of Example A.11.]

The process of expressing a polynomial as a product of other polynomials is called **factoring**. In many cases, the procedure used in factoring is based on the distributive property. Although we presented this property in Section A.1, we restate it here in a slightly modified form.

Distributive Property

If a, b, and c are real numbers,

$$ab + ac = a(b + c)$$

We use the distributive property in Example A.13.

Example A.13

Factor the following polynomials:

a. $5x^3 - 20x^2 + 35x$
b. $3x^3 - 24x^2 + 2x - 16$

Solutions

a. $5x^3 - 20x^2 + 35x = (5x)x^2 - (5x)4x + (5x)7$
$= 5x(x^2 - 4x + 7)$

b. $3x^3 - 24x^2 + 2x - 16 = (3x^3 - 24x^2) + (2x - 16)$
$= 3x^2(x - 8) + 2(x - 8)$
$= (x - 8)(3x^2 + 2)$

The proof of Theorem A.3 is based on the distributive property.

Theorem A.3

Suppose b and c are real numbers and x is a variable. If p and q are real numbers such that

$$pq = c$$

and

$$p + q = b$$

then

$$x^2 + bx + c = (x + p)(x + q)$$

We use Theorem A.3 in Example A.14.

Example A.14

Factor the following polynomials:

a. $x^2 + 2x - 15$
b. $4x^3 + 32x^2 + 48x$
c. $3x^2 + 30x + 75$
d. $x^2 - 2$

Solutions

a. The polynomial $x^2 + 2x - 15$ has the form $x^2 + bx + c$ with $b = 2$ and $c = -15$. To use Theorem A.3, we must find two real numbers whose product is -15 and whose sum is 2. Note that these two requirements are satisfied by the pair consisting of 5 and -3. So, using Theorem A.3,

$$x^2 + 2x - 15 = (x + 5)(x + (-3))$$
$$= (x + 5)(x - 3)$$

b. First we use the distributive property:

$$4x^3 + 32x^2 + 48x = (4x)x^2 + (4x)8x + (4x)12$$
$$= 4x(x^2 + 8x + 12)$$

Then we try to apply Theorem A.3 to $x^2 + 8x + 12$. So we search for two real numbers whose product is 12 and whose sum is 8. These two requirements are met by the pair consisting of 2 and 6. Therefore, using Theorem A.3,

$$x^2 + 8x + 12 = (x + 2)(x + 6)$$

and so

$$4x^3 + 32x^2 + 48x = 4x(x^2 + 8x + 12)$$
$$= 4x(x + 2)(x + 6)$$

c. First we use the distributive property:
$$3x^2 + 30x + 75 = 3(x^2 + 10x + 25)$$
Then we note that
$$5 \cdot 5 = 25$$
and
$$5 + 5 = 10$$
So, using Theorem A.3,
$$x^2 + 10x + 25 = (x + 5)(x + 5)$$
Therefore,
$$3x^2 + 30x + 75 = 3(x^2 + 10x + 25)$$
$$= 3(x + 5)(x + 5)$$
$$= 3(x + 5)^2$$

d. The polynomial $x^2 - 2$ has the form $x^2 + bx + c$ with $b = 0$ and $c = -2$. Thus, we want to find two real numbers whose product is -2 and whose sum is zero. Such a pair consists of $\sqrt{2}$ and $-\sqrt{2}$. So, using Theorem A.3,
$$x^2 - 2 = (x + \sqrt{2})(x - \sqrt{2})$$

Part (d) of Example A.14 illustrates Theorem A.4, which is actually a special case of Theorem A.3.

Theorem A.4

If c is a positive real number, then
$$x^2 - c = (x + \sqrt{c})(x - \sqrt{c})$$

Example A.15 makes use of Theorem A.4.

Example A.15

Factor the following polynomials:

a. $x^2 - 49$
b. $x^2 - 0.25$
c. $5x^2 - \frac{5}{9}$
d. $x^3 + 3x^2 - 4x - 12$

Solutions

a. $x^2 - 49 = (x + 7)(x - 7)$
b. $x^2 - 0.25 = (x + 0.5)(x - 0.5)$

c. $5x^2 - \frac{5}{9} = 5(x^2 - \frac{1}{9})$ distributive property
$= 5(x + \frac{1}{3})(x - \frac{1}{3})$ Theorem A.4

d. $x^3 + 3x^2 - 4x - 12 = (x^3 + 3x^2) - (4x + 12)$
$= x^2(x + 3) - 4(x + 3)$ distributive property
$= (x + 3)(x^2 - 4)$ distributive property
$= (x + 3)(x + 2)(x - 2)$ Theorem A.4

Sometimes Theorems A.3 and A.4 help us factor polynomials of the form

$$x^4 + bx^2 + c$$

Example A.16 involves such a polynomial.

Example A.16

Factor $x^4 - 5x^2 - 36$.

Solution

$x^4 - 5x^2 - 36 = (x^2)^2 - 5x^2 - 36$ a property of exponents
$= y^2 - 5y - 36$ letting $y = x^2$
$= (y + 4)(y - 9)$ Theorem A.3
$= (x^2 + 4)(x^2 - 9)$ since $y = x^2$
$= (x^2 + 4)(x + 3)(x - 3)$ Theorem A.4

Polynomials of degree 2 are often called **quadratic expressions**. Theorem A.3 provides us with a method for factoring quadratic expressions whose leading coefficient is 1. However, the method is often difficult to apply. Theorem A.5 provides us with a more versatile method for factoring quadratic expressions.

Theorem A.5

Suppose a, b, and c are real numbers with $a \neq 0$. If $b^2 - 4ac \geq 0$, then

$$ax^2 + bx + c = a(x - p)(x - q)$$

where

$$p = \frac{-b + \sqrt{b^2 - 4ac}}{2a}$$

and

$$q = \frac{-b - \sqrt{b^2 - 4ac}}{2a}$$

The value of $b^2 - 4ac$ is the **discriminant** of the quadratic expression $ax^2 + bx + c$, and the formula

$$\frac{-b \pm \sqrt{b^2 - 4ac}}{2a}$$

is the **quadratic formula**. We use this formula in Example A.17.

Example A.17

Factor the following polynomials:

a. $9x^2 - 6x - 8$
b. $8x^5 + 12x^4 - 8x^3$
c. $x^2 - 8x + 16$

Solutions

a. The given polynomial has the form $ax^2 + bx + c$ with $a = 9$, $b = -6$, and $c = -8$. Note that

$$b^2 - 4ac = (-6)^2 - 4(9)(-8)$$
$$= 324$$

So the discriminant is nonnegative. Note also that

$$\frac{-b + \sqrt{b^2 - 4ac}}{2a} = \frac{-(-6) + \sqrt{324}}{2(9)}$$
$$= \frac{4}{3}$$

and

$$\frac{-b - \sqrt{b^2 - 4ac}}{2a} = \frac{-(-6) - \sqrt{324}}{2(9)}$$
$$= -\frac{2}{3}$$

Therefore, using Theorem A.5,

$$9x^2 - 6x - 8 = 9(x - \tfrac{4}{3})[x - (-\tfrac{2}{3})]$$
$$= 9(x - \tfrac{4}{3})(x + \tfrac{2}{3})$$

b. Using the distributive property,

$$8x^5 + 12x^4 - 8x^3 = 4x^3(2x^2 + 3x - 2)$$

The polynomial $2x^2 + 3x - 2$ has the form $ax^2 + bx + c$ with $a = 2$, $b = 3$, and $c = -2$. Note that

$$b^2 - 4ac = 3^2 - 4(2)(-2)$$
$$= 25$$

So the discriminant is nonnegative. Note also that

$$\frac{-b + \sqrt{b^2 - 4ac}}{2a} = \frac{-3 + \sqrt{25}}{2(2)}$$
$$= 0.5$$

and
$$\frac{-b - \sqrt{b^2 - 4ac}}{2a} = \frac{-3 - \sqrt{25}}{2(2)}$$
$$= -2$$

Therefore, using Theorem A.5,

$$2x^2 + 3x - 2 = 2(x - 0.5)(x - (-2))$$
$$= 2(x - 0.5)(x + 2)$$

So
$$8x^5 + 12x^4 - 8x^3 = 4x^3(2x^2 + 3x - 2)$$
$$= (4x^3)2(x - 0.5)(x + 2)$$
$$= 8x^3(x - 0.5)(x + 2)$$

c. The given polynomial has the form $ax^2 + bx + c$ with $a = 1$, $b = -8$, and $c = 16$. Note that

$$b^2 - 4ac = (-8)^2 - 4(1)(16)$$
$$= 0$$

So the discriminant is nonnegative. Note also that

$$\frac{-b + \sqrt{b^2 - 4ac}}{2a} = \frac{-(-8) + \sqrt{0}}{2(1)}$$
$$= 4$$

and
$$\frac{-b - \sqrt{b^2 - 4ac}}{2a} = \frac{-(-8) - \sqrt{0}}{2(1)}$$
$$= 4$$

Therefore, using Theorem A.5,

$$x^2 - 8x + 16 = 1(x - 4)(x - 4)$$
$$= (x - 4)^2$$

Example A.18 illustrates how Theorems A.4 and A.5 help us factor polynomials of the form

$$ax^4 + bx^2 + c$$

Example A.18

Factor $5x^4 + 30x^2 - 35$.

Appendix A Algebra Review

Solution

$$\begin{aligned}
5x^4 + 30x^2 - 35 &= 5(x^2)^2 + 30x^2 - 35 && \text{a property of exponents}\\
&= 5y^2 + 30y - 35 && \text{letting } y = x^2\\
&= 5(y+7)(y-1) && \text{Theorem A.5}\\
&= 5(x^2+7)(x^2-1) && \text{since } y = x^2\\
&= 5(x^2+7)(x+1)(x-1) && \text{Theorem A.4}
\end{aligned}$$

Definition A.11

A polynomial is **irreducible** if it cannot be expressed as a product of two or more polynomials of degree 1 or greater.

All polynomials of degree less than 2 and some polynomials of degree 2 are irreducible. Polynomials of degree greater than 2 are *not* irreducible. However, it is usually difficult to express such a polynomial as a product of two or more polynomials of degree 1 or greater.

Theorem A.5 tells us that if a quadratic expression has a nonnegative discriminant, the expression is *not* irreducible. Theorem A.6 tells us that if a quadratic expression has a negative discriminant, the expression is irreducible.

Theorem A.6

Suppose a, b, and c are real numbers with $a \neq 0$. If

$$b^2 - 4ac < 0$$

then $ax^2 + bx + c$ is irreducible.

Example A.19

Determine which of the following quadratic expressions are irreducible:

a. $x^2 - 8.3x + 2.4$
b. $3x^2 - 5x + 9$
c. $x^2 + 36$
d. $x^2 + 6x + 9$

Solutions

a. The discriminant of $x^2 - 8.3x + 2.4$ is

$$\begin{aligned}
b^2 - 4ac &= (-8.3)^2 - 4(1)(2.4)\\
&= 59.29
\end{aligned}$$

So $x^2 - 8.3x + 2.4$ is not irreducible. In fact,
$$x^2 - 8.3x + 2.4 = (x - 8)(x - 0.3)$$

b. The discriminant of $3x^2 - 5x + 9$ is
$$b^2 - 4ac = (-5)^2 - 4(3)(9)$$
$$= -83$$

So $3x^2 - 5x + 9$ is irreducible.

c. The discriminant of $x^2 + 36$ is
$$b^2 - 4ac = 0^2 - 4(1)(36)$$
$$= -144$$

So $x^2 + 36$ is irreducible.

d. The discriminant of $x^2 + 6x + 9$ is
$$b^2 - 4ac = 6^2 - 4(1)(9)$$
$$= 0$$

So $x^2 + 6x + 9$ is not irreducible. In fact,
$$x^2 + 6x + 9 = (x + 3)^2$$

Theorem A.7 helps us factor polynomials of the form $x^3 + c$. The proof of the theorem is based on the distributive property.

Theorem A.7

If c is a real number,
$$x^3 + c = (x + p)(x^2 - px + p^2)$$
where $p = \sqrt[3]{c}$.

In Theorem A.7, the discriminant of the quadratic expression $x^2 - px + p^2$ is a negative number, namely,
$$(-p)^2 - 4(1)(p^2) = p^2 - 4p^2$$

Thus, $x^2 - px + p^2$ is irreducible.

We use Theorem A.7 in Example A.20.

Example A.20

Factor the following polynomials:

a. $x^3 + 64$
b. $5x^4 - 40x$

Solutions

a. Since $\sqrt[3]{64} = 4$,
$$x^3 + 64 = (x + 4)(x^2 - 4x + 16)$$

b. First we use the distributive property:
$$5x^4 - 40x = 5x(x^3 - 8)$$
$$= 5x(x^3 + (-8))$$

Then we apply Theorem A.7 to $x^3 + (-8)$. Since $\sqrt[3]{-8} = -2$,
$$x^3 + (-8) = (x + (-2))(x^2 - (-2)x + (-2)^2)$$
$$= (x - 2)(x^2 + 2x + 4)$$

Thus,
$$5x^4 - 40x = 5x(x^3 + (-8))$$
$$= 5x(x - 2)(x^2 + 2x + 4)$$

Theorems A.5 and A.7 help us factor polynomials of the form
$$ax^6 + bx^3 + c$$

Example A.21 involves such a polynomial.

Example A.21

Factor $2x^6 - 38x^3 - 432$.

Solution

$$\begin{aligned}
2x^6 - 38x^3 - 432 &= 2(x^3)^2 - 38x^3 - 432 &&\text{a property of exponents} \\
&= 2y^2 - 38y - 432 &&\text{letting } y = x^3 \\
&= 2(y + 8)(y - 27) &&\text{Theorem A.5} \\
&= 2(x^3 + 8)(x^3 - 27) &&\text{since } y = x^3 \\
&= 2(x + 2)(x^2 - 2x + 4)(x - 3)(x^2 + 3x + 9)
\end{aligned}$$

(We used Theorem A.7 in the last step.)

The Zeros of a Polynomial

A **zero** of an algebraic expression with one variable is a real number that makes the value of the expression zero if it is substituted for the variable. Thus, 8 is a zero of the polynomial $x^2 + 4x - 96$, because
$$8^2 + 4(8) - 96 = 0$$

Theorem A.8

If c is a zero of a polynomial, then $x - c$ is a factor of the polynomial.

The converse of Theorem A.8 is also valid. So, if $x - c$ is a factor of a polynomial, then c is a zero of the polynomial.

Theorem A.8 implies that irreducible quadratic expressions have no zeros. It also implies that a polynomial has no zeros other than those revealed by expressing the polynomial as a product of irreducible polynomials. We use these results in Example A.22.

Example A.22

Find the zeros of the following polynomials:

a. $3x^3 - 24x^2 + 2x - 16$
b. $x^2 + 2x - 15$
c. $8x^5 + 12x^4 - 8x^3$
d. $x^2 - 8x + 16$
e. $2x^6 - 38x^3 - 432$
f. $3x^2 - 5x + 9$

Solutions

a. In part (b) of Example A.13, we showed that
$$3x^3 - 24x^2 + 2x - 16 = (x - 8)(3x^2 + 2)$$
Thus, since $3x^2 + 2$ is irreducible (why?), the only zero is 8.

b. In part (a) of Example A.14, we showed that
$$x^2 + 2x - 15 = (x + 5)(x - 3)$$
Thus, the zeros are -5 and 3.

c. In part (b) of Example A.17, we showed that
$$8x^5 + 12x^4 - 8x^3 = 8x^3(x - 0.5)(x + 2)$$
Thus, the zeros are 0, 0.5, and -2.

d. In part (c) of Example A.17, we showed that
$$x^2 - 8x + 16 = (x - 4)^2$$
Thus, the only zero is 4.

e. In Example A.21, we showed that
$$2x^6 - 38x^3 - 432 = 2(x + 2)(x^2 - 2x + 4)(x - 3)(x^2 + 3x + 9)$$
Thus, since $x^2 - 2x + 4$ and $x^2 + 3x + 9$ are irreducible (why?), the only zeros are -2 and 3.

f. In part (b) of Example A.19, we showed that $3x^2 - 5x + 9$ is irreducible. Thus, this quadratic expression has no zeros.

A.3 Problems

In Problems 1 to 10, perform the computations.

1. $(x^5 - 4x^3 + 9) + (3x^5 + 8x^2 - 5)$
2. $(-4x^6 + x^2 - 7) + (x^6 - 2x^3 - 6x^2)$
3. $(2x^3 - 5x + 1) - (-8x + 3)$
4. $(x^3 - 4x + 11) - (x^4 + 2x^3 - 4)$
5. $(x - 2)(x^3 + 8x - 4)$
6. $(3x + 1)(2x^4 - x^2 + 5x)$
7. $\dfrac{12x^5 + 9x^3 - 6x}{3x^3}$
8. $\dfrac{2x^7 - 3x^5 + 4x}{x^4}$
9. $\dfrac{8x^3 - 4x^2 - 5}{2x + 3}$
10. $\dfrac{6x^5 + x^3 - 8x}{3x^2 - 4}$

In Problems 11 to 42, express the given polynomial as a product of irreducible polynomials and list the zeros.

11. $x^3 + 3x^2 + 4x + 12$
12. $2x^3 - 2x^2 + 5x - 5$
13. $6x^4 - 9x^3 + 12x^2$
14. $3x^5 + x^4 + 2x^3$
15. $x^3 - 7x^2 - 9x + 63$
16. $x^3 - 5x^2 - 16x + 80$
17. $x^2 + 11x + 28$
18. $x^2 + 9x + 18$
19. $7x^4 - 35x^3 + 28x^2$
20. $2x^5 - 12x^4 - 54x^3$
21. $2x^2 - 24x + 72$
22. $3x^2 - 6x + 3$
23. $x^2 - 81$
24. $x^2 - 64$
25. $3x^2 - 0.75$
26. $7x^2 - 0.28$
27. $x^4 - 29x^2 + 100$
28. $x^4 - 17x^2 + 16$
29. $25x^2 - 20x + 3$
30. $20x^2 + 11x - 3$
31. $36x^2 + 12x + 1$
32. $4x^2 - 4x + 1$
33. $45x^6 + 15x^5 - 10x^4$
34. $24x^4 - 22x^3 + 3x^2$
35. $4x^4 + 11x^2 - 3$
36. $25x^4 + 21x^2 - 4$
37. $x^4 + 125x$
38. $x^5 + 8x^2$
39. $2x^3 - 54$
40. $4x^3 - 256$
41. $x^6 + 936x^3 - 64{,}000$
42. $x^6 - 124x^3 - 125$

A.4 Rational Expressions

A **rational expression** is a quotient of two polynomials. Although it is possible for a rational expression to have more than one variable, in this section we consider only rational expressions with one variable. The following algebraic expressions are examples of such rational expressions:

$$\frac{x^2 - 3x + 9}{7x - 5} \qquad \frac{3}{x^2 - 4} \qquad \frac{2x^3 - 5}{x^3 + 2x^2 + 1}$$

Recall that nonzero real numbers are polynomials of degree 0. Thus, since any polynomial can be expressed as the quotient of itself and 1, every polynomial is a rational expression.

Since polynomials represent real numbers, the following properties of quotients of real numbers apply also to rational expressions:

Properties of Quotients of Real Numbers

Q1 $\dfrac{ad}{bd} = \dfrac{a}{b}$

Q2 $\dfrac{a}{b} \cdot \dfrac{c}{d} = \dfrac{ac}{bd}$

Q3 $\dfrac{a}{b} \div \dfrac{c}{d} = \dfrac{a}{b} \cdot \dfrac{d}{c}$

Q4 $\dfrac{a}{b} + \dfrac{c}{b} = \dfrac{a+c}{b}$

Q5 $\dfrac{a}{b} - \dfrac{c}{b} = \dfrac{a-c}{b}$

In the above properties, we assume the denominators are not zero.

Simplifying Rational Expressions

A rational expression is **reduced to lowest terms** if the numerator and denominator have no common factors. Property Q1 makes it possible for us to use the following procedure for reducing a rational expression to lowest terms.

> Express the numerator and denominator as products of irreducible polynomials. Then eliminate the common factors.

We use the above procedure in Example A.23.

Example A.23

Reduce the following rational expressions to lowest terms:

a. $\dfrac{3-x}{x^2 - x - 6}$

b. $\dfrac{3x^2 + 24}{x^3 + 5x^2 + 8x + 40}$

c. $\dfrac{3x^2 + 16x - 12}{3x^3 - 2x^2}$

d. $\dfrac{x^2 - 16}{2x^3 + 128}$

Solutions

a. $\dfrac{3-x}{x^2 - x - 6} = \dfrac{(-1)(x-3)}{(x-3)(x+2)}$

$= \dfrac{-1}{x+2}$ Q1

b. $\dfrac{3x^2 + 24}{x^3 + 5x^2 + 8x + 40} = \dfrac{3(x^2 + 8)}{x^2(x+5) + 8(x+5)}$

$= \dfrac{3(x^2 + 8)}{(x+5)(x^2 + 8)}$

$= \dfrac{3}{x+5}$ Q1

c. $\dfrac{3x^2 + 16x - 12}{3x^3 - 2x^2} = \dfrac{3(x - \frac{2}{3})(x + 6)}{x^2(3x - 2)}$

$= \dfrac{(3x - 2)(x + 6)}{x^2(3x - 2)}$

$= \dfrac{x + 6}{x^2}$ Q1

d. $\dfrac{x^2 - 16}{2x^3 + 128} = \dfrac{(x + 4)(x - 4)}{2(x^3 + 64)}$

$= \dfrac{(x + 4)(x - 4)}{2(x + 4)(x^2 - 4x + 16)}$

$= \dfrac{x - 4}{2x^2 - 8x + 32}$ Q1

Example A.24 illustrates a method for reducing a rational expression to lowest terms if either the numerator or denominator (not both) is too difficult to factor.

Example A.24

Reduce the following rational expression to lowest terms:

$$\dfrac{x^4 - 5x^3 - 23x^2 - 6x - 27}{x^2 + x - 6}$$

Solution

The numerator is difficult to factor. The denominator, however, can be factored using Theorem A.3 in Section A.3:

$$x^2 + x - 6 = (x + 3)(x - 2)$$

To determine whether $x + 3$ is a factor of the numerator, we divide the numerator by $x + 3$:

$$
\begin{array}{r}
x^3 - 8x^2 + x - 9 \\
x + 3 \overline{) x^4 - 5x^3 - 23x^2 - 6x - 27} \\
\underline{x^4 + 3x^3} \\
-8x^3 - 23x^2 \\
\underline{-8x^3 - 24x^2} \\
x^2 - 6x \\
\underline{x^2 + 3x} \\
-9x - 27 \\
\underline{-9x - 27} \\
0
\end{array}
$$

Since the remainder is zero, $x + 3$ is a factor of the numerator. In fact,
$$x^4 - 5x^3 - 23x^2 - 6x - 27 = (x + 3)(x^3 - 8x^2 + x - 9)$$
Thus,
$$\frac{x^4 - 5x^3 - 23x^2 - 6x - 27}{x^2 + x - 6} = \frac{(x + 3)(x^3 - 8x^2 + x - 9)}{(x + 3)(x - 2)}$$
$$= \frac{x^3 - 8x^2 + x - 9}{x - 2}$$

The polynomial $x^3 - 8x^2 + x - 9$ is difficult to factor. So, to determine whether $x - 2$ is a factor of $x^3 - 8x^2 + x - 9$, we again use division:

$$\begin{array}{r}
x^2 - 6x - 11 \\
x - 2 \overline{)x^3 - 8x^2 + x - 9} \\
\underline{x^3 - 2x^2 } \\
-6x^2 + x \\
\underline{-6x^2 + 12x } \\
-11x - 9 \\
\underline{-11x + 22 } \\
31
\end{array}$$

Since the remainder is *not* zero, $x - 2$ is not a factor of $x^3 - 8x^2 + x - 9$. Therefore,
$$\frac{x^3 - 8x^2 + x - 9}{x - 2}$$
is reduced to lowest terms.

Multiplication and Division

We use properties Q2 and Q3 to multiply and divide rational expressions.

Example A.25

Perform the following computations and simplify:

a. $\dfrac{15 - 3x}{x^2 + 7x} \cdot \dfrac{x^3}{x^2 - 25}$

b. $\dfrac{4x^2 - 12x + 20}{x^4 + 5x^2 - 36} \div \dfrac{12x^5}{x^2 + 9}$

Solutions

a. $\dfrac{15 - 3x}{x^2 + 7x} \cdot \dfrac{x^3}{x^2 - 25} = \dfrac{(15 - 3x)x^3}{(x^2 + 7x)(x^2 - 25)}$ Q2

$= \dfrac{3(5 - x)(x)(x^2)}{x(x + 7)(x + 5)(x - 5)}$

$$= \frac{-3(x-5)(x)(x^2)}{x(x+7)(x+5)(x-5)}$$

$$= \frac{-3x^2}{x^2+12x+35} \quad \text{Q1}$$

b. $\dfrac{4x^2-12x+20}{x^4+5x^2-36} \div \dfrac{12x^5}{x^2+9} = \dfrac{4x^2-12x+20}{x^4+5x^2-36} \cdot \dfrac{x^2+9}{12x^5}$ Q3

$$= \frac{(4x^2-12x+20)(x^2+9)}{(x^4+5x^2-36)12x^5} \quad \text{Q2}$$

$$= \frac{4(x^2-3x+5)(x^2+9)}{(x^2+9)(x+2)(x-2)(4)(3x^5)}$$

$$= \frac{x^2-3x+5}{3x^7-12x^5} \quad \text{Q1}$$

Addition and Subtraction

To add or subtract rational expressions that have a common denominator, we use property Q4 or property Q5. If the denominators are not the same, we first rewrite the expressions so that the denominators are the same. The rewriting is based on property Q1. This property tells us, when read from right to left, that multiplying the numerator and denominator of a rational expression by the same polynomial produces an equivalent rational expression.

Example A.26

Perform the following computations and simplify:

a. $\dfrac{x}{x^2+4x-12} - \dfrac{14-6x}{x^2+4x-12}$

b. $\dfrac{7x^3}{x-6} + \dfrac{3x^3}{6-x}$

c. $\dfrac{x^2}{x^2+4x+4} + \dfrac{x-1}{3x^2+6x}$

Solutions

a. $\dfrac{x}{x^2+4x-12} - \dfrac{14-6x}{x^2+4x-12} = \dfrac{x-(14-6x)}{x^2+4x-12}$ Q5

$$= \frac{7x-14}{x^2+4x-12}$$

$$= \frac{7(x-2)}{(x+6)(x-2)}$$

$$= \frac{7}{x+6}$$

b. $\dfrac{7x^3}{x-6} + \dfrac{3x^3}{6-x} = \dfrac{7x^3}{x-6} + \dfrac{(-1)3x^3}{(-1)(6-x)}$ Q1

$= \dfrac{7x^3}{x-6} + \dfrac{-3x^3}{x-6}$

$= \dfrac{7x^3 + (-3x^3)}{x-6}$ Q4

$= \dfrac{4x^3}{x-6}$

c. $\dfrac{x^2}{x^2+4x+4} + \dfrac{x-1}{3x^2+6x} = \dfrac{x^2}{(x+2)^2} + \dfrac{x-1}{3x(x+2)}$

$= \dfrac{3x(x^2)}{3x(x+2)^2} + \dfrac{(x-1)(x+2)}{3x(x+2)^2}$ Q1

$= \dfrac{3x^3}{3x(x+2)^2} + \dfrac{x^2+x-2}{3x(x+2)^2}$

$= \dfrac{3x^3+x^2+x-2}{3x(x+2)^2}$ Q4

The remainder is not zero if $3x^3 + x^2 + x - 2$ is divided by $3x$ or $x + 2$. So $3x$ and $x + 2$ are not factors of $3x^3 + x^2 + x - 2$. Therefore, the above result is already reduced to lowest terms.

Equations Involving Rational Expressions

An **algebraic equation** is a statement that two algebraic expressions are equal. The following is an example of an algebraic equation with one variable:

$$3x^2 = 6x + 24$$

In this section, we deal only with algebraic equations with one variable. So, for the sake of simplicity, we will use the term "equation" to mean "algebraic equation with one variable."

A **solution** of an equation is a number that makes the two sides of the equation equal when the number replaces the variable. For example, the above equation has two solutions, namely, -2 and 4. (Verify this claim.)

Two equations are **equivalent** if they have the same solutions. For example, the following sequence of equations consists of equivalent equations. Each equation has -2 and 4 as solutions.

$$3x^2 = 6x + 24$$
$$3x^2 - 6x - 24 = 0$$
$$x^2 - 2x - 8 = 0$$
$$(x+2)(x-4) = 0$$

If we add the same polynomial to both sides of an equation, we produce an equivalent equation. (Keep in mind that addition includes subtraction and that nonzero real numbers are polynomials of degree 0.) But if we multiply both sides of an equation by the same polynomial, we may produce a nonequivalent equation. Nevertheless, the solutions of the original equation will be among the solutions of the resulting equation.

To find the solutions of an equation involving rational expressions, we proceed as follows:

1. Multiply both sides by a polynomial that eliminates the denominators.
2. Find the solutions of the resulting equation.
3. From the set of solutions of the resulting equation, eliminate any number that makes any of the denominators zero when the number replaces the variable. The remaining numbers are the solutions of the original equation.

Example A.27

Find the solutions of the following equations:

a. $\dfrac{5}{x-3} = \dfrac{15}{x^2 - 3x} - 1$

b. $\dfrac{x}{x^2 + 4x - 12} + \dfrac{4}{x-2} = \dfrac{7}{x+6}$

Solutions

a.
$\dfrac{5}{x-3} = \dfrac{15}{x(x-3)} - 1$ factoring $x^2 - 3x$

$5x = 15 - x(x-3)$ multiplying by $x(x-3)$

$5x = 15 - x^2 + 3x$

$x^2 + 2x - 15 = 0$

$(x + 5)(x - 3) = 0$

The solutions of the resulting equation are -5 and 3. Among these two numbers, 3 is the only number that makes a denominator zero when it replaces x. So the original equation has one solution, namely, -5.

b. $\dfrac{x}{(x+6)(x-2)} + \dfrac{4}{x-2} = \dfrac{7}{x+6}$ factoring $x^2 + 4x - 12$

$x + 4(x + 6) = 7(x - 2)$ multiplying by $(x + 6)(x - 2)$

$x + 4x + 24 = 7x - 14$

$38 = 2x$

$19 = x$

The number 19 does not make any of the denominators zero when it replaces x. So the original equation has one solution, namely, 19.

A.4 Problems

In Problems 1 to 12, reduce the given rational expression to lowest terms.

1. $\dfrac{3x + 15}{x^2 + 7x + 10}$

2. $\dfrac{7x - 28}{x^2 - 3x - 4}$

3. $\dfrac{x^2 - 9}{x^3 - 3x^2 + 4x - 12}$

4. $\dfrac{x^2 - 36}{x^3 + 6x^2 + 9x + 54}$

5. $\dfrac{5 - x}{2x^2 - 9x - 5}$

6. $\dfrac{1 - 3x}{3x^2 + 5x - 2}$

7. $\dfrac{x^2 - 11x - 26}{x^3 + 8}$

8. $\dfrac{x^2 - 10x + 21}{x^3 - 27}$

9. $\dfrac{x^3 + 11x^2 + 20x - 32}{x^2 + 6x - 16}$

10. $\dfrac{x^3 - 5x^2 - 2x + 24}{x^2 - 3x - 4}$

11. $\dfrac{x^2 - 5x + 25}{3x^6 + 351x^3 - 3000}$

12. $\dfrac{x^2 + x + 1}{5x^6 + 130x^3 - 135}$

In Problems 13 to 26, perform the indicated computation and simplify.

13. $\dfrac{x^2 - 1}{x^2 + 4x} \cdot \dfrac{5x}{3x + 3}$

14. $\dfrac{x^3 + 5x}{7x - 21} \cdot \dfrac{x^2 - 9}{13x^3}$

15. $\dfrac{x^2 + 3x - 10}{15x^2} \div \dfrac{4x^2 - 7x - 2}{3x}$

16. $\dfrac{18x^3}{2x^2 - 23x + 63} \div \dfrac{6x^2}{x^2 - 49}$

17. $\dfrac{x^2 - 7x}{x^3 - 3x^2 - 4x} + \dfrac{12}{x^3 - 3x^2 - 4x}$

18. $\dfrac{x^2}{x^3 - 3x^2 - 10x} - \dfrac{5x + 14}{x^3 - 3x^2 - 10x}$

19. $\dfrac{x^2}{x^4 - 5x^2 - 36} - \dfrac{8x - 15}{x^4 - 5x^2 - 36}$

20. $\dfrac{x^2 + 5x}{x^4 - 15x^2 - 16} + \dfrac{4}{x^4 - 15x^2 - 16}$

21. $\dfrac{13}{x^4 - 5} + \dfrac{7}{5 - x^4}$

22. $\dfrac{3x^2}{x - 9} - \dfrac{x}{9 - x}$

23. $\dfrac{3x - 5}{2x^2 + 6x} - \dfrac{4}{x^2 + 6x + 9}$

24. $\dfrac{x + 1}{x^2 - 8x + 16} + \dfrac{2}{5x^2 - 20}$

25. $\dfrac{5}{x^3} + \dfrac{x - 6}{x + 1}$

26. $\dfrac{2}{x - 2} - \dfrac{x + 7}{x^2}$

In Problems 27 to 32, find the solutions of the given equation.

27. $\dfrac{4}{x^2 - 3x + 2} = \dfrac{4}{x - 2} + 1$

28. $\dfrac{3}{x} = \dfrac{1}{x + 4} + 2$

29. $\dfrac{3x}{x - 4} = x + 3$

30. $\dfrac{4x}{x^2 - 25} + \dfrac{7}{x + 5} + 1 = 0$

31. $\dfrac{2}{x} = \dfrac{5}{x + 7} + \dfrac{35}{x^2 + 7x}$

32. $\dfrac{7}{x^3 - 50} = \dfrac{1}{2}$

APPENDIX

B
Integration by Tables

Many mathematical handbooks contain hundreds of integral rules. Such rules often can help us find indefinite integrals which cannot be found using the rules and techniques presented in Chapter 5.

The integral rules listed below are samples of the integral rules contained in mathematical handbooks. In each rule, u and du are abbreviations for $u(x)$ and $u'(x)\,dx$, respectively. The letters a, b, c, and d represent constants. We use a^2 to represent positive constants and n to represent positive integers.

A Short Table of Integrals

1. $\displaystyle\int \frac{u}{au+b}\,du = \frac{u}{a} - \frac{b}{a^2}\ln|au+b| + C$

2. $\displaystyle\int \frac{u}{(au+b)^2}\,du = \frac{1}{a^2}\left[\frac{b}{au+b} + \ln|au+b|\right] + C$

3. $\displaystyle\int \frac{u^2}{au+b}\,du = \frac{1}{a^3}\left[\frac{1}{2}(au+b)^2 - 2b(au+b) + b^2\ln|au+b|\right] + C$

4. $\displaystyle\int \frac{u^2}{(au+b)^2}\,du = \frac{1}{a^3}\left[(au+b) - \frac{b^2}{au+b} - 2b\ln|au+b|\right] + C$

5. $\displaystyle\int \frac{1}{u(au+b)}\,du = \frac{1}{b}\ln\left|\frac{u}{au+b}\right| + C$

6. $\displaystyle\int \frac{1}{(au+b)(cu+d)}\,du = \frac{1}{bc-ad}\ln\left|\frac{cu+d}{au+b}\right| + C$

7. $\displaystyle\int \frac{u}{(au+b)(cu+d)}\,du = \frac{1}{bc-ad}\left(\frac{b}{a}\ln|au+b| - \frac{d}{c}\ln|cu+d|\right) + C$

8. $\displaystyle\int \frac{1}{u^2-a^2}\,du = \frac{1}{2a}\ln\left|\frac{u-a}{u+a}\right| + C \qquad a>0$

9. $\displaystyle\int \frac{1}{a^2-u^2}\,du = \frac{1}{2a}\ln\left|\frac{u+a}{u-a}\right| + C \qquad a>0$

10. $\displaystyle\int \frac{1}{\sqrt{u^2 \pm a^2}}\,du = \ln|u + \sqrt{u^2 \pm a^2}| + C$

11. $\displaystyle\int \sqrt{u^2 \pm a^2}\,du = \frac{u}{2}\sqrt{u^2 \pm a^2} \pm \frac{a^2}{2}\ln|u + \sqrt{u^2 \pm a^2}| + C$

12. $\displaystyle\int \frac{\sqrt{u^2 \pm a^2}}{u^2}\,du = \frac{-\sqrt{u^2 \pm a^2}}{u} + \ln|u + \sqrt{u^2 \pm a^2}| + C$

13. $\int \dfrac{1}{u^2\sqrt{u^2 \pm a^2}}\, du = \dfrac{-\sqrt{u^2 \pm a^2}}{\pm a^2\, u} + C$

14. $\int u^n \sqrt{au + b}\, du = \dfrac{2u^n(au + b)^{3/2}}{a(2n + 3)} - \dfrac{2bn}{a(2n + 3)} \int u^{n-1}\sqrt{au + b}\, du$

15. $\int \dfrac{u^n}{\sqrt{au + b}}\, du = \dfrac{2u^n\sqrt{au + b}}{a(2n + 1)} - \dfrac{2bn}{a(2n + 1)} \int \dfrac{u^{n-1}}{\sqrt{au + b}}\, du$

16. $\int u^n e^u\, du = u^n e^u - n \int u^{n-1} e^u\, du$

17. $\int (\ln u)^n\, du = u(\ln u)^n - n \int (\ln u)^{n-1}\, du$

We use the above rules in the examples that follow. In Example B.1, we use rule 8 in combination with the constant multiple rule for integrals (Theorem 5.5 in Section 5.1).

Example B.1

Find

$$\int \dfrac{1}{9x^2 - 16}\, dx$$

Solution

First we rewrite the integrand:

$$\int \dfrac{1}{9x^2 - 16}\, dx = \int \dfrac{1}{(3x)^2 - 16}\, dx$$

Now we see that we can use rule 8. We let

$$u = 3x$$

Then,

$$du = 3\, dx$$

So

$$\tfrac{1}{3}\, du = dx$$

Substituting, we obtain

$$\int \dfrac{1}{(3x)^2 - 16}\, dx = \int \dfrac{1}{u^2 - 16}\left(\tfrac{1}{3}\right) du$$

$$= \int \left(\tfrac{1}{3}\right)\dfrac{1}{u^2 - 16}\, du$$

$$= \left(\tfrac{1}{3}\right)\left(\tfrac{1}{8}\right) \ln\left|\dfrac{u - 4}{u + 4}\right| + C \quad \text{rule 8 and Theorem 5.5}$$

$$= \dfrac{1}{24} \ln\left|\dfrac{3x - 4}{3x + 4}\right| + C \quad \text{since } u = 3x$$

Example B.2

Find

$$\int \frac{2}{x^3\sqrt{x^4 - 7}} \, dx$$

Solution

We begin by rewriting the integrand:

$$\int \frac{2}{x^3\sqrt{x^4 - 7}} \, dx = \int \frac{2x}{x^4\sqrt{x^4 - 7}} \, dx$$

$$= \int \frac{1}{(x^2)^2 \sqrt{(x^2)^2 - 7}} \, 2x \, dx$$

The rewriting shows that we can use rule 13. We let

$$u = x^2$$

Then,
$$du = 2x \, dx$$

Substituting, we get

$$\int \frac{1}{(x^2)^2 \sqrt{(x^2)^2 - 7}} \, 2x \, dx = \int \frac{1}{u^2 \sqrt{u^2 - 7}} \, du$$

$$= \frac{\sqrt{u^2 - 7}}{7u} + C \qquad \text{rule 13}$$

$$= \frac{\sqrt{(x^2)^2 - 7}}{7x^2} + C \qquad \text{since } u = x^2$$

$$= \frac{\sqrt{x^4 - 7}}{7x^2} + C$$

Example B.3

Find

$$\int \frac{10}{3x^2 + 14x + 16} \, dx$$

Solution

As in Examples B.1 and B.2, we begin by rewriting the integrand:

$$\int \frac{10}{3x^2 + 14x + 16} \, dx = \int 10 \frac{1}{(3x + 8)(x + 2)} \, dx$$

The result shows that we can use rule 6 in combination with Theorem 5.5. We *mentally* substitute u for x and du for dx and obtain

$$\int 10 \frac{1}{(3x+8)(x+2)} \, dx = \frac{10}{8-6} \ln\left|\frac{x+2}{3x+8}\right| + C \qquad \text{rule 6 and Theorem 5.5}$$

$$= 5 \ln\left|\frac{x+2}{3x+8}\right| + C$$

Example B.4

Find

$$\int \frac{\sqrt{x^2 - 6x + 14}}{x^2 - 6x + 9} \, dx$$

Solution

Note that

$$x^2 - 6x + 9 = (x-3)^2$$

So we are motivated to rewrite the integrand as follows:

$$\int \frac{\sqrt{x^2 - 6x + 14}}{x^2 - 6x + 9} \, dx = \int \frac{\sqrt{(x^2 - 6x + 9) - 9 + 14}}{x^2 - 6x + 9} \, dx$$

$$= \int \frac{\sqrt{(x-3)^2 + 5}}{(x-3)^2} \, dx$$

The rewriting shows that we can use rule 12. We let

$$u = x - 3$$

Then, $\qquad du = 1 \, dx = dx$

Substituting, we obtain

$$\int \frac{\sqrt{(x-3)^2 + 5}}{(x-3)^2} \, dx = \int \frac{\sqrt{u^2 + 5}}{u^2} \, du$$

$$= \frac{-\sqrt{u^2 + 5}}{u} + \ln\left|u + \sqrt{u^2 + 5}\right| + C \qquad \text{rule 12}$$

$$= \frac{-\sqrt{(x-3)^2 + 5}}{x - 3} + \ln\left|x - 3 + \sqrt{(x-3)^2 + 5}\right| + C \qquad \text{since } u = x - 3$$

$$= \frac{-\sqrt{x^2 - 6x + 14}}{x - 3} + \ln\left|x - 3 + \sqrt{x^2 - 6x + 14}\right| + C$$

In Example B.5, we apply rule 16 twice to express the indefinite integral

$$\int x^2 e^x \, dx$$

in terms of the simpler indefinite integral

$$\int e^x \, dx$$

Then we apply the exponential rule for integrals (Theorem 5.4 in Section 5.1).

Example B.5

Find

$$\int x^2 e^x \, dx$$

Solution

We apply rule 16 while mentally substituting u for x and du for dx:

$$\int x^2 e^x \, dx = x^2 e^x - 2 \int x e^x \, dx \qquad \text{rule 16 with } n = 2$$

$$= x^2 e^x - 2 \left(x e^x - \int e^x \, dx \right) \qquad \text{rule 16 with } n = 1$$

$$= x^2 e^x - 2 x e^x + 2 \int e^x \, dx$$

$$= x^2 e^x - 2 x e^x + 2 e^x + C \qquad \text{Theorem 5.4}$$

Rules 14 through 17 are examples of **reduction rules**. As we illustrated in Example B.5, we use reduction rules to express complicated indefinite integrals in terms of less complicated indefinite integrals.

Appendix B Problems

Use the table of integrals included in this appendix to find the following indefinite integrals.

1. $\displaystyle \int \frac{5x}{x^2 + 6x + 9} \, dx$

2. $\displaystyle \int \frac{12}{3x^2 - 4x} \, dx$

3. $\displaystyle \int \frac{2x^3}{x^2 - 6} \, dx$

4. $\displaystyle \int \frac{x^2 - 8x + 16}{x + 9} \, dx$

5. $\displaystyle \int \frac{x^2}{4x^2 + 12x + 9} \, dx$

6. $\displaystyle \int \frac{x}{16x^2 - 9} \, dx$

7. $\displaystyle \int \frac{1}{36 - 25x^2} \, dx$

8. $\displaystyle \int \sqrt{x^2 - 12x} \, dx$

9. $\displaystyle \int \frac{18}{\sqrt{81x^2 + 13}} \, dx$

10. $\displaystyle \int x^2 \sqrt{x + 3} \, dx$

11. $\displaystyle \int (\ln x)^2 \, dx$

12. $\displaystyle \int \frac{x}{\sqrt{2x + 1}} \, dx$

Answers to Odd Numbered Section Problems and All Chapter Review Problems*

CHAPTER ONE

Section 1.1

1. 71 3. 3 5. 3 7. 237
9. −9 11. $\frac{1}{3}$ 13. 21 15. $-\frac{1}{12}$
17. 9 19. $x \neq 9$ 21. All reals
23. $x \neq -3, 3$ 25. $(-\infty, 7]$ 27. $(4, +\infty)$
29. $[-4, 4]$ 31. $(-2, 2)$
33. 0.7 thousand per day 35. 0.5 million
37. $12 thousand 39. 112 feet per second
41. 2.625 thousand per month

45. 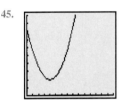 The population will be least approximately 20 years after the beginning of 1995. The population will be approximately 10 thousand when it is least.

47. 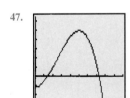 The firm is close to breaking even when its monthly sales level is either 1.4 tons or 7.3 tons.

49. $f(0) = 5.00, f(1) = 10.32, f(2) = 13.64, f(3) = 15.20,$
$f(4) = 15.24, f(5) = 14.00, f(6) = 11.72, f(7) = 8.64,$
$f(8) = 5.00, f(9) = 1.04, f(10) = -3.00, f(11) = -6.88,$
$f(12) = -10.36, f(13) = -13.20, f(14) = -15.16,$
$f(15) = -16.00, f(16) = -15.48, f(17) = -13.36,$
$f(18) = -9.40, f(19) = -3.36, f(20) = 5.00$

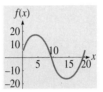

51. *a.* 4.50 *b.* 3.43 *c.* 1.28 *d.* −20.25

Section 1.2

1. $A(x) = 0.02x + \dfrac{18}{x}$ 3. $R(x) = \dfrac{2000x}{x + 500}$
5. $P(x) = -0.002x^2 + 4.8x - 1400$
7. $C(x) = 9x + \dfrac{3600}{x}$; domain: integers from 1 to 25
9. $V(x) = 4x^3 - 64x^2 + 240x$ 11. $A(r) = 2\pi r^2 + \dfrac{40\pi}{r}$
domain: (0, 12) domain: $(0, +\infty)$
13. $g[u(x)] = (x - 9)^5$ 15. $g[u(x)] = \dfrac{1}{\sqrt{3x + 5}}$
17. $g[u(x)] = \dfrac{2}{(4x + 6)^3}$
19. $g(x) = x^5; u(x) = 3x + 5$
21. $g(x) = \dfrac{1}{x}$ 23. $g(x) = \dfrac{9}{\sqrt{x}}$
$u(x) = x^2 + 7x$ $u(x) = x - 11$

25.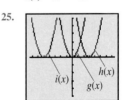

27. 1800 shirts; $32,400 29. 20 molds

*Answers to verbalization problems not included.

Section 1.3

1. Slope: 3; crosses vertical axis: (0, 1)

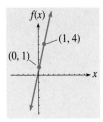

3. Slope: −2.5; crosses vertical axis: (0, 4)

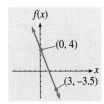

5. Slope: 0; crosses vertical axis: (0, 2)

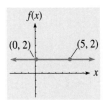

7. Slope: −1; crosses vertical axis: (0, 0)

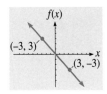

9. Slope: $-2; f(x) = -2x + 7$
11. Slope: $4; f(x) = 4x - 5$
13. Slope: $\frac{1}{2}; f(x) = \frac{1}{2}x + 1$
15. Slope: $0; f(x) = 2$
17. Slope: $\frac{1}{4}; f(x) = \frac{1}{4}x$
19. $f(x) = -4x + 15; f(1) = 11$
21. $f(x) = 5x + 2; f(9) = 47$
23. $S(x) = \frac{2}{3}x - 10$; 30 cartons
25. $C(t) = -\frac{4}{5}t + 70$; 7 years from now
27. $R(t) = 4.7t + 146.2$; 226.1 pounds per year
29. $R(x) = -0.02x + 6$; 1 thousand per day
31. $M(x) = -0.15x + 17$; 7.7 minutes
33. $y = -\frac{3}{5}x + 9$; domain: (0, 15)

37. a.

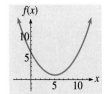

The monthly demand is approximately 4.8 million liters when the price is $4.25 per liter.

b. $m = -0.4315789474, b = 6.635087719$; $D(4.25) = 4.800877193$ million liters

Section 1.4

1. Vertex: (5, 1)

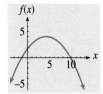

3. Vertex: (4, 4)

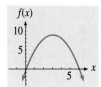

5. Vertex: (3, 9)

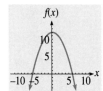

7. Vertex: (0, 12)

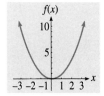

9. Vertex: (0, 0)

11. 3   13. 152   15. $\frac{2}{3}, -\frac{1}{4}$
17. $\frac{3}{8}$ 19. No real number x for which $f(x) = 0$
21. 5 years from now

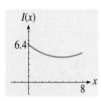

23. *a.* 2 tons, 11 tons   *b.* 6.5 tons; $20.25 hundred
25. *a.* $P(x) = -0.02x^2 + 2x - 3$ *b.* 50 tons; $47 thousand
   c. 1.5 tons; 98.5 tons
27. $8.50 per pound
29. $C(x) = -10x^2 + 320x$; domain: integers from 12 to 20; 16 accounts
31. $A(x) = -2x^2 + 207x$; 51.75 feet

33.

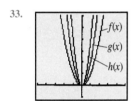

Section 1.5
1. $\frac{4}{3}$ 3. 0   5. 4   7. Does not exist
9. 0   11. 74 years   13. 9 percent
15. 30 thousand barrels

17.

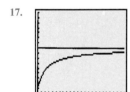

Section 1.6
1. 9   3. 50   5. 45   7. Does not exist
9. $\frac{14}{3}$ 11. 0   13. Does not exist
15. 12   17. $\frac{11}{3}$ 19. Does not exist   21. 2
23. Not continuous; (1) and (3) not met
25. Continuous   27. Continuous
29. Not continuous; (1) and (3) not met
31. Not continuous; (2) and (3) not met
33. Continuous   35. Not continuous; (2) and (3) not met
37. Continuous   39. Not continuous; (3) not met

41.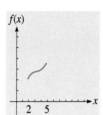

43.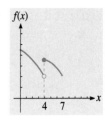

45.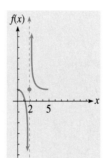

47.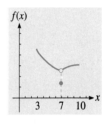

49. *a.* $C(x) = \begin{cases} 400x + 9000 & \text{when } 0 \leq x \leq 20 \\ 800x + 1000 & \text{when } x > 20 \end{cases}$

b.

c. Continuous

Chapter 1   Review Problems
1. -1.3 2. 3.16
3. *a.* $x \neq -6, 6$ *b.* $[4, +\infty)$ *c.* $(-4, 4)$
   d. All reals
4. 1.8 liters per minute   5. $12 thousand per year
6. *a.* $R(x) = \frac{2100x}{x + 260}$ *b.* $P(x) = \frac{2100x}{x + 260} - 3x - 20$
   c. $A(x) = \frac{3x + 20}{x}$
7. $C(x) = 30x + \frac{16{,}250}{x}$; domain: integers from 1 to 40
8. $g[u(x)] = \sqrt{5x - 1}$; $u[g(x)] = 5\sqrt{x} - 1$
9. $g(x) = \frac{8}{x^3}$; $u(x) = x^2 + 1$

10. Slope: −3; crosses vertical axis: (0, 2)

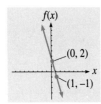

11. Slope: 0.5; crosses vertical axis: (0, 0)

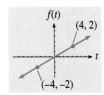

12. Slope: 1; crosses vertical axis: (0, −4)

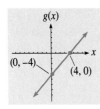

13. Slope: 0; crosses vertical axis: (0, 6)

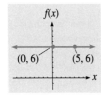

14. Slope: $-\frac{1}{2}$; $f(x) = -\frac{1}{2}x + 10$
15. Slope: 1.6; $f(x) = 1.6x + 2$ 16. Slope: 0; $f(x) = 5$
17. Slope: 1; $f(x) = x$ 18. $f(x) = 1.5x + 3$; $f(8) = 15$
19. −3.6   20. $D(x) = -0.8x + 124$; 52 pairs
21. $E(x) = 2x + 28$; 62 percent
22. *a.* Vertex: (5, 3)

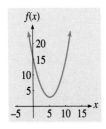

b. Vertex: (8, 10)

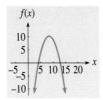

23. 12   24. 6
25. *a.* $\frac{3}{8}$, 5   *b.* 4.5   *c.* $f(x)$ is never 0
26. 20 thousand
27. *a.* $P(x) = -0.08x^2 + 2.4x - 6$
    b. 15 barrels; $12 hundred
    c. 2.8 barrels, 27.2 barrels
28. $E(x) = -4x^2 + 72x$; domain: integers from 5 to 12; 9 tables
29. *a.* 12   *b.* 0   *c.* Does not exist
30. 25 billion cubic centimeters per year
31. *a.* 11   *b.* 11   *c.* $\frac{6}{11}$ *d.* 15   *e.* $\frac{8}{5}$
    f. 0   *g.* Does not exist   *h.* Does not exist
    i. 8   *j.* 12   *k.* Does not exist   *l.* 14.8
32. *a.* Continuous
    b. Not continuous; (1) and (3) not met
    c. Continuous
    d. Not continuous; (1), (2), and (3) not met
    e. Continuous
    f. Continuous
    g. Not continuous; (3) not met
    h. Not continuous; (2) and (3) not met
    i. Continuous
33. *a.* 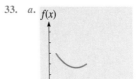   *b.*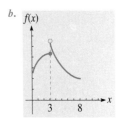

c. 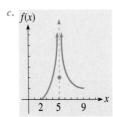   *d.*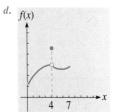

34. *a.* $R(x) = \begin{cases} 4 & \text{when } 0 < x \leq 3 \\ 2x - 2 & \text{when } 3 < x \leq 7 \end{cases}$

b.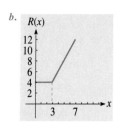

c. Continuous

27. Note that $f(4) = 3.08$. So the point $(4, 3.08)$ is on the graph of $f(x)$. The point with coordinates
$x = 3.9604255$
$y = 3.1178036$
and the point with coordinates
$x = 4.0319149$
$y = 3.0492395$
are two points on the graph of $f(x)$ near the point $(4, 3.08)$. The slope of the regression line for these two points is -0.959 (truncated at three decimal places). Therefore, the slope of the tangent in question is approximately -0.959.

CHAPTER TWO

Section 2.1

1.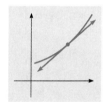
3.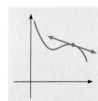

5.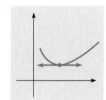
7.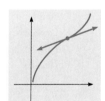

9.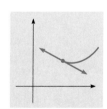

11. 24   13. 48   15. 20   17. -2
19. 52   21. -1.25 23. $y = 3x - 5$

25.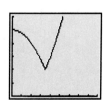

It appears that the point where the graph has no tangent is the point $(3.5, 2)$.

Section 2.2

1. $f'(x) = 16x; 4$
3. $f'(x) = 3x^2; 3$
5. $f'(x) = 6x; 30$
7. $f'(x) = 2x - 6; -2$
9. $f'(x) = 6x^2 + 2x; \frac{1}{2}$
11. $f'(x) = 2x - 3; 0$
13. $f'(x) = 12x^2 + 1; 13$
15. $f'(x) = 4x^3; -32$
17. $f'(x) = \frac{1}{\sqrt{x}}; \frac{1}{3}$
19. $f'(x) = \frac{-1}{x^2}; \frac{-1}{25}$
21. 3   23. $-6, 6$ 25. 3   27. 4
29. 8 years from now; 28   31. 21 barrels per day
33. 6.5 feet
39. Note that $f(6) = 9$. So the point $(6, 9)$ is on the graph of $f(x)$. The point with coordinates
$x = 5.9514894$
$y = 8.9816975$
and the point with coordinates
$x = 6.0446809$
$y = 9.0166623$
are two points on the graph of $f(x)$ near the point $(6, 9)$. The slope of the regression line for these two points is 0.375 (truncated at three decimal places). So, since this regression line approximates the tangent to the graph of $f(x)$ at the point $(6, 9)$, the value of $f'(6)$ is approximately 0.375.

41. It appears that the x value where $f(x)$ is not differentiable is $x = 6$.

Section 2.3

1. $f'(x) = -4$
3. $f'(x) = 0$
5. $f'(x) = 15x^2$
7. $f'(x) = -5x^4$
9. $f'(x) = 2x$
11. $f'(x) = \frac{-1}{x^2}$
13. $f'(x) = \frac{8}{x^3}$
15. $f'(x) = \frac{1}{\sqrt{x}}$
17. $f'(x) = \frac{1}{3(\sqrt[3]{x})^4}$
19. $f'(x) = 3$
21. $f'(x) = 2x - 9$
23. $f'(x) = 6x^2 + 0.5$

25. $f'(x) = 3x^2 + 14x - 5$
27. $f'(x) = 1 + \dfrac{2}{x^2}$
29. $f'(x) = 8 + \dfrac{1}{2\sqrt{x}}$
31. $f'(x) = \dfrac{1}{(\sqrt[3]{x})^2} - \dfrac{2}{x^3}$
33. $f'(x) = 3x^2 - \dfrac{1}{2(\sqrt{x})^3}$
35. $f'(x) = 6x^5 - 30x^2$
37. 6   39. $-4, 5$ 41. $\tfrac{2}{3}, 6$ 43. $-4, 4$
45. 16   47. 3   49. 4
51. 50 miles per hour; 7 miles per gallon
53. 9 months from now
55. After working 6 hours; 261 packages per hour
57. The derivative of 7 is 0, and the derivative of x is 1. So, using the erroneous assumption that the derivative of the product of two functions is the product of their derivatives, we would conclude that $f'(x) = 0 \cdot 1 = 0$. But, according to the power rule, $f'(x) = 7$.
59. $f'(x) = \lim\limits_{h \to 0} \dfrac{[u(x+h) + v(x+h)] - [u(x) + v(x)]}{h}$

$= \lim\limits_{h \to 0} \left(\dfrac{u(x+h) - u(x)}{h} + \dfrac{v(x+h) - v(x)}{h} \right)$

$= \lim\limits_{h \to 0} \dfrac{u(x+h) - u(x)}{h} + \lim\limits_{h \to 0} \dfrac{v(x+h) - v(x)}{h}$

$= u'(x) + v'(x)$

61.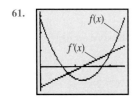
   a. $f(x)$ decreases
   b. $f(x)$ increases

63.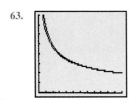

Section 2.4

1. Increases approximately 2 units
3. Decreases approximately 3 units
5. Increases approximately $\tfrac{1}{4}$ of a unit
7. Decreases approximately $\tfrac{1}{64}$ of a unit
9. Decreases approximately $\tfrac{4}{27}$ of a unit
11. The probability increases by approximately 0.0025.
13. The monthly demand decreases by approximately $\tfrac{7}{125}$ of a million bars.
15. 400; The total weekly production cost increases approximately $400 if the weekly production level is increased from 12 liters to 13 liters.
17. 2; The weekly production level increases by approximately 2 calculators if the weekly labor force is increased from 12 units to 13 units.
19. 20 units   21. 60 units

23.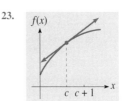

25.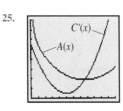
The average cost of producing each unit appears to be equal to the marginal cost when the monthly production level is 30.5 units. At this production level, it appears that the average cost of producing each unit is least.

27.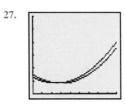

Section 2.5

1. $140x^4(4x^5 + 9)^6$ 3. $3(x^4 - 5x)^2(4x^3 - 5)$
5. $60(2x - 7)^5$
7. $2(x^3 - 4x^2 + 5x)(3x^2 - 8x + 5) - 2x$
9. $\dfrac{-18x^2}{(x^3 + 7)^2}$ 11. $\dfrac{15(8x^3 - 9)}{(2x^4 - 9x)^6}$
13. $\dfrac{8x^3 + 5}{3(\sqrt[3]{2x^4 + 5x - 2})^2}$ 15. $\dfrac{-9}{2(\sqrt{9x + 8})^3}$
17. $30x^5 + 28x^3 + 45x^2 - 90x + 21$
19. $4x^7(5x + 1)^6(75x + 8)$
21. $(6x + 3)^4(42x^2 + 6x) - \dfrac{7}{x^2}$
23. $(x^2 + 9x)^4(60x^3 + 282x^2 - 162x)$
25. $\dfrac{2x^4}{\sqrt{x^4 - 7}} + \sqrt{x^4 - 7}$
27. $3(5x - 2)^2(x^2 + 1)^2(15x^2 - 4x + 5)$
29. $(x^3 + 2x + 4)^4(7x + 3)(119x^3 + 45x^2 + 98x + 86)$
31. $\dfrac{12x^5 + 45x^4}{(3x + 9)^2}$ 33. $\dfrac{17}{(5x + 1)^2} + \dfrac{1}{2\sqrt{x}}$
35. $\dfrac{-(2x^5 + 6x^4 - 3x^2 + 10x + 10)}{(x^4 + 3x - 5)^2}$
37. $\dfrac{-(21x^2 + 4x + 6)}{(7x^2 - 2)^2}$ 39. $\dfrac{(2x + 1)^2(4x + 41)}{(x + 7)^2}$

41. $4\left(\dfrac{x^3-7}{x^3+5}\right)^3 \dfrac{36x^2}{(x^3+5)^2}$ 43. $\dfrac{7x+37}{2(\sqrt{x+3})^3}$
45. 4   47. $\dfrac{25}{3}$
49. Increases approximately 32 units
51. Decreases approximately $\dfrac{45}{289}$ of a unit
53. -120; The weekly revenue decreases approximately \$120 if the weekly production level is increased from 7 tons to 8 tons.
55. By approximately $\dfrac{1}{8}$ of a ton
57. It will decrease by approximately 160 words.
59. a. An additional 5.76 percent (approximately) can be removed.
    b. \$12.75 million
61. 18 tons; \$48 thousand
63. 5 months from now; 12 percent
67.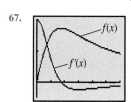
    a. $f(x)$ increases.
    b. $f(x)$ decreases.

Section 2.6

1. Highest point: (1, 39); lowest point: (5, 7)
3. Highest point: (5, 100); lowest point: (1, 20)
5. Highest point: (25, 100); lowest point: (16, 16)
7. Highest point: (4, 81); lowest point: (1, 9)
9. Highest point: (6, 15); lowest point: $(0, \tfrac{15}{37})$
11. Highest point: $(2, \sqrt{20})$; lowest point: (6, 2)
13. Highest point: (2, 2.25); lowest point: (6, 1.35)
15. 7 months from now; 6 percent
17. 7 hours after the patient receives the insulin; 102 milligrams per deciliter
19. 9 months from now; 17.4 percent   21. 45 workers
23. a. 200 recorders
    b. Production level: 190 recorders; price: \$839; profit: \$87,180
    c. Production level: 189 recorders; price: \$842.79; profit: \$85,025.38
25. 8 stands   27. 41 crates   29. 30 miles
31. Approximately 13.6 miles
33. $f(x)$ is minimum at about $x = 3.80500491$.
35. Approximately 6 percent; about 7 months from now
37. Approximately 2.5 hours

Section 2.7

1. Increases on $(-\infty, 3)$ and $(7, +\infty)$; decreases on (3, 7); high point: (3, 72); low point: (7, 40)
3. Increases on $(-5, 1)$; decreases on $(-\infty, -5)$ and $(1, +\infty)$; high point: (1, 8); low point: $(-5, -100)$
5. Increases on $(-\infty, 1)$ and $(1, +\infty)$; no high or low points
7. Increases on $(-\infty, 0)$ and (5, 7); decreases on (0, 5) and $(7, +\infty)$; high points: (0, 1100) and (7, 757); low point: (5, 725)
9. Decreases everywhere; no high or low points
11. Increases on (0, 3) and $(3, +\infty)$; decreases on $(-\infty, 0)$; no high points; low point: (0, 9)
13. Increases on (0, 8); decreases on $(-\infty, 0)$ and $(8, +\infty)$; high point: (8, 1.1875); no low points
15. Increases on $(9, +\infty)$; decreases on $(-\infty, 0)$, (0, 6), and (6, 9); no high points; low point: (9, 486)
17. 3 years from now; 4   19. 6 sacks
21. 16 months from now   23. a. 120 gallons
                                b. 80 gallons
25. a. The derivative of the average cost function is $-600/x^2$, which is negative for each $x > 0$.
    b. The derivative of the revenue function is $1600/(x+2)^2$, which is positive for each $x \geq 0$.
    c. 18 tons per month; price: \$40 thousand per ton; profit: \$48 thousand
    d. 14 tons per month; price: \$50 thousand per ton; profit: \$12.5 thousand
27. 512 cases   29. 12 centimeters
31. 20 tons per week   33. 60 units per day
37.

Section 2.8

1. Increases approximately 2 units
3. Decreases approximately 0.05 of a unit
5. Increases approximately 1.8 units
7. Decreases approximately 3 units
9. Increases approximately 10.86 units
11. Increases approximately 2 percent
13. Decreases approximately 2.5 percent
15. $E(x) = \dfrac{2x}{2x - 500}$; decreases approximately 4 percent

17. $E(x) = \dfrac{6x^2}{3x^2 - 4800}$; decreases approximately $\tfrac{2}{3}$ percent
19. Elastic   21. Inelastic   23. Has unit elasticity
25. Drops approximately \$0.5 billion
27. Decreases by approximately 10 thousand
29. Decreases approximately 80 percent
31. Increases approximately 9 percent
33. Approximately \$1.5 million
35. Approximately \$0.35 million
37. Inelastic at prices less than \$9; elastic at prices greater than \$9
39. $E(p) = \dfrac{-p}{p + 62}$; so $E(p) > -1$ for every $p > 0$.
41. $E(x) = \dfrac{-15x}{x^2 + 25x + 100}$; decreases approximately $\tfrac{1}{3}$ percent
43. Increases approximately 0.5 percent
45. At most 6 percent
47. Approximately 360π square feet ($360\pi \approx 1131$)
49. The value of
$$\dfrac{f(c + \Delta x) - f(c)}{\Delta x}$$
is the slope of the line that represents $f(x)$. Since this slope is also the value of $f'(c)$, it follows that
$$f'(c) = \dfrac{f(c + \Delta x) - f(c)}{\Delta x}$$
Multiplying both sides by Δx,
$$f'(c)\Delta x = f(c + \Delta x) - f(c)$$
The right side of this equality is the exact amount $f(x)$ changes if x changes from c to $c + \Delta x$. So the proof is complete.

53. 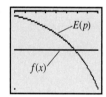 It appears that the demand is inelastic at prices less than \$49 and elastic at prices greater than \$49.

Chapter 2   Review Problems

1. a. $f'(x) = 10x$ b. $f'(x) = 3x^2 - 2$
   c. $f'(x) = \dfrac{7}{2\sqrt{x}}$ d. $f'(x) = \dfrac{-12}{x^2}$
2. a. $f'(x) = 28x^3$ b. $g'(x) = -9$
   c. $\dfrac{dy}{dx} = 0$ d. $f'(x) = \dfrac{-4}{x^5}$
   e. $f'(x) = \dfrac{7}{\sqrt{x}}$ f. $\dfrac{dy}{dx} = \dfrac{-5}{(\sqrt[3]{x})^4}$
   g. $f'(x) = 6x^2 - 1$ h. $g'(x) = 2x - \dfrac{3}{x^2}$
   i. $\dfrac{dy}{dx} = 6 + \dfrac{1}{2(\sqrt{x})^3}$ j. $f'(x) = \dfrac{4}{\sqrt{x}} + \dfrac{3}{x^4}$
3. a. 2, 6   b. No horizontal tangents   c. 36
4. 9
5. a. Decreases approximately 1.5 units; actually decreases 1.3 units (rounded to 1 decimal place)
   b. Increases approximately 0.28 of a unit; actually increases 0.3 of a unit
6. a. 11.5; The weekly revenue increases approximately 11.5 units of money if the weekly production level is increased from 954 units to 955 units.
   b. -15.5; The weekly revenue decreases approximately 15.5 units of money if the weekly production level is increased from 1062 units to 1063 units.
7. a. 290; The total weekly production cost increases approximately \$290 if the weekly production level is increased from 34 tons to 35 tons.
   b. 42 tons; marginal cost: \$2; total weekly production cost: \$116,216
8. a. 20 units
   b. 20 units; The two production levels are equal.
9. a. $f'(x) = 42x(3x^2 + 5)^6$ b. $\dfrac{dy}{dx} = \dfrac{12}{\sqrt{3x + 4}}$
   c. $g'(x) = \dfrac{-(2x + 1)}{(x^2 + x - 8)^2}$ d. $f'(x) = \dfrac{4x + 18}{(\sqrt[3]{x^2 + 9x})^4}$
   e. $\dfrac{dy}{dx} = \dfrac{6 - 72x^2}{(4x^3 - x)^4}$ f. $\dfrac{dy}{dx} = 5(2x^3 - 4)^5(38x^3 - 4)$
   g. $f'(x) = \dfrac{3x^3 - 2x}{2\sqrt{x^3 - 2x}} + \sqrt{x^3 - 2x}$
   h. $g'(x) = (7x + 3)^4(84x - 29)$
   i. $\dfrac{dy}{dx} = (x + 7)^2(x^2 - 4x)^3(11x^2 + 28x - 112)$
   j. $f'(x) = \dfrac{72x}{(x^2 + 4)^2}$ k. $f'(x) = \dfrac{-17}{(5x + 1)^2}$
   l. $g'(x) = \dfrac{-8(x^3 + 4)}{(x^3 - 8)^2}$ m. $f'(x) = \dfrac{3 - 6x}{(x + 2)^4}$
   n. $\dfrac{dy}{dx} = \dfrac{(2x - 8)^3(42x + 64)}{(7x + 1)^2}$
   o. $g'(x) = 4\left(\dfrac{5x - 1}{7x + 2}\right)^3 \dfrac{17}{(7x + 2)^2}$
   p. $f'(x) = \dfrac{6x + 44}{(\sqrt{3x + 11})^3}$ q. $f'(x) = \dfrac{-8}{(5x - 1)^2}\sqrt{\dfrac{5x - 1}{x + 3}}$
   r. $f'(x) = \dfrac{-x^4 + 10x^3 - 45x^2 + 16x - 40}{(x^3 + 8)^2}$
10. 2.25; The monthly production level increases approximately 2.25 tons if the monthly labor force is increased from 20 hundred hours to 21 hundred hours.
11. Decreases approximately 0.25 million gallons
12. a. Highest point: (10, 165); lowest point: (6, 5)

b. Highest point: (1, 4); lowest point: (4, −50)
c. Highest point: (4, 30); lowest point: (10, 3)
d. Highest point: (0, 8); lowest point: (6, 4)
13. At the beginning of 1999; 78.7 percent
14. 4 weeks from now; 67 percent   15. 20 painters
16. a. Increases on $(-\infty, 2)$ and $(6, +\infty)$; decreases on $(2, 6)$; high point: (2, 70); low point: (6, 6)
  b. Increases on $(-\infty, 2)$ and $(2, +\infty)$; no high or low points
  c. Increases on $(-\infty, -3)$ and $(0, 4)$; decreases on $(-3, 0)$ and $(4, +\infty)$; high points: $(-3, 599)$ and $(4, 1285)$; low point: $(0, 5)$
  d. Increases everywhere; no high or low points
  e. Increases on $(0.25, 2.5)$; decreases on $(-\infty, 0.25)$ and $(2.5, +\infty)$; high point: (2.5, 43.75); low point: $(0.25, -1.8125)$
  f. Increases on $(-\infty, -3)$ and $(3, +\infty)$; decreases on $(-3, 3)$; high point: $(-3, 252)$; low point: $(3, -252)$
  g. Increases on $(-\infty, -1)$ and $(5, +\infty)$; decreases on $(-1, 5)$; high point: $(-1, 677)$; low point: $(5, -61855)$
  h. Increases on $(0, 9)$ and $(9, +\infty)$; decreases on $(-\infty, 0)$; no high points; low point: $(0, -5)$
  i. Increases on $(-\infty, 7)$; decreases on $(7, +\infty)$; no high or low points
  j. Increases on $(-\infty, 0)$ and $(4, +\infty)$; decreases on $(0, 2)$ and $(2, 4)$; high point: $(0, 0)$; low point: $(4, 24)$
  k. Increases on $(-7, 7)$; decreases on $(-\infty, -7)$ and $(7, +\infty)$; high point: $(7, \frac{4}{7})$; low point: $(-7, -\frac{4}{7})$
  l. Increases on $(-\infty, 6)$; decreases on $(6, +\infty)$; high point: $(6, 5)$; no low points
  m. Increases on $(1, 5)$ and $(5, 25)$; decreases on $(-\infty, -5)$, $(-5, 1)$, and $(25, +\infty)$; high point: $(25, 0.2)$; low point: $(1, 0.5)$
17. 60 thousand people; 9 thousand per week
18. At the beginning of 1996; 3.3125 million tons per year
19. 800 gallons
20. a. The derivative of the average cost function is $-187{,}500/x^2$, which is negative for each $x > 0$.
  b. The derivative of the revenue function is $10^{10}/(x + 2500)^2$, which is positive for each $x \geq 0$.
  c. 17,500 cases; price: $200 per case; profit: $2,875,000
21. a. Increases approximately 6.792 units; actually increases 6.8784 units
  b. Decreases approximately 0.645 of a unit; actually decreases 258/443 of a unit
  c. Decreases approximately 5.6 units; actually decreases 5.58 units (rounded to two decimal places)
  d. Increases approximately 1.35 units; actually increases 1.63 units (rounded to two decimal places)
22. a. Increases approximately 8 percent; actually increases 7.76 percent
  b. Decreases approximately 4.8 percent; actually decreases 4.72 percent (rounded to two decimal places)
  c. Increases approximately 7 percent; actually increases 7.08 percent (rounded to two decimal places)
  d. Decreases approximately 5.76 percent; actually decreases 5.52 percent (rounded to two decimal places)
23. a. $E(x) = \frac{-2x}{x + 7}$; decreases approximately 1.3 percent
  b. $E(x) = \frac{2}{x + 2}$; increases approximately 0.25 percent
24. a. Inelastic   b. Elastic   c. Has unit elasticity
25. a. Will increase approximately 1.5 million
  b. Will increase approximately $3\frac{4}{7}$ percent
26. a. Approximately 0.7 of a unit increase
  b. Approximately 0.1 of a unit decrease
27. Inelastic at prices less than 52 cents; elastic at prices greater than 52 cents
28. $E(x) = \frac{3x^3}{2x^3 + 4000}$; increases approximately 1.2 percent

CHAPTER THREE

Section 3.1

1. $f'(x) = x + 2$; 7
3. $f'(x) = \frac{-32}{x^3}$; -0.5
5. $f'(x) = \frac{3}{\sqrt{x}}$; $\frac{3}{4}$
7. $f'(x) = \frac{1}{20}(x + 2)^2$; 5
9. $f'(x) = \frac{-2000}{(2x + 3)^3}$; -16
11. $f'(x) = \frac{14}{(x + 2)^2}$; $\frac{2}{7}$
13. $f'(x) = (2x + 1)^2 (8x + 1)$; 20
15. $V'(x) = \frac{-170}{(x + 2)^2}$; $6.8 thousand per year; $49 thousand
17. $f'(t) = \frac{-10}{(t + 2)^2}$; 0.625 million per year
19. a. Growing at the rate of 1.5 thousand per year (rounded to one decimal place)
  b. Decreasing at the rate of 0.26 thousand per year (rounded to two decimal places)
21. $D'(x) = \frac{50}{\sqrt{5x + 1600}}$; $1\frac{1}{9}$ tons per $1 thousand increase in advertising expenditure
23. a. $1.875 thousand per month
  b. $1.2 thousand per month
25. a. $\frac{2}{3}$ of a foot per second
  b. $\frac{3}{8}$ of a foot per second

c. No

27. 0.77 parts per billion per year; at the beginning of the year 2025; the tangent to the graph of $C(x)$ at the point where $x = 35$ is parallel to the line through the points $(10, C(10))$ and $(60, C(60))$

29. a. 2 feet per second; upward
    b. -46 feet per second; downward
    c. 1.5625 seconds after it is thrown; 45.5625 feet high
    d. 54 feet per second

31. 20π square meters per meter; this rate is equal to the circumference, except that the circumference is measured in meters

35. 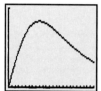 Most rapid increase will occur in about 9 years. At that time, the increase rate will be approximately 0.8 of a million per year.

21.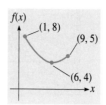

23.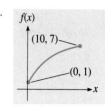

25.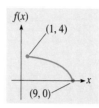

27.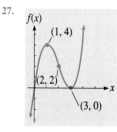

29.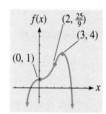

31.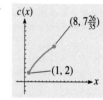

33.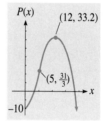

35. a. Increases on [0, 3) and (6, 10); decreases on (3, 6) and (10, 12]
    b. Concave upward on (5, 8); concave downward on [0, 5) and (8, 12]

37.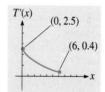

39. (9, 0.5)

Section 3.2

1. Concave downward on $(-\infty, \frac{8}{3})$; concave upward on $(\frac{8}{3}, +\infty)$; inflection point: $(\frac{8}{3}, -\frac{377}{27})$

3. Concave downward on $(-\infty, 3)$ and $(6, +\infty)$; concave upward on $(3, 6)$; inflection points: $(3, -47.25)$ and $(6, -108)$

5. Concave downward on $[0, +\infty)$; no inflection points

7. Concave upward everywhere; no inflection points

9. Concave downward on $(-3, -1)$ and $(1, 3)$; concave upward on $(-\infty, -3)$, $(-1, 1)$, and $(3, +\infty)$; inflection points: $(-3, 0)$, $(-1, -32768)$, $(1, -32768)$, and $(3, 0)$

11. Concave downward on $(-4, 4)$; concave upward on $(-\infty, -4)$ and $(4, +\infty)$; inflection points: $(-4, 3.015625)$ and $(4, 3.015625)$

13.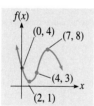

15.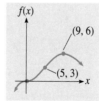

17.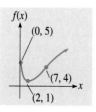

19.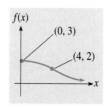

41.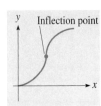

43.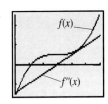

39.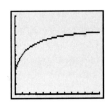

The number of frequent users will increase at a decreasing rate.

Section 3.3

1. $f(x)$ increases at a decreasing rate.
3. $f(x)$ increases at an increasing rate.
5. $f(x)$ increases at an increasing rate on [0, 5) and at a decreasing rate on (5, +∞).
7. $f(x)$ increases at a decreasing rate on [0, 3) and at an increasing rate on (3, +∞).
9. $f(x)$ decreases at a decreasing rate.
11. $f(x)$ decreases at an increasing rate.
13. $f(x)$ decreases at a decreasing rate on [0, 3) and at an increasing rate on (3, +∞).
15. $f(x)$ decreases at an increasing rate on [0, 4) and at a decreasing rate on (4, +∞).

17.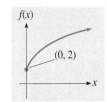

19.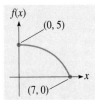

21.

23.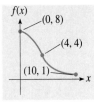

25. The necessary income will increase because $I'(x) > 0$ for $x \geq 0$. It will increase at a decreasing rate.
27. The man will gain weight because $W'(x) > 0$ for $x \geq 0$. His weight will increase at an increasing rate during the next 6 weeks and at a decreasing rate thereafter.
29. $A(x)$ decreases if x is increased because $A'(x) < 0$ for $x > 0$. It will decrease at a decreasing rate.
31. 3 weeks from now; 12.5 percentage points per week
33. 20 units per day
35. 10 tons of fertilizer

Section 3.4

1.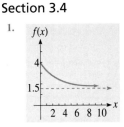

$f(x)$ approaches 1.5 from above while decreasing at a decreasing rate.

3.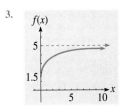

$f(x)$ approaches 5 from below while increasing at a decreasing rate.

5.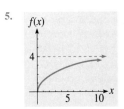

$f(x)$ approaches 4 from below while increasing at a decreasing rate.

7.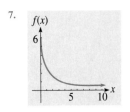

$f(x)$ approaches zero from above while decreasing at a decreasing rate.

9.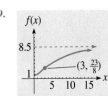

$f(x)$ approaches 8.5 from below while increasing at an increasing rate as x approaches 3 and at a decreasing rate thereafter.

11. 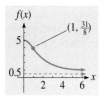 $f(x)$ approaches 0.5 from above while decreasing at an increasing rate as x approaches 1 and at a decreasing rate thereafter.

13. 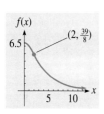 $f(x)$ approaches zero from above while decreasing at an increasing rate as x approaches 2 and at a decreasing rate thereafter.

15. $f(x)$ approaches 7 from below while increasing at an increasing rate as x approaches 5 and at a decreasing rate thereafter.

17. $f(x) = \dfrac{12x + 40}{4x + 5}$

19. $f(x) = \dfrac{7x + 2}{x + 2}$

21. $f(x) = \dfrac{4x}{x + 3}$

23. $f(x) = \dfrac{40}{7x + 5}$

25. $f(x) = \dfrac{2x^2 + 4}{x^2 + 1}$

27. $f(x) = \dfrac{224}{5x^2 + 32}$

29. $f(x) = \dfrac{8x^2 + 144}{x^2 + 48}$

31. $f(x) = \dfrac{6x^2}{x^2 + 75}$

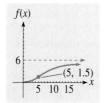

33. 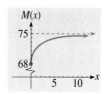 The utilization will approach 75 percent from below while increasing at a decreasing rate.

35. $P(x)$ approaches 30 from above while decreasing at a decreasing rate.

37. 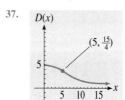 The demand approaches zero as the price increases. At first, the demand decreases at an increasing rate. After the price passes 5 cents, the demand decreases at a decreasing rate.

39. The percent will approach 40 while increasing at an increasing rate during the next 20 years and at a decreasing rate thereafter.

41. 152.4 pounds (rounded to 1 decimal place)
43. $6 thousand   45. 4 years from now
47. Approximately 1 year from now
49. About $334.74 per troy ounce
51. $4 million   53. 8 hours
55.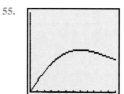

Section 3.5

1. Horizontal asymptote: x axis; vertical asymptote: $x = 2$
3. Slant asymptote: $y = 2.5x + 10$; vertical asymptote: $x = 4$
5. Horizontal asymptote: $y = \frac{3}{4}$; vertical asymptote: $x = -6$
7. Horizontal asymptote: x axis
9. Horizontal asymptote: $y = 3$; vertical asymptotes: $x = -5$ and $x = 5$
11. Slant asymptote: $y = x - 7$; vertical asymptotes: $x = -7$ and $x = 0$
13. Horizontal asymptote: $y = 1$; vertical asymptotes: $x = -1$ and $x = 3$
15. Horizontal asymptote: $y = \frac{1}{5}$
17. No linear asymptotes
19. Horizontal asymptote: $y = \frac{5}{2}$; vertical asymptote: $x = 3$
21. Horizontal asymptote: $y = 4$; vertical asymptote: y axis

23. Horizontal asymptote: x axis; vertical asymptote: $x = 6$
25. Horizontal asymptote: $y = \frac{1}{4}$; vertical asymptote: $x = -3$

27.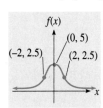
29.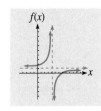
31.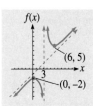
33.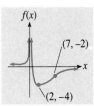
35.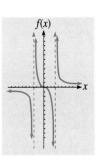
37.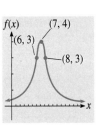
39.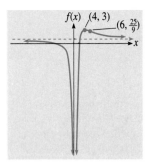
41.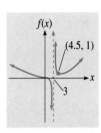
43.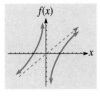

45.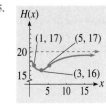

Starting at about 18, the percent will decrease during the next 3 months, reaching a low of 16. Thereafter, the percent will increase while approaching 20.

47.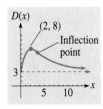

Starting at 3 million barrels, the demand will increase during the next 2 months, reaching a high of 8 million barrels. Thereafter, the demand will decrease while approaching 3 million barrels.

49.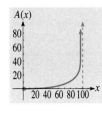

Section 3.6

1. $P(x) = \dfrac{50}{0.5x - 12}$; $-5\frac{5}{9}$

3. $P(x) = \dfrac{200x}{x^2 + 90}$; $10\frac{10}{19}$

5. $P(x) = \dfrac{100x}{x^2 + 100}$; 4

7. $P(x) = \dfrac{-400}{7x^2 + 32x + 36}$; $-1\frac{29}{371}$

9. $P(x) = \dfrac{-200}{x + 1}$; -2

11. $P(x) = \dfrac{200x^2 + 1000}{x^3 + 10x}$; $19\frac{1}{11}$

13. a. $P(t) = \dfrac{5}{0.05t + 1.1}$

    b. 4 percent per week; $3\frac{1}{3}$ percent per week

    c. $P'(t) = -0.25/(0.05t + 1.1)^2$; so $P'(t) < 0$ for t in $[0, 10]$. Thus, $P(t)$ will decrease.

15. $P(t) = -100/(t + 10)$; since $P(t) < 0$ for $t \geq 0$, the supply will decrease. $P'(t) = 100/(t + 10)^2$. Since $P'(t) > 0$ for $t \geq 0$, the absolute value (magnitude) of $P(t)$ will decrease. Thus, the supply will decrease at a decreasing percentage rate.

17. At the beginning of 2001; 12.5 percent per year

19. Invest the $1 million in plan A. After 7.5 years, transfer the balance to plan B.

21. a. 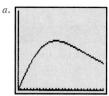 Approximately 12 months from now; approximately $0.3 billion per month

b. 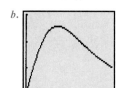 Approximately 10 months from now; approximately 2.6 percent per month

Section 3.7
1. 2.5712
3. 3.0579
5. 2.0912
7. 5.9276
9. 1.1705
11. 4.1617
13. 14.1421
15. 10.44365
17. 0.4006431, 9.793658

Chapter 3   Review Problems
1. a. $f'(x) = -9\sqrt{x}$; -18

   b. $f'(x) = \dfrac{-12}{(x+3)^2}$; $-\dfrac{12}{25}$

   c. $f'(x) = \dfrac{6}{\sqrt{3x+1}}$; 1.5

   d. $f'(x) = 5 + \dfrac{72}{x^2}$; 13

   e. $f'(x) = (0.25x + 1)^2 (x + 1)$; 20

2. $p'(x) = -\frac{1}{16}x + \frac{4}{3}$; $1\frac{1}{12}$ cents per week; 61.5 cents

3. $D'(x) = \dfrac{-192}{(x+2)^2}$; decreases at the rate of 0.1875 million bars per 1 cent increase in price

4. $1/14 billion per year; at the beginning of April in the year 2002; the tangent to the graph of $D(x)$ at the point where $x = 12.25$ is parallel to the line through the points $(4, D(4))$ and $(25, D(25))$.

5. a. Decreasing at the rate of 1.6 thousand bobcats per year
   b. Decreasing at the rate of 0.5 thousand bobcats per year

6. a. Concave downward on $(-3, 6)$; concave upward on $(-\infty, -3)$ and $(6, +\infty)$; inflection points: $(-3, -724)$ and $(6, -3883)$
   b. Concave upward everywhere; no inflection points
   c. Concave downward on $(-\infty, -5)$ and $(5, +\infty)$; concave upward on $(-5, 5)$; inflection points: $(-5, 0.95)$ and $(5, 0.95)$
   d. Concave downward on $(-\infty, 2)$; concave upward on $(2, +\infty)$; inflection point: $(2, 12)$

7. a.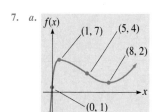

   b. 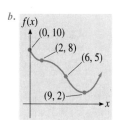   c.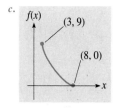

   d. 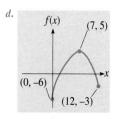   e.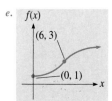

8. a. 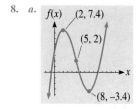   b.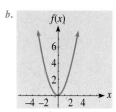

9.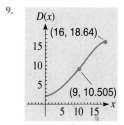

10. $f(x)$ increases at a decreasing rate on $[0, 5)$ and at an increasing rate on $(5, +\infty)$.

11. $f(x)$ decreases at a decreasing rate.

12. a. 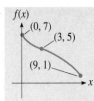 b.

c.   d.

13. The rainwater's pH will increase during the next 10 years because $A'(x) > 0$ for x in $[0, 10]$. It will increase at a decreasing rate.

14. The rabbit population will decline during the next 8 years because $P'(x) < 0$ for x in $(0, 8)$. It will decline at an increasing rate during the next 4 years and at a decreasing rate thereafter.

15. After 4 hours; 128 units per hour.

16. a. 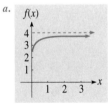 $f(x)$ approaches 4 from below while increasing at a decreasing rate.

b. 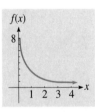 $f(x)$ approaches zero from above while decreasing at a decreasing rate.

c. 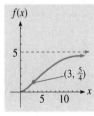 $f(x)$ approaches 5 from below while increasing at an increasing rate as x approaches 3 and at a decreasing rate thereafter.

d. 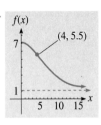 $f(x)$ approaches 1 from above while decreasing at an increasing rate as x approaches 4 and at a decreasing rate thereafter.

17. a. $f(x) = \dfrac{2x + 6}{x + 1}$ b. $f(x) = \dfrac{6x}{2x + 5}$

c. $f(x) = \dfrac{182x + 80}{35x + 40}$ d. $f(x) = \dfrac{26}{x + 2}$

18. a. $f(x) = \dfrac{27x^2 + 6}{3x^2 + 2}$ b. $f(x) = \dfrac{25}{x^2 + 4}$

19. a. $f(x) = \dfrac{15x^2}{2x^2 + 96}$ b. $f(x) = \dfrac{x^2 + 162}{x^2 + 27}$

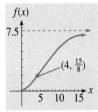 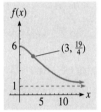

20. 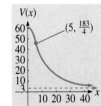 The machine's value will approach $3 thousand while decreasing at an increasing rate during the next 5 years and at a decreasing rate thereafter.

21. 159.6 pounds   22. $15 thousand
23. $6 million
24. a. Horizontal asymptote: $y = 2$; vertical asymptote: $x = 5$
    b. Slant asymptote: $y = 5x$; vertical asymptotes: $x = -7$ and $x = 7$
    c. Horizontal asymptote: x axis
    d. Horizontal asymptote: x axis; vertical asymptotes: $x = 1$ and $x = 4$

25. a. 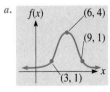 b.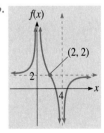

c. 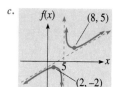 *d.* 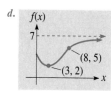   13. 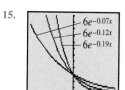   *a.* 12.4 percent   *b.* 14.9 percent

26. *a.* 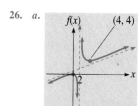 *b.*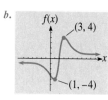

Section 4.2

1. $P(t) = 8000e^{0.035t}$; approximately 32,442 bacteria
3. $V(x) = 75,000e^{-0.12x}$; $22,589.57
5. Approximately 56 milligrams
7. Approximately 79.8 pounds
9. Approximately 82.9 tons
11. Approximately 71,121
13. $f(x) = e^5 e^{0.03x}$ $(e^5 \approx 148)$

27. *a.* $P(x) = \dfrac{600x}{3x^2 + 50}$; 9.6

    b. $P(x) = \dfrac{100x}{x^2 + 11}$; 3

    c. $P(x) = \dfrac{-500}{x^2 + 15x + 50}$; $-1\tfrac{2}{3}$

    d. $P(x) = \dfrac{-200}{x + 4}$; -5

28. At the beginning of 1998; 25 percent per year
29. *a.* 2.8930   *b.* 4.0703
30. *a.* 3.1530   *b.* 2.1285

15. $6e^{-0.07x}$, $6e^{-0.12x}$, $6e^{-0.19x}$ The graph becomes steeper.

Section 4.3

1. After about 13.7 years
3. In about 9.2 weeks
5. Approximately $21.6 thousand
7. About 17 weeks
9. Approximately 13.9 weeks
11. About 23.1 years
13. Approximately every 2.3 hours
15. Approximately 13.9 percent per month
17. About 2.6 percent per minute
19. Approximately 16.9 pounds
21. 36 percent
23. Approximately 5.8 percent per hour
25. *a.* Yes, in about 38.8 months   *b.* No
27. Approximately 14,812 years ago
29. Approximately 5.9 months from now

CHAPTER FOUR

Section 4.1

1. *a.* $B(t) = 300(1.0225)^{4t}$; $383.19
   b. $B(t) = 300(1.00025)^{360t}$; $384.23
   c. $B(t) = 300e^{0.09t}$; $384.25
3. *a.* $14,793.20   *b.* $14,182.70
5. *a.* 8.16 percent   *b.* 8.33 percent
7. Plan A   9. After 8 years
11. 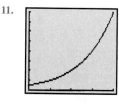   *a.* $16,459.90   *b.* After 29 years

33.   About 16.6 years

35. *a.*

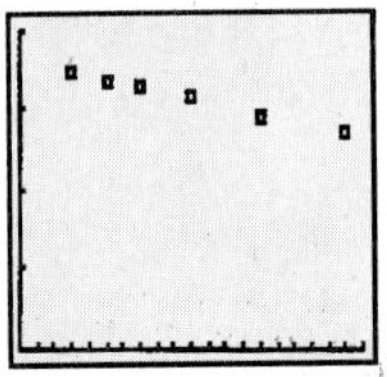

b.

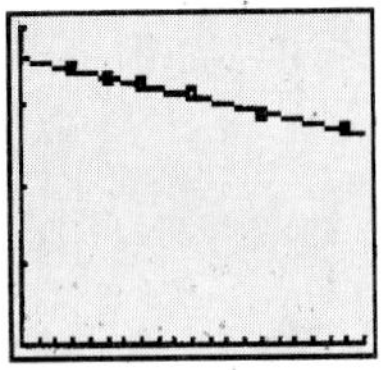

c. $\ln V(x) = -0.048x + 3.6$

d. $V(x) = 37e^{-0.048x}$

e. Approximately \$9.6 million

Section 4.4

1. $3x + 6x \ln x$
3. $\dfrac{9}{x}$
5. $\dfrac{1 - \ln x}{8x^2}$
7. $\dfrac{2 \ln x}{x}$
9. $\dfrac{3}{3x - 4}$
11. $\dfrac{6x}{3x^2 + 7}$
13. $\dfrac{20x}{x^2 + 3}$
15. $\dfrac{3x^2 + 5}{x^3 + 5x}$
17. $\dfrac{1}{x}$
19. $\dfrac{16}{x}$
21. $\dfrac{12}{3x + 1}$
23. $\dfrac{1}{2x + 6}$
25. $\dfrac{1}{x + 3} - \dfrac{2}{2x + 1}$
27. $\dfrac{1}{x} + \dfrac{4}{x + 5}$
29. $\dfrac{2}{4x - 1} - \dfrac{3}{6x + 4}$
31. Approximately \$2.7 thousand
33. 7.5 thousand per year
35. 9 months from now; about \$3.7 billion
37.

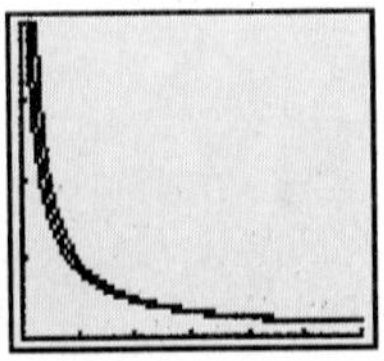

9. $\dfrac{5 + 5e^x - 5xe^x}{(1 + e^x)^2}$
11. $1.92\,(1 - 0.2e^{-0.08t})^2\, e^{-0.08t}$
13. $\dfrac{-2e^x}{(\sqrt{2 + 4e^x})^3}$
15. $(0.03x - 0.85)e^{0.03x}$
17. $\dfrac{60e^{-0.3t}}{(1 + 8e^{-0.3t})^2}$
19. $\dfrac{12e^{-0.5t}}{(1 + 4e^{-0.5t})^2}$
21. $1.5e^{-0.3t}$
23. $\left(\dfrac{1}{x} + \ln x\right)e^x$
25. $\dfrac{2x - (x^2 + 9)\ln(x^2 + 9)}{(x^2 + 9)e^x}$
27. $\dfrac{0.05e^{0.05x}}{e^{0.05x} + 1}$
29. 5
31. $\dfrac{\ln 2}{2}$
33. 8
35.

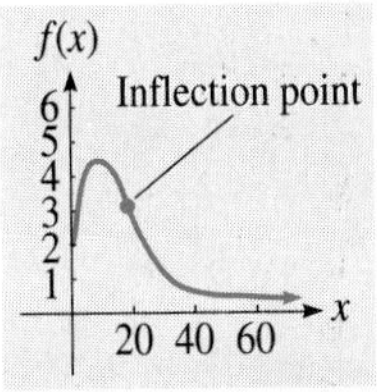

37. A formula for the percentage rate of change of $f(x)$ is $8x$. So the percentage rate of change of $f(x)$ does not remain constant as x increases. Therefore, according to Theorem 4.11, $f(x)$ is not an exponential function.
39. Approximately 1.624 million
41. Approximately 4436 rodents per month
43. At the beginning of September in 1998
45. *a.*

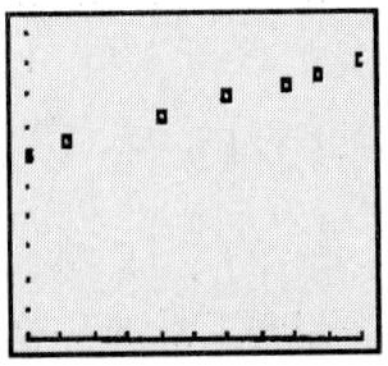

b.

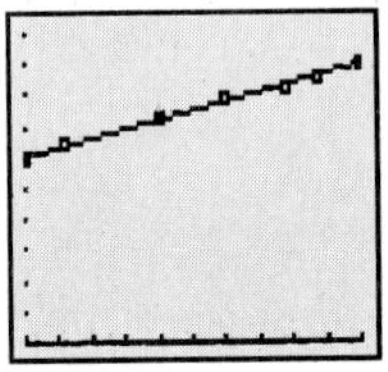

c. $P'(t) = 0.03P(t) - 0.004$

d. $P(t) = 20e^{0.03t}$

e. Approximately 42 million; 1.3 million per year

Section 4.5

1. $(3x^2 + 5)e^{x^3 + 5x}$
3. $-2.2e^{-0.04t}$
5. $0.5e^x - 0.5e^{-x}$
7. $(x^3 + 3x^2)e^x$

Section 4.6

1. 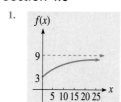 $f(x)$ approaches 9 from below while increasing at a decreasing rate.

3. 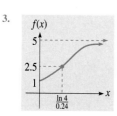 $f(x)$ approaches 5 from below while increasing at an increasing rate as x increases toward $(\ln 4)/0.24$ and at a decreasing rate thereafter.

5. 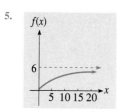 $f(x)$ approaches 6 from below while increasing at a decreasing rate.

7. 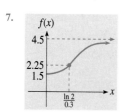 $f(x)$ approaches 4.5 from below while increasing at an increasing rate as x increases toward $(\ln 2)/0.3$ and at a decreasing rate thereafter.

9. $f(x) = 8 - 6e^{-0.23x}$

11. $f(x) = \dfrac{8}{1 + 7e^{-0.17x}}$

13. 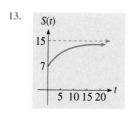 It will approach 15 million while increasing at a decreasing rate.

15. 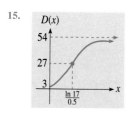 It will approach 54 thousand while increasing at an increasing rate for slightly longer than 5.5 months and at a decreasing rate thereafter.

17. About 9 hours of training
19. Approximately 14 weeks from now
21. About 166 beats per minute
23. Slightly more than 207 thousand copies

25. a.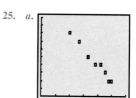

b.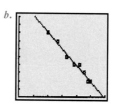

c. $G'(t) = -0.21G(t) + 1.2$
d. $G'(t) = 0.21(5.7 - G(t))$; $G(t) = 5.7 - 3.7e^{-0.21t}$
e. Approximately 5.2 grams; approximately 0.1 gram per hour

Chapter 4 Review Problems

1. a. $B(t) = 8000(1.06)^{2t}$; $19,172.47$
   b. $B(t) = 8000e^{0.12t}$; $19,676.82$
2. a. $27,430.40 b. $27,107.48$
3. a. 6.17 percent b. 6.18 percent
4. Investment B
5. $P(x) = 6000e^{0.042x}$; approximately 7402 microorganisms
6. About 58 milligrams per liter
7. Approximately 8.7 years from now
8. Approximately 4.6 years
9. Approximately 5.8 percent per year
10. About $455,000
11. About 8.3 percent per year
12. Approximately 14,176 years ago
13. a. $5x^2(1 + 3\ln x)$ b. $\dfrac{1 - \ln x}{8x^2}$
    c. $\dfrac{4(\ln x)^3}{x}$ d. $\dfrac{7}{7x + 12}$
    e. $\dfrac{7}{x}$ f. $\dfrac{5x^4 + 3}{x^5 + 3x}$
    g. $\dfrac{16x + 48}{x^2 + 6x}$ h. $\dfrac{4}{4x + 1} - \dfrac{3x^2}{x^3 + 5}$
    i. $\dfrac{1}{x} + \dfrac{3}{6x + 2}$ j. $\dfrac{8}{2x + 5} - \dfrac{24}{6x + 7}$
14. Approximately 8 tons
15. $4.5 thousand per year

16. a. $(2x + 7)e^{x^2 + 7x}$ b. $-2.08e^{-0.08x}$ 21. 7.352 grams   22. Approximately 24 years from now

c. $x^4(x + 5)e^x$ d. $\dfrac{6 + 2e^x - 2xe^x}{(3 + e^x)^2}$

e. $(2 - 0.5e^{-x})^3 \, 2e^{-x}$ f. $\dfrac{0.4e^{-0.2x}}{(\sqrt{1 + e^{-0.2x}})^3}$

g. $3e^{-0.5x}$ h. $\dfrac{48e^{-0.4x}}{(1 + 6e^{-0.4x})^2}$

17. 18 sacks

18.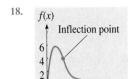

19. A formula for the percentage rate of change of $f(x)$ is $24e^{-0.02x}/(21 - 12e^{-0.02x})$. So the percentage rate of change does not remain constant as x increases. Therefore, according to Theorem 4.11, $f(x)$ is not an exponential function.

20. a. 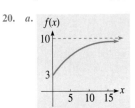 $f(x)$ approaches 10 from below while increasing at a decreasing rate.

b. 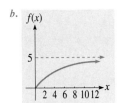 $f(x)$ approaches 5 from below while increasing at a decreasing rate.

c. 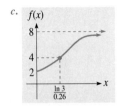 $f(x)$ approaches 8 from below while increasing at an increasing rate as x increases toward (ln 3)/0.26 and at a decreasing rate thereafter.

d. 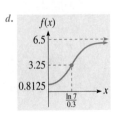 $f(x)$ approaches 6.5 from below while increasing at an increasing rate as x increases toward (ln 7)/0.3 and at a decreasing rate thereafter.

CHAPTER FIVE

Section 5.1

1. $x^3 + C$ 3. $\tfrac{1}{5}x^5 + C$ 5. $\dfrac{-2}{x^3} + C$

7. $4(\sqrt{x})^3 + C$ 9. $\tfrac{2}{3}x^3 + 4x + C$

11. $2x^4 - x^3 + C$

13. $\tfrac{1}{2}x^2 + \dfrac{1}{x} - x + C$ 15. $3x^2 + \tfrac{3}{4}(\sqrt[3]{x})^4 + C$

17. $\tfrac{1}{6}x + 5\ln|x| + C$ 19. $e^x + 3\ln|x| + C$

21. $3x^2 + 2\sqrt{x} + 3x + C$ 23. $F(x) = 2x^4 + 3$

25. $F(x) = 2(\sqrt{x})^3 + 4$ 27. $F(x) = \dfrac{1}{x^4}$

29. $F(x) = 2\ln|x| - 3$ 31. $F(x) = e^x - 7$

33. $F(x) = \tfrac{1}{3}x^3 + 3x^2 + 9x + 5$

35. $F(x) = 2x^2 + \dfrac{3}{x} + 2$

37. $F(x) = x^3 + 6(\sqrt{x})^3 - 12$

39. $F(x) = 2x - \ln|x| + 5$

41. $P(t) = 0.025t^2 + 4t + 175$; 265 thousand

43. $C(x) = \tfrac{1}{3}x^3 - 18x^2 + 325x + 300$; $2250

45. $F(x) = 2\sqrt{x} + 30$; 16 barrels

47. 61.8 million   49. $F(x) = 6\sqrt[3]{x} + 4$; 22 days

51. $S(x) = \tfrac{1}{3}x - \tfrac{1}{3}\sqrt{x}$; $4 million

55. a. $V'(x) = 0.56x - 7.9$
    b. $2.3 thousand per year
    c. $V(x) = 0.28x^2 - 7.9x + 68.3$
    d. $17.3 thousand

Section 5.2

1. 484   3. 0   5. -3.75 7. -0.117

9. $\ln 3 \approx 1.1$ 11. $e^3 - 1 \approx 19.1$ 13. 1107

15. $\displaystyle\int_2^5 (3x^2 + 2x)\,dx$; increases 138 units

17. $-\displaystyle\int_4^9 -\sqrt{x}\,dx$; increases $12\tfrac{2}{3}$ units

19. $\displaystyle\int_8^{16} \dfrac{3}{x}\,dx$; increases $3\ln 2$ units

21. $\displaystyle\int_{20}^{30} (-0.1x + 5)\,dx$; increases 25 barrels

23. $\displaystyle\int_9^{16} -\dfrac{5}{\sqrt{x}}\,dx$; will decrease $10 thousand

25. $-\displaystyle\int_{10}^{20} (-3x^2 + 90x)\,dx$; decreases 6500 sacks

27. $\int_2^6 \frac{10}{x}\,dx$; approximately 11 million boxes

29. Increases $6\frac{2}{3}$ million

31. $\int_0^5 (0.036t - 0.36)\,dt$; will decrease 1.35 units

33. 20 years

35. a. $\int_a^b k \cdot f(x)\,dx = k \cdot F(b) - k \cdot F(a)$
$= k[F(b) - F(a)]$
$= k\int_a^b f(x)\,dx$

b. $\int_a^b [f(x) + g(x)]\,dx = [F(b) + G(b)] - [F(a) + G(a)]$
$= [F(b) - F(a)] + [G(b) - G(a)]$
$= \int_a^b f(x)\,dx + \int_a^b g(x)\,dx$

37. a. 200.75

b. Suppose $f(x) = mx + k$. Then
$\int_a^b f(x)\,dx = \int_a^b (mx + k)\,dx$
$= \left(\frac{m}{2}b^2 + kb\right) - \left(\frac{m}{2}a^2 + ka\right)$
$= \frac{m}{2}(b^2 - a^2) + k(b - a)$
$= \frac{m}{2}(b + a)(b - a) + k(b - a)$

So
$\frac{\int_a^b f(x)\,dx}{b - a} = \frac{m}{2}(b + a) + k$
$= \frac{(ma + k) + (mb + k)}{2}$
$= \frac{f(a) + f(b)}{2}$

41. a. 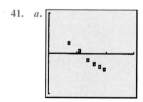 b.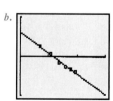

c. $y = -0.528x + 0.613$

d. $W'(t) = \frac{1.8}{\sqrt{t}}$

e. $\int_8^{13} \frac{1.8}{\sqrt{t}}\,dt$

f. Approximately 2.8 tons

Section 5.3

1. 18.75   3. 32   5. 3   7. 54

9. 261   11. $e - 1 \approx 1.7$ 13. 4   15. 9
17. 36   19. $54\frac{1}{3}$ 21. $30\frac{11}{12}$
23. a. 0   b. 5.5
25. The region is bounded by the graph of
$y = \frac{1}{225}x^2 - \frac{4}{15}x + 5$, the x axis, and the vertical lines $x = 20$ and $x = 25$.

29. a.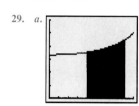

The area of the shaded region is $\int_4^8 \frac{40}{\sqrt{144 - x^2}}\,dx$, which is approximately 15.711. So $F(8)$ is approximately $30 + 15.711 = 45.711$.

b.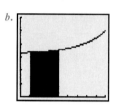

The area of the shaded region is $\int_1^4 \frac{40}{\sqrt{144 - x^2}}\,dx$, which is approximately 10.154. So $F(1)$ is approximately $30 - 10.154 = 19.846$.

31. -62.898 (using 30,000 points)

Section 5.4

1. $2(x^3 + 7)^6 + C$ 3. $2\sqrt{x^4 - 5x} + C$
5. $\frac{1}{40}(8x - 9)^5 + C$ 7. $\frac{1}{9}(\sqrt{6x^2 + 18x})^3 + C$
9. $\frac{-2}{18x^2 + 15} + C$
11. $\frac{1}{3}(x + 2)^9 - \frac{3}{4}(x + 2)^8 + C$
13. $\frac{1}{2}(\ln 8x)^2 + C$ 15. $4\ln(3x^2 + 7) + C$
17. $\ln|8x - 1| + C$
19. $\frac{1}{2}\ln|4x^2 + 10x - 7| + C$
21. $\frac{2}{3}(\sqrt{2x + 5})^3 - 10\sqrt{2x + 5} + C$
23. $\ln(5 + e^{4x}) + C$ 25. $5\ln|\ln x| + C$
27. $6\ln|x + 4| + \frac{24}{x + 4} + C$
29. $\sqrt{x} - 3\ln(\sqrt{x} + 3) + C$
31. $\frac{1}{4}e^{6x^2} + C$ 33. $-25e^{-0.2x} + C$
35. $e^{0.06x} + C$ 37. $12.5e^{0.08x} + C$
39. $6e^{x^2 - 3x} + C$ 41. $x - 7\ln|x + 7| + C$
43. $\frac{1}{5}(\ln x)^5 + C$ 45. 4   47. $\ln 5 \approx 1.6$
49. $4e^2 - 4e \approx 18.7$ 51. 10   53. 57,186
55. Decreases 12 units   57. Decreases ln 5 units
59. $F(x) = \ln|3x - 5| + 9$ 61. $F(x) = \frac{-1}{16x + 12} + 1$
63. a. Approximately 1.1 units
b. $F(t) = 1.7e^{-0.2t} + 3.8$
c. Approximately 4.9
65. a. Approximately 12.5 thousand barrels
b. $F(t) = 9\ln(3t + 1)$

67. a. By approximately 5.6 thousand Teddy Bears
    b. $Q(x) = -24e^{(-1/3)x} + 50$; approximately 48.8 thousand
69. Approximately $79,000
71. a. 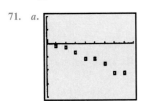 b.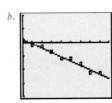

c. $y = -0.32x + 0.2$
d. $G'(t) = 1.2e^{-0.32t}$
e. $G(t) = -3.75e^{-0.32t} + 6.75$
f. Approximately 6.7 grams; approximately 0.03 gram per hour

Section 5.5

1. $\frac{1}{5}x(x+3)^5 - \frac{1}{30}(x+3)^6 + C$
3. $2x^4 \ln x - \frac{1}{2}x^4 + C$
5. $\frac{2}{3}x^3(\sqrt[4]{x^3-1})^3 - \frac{4}{15}(\sqrt[4]{x^3-1})^5 + C$
7. $\frac{-4x}{x+6} + 4 \ln|x+6| + C$
9. $\frac{1}{3}x^3 e^x - x^2 e^x + 2xe^x - 2e^x + C$
11. $250xe^{0.02x} - 12{,}500e^{0.02x} + C$
13. $\frac{-x}{(7x+4)^2} - \frac{1}{49x+28} + C$
15. $x \ln(2x+1) - x + \frac{1}{2}\ln(2x+1) + C$
17. $\frac{64}{9}$
19. $8 \ln 8 - 7 \approx 9.6$
21. -549
23. a. 1024 b. 2526 c. 2582 d. 16,418
25. Decreases $5\frac{7}{9}$ units
27. $F(x) = \frac{x}{3x-5} - \frac{1}{3}\ln|3x-5| + C$
29. $F(x) = 2xe^{0.5x} - 4e^{0.5x} + C$
31. a. $\frac{43}{72}$ of a ton
    b. $Q(x) = \frac{-8x}{(x+1)^2} - \frac{8}{x+1} + 8$

Section 5.6

1. 51   3. 27.4   5. 546   7. 18
9. $\frac{7}{18}$ 11. 41
13. $\sum_{k=1}^{25} 4\left(5 + k\frac{4}{25}\right)^3 \frac{4}{25} \approx 5936$
15. $\sum_{k=1}^{10}\left[3\left(k\frac{7}{10}\right)^2 + 2\left(k\frac{7}{10}\right)\right]\frac{7}{10} \approx 392$
17. Present value $\approx \int_0^7 2040e^{-0.085t}\,dt = \$10{,}762.50$

Future value $\approx \int_0^7 2040e^{0.085(7-t)}\,dt = \$19{,}512.74$

19. $83,470.11   21. $54,691.85
23. $29,485.79   25. Approximately 826 milligrams
27. 632   29. 36   31. Approximately 59.1089

Chapter 5 Review Problems

1. a. $3x^5 + C$ b. $3x^2 + C$
   c. $\frac{1}{4}x^4 + \frac{7}{2}x^2 - 5x + C$ d. $9 \ln|x| + \frac{1}{x^3} + C$
   e. $5e^x + 6x + C$ f. $e^x + 12(\sqrt[4]{x})^3 + C$
2. a. $F(x) = x^5 + 3x - 8$
   b. $F(x) = 4(\sqrt{x})^3 + 13$
   c. $F(x) = \ln|x| - \frac{3}{x}$
   d. $F(x) = 8e^x - \frac{1}{5}x^5 + x - 6$
3. $P(x) = 0.25x - 0.001x^2 + 3$; 3.984 million
4. $C(x) = \frac{1}{3}x^3 - 12x^2 + 210x + 2120$; $3956
5. a. 272   b. -112 c. 4
   d. 16   e. $\ln 7$ f. Approximately 21.06
6. a. Increases 6 units
   b. Increases 109.2 units
   c. Decreases 19.46 units
7. $\int_0^{20}(0.28x - 52)\,dx$; decreases 984 million acres
8. $-\int_{18}^{22}(0.03x^2 - 25)\,dx$; increases 51.84 units
9. a. 4   b. 27   c. 44
10. a. 4.9   b. 9.7
11. a. $\frac{14}{9}(\sqrt{x^3+4})^3 + C$
    b. $\frac{-1}{x^4-5x} + C$
    c. $\frac{1}{2}\ln|2x^3 + 4x - 1| + C$
    d. $\ln|4x+1| + C$
    e. $150e^{0.16x} + C$
    f. $e^{x^2-6} + C$
    g. $2\ln|2x-3| - \frac{6}{2x-3} + C$
    h. $5x - 35\ln|x+7| + C$
    i. $\frac{1}{7}(x+8)^7 - \frac{4}{3}(x+8)^6 + C$
    j. $6\sqrt{x} + 48 \ln|\sqrt{x}-8| + C$
12. a. $-160xe^{-0.05x} + 3200e^{-0.05x} + C$
    b. $(2x^2 + 5x)\ln x - x^2 - 5x + C$
    c. $\frac{3}{4}x^2 e^{4x} - \frac{3}{8}xe^{4x} + \frac{3}{32}e^{4x} + C$
    d. $\frac{-2x}{(2x+5)^2} - \frac{1}{2x+5} + C$
    e. $x^5(x^5+1)^8 - \frac{1}{9}(x^5+1)^9 + C$
    f. $x \ln(4x+8) - x + 2\ln(4x+8) + C$

Section 6.3

1. The denominator is different from zero for each x in $[1, 6]$. So, since $f(x)$ is a rational function, $f(x)$ is continuous on $[1, 6]$. The value of $1.2/x^2$ is nonnegative for each x in $[1, 6]$. Thus, $f(x)$ satisfies property 1 in Definition 6.3. Note that $\int_1^6 \frac{1.2}{x^2}\,dx = \left.\frac{-1.2}{x}\right|_1^6 = 1$. Therefore, $f(x)$ also satisfies property 2 in Definition 6.3.

3. The denominator is different from zero for each x in $[0, +\infty)$. Thus, since $f(x)$ is a rational function, $f(x)$ is continuous on $[0, +\infty)$. The value of $4/(x+4)^2$ is nonnegative for each x in $[0, +\infty)$. So $f(x)$ satisfies property 1 in Definition 6.3. The area of the region between the graph of $f(x)$ and $[0, +\infty)$ is

$$\lim_{b \to +\infty} \int_0^b \frac{4}{(x+4)^2}\,dx$$

where $b > 0$. This limit is equal to

$$\lim_{b \to +\infty}\left(1 - \frac{4}{b+4}\right)$$

which, in turn, is equal to 1. Therefore, $f(x)$ also satisfies property 2 in Definition 6.3.

5. *a.* $\frac{7}{27}$ *b.* $\frac{23}{54}$ 7. 0.9375
9. $\frac{1}{3}$ 11. $\frac{4}{9}$ 13. *a.* $1 - \frac{5}{e^4} \approx 0.9$
   b. $1 - \frac{3}{e^2} + \frac{7}{e^6} \approx 0.6$
15. *a.* Approximately 0.7   17. 43.75 percent
    b. 0.216
19. Since the probability is approximately 0.47, the owner will not replace the truck.
21. $\frac{3}{117}$

Section 6.4

1. $1\frac{2}{3}$ 3. 4   5. 1
7. Approximately 1.84
9. Approximately 0.75   11. 56.25 minutes
13. $32 thousand   15. Approximately 61 percent
17. *a.* The random variable x distributed on $[0, 6]$ according to the probability density function $f(x) = -\frac{1}{36}x^2 + \frac{1}{6}x$ has a mode and an expected value that are both equal to 3.
    b. The random variable x distributed on $[0, 4]$ according to the probability density function $f(x) = \frac{3}{64}x^2$ has a mode of 4 and an expected value of 3.
19. *a.*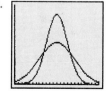

---

13. *a.* 632   *b.* $\frac{128}{3}$ *c.* 80
    d. $-\frac{1}{3}$ *e.* 74.2   *f.* 60.9

14. *a.* $\sum_{k=1}^{120}[480e^{-0.08(k10/120)}]\frac{10}{120} \approx 3304.03$

    b. $\sum_{k=1}^{50} \frac{1}{2 + k\frac{7}{50}} \cdot \frac{7}{50} \approx 1.504$

15. Present value $\approx \int_0^4 1920e^{-0.06t}\,dt = \$6827.91$

    Future value $\approx \int_0^4 1920e^{0.06(4-t)}\,dt = \$8679.97$

16. $64,011.64   17. $13,639.49

CHAPTER SIX

Section 6.1

1. 10.75   3. 8.883   5. 26.491
7. 1.642   9. *a.* 8.778   *b.* 21.778
11. 17.325   13. $34,845.33
15. *a.* Greater than 40   *b.* Greater than 71
19. 235,242.2   21. 11.40157   23. 7.722014

Section 6.2

1. *a.* 10 units   3. *a.* 16 units
   b. $66\frac{2}{3}$ thousand   *b.* $65\frac{1}{3}$ hundred
   c. $40 thousand   *c.* $80 hundred
   d. $26\frac{2}{3}$ thousand   *d.* $14\frac{2}{3}$ hundred
5. *a.* 8 units
   b. $17 thousand
   c. Consumers' surplus: $10\frac{2}{3}$ thousand
      Producers' surplus: $80 thousand
7. $32 hundred
9. $109.5 thousand (rounded to one decimal place)
11. *a.* $8.30 hundred (rounded to two decimal places)
    b. $3.14 hundred
15. *a.* $23.93 thousand (rounded to two decimal places)

    b. $7.68 thousand

Section 6.5

1. $f(x) = Ce^{-0.15x}$
   $f(x) = 45e^{-0.15x}$
3. $f(x) = Ce^{0.7x}$
   $f(x) = 60e^{0.7x}$
5. $f(x) = Ce^{-0.04x}$
   $f(x) = 103.28e^{-0.04x}$ (*C* rounded to two decimal places)
7. $f(x) = 25e^{0.047x}$ (*r* rounded to three decimal places)

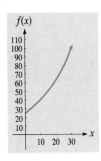

9. $f(x) = 15e^{-0.2x}$

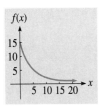

11. Approximately 11 billion
13. A little more than $26,000
15. 10 milligrams per liter
17. About 33 million
19. Approximately 6.4 hours from now

Section 6.6

1. $f(x) = 6 - ce^{-0.24x}$
   $f(x) = 6 - 4e^{-0.24x}$
3. $f(x) = 2 - ce^{-0.18x}$
   $f(x) = 2 + 6e^{-0.18x}$
5. $f(x) = 3 - ce^{-0.26x}$
   $f(x) = 3 + 3.7e^{-0.26x}$ (*c* rounded to one decimal place)
7. $f(x) = 9 - 6e^{-0.2x}$

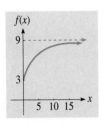

9. $f(x) = 4 + 5e^{-0.4x}$

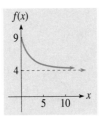

11. $f(x) = 11 - 9e^{-0.24x}$ (*r* rounded to two decimal places)

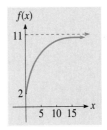

13. Approximately $25,800
15. Approximately 8 weeks from now
17. $V(t) = 800 - 800e^{-0.04t}$
19. Approximately 9.5 years from now
21. Approximately 17.7 years from now
23. $A'(t) = 0.04[300{,}000 - A(t)]$
    $A(t) = 300{,}000 - 300{,}000e^{-0.04t}$
    Approximately 12.8 hours

Chapter 6 Review Problems

1. a. 1.022   b. 2.094   c. 21.499
   d. 0.432
2. a. 0.022   b. 9.022
3. a. $12 thousand   b. $4.088 thousand
4. The denominator is different from zero for each *x* in [5, 13]. So, since $f(x)$ is a rational function, $f(x)$ is continuous on [5, 13]. The value of $30/(x+7)^2$ is nonnegative for each *x* in [5, 13]. Thus, $f(x)$ satisfies property 1 in Definition 6.3. Note that $\int_5^{13} \frac{30}{(x+7)^2} dx = \frac{-30}{x+7} \Big|_5^{13} = 1$. Therefore, $f(x)$ also satisfies property 2 in Definition 6.3.
5. The denominator is different from zero for each *x* in $[0, +\infty)$. Thus, since $f(x)$ is a rational function, $f(x)$ is continuous on $[0, +\infty)$. The value of $8x/(x^2+4)^2$ is nonnegative for each *x* in $[0, +\infty)$. So $f(x)$ satisfies property 1 in Definition 6.3. The area of the region between the graph of $f(x)$ and $[0, +\infty)$ is

$$\lim_{b \to +\infty} \int_0^b \frac{8x}{(x^2+4)^2} dx$$

where $b > 0$. This limit is equal to

$$\lim_{b \to +\infty} \left(1 - \frac{4}{b^2+4}\right)$$

which, in turn, is equal to 1. Therefore, $f(x)$ also satisfies property 2 in Definition 6.3.

6. *a.* Approximately 0.69 *b.* Approximately 12.6 percent
7. 2.4   8. 3.125   9. $2.9 thousand
10. *a.* $f(x) = Ce^{-0.25x}$ *b.* $f(x) = Ce^{0.16x}$
    $f(x) = 38e^{-0.25x}$ $f(x) = 30.06e^{0.16x}$ (*C* rounded to two decimal places)
11. $f(x) = 32e^{0.045x}$ (*r* rounded to three decimal places)
12. Approximately 1.79 million
13. *a.* $f(x) = 5 - ce^{-0.14x}$ *b.* $f(x) = 3 - ce^{-0.19x}$
    $f(x) = 5 - 3e^{-0.14x}$ $f(x) = 3 + 4e^{-0.19x}$
    c. $f(x) = 4 - ce^{-0.22x}$
    $f(x) = 4 + 9.67e^{-0.22x}$ (*c* rounded to two decimal places)
    d. $f(x) = 6 - ce^{-0.34x}$
    $f(x) = 6 - 6e^{-0.34x}$
14. *a.* $f(x) = 1 + 5e^{-0.225x}$

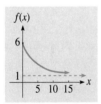

b. $f(x) = 7 - 6e^{-0.5x}$

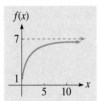

c. $f(x) = 2 + 3e^{-0.275x}$ (*r* rounded to three decimal places)

15. In about 4 years

CHAPTER SEVEN

Section 7.1

1. 75   3. 2   5. 450   7. 26
9. Increases 4.8 units   11. Increases 40 units
13. Increases 17 units

15.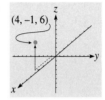

17.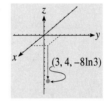

19.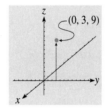

21.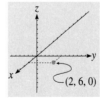

23.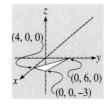

25.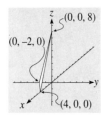

27. a. 18 million barrels
    b. Decreases by 3 million barrels
    c. Increases by 1.6 million barrels
29. a. 44,100 gallons
    b. $P(2x, 2y) = 100(2x)^{2/3}(2y)^{1/3}$
    $= 100(2^{2/3})(x^{2/3})(2^{1/3})(y^{1/3})$
    $= 2 \cdot 10x^{2/3}y^{1/3}$
    $= 2 \cdot P(x, y)$
31. a. $z = -0.5x - 0.75y + 2000$
    b. 900 pounds
    c. 720 pounds

Section 7.2

1. $f_x(x, y) = 6x^2y; f_y(x, y) = 2x^3 + 14y$
3. $f_x(x, y) = 4x^3y^3 - 3; f_y(x, y) = 3x^4y^2$
5. $f_x(x, y) = \dfrac{25y^{3/4}}{x^{3/4}}; f_y(x, y) = \dfrac{75x^{1/4}}{y^{1/4}}$
7. $f_x(x, y) = 3; f_y(x, y) = -7$
9. $f_x(x, y) = \dfrac{-9y^2}{x^4} - \dfrac{7}{y}; f_y(x, y) = \dfrac{6y}{x^3} + \dfrac{7x}{y^2}$
11. $f_x(x, y) = 42x(3x^2 + y^3)^6; f_y(x, y) = 21y^2(3x^2 + y^3)^6$
13. $f_x(x, y) = \dfrac{6x^2y + 1}{2\sqrt{2x^3y + x}}; f_y(x, y) = \dfrac{x^3}{\sqrt{2x^3y + x}}$
15. $f_x(x, y) = \dfrac{-y}{(xy + 5y)^2}; f_y(x, y) = \dfrac{-(x + 5)}{(xy + 5y)^2}$
17. $f_x(x, y) = 6y^5(2xy^3 - 4)^2$
    $f_y(x, y) = 8y(xy^3 - 2)^2(11xy^3 - 4)$
19. $f_x(x, y) = \dfrac{5y^2 - 2x^3y^2}{(x^3 + 5)^2}$
    $f_y(x, y) = \dfrac{2xy}{x^3 + 5}$
21. $f_x(x, y) = \dfrac{3x^2}{x^3 + 5y}; f_y(x, y) = \dfrac{5}{x^3 + 5y}$
23. $f_x(x, y) = \dfrac{3y^2}{3x + y}$
    $f_y(x, y) = \dfrac{y^2}{3x + y} + 2y \ln(3x + y)$
25. $f_x(x, y) = \dfrac{2xy^3}{2x + 1} + y^3 \ln(2x + 1)$
    $f_y(x, y) = 3xy^2 \ln(2x + 1)$
27. $f_x(x, y) = ye^{xy+5y}; f_y(x, y) = (x + 5)e^{xy+5y}$
29. $f_x(x, y) = x^2(8x + 3)e^{8x-y^2}$

$f_y(x, y) = -2x^3ye^{8x-y^2}$
31. $f_x(x, y) = 4(e^{xy} + 2y)^3 ye^{xy}$
    $f_y(x, y) = 4(e^{xy} + 2y)^3(xe^{xy} + 2)$
33. $\dfrac{\partial w}{\partial x} = 2xz^3 - 7y$

$\dfrac{\partial w}{\partial y} = -7x + z^2$

$\dfrac{\partial w}{\partial z} = 3x^2z^2 + 2yz$

35. Decreases approximately 3 units
37. Increases approximately 4 units
39. Decreases approximately 0.18 of a unit
41. $f_x(5, 10) = 1$; so $f(x, y)$ increases approximately 1 unit if x increases from 5 to 6 and y remains fixed at 10.
    $f_y(5, 10) = -0.25$; so $f(x, y)$ decreases approximately 0.25 of a unit if y increases from 10 to 11 and x remains fixed at 5.
43. A 1-hour increase in the weekly time spent jogging
45. Approximately $\$1\frac{5}{7}$ thousand
47. About 609 barrels greater
49. The marginal productivity of labor is 0.65. So the daily production level increases approximately 0.65 of a unit if the labor force increases 1 unit. The marginal productivity of capital is 1.1. So the daily production level increases approximately 1.1 units if the capital outlay increases 1 unit.
51. The marginal productivity of labor is 21. So the daily production level increases approximately 21 units if the labor force increases 1 unit. The marginal productivity of capital is 15. So the daily production level increases approximately 15 units if the capital outlay increases 1 unit.
53. The marginal productivity of labor is $5\frac{5}{24}$. So the daily production level increases approximately $5\frac{5}{24}$ units if the labor force increases 1 unit. The marginal productivity of capital is $5\frac{1}{3}$. So the daily production level increases approximately $5\frac{1}{3}$ units if the capital outlay increases 1 unit.
55. Substitutes   57. Complementary   59. Substitutes

Section 7.3

1. $(3, 5, 29)$ 3. $(5, 7, \ln 104.5)$
5. $(7, 2, \frac{-121}{3}), (7, 5, \frac{-269}{6})$
7. $(0, 0, 3), (\frac{1}{6}, \frac{1}{3}, \frac{161}{54})$
9. $(3, 1, \frac{119}{6}), (3, 4, \frac{46}{3}), (5, 1, \frac{21}{2}), (5, 4, 14)$
11. $(1, 3, 3)$ 13. $x = 5$ and $y = 10$; 75
15. $x = 14$ and $y = 6$ 17. $x = 8$ and $y = 10$
19. 30 workers and 20 machines
21. 90 tons of beef and 30 tons of pork
23. a. $P(x, y) = (166 - 2x + y)(x - 35)$
    $+ (191 - 3y + 2x)(y - 28) - 7000$
    b. Sell product G at $120 per ton and product H at $100 per ton

Section 7.4
1. (0, 0, 0) is a saddle point; (1, 3, 3) is a high point.
3. (5, 2, 3) is a saddle point; (5, 3, 2) is a low point.
5. (14, 4, 36) is a saddle point; (17, 5, 37) is a high point.
7. (1, 6, 220) and (7, 2, 36) are saddle points; (1, 2, 252) is a high point; (7, 6, 4) is a low point.
9. $(-3, 3, 135)$ is a saddle point.
11. (5, 3, 93) is a high point. 13. $(\frac{1}{3}, \frac{1}{3}, 9)$ is a low point.
15. Temperature: 70 degrees; humidity: 80 percent
17. 300 units of product A and 200 units of product B
19. $C(x, y) = 7xy + \frac{56}{x} + \frac{56}{y}$; $x = 2$ and $y = 2$

Section 7.5
1. $f(x) = 0.289x + 1.868$; 4.758

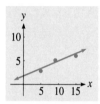

3. $f(x) = -0.405x + 9.092$; 3.017

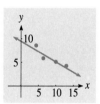

5. $f(x) -0.472x + 12.163$; 7.443

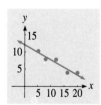

7. $f(x) = 0.036x^2 + 2.051$

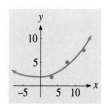

9. $f(x) = -0.031x^2 + 11.98$

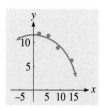

11. $f(x) = -0.104x^2 + 2.047x$

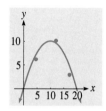

13. $f(x) = -0.097x^2 + 2.335x$

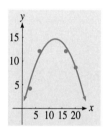

15. a. $D(x) = -0.218x + 6.905$
    b. 4.507 million belts per month
    c.

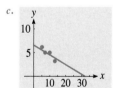

17. a. $P(x) = 0.023x^2 + 5.158$
    b. Increasing at the rate of 0.828 of a million per year
19. a. $P(x) = -0.29x^2 + 5.258x$
    b. Approximately 18 weeks after the epidemic started
21. a. $f(x) = 0.714x + 2.286$
    b. $f(x) = 0.037x^2 + 5.291$
    c. The function in part (b) is the better approximation because $S(0.037, 5.291)$ is less than $S(0.714, 2.286)$.
25. $y = 3.09x + 4.46$
27. $y = 0.51x - 0.16$

Section 7.6
1. $x = 6$ and $y = 3$
3. $x = 3$ and $y = 6$
5. $(2, 1)$
7. $x = 7$ and $y = 49$
9. $x = 6$ and $y = 1$
11. $x = 13$ and $y = 5$
13. $x = 1, y = 1$, and $z = 1$
15. $x = 4, y = \frac{1}{2}$, and $z = 4$
17. 25 units of brand A, 15 units of brand B, and 45 units of brand C
19. a. $24,000 on unskilled labor and $12,000 on skilled labor
    b. $30,000 on unskilled labor and $15,000 on skilled labor
21. 54,000 square feet
23. a. Base: 4 feet by 4 feet; height: 6 feet
    b. Base: 5 feet by 5 feet; height: 7.5 feet
25. 4 feet by 4 feet by 4 feet

Section 7.7
1. $x = 6$ and $y = 3$
3. $x = 3$ and $y = 6$
5. $(2, 1)$
7. $x = 7$ and $y = 49$
9. $x = 6$ and $y = 1$
11. $x = 13$ and $y = 5$
13. 20 units of product A and 6 units of product B
15. 60 experienced workers and 40 inexperienced workers

Chapter 7 Review Problems
1. 6
2.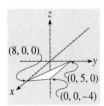
3. a. $P(x, y) = 3300x + 2400y - 2x^2 - y^2 - xy$
   b. The production level increase of product A
4. $f(x, y)$ increases approximately 8 units.
5. $f(x, y)$ decreases approximately $\frac{35}{256}$ of a unit.
6. w increases approximately 18.7055 units (rounded to 4 decimal places).
7. The marginal productivity of capital is 2. So the daily production level increases by approximately 2 tons if the capital outlay increases $1000.
8. Complementary
9. $x = 4$ and $y = 6$; 8
10. 600 tons of product A and 900 tons of product B; $2,070,000
11. $(0, 0, 0)$ is a saddle point; $(2, 8, \frac{640}{3})$ is a high point.
12. $C(x, y) = 6xy + \frac{48}{x} + \frac{48}{y}$; $x = 2$ and $y = 2$
13. $f(x) = -0.403x + 10.019$; $y \approx 8.004$ when $x = 5$; $y \approx 3.168$ when $x = 17$

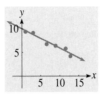

14. a. $D(x) = -2.98x^2 + 768$
    b. Around the beginning of the year 2004
15. 63,281.25 square feet
16. 63,281.25 square feet

APPENDIX A

Section A.1
1. True
3. True
5. False
7. False
9. True
11. $-8 < 3$
13. $-4.7 < -4$
15. $7.46 > 7.439$
17. $-12.3 > -20$
19. $-5 < -4.1 < -2$
21. $x \leq 5$
23. $x < 0$
25. $x \geq 0$
27. $x \leq 7.8$
29. $c < d$
31. Open; $(2, 7)$

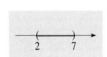

33. Closed; $[0, 4]$

35. Half-open; $(-\infty, 0]$

37. Open; $(-4, +\infty)$

39. Half-open; (0, 12.4]

41. Open; (4.2, 8)

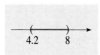

43. Open; (−∞, 3)

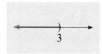

45. Open; (−4, 4)

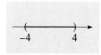

47. −16   49. 8   51. −26   53. −18   55. 21
57. 9   59. 5   61. 4

Section A.2

1. 512   3. 64   5. −64   7. −32
9. 2   11. −3   13. 7   15. 16
17. $\frac{1}{9}$ 19. $\frac{1}{4}$ 21. 1   23. 0.04
25. 256   27. 62,748,517
29. 9,682,651,996,000   31. 0.000000017210368
33. −279936   35. 0.015625   37. 13
39. 6.782329983   41. −1.729140094
43. 3.314454017   45. 0.3801685654
47. 0.005044076   49. 31.30495168
51. $2x\sqrt[4]{3x^3}$ 53. $\frac{8\sqrt[4]{x}}{x}$ 55. $\frac{2\sqrt[5]{x^2}}{x}$
57. $\sqrt[4]{x^3}$ 59. $\frac{x^6}{4}$ 61. $\frac{\sqrt[6]{x^5}}{x}$ 63. $\sqrt[4]{x^3}$
65. $5\sqrt[3]{3x^2}$ 67. $\frac{3\sqrt[6]{x^5}}{x}$

Section A.3

1. $4x^5 - 4x^3 + 8x^2 + 4$
3. $2x^3 + 3x - 2$
5. $x^4 - 2x^3 + 8x^2 - 20x + 8$

7. $4x^2 + 3 - \frac{2}{x^2}$
9. $4x^2 - 8x + 12 - \frac{41}{2x + 3}$
11. $(x^2 + 4)(x + 3)$; −3   13. $3x^2(2x^2 - 3x + 4)$; 0
15. $(x + 3)(x - 3)(x - 7)$; −3, 3, and 7
17. $(x + 7)(x + 4)$; −7 and −4
19. $7x^2(x - 1)(x - 4)$; 0, 1, and 4
21. $2(x - 6)^2$; 6
23. $(x + 9)(x - 9)$; −9 and 9
25. $3(x + 0.5)(x - 0.5)$; −0.5 and 0.5
27. $(x + 2)(x - 2)(x + 5)(x - 5)$; −5, −2, 2, and 5
29. $25(x - \frac{1}{5})(x - \frac{3}{5})$; $\frac{1}{5}$ and $\frac{3}{5}$
31. $36(x + \frac{1}{6})^2$; $-\frac{1}{6}$
33. $45x^4(x + \frac{2}{3})(x - \frac{1}{3})$; $-\frac{2}{3}$, 0, and $\frac{1}{3}$
35. $4(x + \frac{1}{2})(x - \frac{1}{2})(x^2 + 3)$; $-\frac{1}{2}$ and $\frac{1}{2}$
37. $x(x + 5)(x^2 - 5x + 25)$; −5 and 0
39. $2(x - 3)(x^2 + 3x + 9)$; 3
41. $(x + 10)(x^2 - 10x + 100)(x - 4)(x^2 + 4x + 16)$; −10 and 4

Section A.4

1. $\frac{3}{x + 2}$ 3. $\frac{x + 3}{x^2 + 4}$ 5. $\frac{-1}{2x + 1}$
7. $\frac{x - 13}{x^2 - 2x + 4}$ 9. $\frac{x^2 + 3x - 4}{x - 2}$
11. $\frac{1}{3x^4 + 15x^3 - 24x - 120}$ 13. $\frac{5x - 5}{3x + 12}$
15. $\frac{3x + 15}{20x^2 + 5x}$ 17. $\frac{x - 3}{x^2 + x}$
19. $\frac{x - 5}{x^3 + 3x^2 + 4x + 12}$ 21. $\frac{6}{x^4 - 5}$
23. $\frac{3x^2 - 4x - 15}{2x^3 + 12x^2 + 18x}$ 25. $\frac{x^4 - 6x^3 + 5x + 5}{x^4 + x^3}$
27. −3   29. −2 and 6   31. No solution

APPENDIX B

1. $\frac{15}{x + 3} + 5\ln|x + 3| + C$
3. $x^2 + 6\ln|x^2 - 6| + C$
5. $\frac{1}{8}\left(2x + 3 - \frac{9}{2x + 3} - 6\ln|2x + 3|\right) + C$
7. $\frac{1}{60}\ln\left|\frac{5x + 6}{5x - 6}\right| + C$
9. $2\ln|9x + \sqrt{81x^2 + 13}| + C$
11. $x(\ln x)^2 - 2x\ln x + 2x + C$

Index

Abscissa, 5
Absolute maximum value, 125–126
Absolute minimum value, 125–126
Absolute value, 465
Algebraic equation, 495
   equivalent, 495
   solution of, 495
Algebraic expression, 477–478
   zero of, 488
Annuity:
   definition of, 341
   future value of, 342–343
   present value of, 341–342
Antiderivative:
   definition of, 279
   general, 279–280
   particular, 280–281
Area of a region:
   above the horizontal axis, 304
   below the horizontal axis, 305
Asymptote:
   horizontal, 177
   slant, 191–192
   vertical, 191
Average cost function, 15
Average rate of change, 7
Average value:
   of a function, 301
   of a random variable, 380
Average velocity, 10, 151

Base, 469

Carbon dating, 244–245
Cartesian plane, 5
Chain rule:
   for derivatives, 108, 155
   for integrals, 313
Change:
   in $f(x)$:
      actual, 6
      approximate, 92, 132
   in $f(x, y)$:
      actual, 405–407

Change (*cont.*):
      approximate, 417
Closed interval, 464, 465
Cobb-Douglas production function, 418
Coefficient, 478
   leading, 38, 478
Complementary products, 420
Composite function:
   definition of, 17
   derivative of, 108, 155
Compound interest (*see* Interest)
Concave:
   downward, 162
   upward, 169
Concavity:
   definition of, 162
   test for, 162
Constant function, 5
Constant multiple rule:
   for derivatives, 105
   for integrals, 285
Constant of antidifferentiation, 280
Constant of integration, 280
Constant rate of change, 24, 26
Constant rule:
   for derivatives, 84
   for integrals, 281
Constrained critical point, 454
Constrained maximum value, 447
Constrained minimum value, 447
Constrained optimization problem, 447
Constrained point, 447
Constraint equation, 447
Consumers' surplus, 361–362
   at market equilibrium, 365–366
Continuity of a function:
   on an interval, 58
   at a point, 57–59
Continuous function, 57–59
Coordinate of a point on a line, 462
Coordinate system on a line, 462
Critical point:
   constrained, 454
   first-order, 164
   of $f(x)$, 124

Critical point (*cont.*):
  on the graph of $f(x)$, 124
  on the graph of $f(x, y)$, 424
  second-order:
    of $f(x)$, 164
    on the graph of $f(x)$, 164
Cube root, 472
Curve sketching procedure, 165

Data point, 28, 435
Decreasing function, 122–123
Decreasing rate:
  of decrease, 172
  of increase, 170
Definite integral:
  as change, 295
  characterized in terms of area, 307
  characterized in terms of limits, 336–337
  definition of, 294
  notation, 294
  used as an approximation:
    for the future value of an annuity, 342–343
    for the present value of an annuity, 341–342
    for a Riemann sum, 338–341
  with a variable limit of integration, 300, 308, 310
Degree:
  of a polynomial expression, 478
  of a polynomial function, 37–38
Demand function, 13–14
Dependent variable, 3, 405
Derivative:
  as approximate change, 92
  of a composite function, 108, 155
  of a constant, 84
  of a constant times a function, 105
  definition of, 76
  of e^x, 258
  first, 161
  of a function of the form $e^{u(x)}$, 256–257
  of a logarithmic function, 252–253
  of the natural logarithmic function, 250
  partial, 414, 419
  of a power, 85, 102
  of a product, 104
  of a quotient, 106
  second, 160–161
  as the slope of a tangent line, 77
  of a sum or difference, 86
Descartes, René, 5
Difference quotient, 76

Differentiable function, 76, 89, 108
Differential equation, 386–388, 393–394
  general solution of, 388, 394–395
  particular solution of, 388–389, 395
Differential of a function, 132
Discontinuity of a function at a point, 57–58
Discontinuous function, 57–58
Discriminant, 36, 484
Distributive property, 480
Domain, 3, 4, 404, 409
Doubling time, 240–241

e, the number, 222–223
Effective interest rate, 230–231
Elastic demand, 138–139
Elasticity, 135–137
Endpoint of an interval, 464
Equivalent equation, 495
Error:
  in Newton's method, 210–211
  in Simpson's rule, 358
Euler, Leonhard, 223
Expected value of a random variable, 380, 382
Exponent, 469
  irrational, 228
  negative, 473
  positive integer, 228, 469
  positive rational, 228, 472–473
  properties of, 228, 474
Exponential decay, 235–236
Exponential function:
  alternative form of, 243
  definition of, 234
  graph of, 234, 235
Exponential growth, 234–235
Exponentially distributed random variable:
  definition of, 381
  expected value of, 382

Factor of a polynomial expression, 480
Factoring, 480
  theorems for, 480–483, 486–488
First derivative, 161
First-order critical point, 164
Fixed cost, 7
Function:
  average cost, 15
  composite, 17–18
  constant, 5
  continuous, 57–59
  decreasing, 122–123
  definition of, 2

Function (cont.):
- demand, 13–14
- dependent variable of, 3
- differentiable, 76
- differential of, 132
- discontinuous, 57–59
- domain of, 3
- exponential, 234, 243
- increasing, 122–123
- independent variable of, 3
- Lagrange, 453, 456
- least-squares, 438
- least-squares linear, 27–29, 440–441
- linear, 23
- logarithmic, 251
- logistic, 272
- marginal cost, 93
- marginal productivity of capital, 94–95
- marginal productivity of labor, 94–95
- natural logarithmic, 238
- notation, 3
- polynomial, 37–38
- power, 101
- probability density, 371
- profit, 14, 15
- quadratic, 32
- range of, 3
- rational, 46
- revenue, 14, 15
- of three variables:
  - definition of, 409
  - domain of, 409
  - notation, 409
- total cost, 6
- of two variables:
  - definition of, 404
  - domain of, 404
  - graph of, 408
  - linear, 408
  - notation, 405
  - partial differentiable, 415
  - range of, 404
- utility, 449

Fundamental theorem of calculus, 295
Future value of an annuity, 342–343

General antiderivative:
- of a constant, 281
- of a constant times a function, 285
- definition of, 279–280
- of the natural logarithmic function, 329–330
- notation, 280
- of a power, 282
- of a sum or difference, 283

General solution, 388, 394–395
Generalized exponential rule, 256–257
Generalized logarithm rule, 252–253
Generalized power rule, 102
Graph:
- of an exponential function, 234, 235
- of a function, 5
- of a function of two variables, 408
- of an interval, 464, 465
- of a linear function, 24
- of a linear function of two variables, 408–409
- of a logistic function, 270, 272
- of the natural logarithmic function, 238
- of a quadratic function, 32–34

Half-life, 240, 241
Half-open interval, 464, 465
Horizontal asymptote, 177
Horizontal axis, 5
Horizontal line, 25

Improper integral:
- convergence of, 373
- definition of, 373
- divergence of, 373

Increasing function, 122–123
Increasing rate:
- of decrease, 172
- of increase, 170

Indefinite integral:
- of a constant, 281
- of a constant times a function, 285
- definition of, 280
- of the natural logarithmic function, 329–330
- notation, 280
- of a power, 282
- of a sum or difference, 283

Independent variable, 3, 405, 409
Index of a radical, 472
Inelastic demand, 138–139
Inequality:
- definition of, 462
- symbols, 462–463

Infinity symbol, 465
Inflection point, 163
Instantaneous rate of change, 152
Instantaneous velocity, 150–152
Integer, 461
Integer exponent, positive, 228, 469
Integral:
- definite, 294
- improper, 373

Integral (*cont.*):
  indefinite, 280
Integrand, 280
Integration:
  by parts, 326–332
  by substitution:
    applied to $h(x)e^{u(x)}$, 321–322
    applied to $h(x)/u(x)$, 316–319
    applied to $[u(x)]^n h(x)$, 314–316
    using a table of integrals, 499–503
Integration by parts rule, 326
Interest:
  compounded continuously, 225–227
  compounded n times per year, 223–227
Interval:
  closed, 464, 465
  endpoints of, 464
  graph of, 464, 465
  half-open, 464, 465
  notation, 464, 465
  open, 464, 465
Irrational exponent, 228
Irrational number, 461
Irreducible polynomial expression, 486

Lagrange, Joseph L., 453
Lagrange function, 453, 456
Lagrange multiplier, 453, 456
Lagrange multipliers, method of, 454
Leading coefficient, 38, 478
Learning curve, 266
Least-squares:
  function, 438
  linear function, 27–29, 440–441
Least squares, method of, 438
Leibniz, Gottfried Wilhelm, 67, 79
Leibniz notation, 79
L'Hôpital, Guillaume François de, 259
L'Hôpital's rule, a special case, 259
Libby, Willard F., 245
Limit:
  at infinity:
    definition of, 42
    nonexistence of, 42–43
    properties of, 44
    of a rational function, 46–47
  as x approaches a number:
    definition of, 49
    nonexistence of, 54–56
    one-sided, 53
    of a polynomial function, 51
    properties of, 50
    of a rational function, 52–53
Limits of integration, 294

Linear function:
  definition of, 23
  graph of, 24
  least-squares, 27–29, 440–441
Linear function of two variables:
  definition of, 408
  graph of, 408
Logarithm rule:
  for derivatives, 250
  for integrals, 287
Logarithmic function:
  definition of, 251
  derivative of, 252–253
Logistic curve, 272
Logistic function:
  definition of, 272
  graph of, 270, 272
Logistic increase, 272

Magnitude, 465
Marginal cost function, 93
Marginal productivity:
  of capital, 94–95, 417
  of labor, 94–95, 417
Marginal propensity:
  to consume, 99, 290
  to save, 290
Marginal revenue function, 93
Market equilibrium, 11, 365
Maximum value:
  absolute, 125–126
  on a closed interval, 112–113
  constrained, 447
  definition of, 31
  of a function of two variables, 425
  on a nonclosed interval, 126
  of a quadratic function, 34
  relative, 125–126
Mean of a random variable, 380
Median of a random variable, 386
Minimum value:
  absolute, 125–126
  on a closed interval, 112–113
  constrained, 447
  definition of, 31
  of a function of two variables, 425
  on a nonclosed interval, 126
  of a quadratic function, 34
  relative, 125–126
Mode of a random variable, 386
Monomial, 478

Natural logarithmic function:
  definition of, 238
  derivative of, 250

Natural logarithmic function (*cont.*):
  general antiderivative of, 229–230
  graph of, 238
  indefinite integral of, 229–230
  properties of, 238
Negative exponent, 473
Newton, Isaac, 67, 207
Newton's method:
  error in, 210–211
  idea behind, 208–209
  statement of, 210
Nominal interest rate, 230–231
Normally distributed random variable, 384
nth power, 469
nth root, 471
Number line, 462

Open interval, 464, 465
Order of operations, 466, 476
Ordered pair, 4
Ordered quadruple, 409
Ordered triple, 407, 409
Ordinate, 5
Origin on a line, 462

Parabola, 32–34
Parabolic rule, 350
Parameter, 437–438
Partial derivative:
  alternative notation for, 416
  as approximate change, 417
  definition of, 414, 419
  geometric interpretation of, 418–419
  second, 430
Partial differentiable function, 415
Particular antiderivative, 280–281
Particular solution, 388–389, 395
Percentage change in $f(x)$:
  actual, 134
  approximate, 134
Percentage rate of change:
  definition of, 200
  of exponential change, 242, 260–261
Point of diminishing returns, 174, 187
Polynomial expression:
  arithmetic operations on, 478–480
  definition of, 478
  degree of, 478
  factors of, 480
  irreducible, 486
  leading coefficient of, 478
Polynomial function:
  continuity of, 59
  definition of, 37
  degree of, 37–38

Polynomial function (*cont.*):
  examples of, 38
  leading coefficient of, 38
Power function, 101
Power rule:
  for derivatives, 85
  for integrals, 282
Present value, 229–230
Present value of an annuity, 341–342
Principal nth root, 471
Probability, 369
Probability density function:
  definition of, 371
  of a random variable, 375
Producers' surplus, 363–364
  at market equilibrium, 365–366
Product rule, 104
Profit function, 14, 15
Proportional, 390, 397

Quadratic expression, 483
Quadratic formula, 36, 484
Quadratic function:
  definition of, 32
  graph of, 32–34
Quotient rule, 106

Radical, 472
  simplest form of, 474
Radicand, 472
Random variable:
  average value of, 380
  definition of, 374
  expected value of, 380
  exponentially distributed, 381
  mean of, 380
  median of, 386
  mode of, 386
  normally distributed, 384
  standard deviation of, 384
Range, 3, 404
Rate of change:
  average, 7
  constant, 24, 26
  instantaneous, 152
  of linear change, 24, 26–27
  percentage, 200
Rational equation, procedure for solving, 496
Rational exponent, positive, 228, 472–473
Rational expression, 490
  addition and subtraction of, 494–495
  multiplication and division of, 493–494
  reduced to lowest terms, 491

Rational function:
 continuity of, 59
 definition of, 46
Rational number, 461
Rationalized denominator, 475
Real number:
 definition of, 461
 properties of, 467, 490
Rectangular coordinate plane, 5
Reduction rule, 503
Regression:
 curve, 438
 line, 27–29
Related rates, 154–156
Relative maximum value, 125–126
Relative minimum value, 125–126
Replacement set, 477
Revenue function, 14, 15
Riemann, Georg Friedrich, 337
Riemann sum, 337–338
Root:
 cube, 472
 nth, 471
 principal nth, 471
 square, 472

Saddle point, 425
Second derivative, 160–161
Second derivative test, 431
Second-order critical point, 164
Second partial derivative, 430
Sign chart, 123
Simplest form of a radical, 474
Simpson, Thomas, 350
Simpson's rule:
 definition of, 354
 error in, 358
Slant asymptote, 191–192
Slope of a line, 24–26
Slope of a tangent line, 69, 77
Square root, 472
Standard deviation, 384
Substitute products, 420
Substitution method:
 for constrained optimization, 449
 for integration, 313–323

Substitution method (*cont.*):
 using a table of integrals, 499–503
Sum and difference rule:
 for derivatives, 86
 for integrals, 283
Summation notation, 334–335
Surface, 408

Table of integrals, 499–500
Tangent line, 67–68
Test for concavity, 162
Three-dimensional coordinate system, 407
Total cost function, 6

Unit elasticity, 138
Utility function, 449

Variable, 477
 dependent, 3, 405
 independent, 3, 405, 409
Velocity:
 average, 10, 151
 instantaneous, 150–152
Verhulst, Pierre-François, 270
Vertex of a parabola, 33–34
Vertical asymptote, 191
Vertical axis, 5
Vertical line, 25

x axis, 5
x coordinate, 5

y axis, 5
y coordinate, 5

Zero:
 of an algebraic expression, 488
 of a function, 35
 of a quadratic function, 35–37
Zero interval, 209
Zero polynomial function, 38
Zeroth power, 474

Index of Selected Applications

Business and Economics

Average outstanding balance, 402
Book sales, 248
Breakeven sales levels, 11
Budget constraints, 411–412
Car affordability, 128
Comparing investment plans, 232
Comparing two deals, 345
Consumer debt, 274
Consumers' and producers' surpluses, 368
Defense spending, 198
Demand for a gasoline additive, 30–31
Depreciation, 264
Effective interest rate, 233
Elasticity of demand, 141
Energy production cost, 29
Factory utilization, 184
Household income, 128
Inflation rate, 39
Investing $1 million, 205
Labor productivity, 185–186
Marginal cost, 96–98
Marginal productivity, 98–99
Market equilibrium, 11
Minimum average cost, 131
Mortgage rates, 111
Necessary income, 236
Optimal delivery batch, 119
Optimal dimensions, 452
Optimal enrollment, 40
Optimal firing batch, 22
Optimal mix of machines, 452
Optimal order size, 126–127
Optimal production levels, 118–119
Optimal time to sell, 40
Optimal transferring time, 233
Optimal work force, 118
Present and future values of an annuity, 344
Prime rate, 48
Propensity to consume, 99
Rental fee, 61
Single payment amount, 345
Substitute and complementary products, 423
Total cost function, 6–7

Life Sciences

Acid rain, 128
Beaver population, 61
Blood sugar level, 118
Carbon dating, 245
Cholesterol levels in an African country, 386
Cigarette smoking, 140
Concentration of EDB in a lake, 134–135
Determining time of death, 397–398
Doubling time of a population, 247
Drug concentration in a bloodstream, 345
Elasticity of blood sugar level, 141–142
Eradication of a disease, 402
Excess weight, 236
Growth of algae, 339–341
Heart disease mortality, 98
Insect population, 206
Intravenous feeding, 325
Life expectancy, 48
Maximal heart rate, 30
Maximum infection rate, 146
Oxygen uptake, 30
Perspiration rate, 185
Poiseuille's law of blood flow, 9
Puberty age, 184